Progress in Colloid and Polymer Science

Recently Published and Forthcoming Volumes

Surface and Colloid Science
Volume Editor: Galembeck, F.
Vol. 128, 2004

Analytical Ultracentrifugation VII
Volume Editors: Lechner, M. D., Börger, L.
Vol. 127, 2004

Trends in Colloid and Interface Science XVII
Volume Editors: Cabuil, V., Levitz, P., Treiner, C.
Vol. 126, 2004

From Colloids to Nanotechnology
Volume Editors: Zrinyi, M., Hórvölgyi, Z. D.
Vol. 125, 2004

Aqueous Polymer Dispersions
Volume Editor: Tauer, K.
Vol. 124, 2004

Trends in Colloid and Interface Science XVI
Volume Editors: Miguel, M., Burrows, H. D.
Vol. 123, 2004

Aqueous Polymer – Cosolute Systems
Volume Editor: Anghel, D. F.
Vol. 122, 2002

Molecular Organisation on Interfaces
Volume Editor: Lagaly, G.
Vol. 121, 2002

Lipid and Polymer-Lipid Systems
Volume Editor: Nylander, T.
Vol. 120, 2002

Progress in Colloid and Polymer Science
Editors: F. Kremer, Leipzig and W. Richtering, Aachen

Volume 128 · 2004

Surface and Colloid Science

Volume Editor:
Fernando Galembeck

Springer

The series Progress in Colloid and Polymer Science is also available electronically (ISSN 1437-8027)

- Access to tables of contents and abstracts is *free* for everybody.
- Scientists affiliated with departments/institutes subscribing to Progress in Colloid and Polymer Science as a whole also have full access to all papers in PDF form. Point your librarian to the LINK access registration form at http://www.springerlink.com

ISSN 0340-255X
ISBN 3-540-21247-7
DOI 10.1007/b12355
Springer Berlin, Heidelberg, New York

Springer is a part of Springer Science + Business Media

http://www.springeronline.com

Typesetting: SPS, Chennai, India

Cover: Estudio Calamar,
F. Steinen-Broo, Pau/Girona, Spain

Printing: Mercedes Druck, Berlin,
Germany

Printed on acid-free paper
Printed in Germany

Progr Colloid Polym Sci (2004) 128: V

PREFACE

The International Conferences on Surface and Colloid Science were initiated in 1975 in Budapest and they are held every three years, under the auspices of the International Association of Colloid and Interface Scientists. In September 2003, 454 researchers from 32 countries gathered at Iguassu Falls in Brazil presenting 538 papers on the following topics: Biocolloids and Biological Applications, Charged Particles and Interfaces, Colloid Stability, Colloidal Dispersions, Environmental Colloidal Science, Interfaces and Adsorption, Nanostructures and Nanotechnology, Self-Assembly and Structured Fluids, Surfactants and Polymers, Technology and Applications, Colloids and Surfaces in Oil Production.

The participants submitted manuscripts for publication in this volume and these went through peer review. Fifty-eight contributions were selected and they are now presented in this volume.

This is a good display of current colloid and interface science, thanks to the contributions of plenary and invited lecturers covering many frontier topics. As expected, latin-american authors are well represented in this volume, including some new researchers presenting high-quality contributions.

The 12th ICSCS will be held in China and it is already being looked forward by researchers from all over the world.

F. Galembeck, Campinas, Brazil

Progr Colloid Polym Sci (2004) 128: VI–IX

CONTENTS

Nanostructures and Nanotechnology

Self-Assembly and Structured Fluids

Biocolloids and Biological Applications

Colloidal Dispersions

Surfactants and Polymers

Technology and Applications

Colloids and Surfaces in Oil Production

Environmental Colloidal Science

Progr Colloid Polym Sci (2004) 128: 1–10
DOI 10.1007/b97058

Robert J. Hunter

More reliable zeta potentials using electroacoustics

R.J. Hunter
School of Chemistry,
University of Sydney,
NSW 2006, Australia
e-mail: hunter_r@chem.usyd.edu.au

Abstract The traditional methods of determining zeta potentials (particle electrophoresis, streaming potential and electroosmosis) all suffer from real drawbacks in the matter of reliability. Measurements on very dilute systems (using, say, microelectrophoresis) are prone to contamination problems, but even when the direct current mobility is accurately assessed, current estimates of zeta potentials can be in error by as much as 100% or 100 mV in the most unfavourable cases. Electroacoustics overcomes the contamination problem by allowing measurements at significant particle concentrations. The simultaneous measurement of both size and zeta potential, using the electroacoustic spectroscopy technique, means that the correct average size is used for estimating zeta in polydisperse systems. Use of the entire mobility spectrum, over a range of frequencies, allows one to avoid the ambiguities which can occur with a single point (direct current) measurement. One can also take account of the effects of stagnant layer conduction, either by undertaking simultaneous measurements of the dielectric relaxation behaviour at high frequency or, more simply, by doing a salt titration from low ionic strength to concentrations where the effect of stagnant layer conduction is negligible.

Keywords Electroacoustic spectrum · Zeta potential · Stagnant layer conduction · Mobility spectrum

Introduction

Accurate assessment of the zeta potential is very important when one seeks to characterise the stability behaviour of a particulate suspension or emulsion. The interaction energy and the force between approaching particles both vary as the square of the zeta potential [1] and that has a profound effect on not only the colloid stability of the sol, but also on its rheological (flow) behaviour, its sedimentation and redispersion after settling, its filtration and flotation behaviour, and on more subtle effects like the shelf life of suspension products.

Johnson et al. [2] give an excellent recent example of the use of electroacoustics to assess the zeta potential of the well-known Sumitomo product AKP 30 alumina, as a function of pH at various volume fractions. They show that the zeta potential is independent of volume fraction from 20 to 30% (v/v) and has the same value as that determined by electrophoresis at high dilution. The shear stress, τ, is a parabolic function of the pH with a maximum at the isoelectric point of the suspension (around pH 9) and all of the data can be reduced to a single curve by normalising the shear stress to the maximum value, τ_{max}. The plot of the reduced shear stress (τ/τ_{max}) of the suspension can then be quantitatively correlated with the square of the zeta potential with the only adjustable parameter being the distance of closest approach of the particles during a collision (which is around 2.3 nm). The measurements of shear yield strength were made with a Vane rheometer and the measurements of the zeta potential as a function of pH were done by electroacoustics so that they could be done at the same concentration as the shear stress measurements.

The range of zeta potentials involved in this case was from 0 to 50 mV and in this region the relation between ζ and the direct current (d.c.) electrophoretic mobility, μ_E, is, to a good approximation, linear with a slope which depends on the ratio of the particle size, a, to the thickness of the electrical double layer, $1/\kappa$, where κ is the Debye–Hückel parameter. For κa approaching infinity (a large particle with a relatively thin double layer), the relation is given by the Smoluchowski equation:

$$\mu_E = \varepsilon\zeta/\eta, \quad (1)$$

where ε and η are the permittivity and viscosity of the suspension medium, respectively. At the other extreme, ($\kappa a < 1$: a small particle and/or a very low salt concentration) the Hückel equation is appropriate:

$$\mu_E = 2\varepsilon\zeta/3\eta. \quad (2)$$

For intermediate values, one must take account of how the electric field lines are disposed around the particle. This was first done by Henry [3] but a much more easily applied solution has since been given by Ohshima [4]:

$$\mu_E = (2\varepsilon\zeta/3\eta)f_1(\kappa a), \quad (3)$$

where

$$f_1(\kappa a) = 1 + 0.5(1 + \delta/\kappa a)^{-3} \text{ and}$$
$$\delta = 2.5[1 + 2\exp(-\kappa a)]. \quad (4)$$

These are the equations used to calculate the data in Fig. 1. Many commercial instruments which purport to measure the zeta potential proceed no further with the analysis than Eq. (1), preferring to leave the calculation of a true zeta potential to the working scientist. It is obvious that this can involve an error in ζ of 50% in an extreme case, if the wrong value of κa is used.

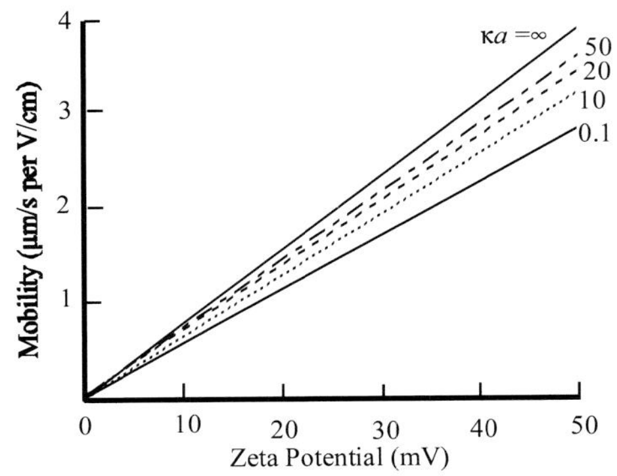

Fig. 1 Relation between direct current (*d.c.*) electrophoretic mobility and zeta potential for particles with relatively small zeta potentials

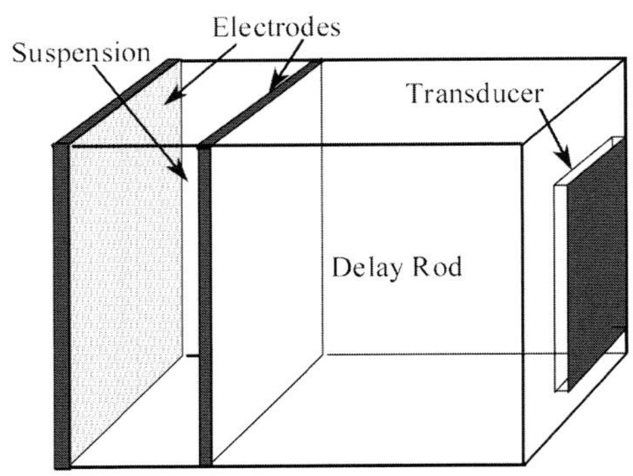

Fig. 2 Schematic arrangement for measuring the electroacoustic spectrum (see text)

The electroacoustic spectrum

A schematic diagram of the general principle involved in the measurement of the electroacoustic spectrum is shown in Fig. 2. The suspension or emulsion is contained between two inert electrodes a few centimetres apart and a high frequency (megahertz) electric field of the order of a hundred volts per centimetre is applied to it in the form of a short (microsecond) pulse (strictly a "tone burst"). The particles move backwards and forwards in response to the field and their joint motion creates a pressure wave or a sound pulse. The pressure wave is actually cancelled in the bulk of the liquid but near the electrodes there is no cancellation and it is from there that the sound waves emanate (one from each electrode). The sound waves travel along the delay rod and are detected by a transducer. The purpose of the delay rod (usually made of glass or plastic) is to separate the electronic noise created by the application of the field from the resulting very small sound signal so that it can be more readily detected.

The relation between the applied field and the resulting response is illustrated in Fig. 3, which defines the dynamic mobility, μ_d. The magnitude of the mobility is the ratio of the velocity amplitude of the sound wave to the peak voltage of the applied field. It is the analogue of the d.c. mobility referred to earlier. At these high frequencies the particles (if they exceed about 70 nm in diameter) begin to lag behind the field and it is this phase

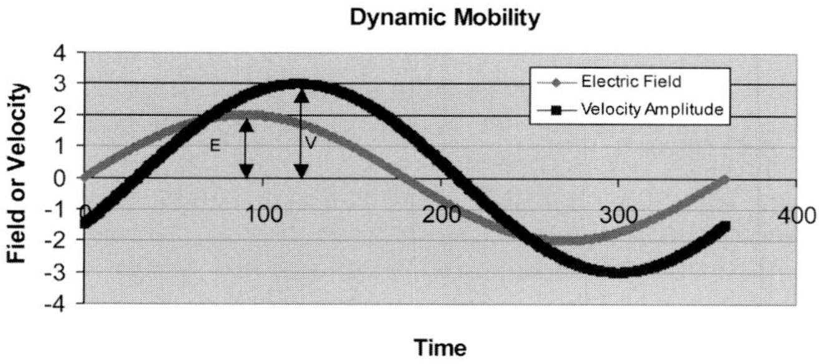

Fig. 3 The applied field here is $E\sin(\omega t)$ and the particle velocity is $V\sin(\omega t-\theta)$. The dynamic mobility μ_d is a complex quantity which has both a magnitude $|\mu_d| = V/E$ and a phase angle $\arg(\mu_d) = -\theta$

lag which, if accurately measured as a function of frequency, can be used to estimate the size of the particles.

The measurement which is made in the AcoustoSizer (Colloidal Dynamics, Warwick, RI) [5] is of the electrokinetic sonic amplitude (ESA) effect. This is the ratio of the voltage output of the transducer compared to the voltage applied across the electrodes, with a correction for the relative impedances of those two systems. O'Brien [6] has shown that this quantity (which is complex) is given by

$$\mathrm{ESA} = A(\omega)\phi(\Delta\rho/\rho)\mu_d[Z_s Z_r/(Z_s + Z_r)]. \qquad (5)$$

Here the Zs are the acoustic impedances of the suspension (s) and the delay rod (r), $\Delta\rho$ is the difference between the density of the suspended particles and the suspension medium of density ρ, ϕ is the volume fraction and $A(\omega)$ is a frequency-dependent calibration function which can be evaluated using a salt solution for which the electroacoustic signal can be calculated from its transport and thermodynamic equilibrium properties. (The usual choice is potassium dodecasilicotungstate because it has a large ESA signal.) This equation is valid for arbitrary concentrations and if the density and volume fraction are known it allows one to determine directly the dynamic mobility from the measured signal since the instrument measures the relevant acoustic impedances directly by sound reflection.

For relatively dilute systems (approximately 5% v/v or less) O'Brien [7] has shown that for spherical particles with a relatively thin double layer ($\kappa a > 20$ preferably but certainly greater than about 13) the dynamic mobility is related to the particle properties as follows:

$$\mu_d = \frac{2\varepsilon\zeta}{3\eta}\left[1 + f\left(\omega, Du, \varepsilon_p\right)\right] G\left(\omega a^2 \rho/\eta\right), \qquad (6)$$

where $Du(=K_s/K^{\infty}a)$ is the Dukhin number, which measures the relative conduction through the surface regions of the particle (K_s) compared with the bulk conductivity (K^{∞}) and ε_p is the particle permittivity. Both f and G are complex quantities: f describes the tangential field around the particle and it is this which drives the motion; G is called the inertial term and it is determined primarily by the particle size. It has a value of unity and zero phase angle at low frequencies for small particles so it has no effect under those conditions. As both ω and a increase, G falls exponentially to zero and the phase angle falls steadily to a minimum valued of −45°. For small Du, ε_p and zeta potential (below 50 mV, say) the value of f is close to 0.5 so Eq. (6) reduces to the Smoluchowski equation (Eq. 1) with a correction for particle inertia which depends largely on ωa^2. The dependence on ε_p can usually be ignored unless one is dealing with a material like titania which has a very high permittivity. The resulting spectra for "normal" systems are shown in Fig. 4, for small a (about 50 nm) and larger (micron-sized) particles. Note that the particle size effect shows up strongly in the phase angle and also in the magnitude, which falls off more slowly for small particles. Both of these characteristics are used to assess the size of the particles, whereas the zeta potential is determined almost entirely by the magnitude of the signal (except for some anomalous situations which will be discussed later).

One important consequence of the simultaneous measurement of zeta potential and size using the same signal is that in polydisperse systems, the average size estimated by this procedure will automatically be the appropriate choice for calculating the value of the parameter κa and hence assuring that the correct formula for ζ is used. In the case of electroacoustic spectrum determination using the AcoustoSizer, the entire calculation is conducted by the instrument software and is invisible to the user.

For suspensions of higher concentration, a more elaborate calculation is required. Rider and O'Brien [8] made the first attack on the problem by calculating the interaction between two particles. This enabled the calculations to be extended up to about 10% volume fraction. The next important step occurred when O'Brien at al. [9] were able to solve the more general problem for near-neutrally buoyant particles ($\Delta\rho$ small compared

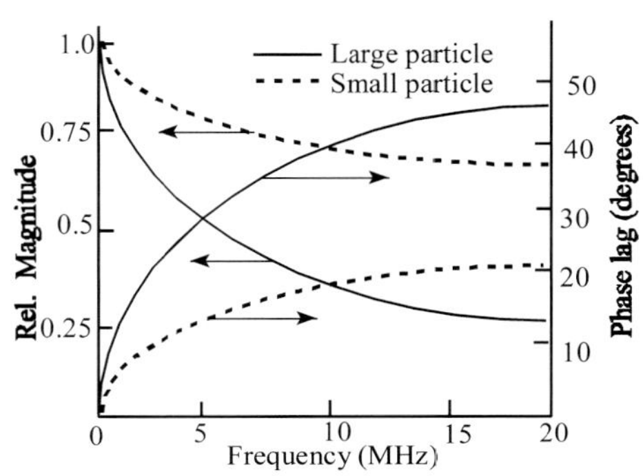

Fig. 4 Typical electroacoustic spectra for small (a~50 nm) and larger (micron-sized) particles

with ρ) by assuming that the interactions could be treated as pairwise-additive. Using the standard Percus–Yevick equation of statistical mechanics to describe the average distance between particles, they were able to provide a very satisfactory solution which could account quantitatively for the effects of concentration for these systems up to over 50% volume fraction. (Fig. 5 [10].) It is important to note that the concentration correction correctly handles both the zeta potential and the size, which means that it gives a good account of both the magnitude and the phase angle over the entire frequency range (1–20 MHz). Some other treatments [11] are concerned only with correcting the value of the zeta potential obtained from a measurement of the magnitude of the electroacoustic signal at a single frequency. That is a good deal easier to do (provided the correct value of κa is used) since the effect on the magnitude can be allowed for by the usual device of replacing the density of the suspension medium with the density of the suspension (the "effective medium" treatment which has proved so effective in sedimentation and rheological studies [12]). Ennis et al. [13] have shown how well this works for the magnitude of the dynamic mobility even at frequencies above 1 MHz (see the comparison in Ref. [14]). Unfortunately it proves quite inadequate to account for the phase-angle behaviour in this case.

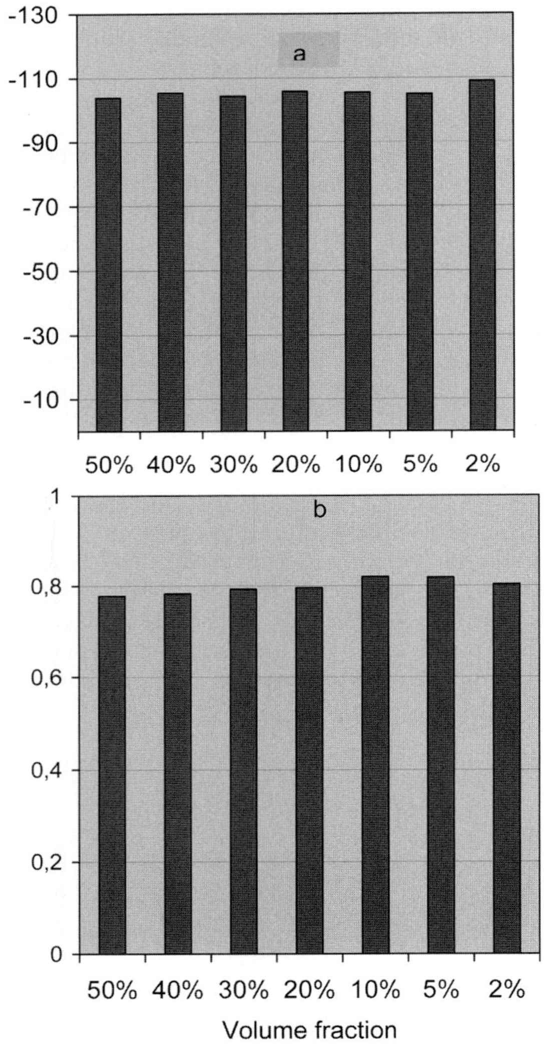

Fig. 5a Zeta potential as a function of dilution for a sunflower oil-in-water emulsion initially at 50% (v/v) and diluted with the ambient fluid. (From Ref. [10]). **b** Particle size of the same emulsion as in **a** [10] calculated using O'Brien's concentration correction [9]

Returning briefly to the determination of the zeta potential from d.c. electrophoresis measurements, we see from Fig. 6 that the numerically calculated relation between the two [15] reveals regions (A) where even the most precise mobility determination could still give rise to significant uncertainties (about 60 mV) in the zeta potential (the horizontal broken lines indicate an error of a few percent in the mobility measurement). Even worse would be a mobility of 6.0 units at a κa value of 20 where one would be presented with alternatives of 90 and 210 mV.

A few years ago we could dismiss the upper value as impossible and settle for the lower one, but that is not always appropriate. An electroacoustic spectrum of an emulsion system which returned a zeta potential of –175 mV at $\kappa a = 43$ [16] is shown in Fig. 7. If we consult the O'Brien and White data [15] for this value of κa (Fig. 6) we find a lower root at 103 mV, but Fig. 7 shows that that value is incorrect.

The electroacoustic spectrum clearly demonstrates that the higher absolute value of the zeta potential is the correct one. The spectrum does not exhibit the ambiguity of the d.c. result, but one can clearly see that both the low and the high zeta value have the same limiting low-frequency magnitude of about 6.0 units.

Stagnant layer conduction

The problem of what has become known as stagnant layer conduction (SLC) (sometimes referred to as anomalous surface conduction or the dynamic Stern layer) occurs most obviously at low electrolyte concentrations or for relatively small particles, when the term Du, introduced in Eq. (6), becomes significant [17]. Since Eq. (6) is derived for thin double-layer systems ($\kappa a > 15$, say) there is a lower limit to the value of the salt content and the particle size which can be dealt with in any case, but it is still possible to study some systems in which SLC is an important factor. It is important to recognise, however, that SLC is but one possible explanation for some of the phenomena for which it has been invoked as an explanation.

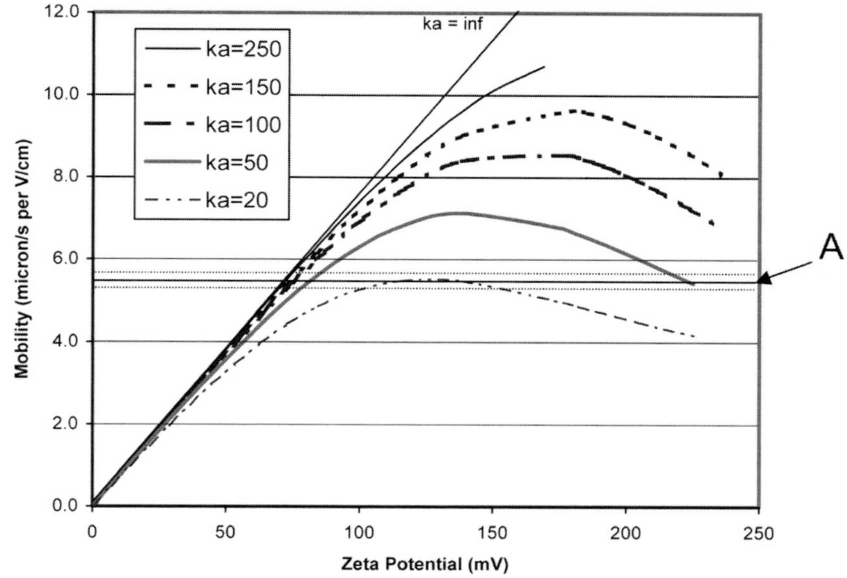

Fig. 6 Relation between d.c. mobility and zeta potential for a sphere (numerical results calculated by O'Brien and White [15]). A mobility of 5.5 μm/s/V/cm (*A*) with the usual experimental error of a few percent could give rise to an uncertainty in the zeta potential of the order of 50 mV or more

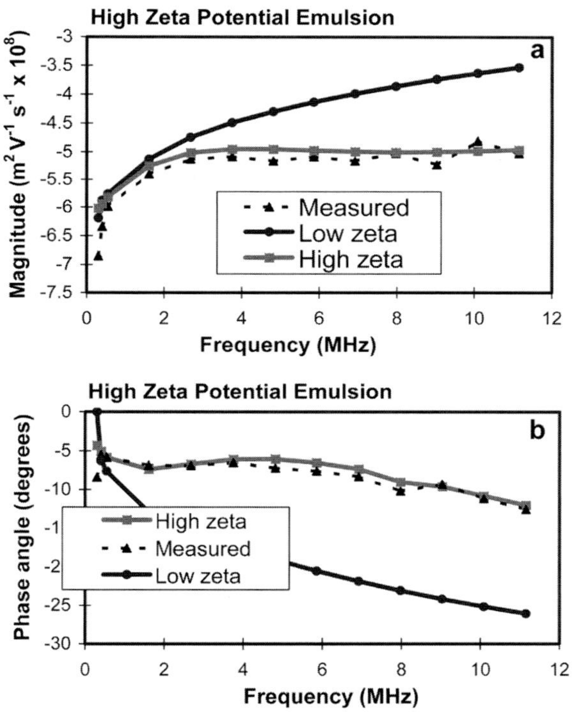

Fig. 7a Comparison of the electrokinetic sonic amplitude (*ESA*) spectrum (magnitude) for low and high zeta potential values. Note that the limiting value at low frequencies is the same for both. **b** Comparison of the ESA spectrum (phase) for low and high zeta potential values

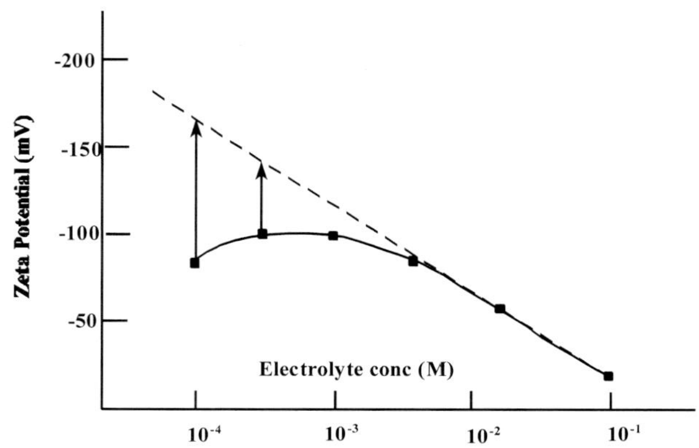

Fig. 8 Dependence of the zeta potential of a monodisperse polystyrene sulfate latex on the concentration of indifferent electrolyte. The *broken line* indicates the anticipated trend for a simple Gouy–Chapman model of the double-layer system

Studies of the phenomenon [18, 19] have been conducted mainly on monodisperse spherical latex particles, chiefly the polystyrene sulfate system prepared by the method of Goodwin et al. [20]. The characteristic feature of these experiments is that the zeta potential is studied as a function of indifferent electrolyte concentration, usually from very low concentrations (about 10^{-5} M) up to 10–20 mM, where the system remains stable. The observed behaviour is illustrated in Fig. 8, which shows the characteristic maximum in the magnitude of the zeta potential in the neighbourhood of 1 mM salt concentration.

Simple models of such a system would suggest that the zeta potential would continue to increase in magnitude as the indifferent salt concentration was reduced (though it need not be log–linear, as indicated in the figure). The decrease in the measured values of $|\zeta|$ at low electrolyte concentration can be explained by invoking the SLC process and we will discuss that subject later. It should be noted, however, that these sorts of studies are customarily conducted using data from electrophoresis and consideration of the results of the previous section would suggest a possible alternate explanation. As indicated in Fig. 8, it may be that in some cases the low value for $|\zeta|$ is simply a consequence of choosing the wrong root of the governing equations. Such a case was examined by Midmore et al. [21], who found that by choosing the upper value for $|\zeta|$, their plot of $|\zeta|$ against indifferent electrolyte concentration was log–linear. They measured a (dimensionless) mobility of 4.39 at 1 mM salt concentration and so had a choice of either 87 or 179 mV. Subsequently we were able to show, using electroacoustics [22], that the original interpretation had been correct (Fig. 9).

Midmore et al. [21] also showed that the charge corresponding to their measured zeta potentials was close to the surface charge (σ_s) determined by titration. That means there can be no (or very little) charge between the surface and the slipping plane. That is also suggested by the slope of the data in Fig. 9, which is close to that required by the Nernst equation (−60 mV per tenfold change in electrolyte concentration). When the zeta potential is measuring all of the countercharge, then the electrokinetic charge (σ_e) remains constant and equal to σ_s as the electrolyte concentration changes. It can then be shown using the standard Gouy–Chapman equations for the diffuse double layer that

$$\left(\frac{d\zeta}{d\log_{10}c}\right)_\sigma = -2.303\frac{kT}{ze}\tanh\left(\frac{ze\zeta}{2kT}\right) \approx -60\,\mathrm{mV}\ \text{for large}\ \zeta \tag{7}$$

when the tanh function approaches unity.

It is not always so, however. Indeed this seems to be the exception rather than the rule. More usually the electrokinetic charge is significantly lower in magnitude than the surface charge and a considerable amount of the countercharge lies behind the shear surface. It is in these cases that the SLC can become significant and Minor et al. [18] give convincing evidence for its importance in the d.c. electrophoresis context and we have been able to verify that it occurs at high frequency in some systems.

The effect is illustrated in a very simple way in Fig. 10. In the classical theory of electrophoresis the back field was calculated from the zeta potential, which inherently assumed that only the diffuse-layer ions outside the shear surface contribute to the conduction current. If SLC is occurring then the back field is much larger so the particle is slowed down more severely and one underestimates the zeta potential magnitude. Since the relative importance of this effect is measured by the Dukhin number ($Du = K_s/K^\infty a$) it is obviously more serious at low salt concentrations and for small particles. This then provides an alternative explanation for the maximum shown in Fig. 8. Furthermore, when one measures the actual conduction behaviour of colloidal dispersions one finds a discrepancy between the values of the zeta potential calculated from the conduction behaviour and that obtained from, say, electrophoresis or streaming current measurements in plugs. The resolution of that discrepancy is discussed in Ref. [24].

The Dukhin number enters directly into the definition of the function f, which defines the tangential electric field in Eq. (6):

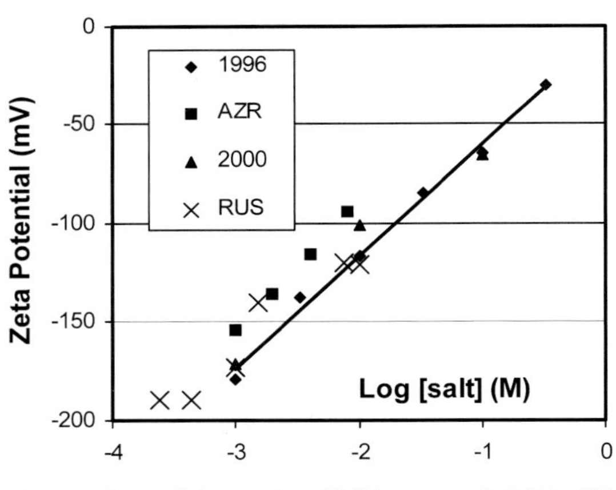

Fig. 9 Comparison of the results of Midmore et al. [21] in 1996 with Midmore's more recent measurements on the same polystyrene sulfate latex system (2000) and electroacoustic (AZR) results on this material. The *crosses* represent data from Russell [23] measured by electrophoresis and high-frequency conduction on a similar system. The slope of the original data is very close to the Nernst value of −60 mV per tenfold change in concentration

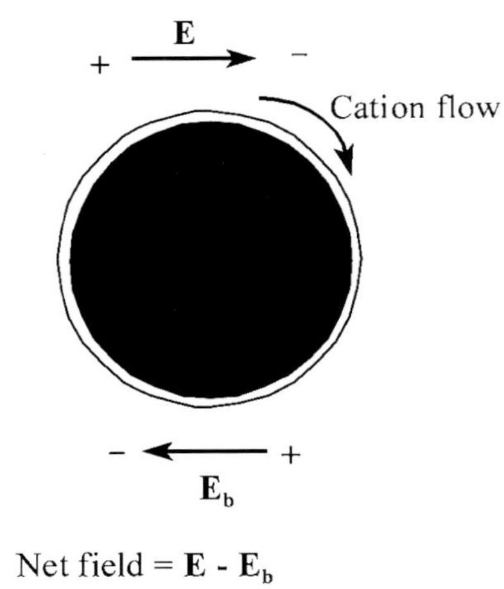

Fig. 10 The effect on a negative particle of surface conduction. The applied field, **E**, causes a flow of cations in the direction shown and this creates a back field, $\mathbf{E_b}$, which slows the particle's mobility (see text)

$$f = \frac{1 + i\omega' - (2Du + i\omega'\varepsilon_p/\varepsilon)}{2(1 + i\omega') + (2Du + i\omega'\varepsilon_p/\varepsilon)} \quad \text{where } \omega' = \omega\varepsilon/K^\infty. \tag{8}$$

The initial procedure used to evaluate f in the AcoustoSizer software was to use the estimated value of the zeta potential to assess the value of Du, using a relation of the type

$$Du^d = Du(\zeta) = \left(\frac{2}{\kappa a}\right)\left[\cosh\left(\frac{ze\zeta}{2kT}\right) - 1\right]\left(1 + \frac{3m}{z^2}\right), \tag{9}$$

where the superscript d refers to the diffuse double layer and $m \approx m_+$ is the ionic mobility constant, which is 0.256 (for the sodium ion) in the present case. (This relation is adequate for high zeta potentials when only the counterion needs to be considered, but a more elaborate expression is used in the instrument software to cover other possibilities.) This procedure implicitly assumes that conduction occurs only in the diffuse layer beyond the shear plane. An alternative procedure is, however, possible if one has either (1) an independent assessment of Du or (2) an alternative procedure for estimating Du from the electroacoustic spectrum. Both methods are now accessible. We will show in the following how to incorporate Du values obtained from high-frequency conductivity (or dielectric relaxation) data and how to correct for SLC by the use of salt titration.

High-frequency conduction (dielectric relaxation) studies

Since the electroacoustic spectrum is measured in the frequency range from 1 to 20 MHz it is natural to assume that the conductivity behaviour which is relevant is in the same frequency range. Although measurement of such high-frequency conductivity presents some experimental challenges, it turns out that the theory to describe the results is a good deal simpler than that for low frequencies. The reason for this is that the alternations of the field occur so rapidly that the ions are limited in their movement to an oscillation within the double layer around the particle. At lower (kilohertz) frequencies there occurs a build up of concentration of salt on one side of the particle during an oscillation and this introduces an additional diffusion process which does not occur at high frequency.

The complex conductivity of a suspension, K^*, has a real (dissipative) part and an imaginary (storage or capacitive) part, as does the suspension medium. O'Brien [25] has shown that the bulk electrical conductivity of a suspension of particles with thin double layers is given by

$$K^*/(K^\infty + i\omega\varepsilon) = 1 + N_p < S > /\varepsilon_0, \tag{10}$$

where N_p is the number of particles per unit volume and $<S>$ is the volume-average value of what is called the dipole strength of the particles. The procedures for measuring K^* for the suspension and for the medium are discussed by Midmore et al. [26]. They also show that

$$K^* = (K^\infty + i\omega\varepsilon)(1 - 2\phi f)/(1 + \phi f), \tag{11}$$

so

$$< S > = -[f/(1 + \phi f)]4\pi a^3\varepsilon_o, \tag{12}$$

where the function f is identical to that in Eq. (6). Measurement of the dielectric relaxation behaviour thus gives us a direct measure of the function f instead of having to rely on the developing assessment of the best-fit zeta potential value (which provides only the outer diffuse layer contribution).

The success of this procedure [27] is illustrated in Fig. 11 and Table 1. The figure shows the fits to the magnitude data for a sample of perfluoromethyldecalin stabilised with 5 mM sodium dodecyl sulfate. The fits to the phase angle (Fig. 12) are also quite satisfactory [27]. Note that the low-frequency limiting values of the magnitude increase with increasing concentration of salt and reach a maximum at about 10 mM and then decrease slightly in this case. This is the same behaviour as is observed in an electrophoresis experiment in cases like this and is an indication of possible SLC. Note also that the phase-angle behaviour is more like the normal theoretical curve (Fig. 4) at the higher concentration when SLC is much less important.

In these systems with rather high zeta potentials we can confidently assume that changing the indifferent electrolyte concentration up to, say, 30 mM should not cause any flocculation and therefore should not cause any change in the particle size. In this case we can see from Table 1 that by using the measured dielectric data we can fit both the magnitude and the phase-angle spectrum with a constant particle size (as determined by electroacoustics at the highest concentration) and a zeta potential which steadily decreases in magnitude with increasing salt concentration as one would expect.

The dielectric data are fitted by matching the theoretical and experimental values of the real and imaginary parts of the function S [27]. One can then obtain f from Eq. (12) and hence an estimate of the Dukhin number (Du_{DR} in Table 1). The Du value estimated from the zeta potential is Du^d and we can also estimate a Du value by simply trying to fit the electroacoustic data using a constant size. In this case the size is 0.55 μm with a spread (15–85%) from 0.52 to 0.58 μm, estimated from the behaviour at the highest salt concentration when SLC is no longer important. The resulting value of Du is listed in Table 1 as Du_{EA} and it is in excellent agreement with the value obtained more directly from the complex conductivity (dielectric relaxation) measurements. In constructing the table it was assumed that all, or most,

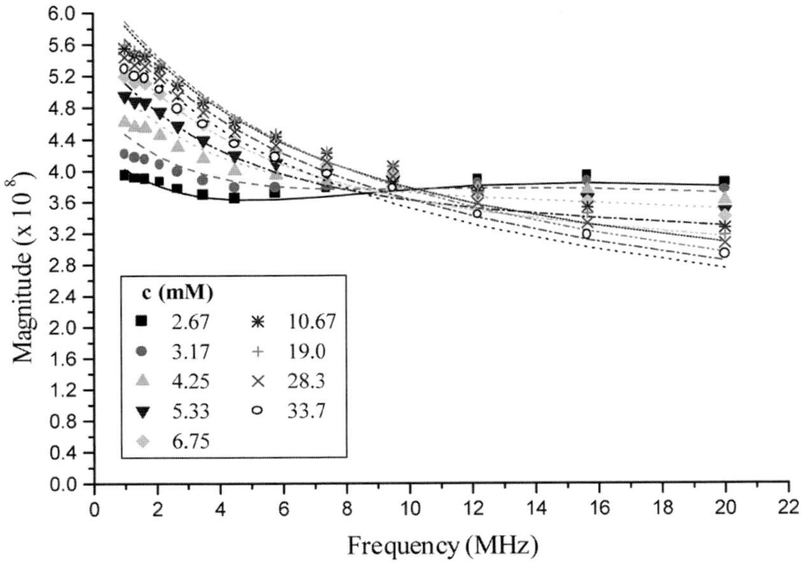

Fig. 11 Magnitude of the dynamic mobility of perfluormethyldecalin emulsion as a function of added salt (NaCl) concentration. Note that the low-frequency limiting values increase with increasing salt concentration and reach a maximum value at around 10 mM in this case [27]

Table 1 Zeta potential and conductivity results for a 5 vol % perfluormethyldecalin emulsion as a function of NaCl concentration [27]

c (mM)	K^{∞} (S/m)	ζ (mV)	q	Du^{d}	Du_{EA} (qDu^{d})	Du_{DR}	$\lvert\psi_{DR}\rvert$ (mV)	$\lvert\psi_{EA}\rvert$ (mV)	σ_{DR} ($\mu C/cm^2$)
2.67	0.032	−140	3.2	0.452	1.45	1.6	202	197	13.1
3.17	0.038	−138	3.0	0.398	1.19	1.3	195	191	12.6
4.25	0.051	−132	3.0	0.303	0.91	0.93	187	186	12.2
5.33	0.064	−125	3.25	0.233	0.757	0.77	183	182	12.6
6.75	0.081	−118	3.3	0.178	0.588	0.60	176	175	12.6
10.67	0.128	−111	3.15	0.122	0.383	0.38	165	165	12.8
19.0	0.228	−98	3.2	0.068	0.218	0.22	153	152	13.5
28.3	0.339	−89	3.4	0.045	0.154	0.15	145	146	14.1
33.7	0.404	−85	3.3	0.038	0.125	0.12	138	139	13.7

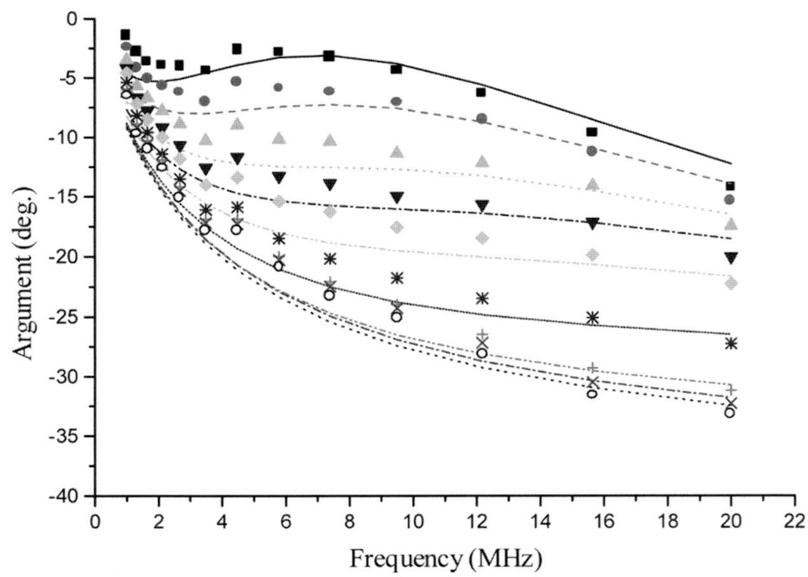

Fig. 12 Phase-angle data for the system described in Fig. 11. The *lines* are fitted with the measured value of *Du*

Table 2 The electroacoustic behaviour of eicosane ($C_{20}H_{42}$) as a function of temperature in the neighbourhood of the melting point (42 °C). The factor q is defined by $Du = qDu^d$ [28]

T (°C)	ρ (g/cm^3)	K^∞ (S/m)	ζ (mV)	Du^d	d_{15}–d_{50}–d_{85}(μm)	Factor q	Phase
43.0	0.79	0.179	−137	0.272	0.29–0.32–0.35	1.80	Liquid
39.3	0.79	0.168	−137	0.292	0.29–0.32–0.35	1.80	Liquid
35.2	0.80	0.156	−138	0.312	0.29–0.33–0.38	1.60	Liquid
25.6	0.80	0.130	−140	0.394	0.29–0.33–0.38	1.30	Liquid
21.4	0.905	0.118	−144	0.78	0.17–0.20–0.27	1.0	Solid
15.9	0.917	0.103	−143	0.865	0.14–0.20–0.29	1.0	Solid

of the counterions behind the shear plane were still in the diffuse layer and the resulting outer Helmholtz plane potentials are listed in the columns headed by ψ. The diffuse layer charge is fairly constant with increasing salt concentration and is presumably determined solely by the adsorption density of the sodium dodecyl sulfate. The plot of ψ against $\log_{10}c$ has a slope of −55 mV [27], which also indicates that with ψ we are measuring most of the diffuse layer charge (from Eq. 7).

At least for these high zeta potential situations, the measurement of the dielectric behaviour is then rather superfluous. One can instead undertake a titration with an indifferent salt, from below 1 mM to a few times 10^{-2} M and assume that the particle size will be the same as that measured at the highest concentration. (The rapid coagulation value of most sols occurs at about 50 mM for a 1:1 indifferent electrolyte.) One can then use the Du value as an additional fitting parameter. The procedure is described in Ref. [27], which also gives the results of a very satisfactory analysis of the electroacoustic spectrum of a hexadecane emulsion where the data were fitted without recourse to the dielectric behaviour.

Rigidity of the liquid–liquid interface

An important aspect of this analysis is that Eq. (6) was derived for the case of a solid spherical particle immersed in a liquid. It has long been assumed that, in d.c. electrophoresis, an emulsion droplet behaves like a solid particle if it is stabilised with a saturated layer of surfactant. No momentum transfer then occurs across the interface and there is no fluid motion inside the droplet. It cannot be assumed that this is necessarily the case at high frequency since the motions involved are so very different. [The distance through which a particle (or a parcel of fluid) moves in the course of a half cycle is of the order of 10 pm at the higher frequencies.] We therefore investigated what happens in these systems when the emulsion droplets change from liquid to solid.

The study [28] was done on the straight-chain alkane eicosane ($C_{20}H_{42}$), which was chosen because it melts at about 38 °C. Cooling the emulsion from an initial temperature of about 42 °C, one can distinguish the crystallisation process at a temperature of about 22 °C. The high degree of supercooling is to be expected since the individual droplets must each crystallise separately. The process does not seem to be assisted by collisions between liquid droplets and ones which have already solidified since we find no evidence of doublet formation. In fact the apparent particle size decreases quite markedly on solidification (from about 0.33 to 0.2 μm). Some of this reduction is due to the increased density of the solid but most of it is an artefact of the measurement. When they solidify, the particles change from spherical to more ellipsoidal in shape. In such a system the electroacoustic size measurement yields an equivalent spherical radius which is intermediate in value between the radii of the major and minor axes [29]. In this case the result can be explained by an axial ratio of about 2.5 to 1 which is not surprising given the rather flat crystal habit of solid alkanes.

The key observation in the study [28] is that the zeta potential changes very little in the crystallisation process as is indicated in Table 2.

The solid shows no evidence of SLC but the liquid has some, which increases with temperature. The increase in the value of Du^d as the liquid crystallises is almost entirely due to the decrease in the particle size.

Conclusions

Because electroacoustics measures both size and charge from the same signal it possesses an inherent advantage over methods which measure the zeta potential alone. The fact that an entire spectrum of values over a range of frequencies must be fitted for both magnitude and phase angle also provides more certainty in the result. The ambiguities which naturally attend the electrophoretic process for moderate to high zeta potentials can be immediately removed using electroacoustics. It can also provide an intrinsic solution to the SLC problem, in the simple case where one is investigating the influence of an indifferent electrolyte.

Acknowledgements I am indebted to my colleagues, Richard O'Brien (for his insights into electroacoustics), James Beattie (for his enthusiastic involvement in our joint research program) and Alex Djerdjev for his careful attention to the experimental program.

References

1. Hunter RJ (2001) Foundations of colloid science, 2nd edn. Oxford University Press, Oxford, p 755
2. Johnson SB, Russell AS, Scales PJ (1998) Colloids Surf A 141:119–130
3. Henry DC (1931) Proc R Soc Lond Ser A 133:106
4. Ohshima H (1994) J Colloid Interface Sci 168:269–71
5. O'Brien RW, Cannon DC, Rowlands WN (1995) J Colloid Interface Sci 173:406
6. O'Brien RW (1988) J Fluid Mech 190:71–86
7. O'Brien RW (1990) J Fluid Mech 212:81–93
8. Rider P, O'Brien RW (1993) J Fluid Mech 257:607–636
9. O'Brien RW, Wade TA, Carasso ML, Hunter RJ, Rowlands WN, Beattie JK (1996) In: Provder T (ed) Proceedings of the American Chemical Society symposium, Orlando, Florida, August 1996. ACS symposium series 693. American Chemical Society Washington, DC, p 311
10. Kong L, Beattie JK, Hunter RJ (2001) J Colloid Interface Sci 238:70–79
11. Dukhin AS, Goetz PJ (2002) In: Mobius D, Miller R (eds) Studies in interfacial science no 15. Elsevier, Amsterdam
12. Hunter RJ (2001) Foundations of colloid science, 2nd edn. Oxford University Press, Oxford, p 197
13. (a) Ennis J, Shugai AA, Carnie SL (2000) J Colloid Interface Sci 223:21–36; (b) Ennis J, Shugai AA, Carnie SL (2000) J Colloid Interface Sci 223:37–53
14. Hunter RJ (2001) Foundations of colloid science, 2nd edn. Oxford University Press, Oxford, p 425
15. O'Brien RW, White LR (1978) J Chem Soc Faraday Trans II 74:1607
16. Hunter RJ, O'Brien RW (1997) Colloids Surf A 126:123–128
17. Hunter RJ (2003) Adv Colloid Interface Sci 100–102:153–167
18. Minor M, van der Linde ABJ, Lyklema J (1998) J Colloid Interface Sci 203:177–188
19. Moncho A, Martinez-Lopez F, Hidalgo-Alvarez R (2001) Colloids Surf 192:215–226
20. Goodwin JW, Hearn J, Ho CC, Ottewill RH (1974) Colloid Polym Sci 252:464
21. Midmore BR, Pratt GV, Herrington TM (1996) J Colloid Interface Sci 184:170
22. Hunter RJ, Midmore BR, Zhang H (2001) J Colloid Interface Sci 237:147–149
23. Russell AS (1999) PhD thesis. University of Melbourne
24. Hunter RJ (2001) Foundations of colloid science, 2nd edn. Oxford University Press, Oxford, pp 426–430
25. O'Brien RW (1986) J Colloid Interface Sci 113:81
26. Midmore BR, Hunter RJ, O'Brien RW (1987) J Colloid Interface Sci 120:210–217
27. Djerdjev AM, Beattie JK, Hunter RJ (2003) J Colloid Interface Sci 265:56–64
28. Djerdjev AM, Beattie JK, Hunter RJ (2003) Langmuir 19:6605–6610
29. Loewenberg M, O'Brien RW (1992) J Colloid Interface Sci 150:158–168

Progr Colloid Polym Sci (2004) 128: 11–13
DOI 10.1007/b97062

Hiroyuki Ohshima

Electrophoretic mobility of colloidal particles in salt-free media

H. Ohshima
Faculty of Pharmaceutical Sciences and Institute of Colloid and Interface Science, Science University of Tokyo, 2641 Yamazaki, Noda, Chiba 278-8510, Japan
e-mail: ohshima@rs.noda.tus.ac.jp
Tel.: +81-4-71213661
Fax: +81-4-71213661

Abstract It is shown that the electrophoretic mobility μ of dilute spherical colloidal particles of radius a, carrying charge Q in a salt-free medium (containing only counterions) is expressed generally as $\mu = Q_{\mathrm{eff}}/D_{\mathrm{H}}$, irrespective of the particle type, where Q_{eff} is the effective particle charge and D_{H} is the drag coefficient of the particle. There is a certain critical value of Q separating two cases: the low-charge case and the high-charge case. In the low-charge case, Q_{eff} equals the real particle charge Q, while in the high-charge case, Q_{eff} becomes constant independent of Q. In the high-charge case, in which case counterion condensation occurs in the vicinity of the particle surface, there are two kinds of counterions: bound and free counterions. Bound counterions condensed at the particle surface do not contribute to the electrophoretic mobility so Q_{eff} and μ become constant independent of Q.

Keywords Electrophoretic mobility · Counterion condensation

Introduction

Electric phenomena in nonaqueous colloidal dispersions, which differ from those in aqueous dispersions owing to the low concentration of electrolyte ions, are of great importance in many applications [1]. Previous theoretical studies have shown that the potential distribution around a charged particle in a salt-free medium containing counterions only is quite different from that in an electrolyte solution [2–6] and that the electrophoretic mobility of particles in a salt-free medium also becomes different from that in an electrolyte solution [7–11]. There is a certain critical value of the particle charge separating two cases: the low-charge case and the high-charge case. In the high-charge case, counterions are accumulated in the vicinity of the particle surface and the electrophoretic mobility of the particles becomes constant independent of the particle charge. Although experimentally there is no salt-free medium, a salt-free medium serves as a good model for a nonaqueous medium. In the present paper we derive the most general expression for the electrophoretic mobility of spherical charged particles in a salt-free medium, which is applicable for all types of particles, including hard particles and liquid drops.

Potential distribution around a particle in a salt-free medium

Consider a dilute suspension of spherical colloidal particles of radius a with charge Q in a salt-free medium containing only counterions. We assume that each sphere is surrounded by a spherical free volume of radius R, within which counterions are distributed so that electrical neutrality as a whole is satisfied. The particle volume fraction ϕ is given by $\phi = (a/R)^3$. We treat the case of dilute suspensions, viz., $\phi \ll 1$. We denote the electric potential at a distance r from the center O of one particle by $\psi(r)$. Let the average number density and the valence of the counterions be n and $-z$, respectively. Then we have $Q = (4\pi/3)(R^3 - a^3)zen$, where e is the elementary electric charge. We set the electric potential $\psi(r)$ to zero at points where the concentration of counterions equals its average value n. We introduce the scaled particle charge $Q^* = Q/\beta$, where $\beta = 4\pi\varepsilon_r\varepsilon_o a(kT/ze)$, ε_r is the relative permittivity of the medium, ε_o is the permittivity of a vacuum, k is the Boltzmann constant, and T is the absolute temperature.

It has been shown [4, 6] that there is a certain critical value of the particle charge Q separating two cases: the low-charge case (case 1): $Q^* \leqq \ln(1/\phi)$ and the high-charge case(case 2): $Q^* > \ln(1/\phi)$. In case 1, the potential distribution is the same as if counterions were absent and is given by the Coulomb potential, so $y(a)–y(R) = Q^*$. In case 2, since counterion condensation occurs, counterions condensed very near the particle surface form a thin layer at $a = r = a^*$, so $y(a^*)–y(R)$ becomes the effective surface potential of the particle. As shown in a previous paper [6], $y(a^*)–y(R)$ is equal to $\ln(1/\phi)$, independent of Q^* (see also the Appendix). By comparing this result with that for case 1 [$y(a)–y(R) = Q^*$], we find that the particle behaves as if the scaled particle charge were $\ln(1/\phi)$. That is, the effective scaled charge of the particle Q^*_{eff} is given by $\ln(1/\phi)$, or the effective particle charge Q_{eff} is $\beta\ln(1/\phi)$. These results may be summarized as follows:

$$Q^*_{\text{eff}} = \begin{cases} Q^*, & [Q^* \leq \ln(1/\phi), \text{case 1}] \\ \ln(1/\phi), & [Q^* > \ln(1/\phi), \text{case 2}] \end{cases} \quad (1)$$

Note that there are two kinds of counterions: bound and free counterions. In case 1, there exist only free counterions. In case 2, there are both types of counterions, i.e., bound counterions in the thin layer between $a \leqq r \leqq a^*$ and free counterions distributed in the region outside the thin layer $a^* \leqq r \leqq R$. As Q increases, the total number of counterions contained in a free volume increases. In case 1 the total number of free counterions always increases linearly as Q increases. In case 2, however, as Q increases, the number of free counterions remains constant and instead the number of bound counterions increases (Fig. 1).

In case 2, the ratio of the number of bound counterions to the total number of counterions, i.e., the binding constant K, is given by $K = 1–\ln(1/\phi)/Q^*$, which tends to 1 in the limit of very high Q. The ratio of the number of free counterions to the total number counterions in case 2 is given by $\ln(1/\phi)/Q^*$, which tends to 0 in the limit of very high Q. For case 1, on the other hand, there are no bound counterions (all counterions are free) and thus $K = 0$.

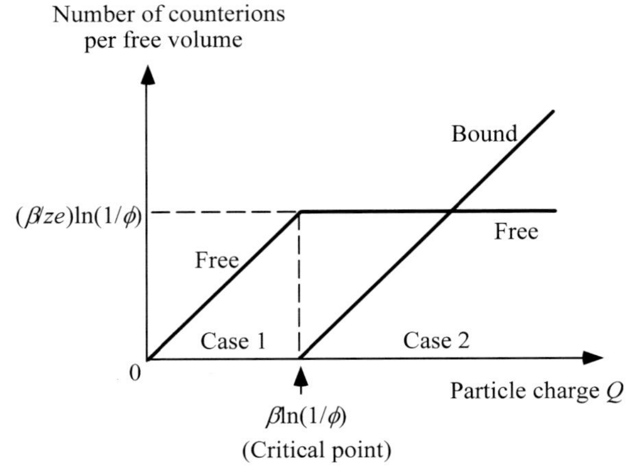

Fig. 1 The number of bound counterions and the number of free counterions per free volume as functions of the particle charge Q, where $\beta = 4\pi\varepsilon_r\varepsilon_o a(kT/ze)$. Only free counterions contribute to the electrophoretic mobility, leading to a constant electrophoretic mobility in case 2

Electrophoretic mobility

Consider a dilute suspension of spherical particles of radius a moving with a velocity $\mathbf{U}$ in a salt-free medium containing counterions only in an applied electric field $\mathbf{E}$. Since the particle behaves as if the scaled particle charge were Q^*_{eff}, which is Q^* for case 1 and $\ln(1/\phi)$ for case 2, the equation of balance of forces acting on the particle is $Q_{\text{eff}}E = D_H U$, where $E = |\mathbf{E}|$, $U = |\mathbf{U}|$ and D_H is the drag coefficient of the particle. The electrophoretic mobility $\mu = U/E$ is thus expressed by

$$\mu = \frac{Q_{\text{eff}}}{D_H} = \begin{cases} \frac{Q}{D_H}, & [Q^* \leq \ln(1/\phi), \text{case 1}] \\ \frac{1}{D_H}\beta\ln(1/\phi), & [Q^* > \ln(1/\phi), \text{case 2}] \end{cases} \quad (2)$$

If one introduces the scaled electrophoretic mobility defined by $E_m = (3\eta ze/2\varepsilon_r\varepsilon_o kT)\mu$, η being the viscosity of the medium, then Eq. (2) can be rewritten as

$$Em = \frac{6\pi\eta a}{D_H} Q^*_{\text{eff}} = \begin{cases} \frac{6\pi\eta a}{D_H} Q^*_{\text{eff}}, & [Q^* \leq \ln(1/\phi), \text{case 1}] \\ \frac{6\pi\eta a}{D_H} \ln(1/\phi), & [Q^* > \ln(1/\phi), \text{case 2}] \end{cases} \quad (3)$$

Equation (2) (or Eq. 3) is the required general expression for the electrophoretic mobility of spherical particles carrying charge Q in a salt-free medium, which is applicable for all particle types. The drag coefficient of the particle in Eq. (4) is given in the following for various types of particles.

1. Spherical particles of radius a:

$$D_H = 6\pi\eta a. \quad (4)$$

2. Liquid drops of radius a:

$$D_H = \frac{2\pi\eta a(3\eta_d + 2\eta)}{(\eta_d + \eta)}. \quad (5)$$

Conclusion

A simple expression for the electrophoretic mobility of particles in a salt-free medium was derived. For the low-charge case the surface potential of the particle is given by the Coulomb potential. For the high-charge case, there are two kinds of counterions: bound and free counterions. Only free counterions contribute to the electrophoretic mobility, leading to a constant electrophoretic mobility, independent of the real particle charge.

Appendix

We start with the Poisson–Boltzmann equation for the electric potential $\psi(r)$ around a sphere in a salt-free medium

$$\frac{d^2y}{dr^2}+\frac{2}{r}\frac{dy}{dr}=\kappa^2 e^y, \tag{6}$$

where $y(r) = ze\psi(r)/kT$ is the scaled potential and

$$\kappa=\left[\frac{3zeQ}{4\pi\varepsilon_r\varepsilon_0 kT(R^3-a^3)}\right]^{1/2}\approx\left(\frac{3Q^* a}{R^3}\right)^{1/2} \tag{7}$$

is the Debye–Hückel parameter in the present system. For the region $a^* \leqq r \leqq R$, Eq. (6) can be approximated by

$$\frac{d^2y}{dr^2}+\frac{2}{r}\frac{dy}{dr}=\kappa^2\exp[y(R)], \tag{8}$$

with the solution

$$y(r)=Q^*\exp[y(R)]\frac{a}{r}\left(1+\frac{r^3}{2R^3}-\frac{3r}{2R}\right)+y(R). \tag{9}$$

We note that as $r \rightarrow a^*$, Eq. (9) tends to a Coulomb potential $y(r) = Q^*\exp(y(R))a/r + y(R)$, which is the solution of the Laplace equation,

$$\frac{d^2y}{dr^2}+\frac{2}{r}\frac{dy}{dr}=0. \tag{10}$$

This implies that for $r \approx a^*$, the condition

$$-\frac{1}{a^*}\frac{dy}{dr}\bigg|_{r=a^*}\gg\kappa^2\exp[y(a^*)] \tag{11}$$

must hold, since under this condition, Eq. (6) becomes Eq. (10). From Eqs. (9) and (11) we have $R^3 \exp[y(R)] >> a^{*3}\exp[y(a^*)]$, which gives

$$y(a^*)-y(R)\leq\ln\left(R^3/a^{*3}\right)\approx\ln(1/\phi). \tag{12}$$

Equation (12) means that the value of $y(a^*)–y(R)$ cannot exceed $\ln(1/\phi)$ for the dilute case. In case 1 [$Q^* \leqq \ln(1/\phi)$], we find that $a^* = a$ and $y(a)–y(R) \leqq \ln(1/\phi)$, whereas in case 2 [$Q^* > \ln(1/\phi)$], $y(a^*)–y(R)$ can reach $\ln(1/\phi)$. For the region very near the particle surface $a < r < a^*$, the potential $y(r)$ decreases very sharply from $y(a)$ to $y(a^*)$, where an approximate expression for $y(a)$ as well as that for $y(R)$ have been derived as [6]

$$y(a)=\ln\left(\frac{Q}{6\phi}\right) \tag{13}$$

and

$$y(R)=-\ln\left[\frac{Q^*}{\ln(1/\phi)}\right] \tag{14}$$

Note that a^* approaches a as Q increases.

References

1. Siffert B, Jada A, J. Eleli Letsango (1951) J Colloid Interface Sci 163:327
2. Fuoss RM, Katchalsky A, Lifson S (1951) Proc Natl Acad Sci USA 37:579
3. Afrey T, Berg PW, Morawetz H (1951) J Polym Sci 7:543
4. Imai N, Oosawa F (1952) Busseiron Kenkyu 52:42
5. Oosawa F (1971) Polyelectrolytes. Dekker, New York
6. Ohshima H (2002) J Colloid Interface Sci 247:18
7. Ohshima H (2002) J Colloid Interface Sci 247:499
8. Ohshima H (2002) J Colloid Interface Sci 255:202
9. Ohshima H (2003) J Colloid Interface Sci 262:294
10. Ohshima H (2003) J Colloid Interface Sci 263:333
11. Ohshima H (2003) Colloids Surf A 222:207

Progr Colloid Polym Sci (2004) 128: 14–20
DOI 10.1007/b97065

Sandor Barany
Ferenc Madai
Vladimir Shilov

Study of nonlinear electrophoresis

Abstract We distinguish two kinds of nonlinear electrophoresis that can be observed in strong electric fields. The first is due to the interaction of the applied field E with the field-induced ionic charges, under conditions when the field-induced variation of the electrolyte concentration remains small compared to its equilibrium value. The second one ("superfast electrophoresis") is related to the interaction of a strong applied field with a secondary diffuse layer of counterions ("space charge") which is induced by the external field outside the primary diffuse electrical double layer because of the concentration polarization. In the first case, with increasing field strength, the processes that are nonlinear with respect to E start playing a more important role that can be characterized by the third power term in the expansion of the electrophoretic velocity in powers of E. An expression is derived for "cubic electrophoresis" of spherical particles with a thin double layer, using the procedure of successive approximation in powers of small-parameter E. In line with the theory at $E = 100$ V/cm and higher substantial nonlinear effects in the electrophoretic velocity of polystyrene latex and aluminum oxide particles as well as yeast cells were observed. In the second case, the interaction of the "space charge" that is formed near the surface of electron-type conducting particles like magnesium or semiconductors like pyrite and molybdenite with a strong electric field (100–500 V/cm) produces electrophoretic velocities that are 1–2 orders of magnitude higher compared with those typical for nonconducting particles. It has been shown that the mobility of particles increases linearly with E and the particle size, in contrast to classical electrophoresis. This can be explained by the Dukhin–Mishchuk theory of superfast electrophoresis. A new method for measuring electrophoresis in strong electric fields has been developed.

Keywords Electrophoresis · Nonlinear · Strong fields · Nonconducting particles · Conducting particles

S. Barany (✉) · F. Madai
University of Miskolc, 3515 Miskolc-Egyetemvaros, Hungary
e-mail: akmbsab@uni-miskolc.hu
Tel.: +36-46-565110
Fax: +36-46-565110

V. Shilov
Institute of Biocolloid Chemistry, National Academy of Sciences of Ukraine, Kiev, Ukraine

Introduction

Electrophoresis arises when charged colloidal particles or macromolecules are subjected to an outer electric field, and under the effect of this field the particles/molecules begin to move with respect to the liquid. It is widely used in colloid chemistry as a method for characterization of the surface of particles, for removal of dispersed particles from liquids, in biology and medicine to separate different biocolloids, macromolecules of proteins, in coating technology, etc.

Till recently electrokinetic investigations were carried out only in linear or slightly nonlinear regimes. Theories of concentration polarization of the electrical double layer (EDL) and the electrophoresis of nonconducting and conducting particles in mainly linear regimes have been developed. These problems are discussed in detail in the literature [1–4]. The electrophoretic mobility of particles of different natures, size and forms, in electric fields from several hundreds volts per meter up to 10^5 V/m, has been studied, both in aqueous and in nonaqueous media. Typically, deviations of the measured electrophoretic mobilities from those predicted by the classical Smoluchowski theory do not exceed a few dozen percent. The situation changed not long ago. It has been shown [5–7] that there are special conditions which ensure the existence of a nonlinear regime with peculiar features: the electroosmosis and electrophoresis velocities may be greater than those predicted by Smoluchowski's theory by 1 or 2 orders of magnitude [7–12]. We called this phenomena "superfast electrophoresis and electroosmosis" or "electrophoresis (electroosmosis) of the second kind" [7–12]. These names highlight a new mechanism behind these phenomena. Whereas classical electrokinetic phenomena are connected with the charges of the usual (primary) double layer, the new phenomena are connected with a secondary diffuse charge cloud which arises in strong electric fields behind the quasi-equilibrium primary EDL and which contains an additional induced space charge [5–13].

Thus, we distinguish two kinds of nonlinear electrokinetic phenomena. The first is due to the interaction of the outer field with field-induced ionic charges of the classical (primary) EDL under conditions when field-induced variations of the electrolyte concentration remain small compared with the equilibrium value. The second one, the electrophoresis of the second kind or superfast electrophoresis is connected with the interaction of a strong outer field with a secondary diffuse layer of counterions (space charge) that is induced outside the primary (classical) diffuse EDL by the external field itself because of concentration polarization. In the following the main theoretical considerations describing this phenomena and our present data on nonlinear electrophoresis of nonconducting dispersed particles and superfast electrophoresis of unipolar conducting particles are described.

Materials and methods

The laws of "cubic" electrophoresis were investigated using monodisperse polystyrene particles of diameter 30 μm, close-to-spherical γ-Al_2O_3 particles of mean diameter 150 μm and ellipsoidal *Saccharomices cerevisiae* yeast cells of size 10×15 μm. For superfast electrophoresis studies as representatives of semiconductors, molybdenite and pyrite mineral particles were used. The conductivities of these minerals are sufficiently high: 10–10^4 S/m for pyrite and 10–10^3 S/m for molybdenite. The minerals were crushed in an agate mortar and the particles obtained were fractionated by sedimentation. The electrophoretic measurements were performed in distilled water and potassium chloride solutions.

A new optical method was developed to measure the electrophoretic velocity of the particles in strong electric fields. The process was observed in a microscope. A Leitz Panphot microscope was remounted and placed so as to have a horizontal axis; thus, the standing (vertical) electrophoresis cell could be observed through this microscope. A black-and-white TV camera was connected to the microscope and its signal was received by a frame grabber card of the computer. The experiments were recorded on AVI files. The particles were allowed to sediment in a vertical planar open cell in which the distance between the electrodes was 2 cm. At the moment when the particles appeared in the visual field, the cell was set under direct current voltage. At the moment of disappearance of the particles from the screen, the voltage was switched off. After recording, still images were taken from the AVI files and were saved separately. Using a macrosequence, it was possible to fix a series of ten images from an experiment, when the interval between two images was only 0.04 s. Combining 7–10 pictures, we were able to trace the path of the particles and their velocity. In such a way the size and the electrophoretic displacement, i.e., the lateral drift of the particles, were measured on the combined image (Fig. 1) with high accuracy by image analyzer software (ImageTool 3.0).

Results

"Classical" nonlinear electrophoresis

As mentioned, electrophoresis is due to the interaction of strong electric fields with the "classical" diffuse double-layer charges of nonconducting and conducting particles.

Electrophoresis is linear when the velocity of electrophoretic movement of the particles is proportional to the strength of the applied field:

$$v_{eph} = \mu_{eph} E. \quad (1)$$

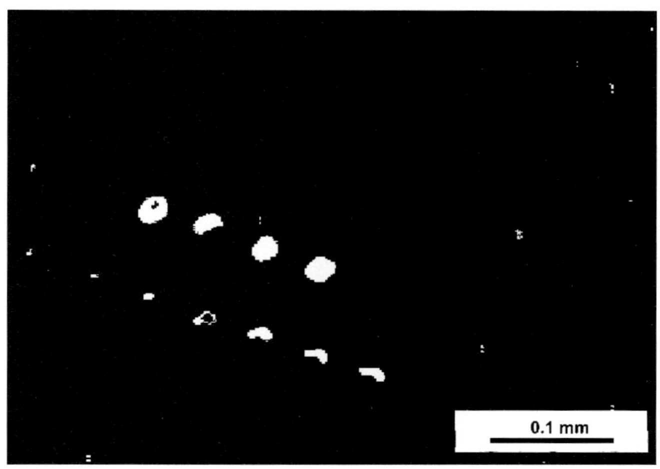

Fig. 1 A combined image of four individual still images from an experiment run. Experimental conditions: *Saccharomices cerevisiae* yeast cells, 50 mmol/l KCl solution, 100 V/cm

This relationship is valid for sufficiently small field strengths only. The field-independent electrophoretic mobility, μ_{eph}, depends on the characteristics of the EDL of the particle, mainly on its zeta potential and surface conductivity, and on the concentration and viscosity of the electrolyte surrounding the particle. The calculation of μ_{eph} is the most common goal of the theory of electrophoresis. As soon as the strength of the applied field increases, the processes that are nonlinear with respect to E start playing an even more important role, determining a noticeable nonlinearity of the dependence $v_{eph}(E)$. The influence of nonlinearity appears in higher than first order terms in the expansion of v_{eph} in powers of E. Electrophoretic velocity v_{eph} is an odd function of E and, correspondingly, even powers of E are absent in the expansion of v_{eph}. It follows from the fact that a change in the sign (direction) of the applied field without changing its absolute value gives rise to a change in the sign of the electrophoretic velocity, without changing its absolute value. Hence, the term that follows the linear electrophoretic term in the expansion of the electrophoretic velocity is cubic with respect to the applied field:

$$v_{eph} = \mu_{eph} E + \mu_{eph}^{(3)} E^3, \tag{2}$$

where the coefficients μ_{eph} and $\mu_{eph}{}^{(3)}$ are field-independent.

An expression for $\mu_{eph}{}^{(3)}$, which characterizes the cubic electrophoresis, for a spherical particle with a thin double layer for the case of a symmetrical electrolyte $z^+ = z^- = 1$ has been derived [14]. This problem is solved following the procedure of successive approximations in powers of the small parameter E [15, 16]. The first step of this procedure consists in considering the action of an external field and of the fields corresponding to the system's linear response on the equilibrium distribution of ions. It leads to the well-known results of linear polarization of the EDL: field-induced deviations of the distributions of the electric potential and electrolyte concentration, both proportional to the first power of the external field. The second step consists in considering the interactions between these field-induced deviations of the electric potential and electrolyte concentration with each other, leading to the appearance of potential and ion concentration distributions that are proportional to the second power of the external field. An important qualitative result of this second approximation is the deviation of the particle's EDL, as a whole, from electroneutrality. This qualitative result was independently confirmed by numerical computations [17].

The appearance of a net charge of the particle together with the surrounding electrolyte volume leads to the appearance of a net force resulting from the action of the field on the total charge. This gives rise to a contribution to the nonlinear electrophoresis that is proportional to the third power of the applied field strength. The fact that this component of the nonlinear electrophoresis arises from the total force distinguishes it from the linear component of the electrophoretic motion that arises owing to the field-induced slip of the external surface of the EDL with respect to the particle, under the condition of zero total force. The new term leads to an unusual distribution of hydrodynamic fields and, correspondingly, to some paradoxical features of nonlinear electrophoresis that we call the "force component" of the nonlinear electrophoresis [14]. It has been shown that the force component of the nonlinear electrophoresis can be described without any reference to the zeta potential, which is the main parameter of the standard model, but rather in terms of a single parameter, namely the Dukhin number Du [14]:

$$Du = \frac{K^{\sigma}}{K_m a}. \tag{3}$$

Here K^{σ} is the surface conductivity, K_m is the conductivity of the medium and a is the particle radius. The Dukhin number is a dimensionless parameter that characterizes the role of the EDL polarization in the electric response of the suspended particle [1, 2, 6].

For the case of a symmetrical electrolyte $z^+ = z^- = 1$ with $D^+ = D^-$ ($t^+ = 1/2$), the final expression is [14]

$$\mu_{eph}^{(3)} = -a^2 \frac{e}{k_B T} \frac{\varepsilon_m}{\eta} \frac{Du[23 + Du(269 + 464Du)]}{252(1 + 2Du)^3} \tag{4}$$

or in simplified form

$$\mu_{eph}^{3} = -a^2 z^+ z^- \frac{e}{3kT} \varepsilon_m \frac{Du}{1 + 2Du}. \tag{5}$$

Equation (4) expresses the value of the cubic electrophoretic mobility $\mu_{eph}{}^{(3)}$ for a spherical particle with a thin EDL in a symmetrical electrolyte through the Du number without any restrictions on other parameters of the thin EDL.

Some preliminary results of our experiments on nonlinear electrophoresis of emulsifier-free polystyrene latices, γ-Al_2O_3 and *S. cerevisiae* yeast cells are described in the following.

Measurements of the electrophoretic velocity in strong electric fields are accompanied by undesirable phenomena like thermoconvection and gas release on electrodes. In order to eliminate them, we measured the deviation of the particle movement from the normal sedimentation trajectory by using electric field pulses, i.e., during a short time after switching on the electric field.

Using the image analyzer equipment, we measured the electrophoretic velocity of polystyrene latex and γ-Al_2O_3 particles, as well as *S.cerevisiae* yeast cells as a function of the electric field strength and electrolyte concentration. The ionic strength was adjusted using KCl solutions. The results are shown in Figs. 2–4.

The common feature of the electrophoretic measurements is the existence of a stepwise, linear increase

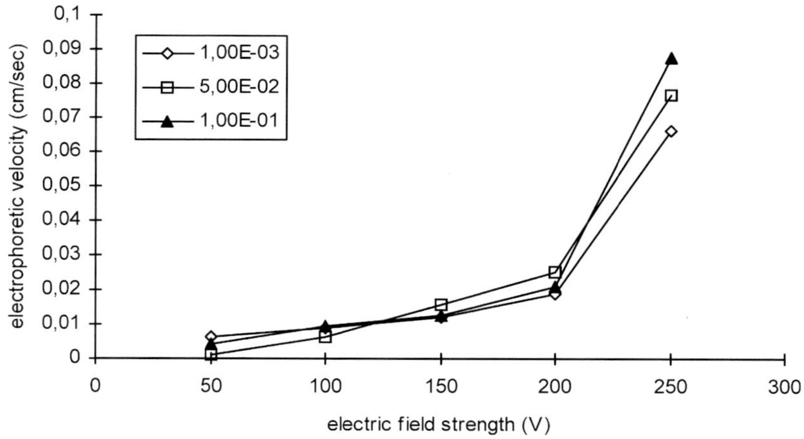

Fig. 2 Dependence of the electrophoretic velocity of *S. cerevisiae* cells on the applied field strength at KCl concentrations of 1, 50 and 100 mmol/l. The distance between the electrodes in the cell was 2 cm

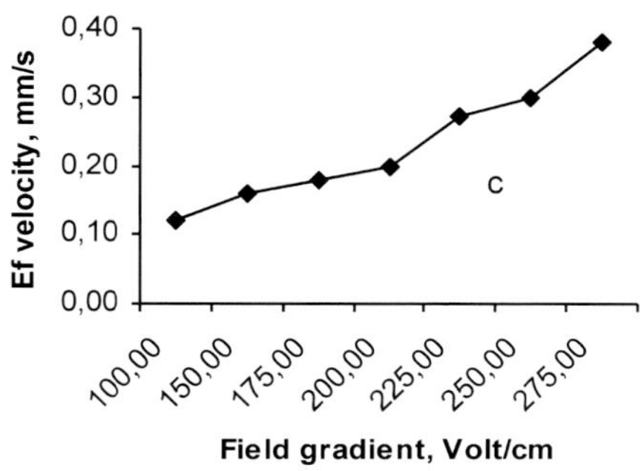

Fig. 3 Dependence of the electrophoretic velocity of γ-Al_2O_3 oxide particles of $d = 250 \mu m$ in distilled water on the applied field gradient

in electrophoretic velocity of particles/cells with the field strength, and a much steeper increase of the electrophoretic movement above 100–200 V/cm, i.e., in high electric fields substantial nonlinear effects were recorded. At these voltages high values of the electrophoretic velocity (millimeters per second) were registered in line with the theoretical predictions (in strong fields $V_{ef} \sim E^3$). The sharp rise in the velocity of electrophoresis in different systems is observed at different voltages: about 100 V/cm for polystyrene particles and cells both in distilled water and in dilute electrolyte solutions and at about 200 V/cm for aluminium oxide particles. The theory developed has been quantitatively checked for polystyrene latex particles. The experimental results are shown by black points in Fig. 4a (pure water) and Fig. 4b (5 mmol/l KCl). The measured values of the linear electrophoretic mobility were 1.1×10^{-8} and 0.8×10^{-8} $m^2/(V\ s)$, correspondingly. The curves represent the results of the theory, corresponding to $Du = 0.194$ (Fig. 4a) and $Du = 0.136$ (Fig. 4b). Good agreement between the theoretical predictions and experiment can be observed.

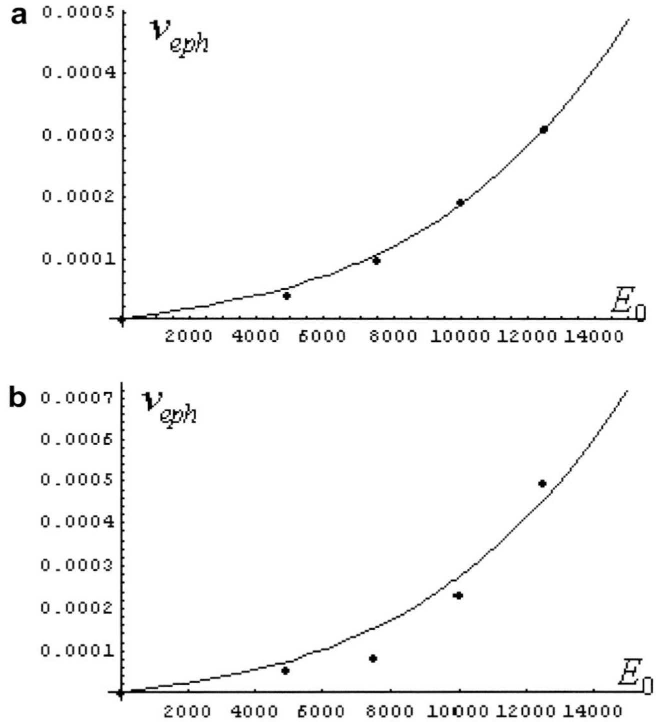

Fig. 4 The field dependence of the electrophoretic velocity of polystyrene latex particles of diameter 30 micrometers in **a** pure water and in **b** 5 mmol/l KCl. Experimental data (*points*), theory (*curves*) (see text)

Superfast electrophoresis

As mentioned, theoretical and experimental electrokinetic investigations in recent years have provided evidence that besides the field-induced disturbance of the classical EDL already described in strong electric fields around conducting dispersed particles under certain circumstances a secondary double layer containing a huge induced space charge can be formed [5–13, 18].

The theory of superfast electrophoresis and electroosmosis was developed by Dukhin and Mishchuk [5, 6, 10, 18, 19]. We discuss here only the main points of the theory necessary for understanding the phenomenon.

It was shown that the necessary conditions for the appearance of superfast electroosmosis and electrophoresis are the following: (1) the existence of a current through the particles; (2) unipolar conductivity of the particle's material, i.e., only one type of ions (or electrons) is mobile; (3) higher conductivity of the particles (K_p) in comparison with the conductivity of the medium (K_e); and (4) high electric field gradients to reach the so-called regime of over-limit current which is required for the formation of the induced space charge near the surface behind the quasi-equilibrium double electric layer, owing to strong concentration polarization [5–10, 18, 19].

The authors [5–9, 18, 19] discussed and described in detail how the space charge near a surface with unipolar conductivity (e.g., an ion-exchange membrane) originates. The reason is that in an external field the equilibrium distribution of the ions in the system is disturbed. This is due to the fact that the electric current through a unipolar conductor is realized by the migration of counterions only. Ions of the same charge (coions) leaving the vicinity of the surface are not replaced by others since these ions do not diffuse in the solid phase. So, the concentration of counterions near the surface considerably exceeds that of coions, leading to a net space charge.

In the case of spherical ion-exchanger particles, the space charge is formed under the influence of not only the perpendicular but also the tangential component of the electric field. The latter induces a motion of the liquid near the interface which contains the space charge. The shear of this layer occurs by the same mechanism as the electroosmotic shear. Dukhin and Mishchuk [18, 19] have shown that the magnitude of this shear can be estimated from the classical Smoluchowski equation by replacing the zeta potential by the potential drop across the region where the space charge exists. This potential drop is approximately $2Ea$, where E is the strength of the external field and a is the particle radius, i.e., the appearance of the space charge leads to an increase of the electrophoretic velocity by a factor of $2Ea/\zeta$ which may be 1 or 2 orders of magnitude.

The superfast electrophoresis of an ion-exchanger granule can be expressed by the following approximate expression [20]:

$$V_{ef} \approx \frac{8}{27}\frac{\varepsilon\varepsilon_0 E}{\eta}2Ea, \qquad (6)$$

where ε is the dielectric permittivity of the liquid and η is its viscosity.

It is seen that the electrophoretic velocity of ion-type conducting particles depends on the square of the external field, and on the particle size.

The situation is more complex in the case of metal or semiconductor particles, since space charges of opposite sign may arise on opposite sides of the particle. The reason is that on opposite sides of a metal particle, different electrochemical reactions take place; as a consequence, ions of opposite sign are transported to these surfaces. Close to that part of the surface where electron transport is accomplished from the metal to the cations that are continually transported from the volume of the electrolyte, a cationic space charge appears. Close to the part of the surface where the anions, continuously transported from the volume, draw off electrons from the metal surface, an anionic space charge is formed. The effect of an external field on the induced space charges, that are localized near the particle surface, leads to the appearance of the superfast electroosmotic and electrophoretic movement.

In the general case, the magnitude of the induced space charges formed on opposite sides of an electron-type conducting particle and the electroosmotic slip provoked by these charges for different reasons are not symmetrical. These reasons may be the different types of ions that are oxidized or reduced on the particle surface, the presence of different admixtures in solution which affect the oxidation/reduction potential of ions and the process of dissociation of water as well as the surface properties of metallic particles. This modifies the density and the thickness of induced space charges. Asymmetry of cationic and anionic induced space charges leads to the asymmetry of electroosmotic slip: a higher induced space charge causes more intensive hydrodynamic flow.

Current through an electron-type conducting particle flows only in the case if the potential drop on it reaches the value necessary for decomposition of the electrolyte Φ_d, i.e., when $3Ea > \Phi_d$ [21]. It has been shown that the electrophoretic velocity of particles with electron-type conductivity obeys the expression [12, 21]

$$V_{ef} \approx \frac{4}{27}\frac{\varepsilon\varepsilon_0}{\eta}\frac{(2Ea - \Phi_d)^2}{a}. \qquad (7)$$

Equation (7) is valid for the situation when the induced space charge is formed only on one side of a particle. The velocity of electrophoresis in the case of the existence of the space charge on both sides of particles (it is valid for electron-type conducting particles) can be evaluated from the formula [12, 21]

$$V_{ef} \approx \frac{4}{27}\frac{\varepsilon\varepsilon_0}{\eta a}(2Ea - \Phi_d)^2(2\gamma - 1), \qquad (8)$$

where γ is a coefficient that characterizes the fraction of the total potential drop $2Ea$–Φ_d on one side of the particle.

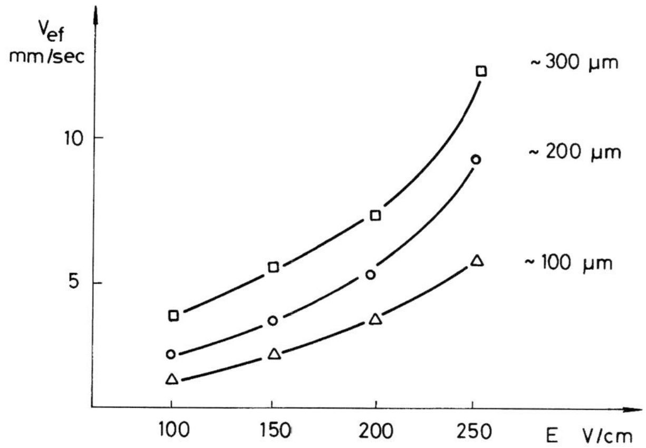

Fig. 5 Dependencies of the electrophoretic velocity on the electric field strength for molybdenite particles of mean diameter 100, 200 and 300 μm

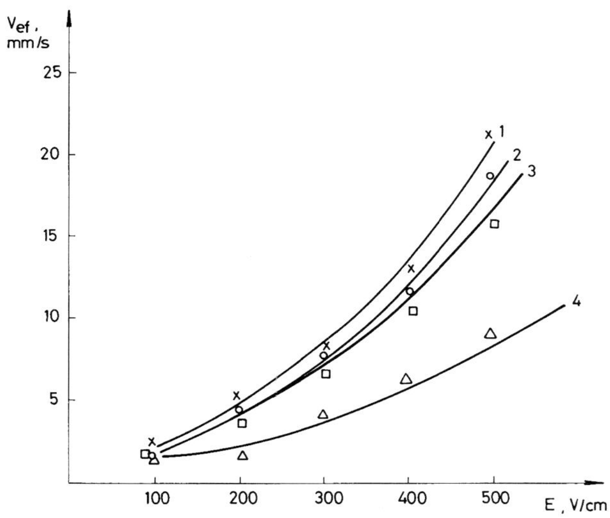

Fig. 6 Dependencies of the electrophoretic velocity of pyrite particles of mean diameter 300 μm on the external field strength in the suspension at KCl concentrations of 0.001 mol/l (*1*), 0.01 mol/l (*2*), 0.1 mol/l (*3*) and 0.5 mol/l (*4*)

The experimental data on electrophoretic movement of different ion-exchanger particles, metals (Mg/Al alloy) and semiconductors (graphite, activated carbon) are described in our previous papers [7–9, 11, 12, 20, 21]. For the illustration of this unusual phenomenon recent data on superfast electrophoresis of pyrite and molybdenite (semiconductors) particles are given in the following. The regularities obtained (Figs. 5–7) can be summarized as follows:

1. In strong electric fields (100–500 V/cm) the electrophoretic velocities of large magnesium, pyrite and molybdenite particles range up to 200 mm/s, which is nearly 2 orders of magnitude greater than the electrophoretic velocities characteristic for nonconducting particles and calculated by the classical Smoluchowski formula (microns per second). Thus, the phenomenon of superfast electrophoresis observed by us earlier for ion-exchange resins and carbonaceous materials also occurs in the case of semiconducting mineral particles. This is evidence for the general character of the new phenomenon which we studied.
2. For the particles under consideration, the electrophoretic velocity increases rapidly with the strength of the external field; this dependence is steeper than the linear dependence predicted by the Smoluchowski theory.
3. Both for metallic (magnesium) and semiconductor (molybdenite, pyrite) particles, the electrophoretic velocity increases with the particle size. These dependencies in some cases are linear, which is readily illustrated by the V_{ef} (*d*) dependence for molybdenite particles at $E = 200$ V/cm (Fig. 7).
4. The electrophoretic velocity of the particles studied decreases with an increase in the electrolyte concentration in the system; in this case, at low values of the ionic strength, the velocity decreases only slightly, whereas in systems with a high electrolyte concentration, the decrease in the velocity becomes significant. This is demonstrated with the example of electrophoresis of pyrite particles (Fig. 6). This regularity can be explained in terms of the Dukhin–Mishchuk theory, according to which the superfast electrophoresis is independent of the EDL parameters (which are very

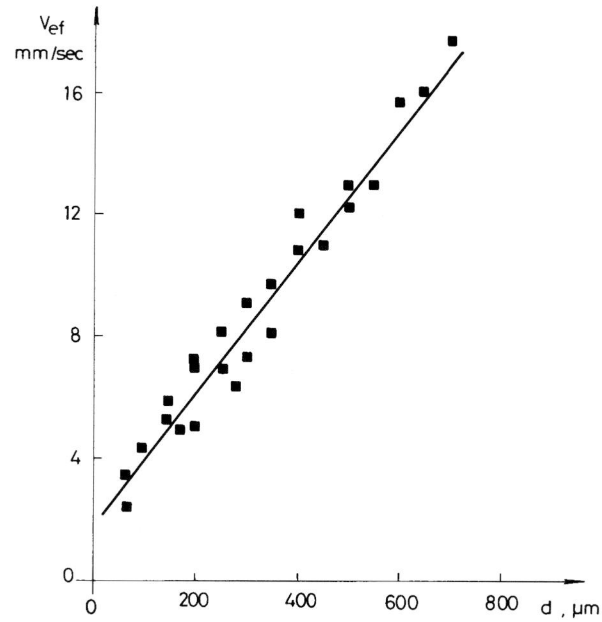

Fig. 7 The dependence of the electrophoretic velocity of molybdenite particles on their size at an electric field gradient 200 V/cm

sensitive to the electrolyte concentration), but it is dependent on the ratio between the conductivity of the dispersed phase and the medium. It is evident that, at low electrolyte concentrations, the conductivity of metallic or mineral particles is greater than that of the equilibrium solution; as a result, the velocity depends only slightly on the KCl concentration in the suspension, in good agreement with the theory [5, 6, 18–21]. At the same time, at sufficiently high ionic strength, the solution conductivity becomes comparable to that for the particles. This is the reason for the significantly decreased velocity of electrophoresis at electrolyte concentrations above 0.1 mol/l.

Conclusion

Nonlinear electrophoresis that can be observed in strong electric fields (several hundred volts per centimeter) is due (1) either to the field-induced disturbances of the EDL neutrality that leads to the appearance of a net force resulting from the action of the field on the total charge or (2) to the interaction of a strong electric field with the space charge near the surface of unipolar (ionically or electronically) conducting particles. This space charge is induced by the strong external field owing to concentration polarization. It has been shown that in case 1 with increasing field strength, the processes that are nonlinear with respect to E can be characterized by the third-power term in the expansion of the electrophoretic velocity in powers of E. An expression is given for "cubic electrophoresis" of spherical particles with a thin double layer, using the procedure of successive approximation in powers of small-parameter E. In line with the theory, at $E = 100$ V/cm and higher substantial nonlinear effects in the electrophoretic velocity of polystyrene latex and aluminum oxide particles as well as yeast cells were observed.

In case 2, the interaction of the space charge that is formed near the surface of electron-type conducting particles (semiconductors) of pyrite and molybdenite, a strong electric field (100–500 V/cm) produces electrophoretic velocities that are 1–2 orders of magnitude higher than electrophoretic velocities typical of nonconducting particles. It has been shown that the mobility of the particles increases linearly with E and the particle size, in contrast to classical electrophoresis. This can be explained by the Dukhin–Mishchuk theory of superfast electrophoresis.

Acknowledgements The authors thank C. Grosse from the University of Tucuman, Argentina, and N. Mishchuk, National Academy of Sciences of Ukraine, for discussing the results, and Y. Babich and Victor Madai for help in some experiments. The study was supported by the Hungarian OTKA Agency (grant T043360) and Hungarian–Argentinean TeT project no. Arg. 2/99; that support is gratefully acknowledged.

References

1. Dukhin SS, Derjaguin BV (1976) Elektroforez. Nauka, Moscow
2. Derjaguin BV, Dukhin SS, Shilov VN (1980) Adv Colloid Interface Sci 13:41
3. Hunter RJ (1981) Zeta potential in colloid science: principles and applications. Academic, New York
4. O'Brian RW (1983) J Colloid Interface Sci 92:204
5. Dukhin SS, Mishchuk NA (1989) Colloid J 51:659
6. Dukhin SS (1991) Adv Colloid Interface Sci 35:173
7. Dukhin SS, Mishchuk NA, Tarovskii AA, Baran AA (1987) Colloid J 49:616
8. Babich YA, Baran AA, Mishchuk NA (1990) Ukr Khim Zh 56:1034
9. Baran AA, Babich YA, Tarovsky AA, Mishchuk NA (1992) Colloids Surf A 68:141
10. Mishchuk NA, Takhistov PV (1995) Colloids Surf A 95:119
11. Barany S, Mishchuk NA, Prieve DC (1998) J Colloid Interface Sci 207:240
12. Barany S (1998) Adv Colloid Interface Sci 75:45
13. Rubinstein I (1981) J Chem Soc Faraday Trans 77:1595
14. Shilov V, Barany S, Grosse C, Shramko O (2003) Adv Colloid Interface Sci 104:159
15. Simonova TS, Dukhin SS (1976) Colloid J 38:79
16. Shilov VN, Shramko OA, Simonova TS (1992) Colloid J 54:619
17. Fixman M, S. Jogannathan S (1983) Macromolecules 16:685
18. Dukhin SS, Mishchuk NA (1990) Colloid J 52:452
19. Mishchuk NA, Dukhin SS (1990) Colloid J 52:497
20. Mishchuk NA, Barany S, Madai F (1997) Colloid J 59:508
21. Mishchuk N, Barany S, Madai F (1998) Colloids Surf A 140:43

Progr Colloid Polym Sci (2004) 128: 21–24
DOI 10.1007/b97068

Claudio L.A. Berli
María V. Piaggio
Julio A. Deiber

Theoretical relation between the tube zeta potential and the background electrolyte composition in capillary electrophoresis

C.L.A. Berli · Julio A. Deiber (✉)
Instituto de Desarrollo Tecnológico para la Industria Química, Universidad Nacional del Litoral, Consejo Nacional de Investigaciones Científicas y Técnicas, Güemes 3450, 3000 Santa Fe, Argentina
e-mail: treoflu@ceride.gov.ar
Tel.: +54-342-4559174
Fax: +54-342-4550944

C.L.A. Berli
Departamento de Física, Facultad de Bioquímica y Ciencias Biológicas, Universidad Nacional del Litoral, Pje. El Pozo, 3000 Santa Fe, Argentina

M.V. Piaggio
Cátedra de Bioquímica Básica de Macromoléculas, FBCB, Universidad Nacional del Litoral, Pje. El Pozo, 3000 Santa Fe, Argentina

Abstract This work presents a theoretical analysis of the relation between the tube zeta potential and the electrolyte solution composition typically found in capillary electrophoresis. A model including the relevant physicochemical parameters of the capillary–solution interface is derived in the framework of Poisson–Boltzmann theory and is then validated experimentally. This model is applied to capillaries containing acidic groups attached to the inner wall as the main source of surface charges. In particular, experimental data of electroosmotic flow in synthetic organic polymer capillaries are considered to prove the validity of the model with less conventional solid materials. It is also found that the model describes successfully experimental data of the zeta potential for a wide range of pH and ionic strength.

Keywords Organic polymer capillaries · Electroosmotic flow · Capillary electrophoresis Zeta potential · Surface potential

Introduction

When an electrical field is applied along the axial direction of a capillary tube filled with an electrolyte solution, the so-called electroosmotic flow (EOF) develops [1–4]. This phenomenon is a manifestation of the presence of electrical charges on the inner capillary surface and the associated double layer of ions in solution. The electroosmotic fluid velocity v_{EO} is given by the Helmholtz–Smoluchowski equation, provided the relevant conditions are satisfied [5–7]. It is expressed

$$v_{EO} \approx -(\varepsilon/\mu)E\zeta, \qquad (1)$$

where ε is the electrical permittivity, μ is the solution viscosity, E is the axial electrical field strength and ζ is the zeta potential at a moving boundary between the capillary wall and the solution. The EOF has critical implications in electrophoresis, mainly in capillary zone electrophoresis (CZE) [8–11], where the tubes used normally contain acidic groups attached to the wall as a source of surface charges. The resulting ζ is thus related to the chemical composition of the background electrolyte (BGE). Therefore, a rational formulation of solutions concerning pH, ionic strength, viscosity and electrical permittivity is required to optimize the separation performance of this technique [12–16]. Further, the knowledge of ζ is of particular interest in simulations of CZE, where "virtual electropherograms" are predicted from physicochemical data of the capillary, the BGE and analytes under separation [17–20].

In this context of analysis, one may observe that theoretical work is required to model the ion exchange between the capillary wall and the BGE, which is the phenomenon determining the surface potential and hence the EOF. Several approaches relating ζ to the BGE composition have been reported [7, 15, 21–23]. Following these works, here we use and discuss a model proposed recently by the authors [7], which was initially aimed at describing silica capillary interfaces, where the main source of surface charge is the dissociation of silanol groups. Derivations were made in the framework of Poisson–Boltzmann theory, with further considerations of surface chemistry to account for surface charge generation. In the present work we extend the model to the case of polymer capillaries, where the ionizable species are carboxylate groups.

The main aspects of the model are outlined in the next section. Then, the results of applying the model to experimental data available in the literature are shown. The data considered involve different systems for which the EOF was measured and the zeta potential was calculated through Eq. (1). Special emphasis is placed on the case of synthetic organic polymer capillaries. In this section we also include examples of silica capillaries for the purposes of visualizing the reach of the model predictions.

Theoretical concepts

In the framework of the standard electrokinetic model [24, 25], the interface composed of a solid surface in equilibrium with an aqueous electrolyte solution is shown schematically in Fig. 1. The dissociation of acid groups attached to the capillary wall generates a negatively charged surface, with charge density per unit area σ_0. The electrostatic potential associated with this charge density decreases away from the surface because of counterion screening. In particular, the region between $x=0$ and $x=d$, the compact layer, is generally assumed to be free of charges owing to the finite size of the hydrated ions. Thus the potential drop across this layer is written [4]

$$\psi_0 - \psi_d = \sigma_0 / C_{in}, \quad (2)$$

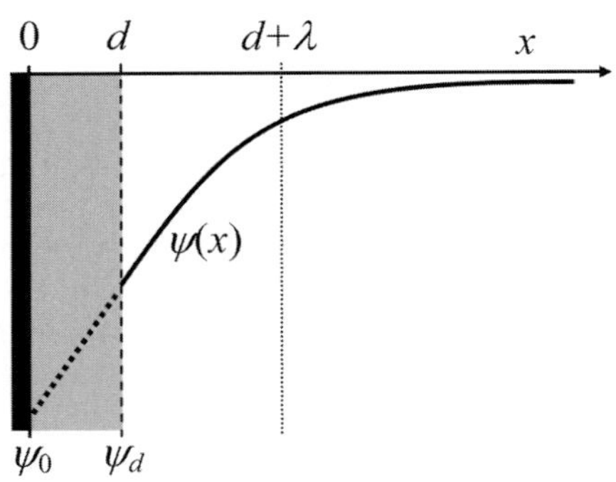

Fig. 1 Schematic representation of the electrostatic potential $\psi(x)$ in the solid–liquid interface of a capillary tube used in electrophoresis. In this scheme, $x=R-r$, where R is capillary radius and r is the radial cylindrical coordinate. Also ψ_0 is the surface potential, ψ_d is the potential at the outer Helmholtz plane, d is the thickness of the compact layer and λ is the Debye screening length. When an electrical field is applied along the capillary, normal to the x-direction, the electric force acting on the ions drags the electrolyte solution through the capillary. The limiting value of the electroosmotic velocity is given by Eq. (1), provided $\zeta \approx \psi_d$ and $\lambda << R$

where C_{in} is the capacitance per unit area of the inner region between the equipotential planes at ψ_0 and ψ_d (see also the legend to Fig. 1). For $x \geq d$, in the diffuse layer, the ion distribution is determined by the balance between entropic and electrostatic forces, and hence $\psi(x)$ is governed by the Poisson–Boltzmann equation. The characteristic screening length illustrated in Fig. 1 is $\lambda = \left(\varepsilon k_B T / e^2 \sum_k z_k^2 n_k^b\right)^{1/2}$, where e is the elementary charge, k_B is the Boltzmann constant, T is the absolute temperature, z_k are the ion valences and n_k^b are the ion number densities in the bulk. In typical conditions of CZE, λ is smaller than the capillary radius R.

The total charge density in the diffuse layer, σ_d, involves the summation over all ion species as follows:

$$\sigma_d = \left\{ 2\varepsilon k_B T \sum_k n_k^b \left[\exp\left(-\frac{e z_k \psi_d}{k_B T} \right) - 1 \right] \right\}^{1/2}. \quad (3)$$

In addition, it is assumed throughout this work that $\zeta \approx \psi_d$ (see also Refs. [4, 24, 26]). Thus Eq. (3) relates ζ to the concentration of the ions in solution. In addition, a suitable relation between the molar concentration of protons in the bulk, $[H^+]$, and the surface charge density is required. For this purpose, we consider a system composed of a solid surface containing n_s weak acid groups per unit area in equilibrium with a large electrolyte reservoir. Consequently, the equilibrium dissociation constant K_s is obtained in the framework of statistical thermodynamics [22]. Thus the charge density as a function of $[H^+]$ is found to be

$$\sigma_0 = \frac{-e n_s}{1 + ([H^+]/K_s)\exp(-e\psi_0/k_B T)}, \quad (4)$$

where $e n_s$ is the total number of surface charges per area unit available. Therefore, through the electroneutrality condition ($\sigma_d + \sigma_0 = 0$), Eqs. (2)–(4) are combined to obtain the following expression

$$\sigma_d\left(n_k^b, \zeta\right) - \frac{e n_s}{1 + 10^{(pK_s - pH)} \exp\left\{ \frac{-e}{k_B T} \left[\zeta - \frac{\sigma_d\left(n_k^b, \zeta\right)}{C_{in}} \right] \right\}} = 0, \quad (5)$$

which establishes a relationship between ζ and the BGE characteristics. Equation (5) is quite general as long as the ion densities in $\sigma_d\left(n_k^b, \zeta\right)$ are appropriately quantified through Eq. (3).

In applying the model, the first step generally consists in the evaluation of parameters n_s K_s and C_{in} from a set of N experimental data $\{pH, n_k^b, \zeta\}$. The best values of these parameters are those that minimize the sum $\sum_{i=1}^{N} (\Delta_i)^2$, where Δ_i represents the left-hand side of Eq. (5) and the subindex i refers to each triad $\{pH, n_k^b, \zeta\}$ placed into Eq. (5). Numerical calculations are carried out with a typical minimizing subroutine. Once the parameters for a given capillary are known, ζ can be predicted for a different pH and ionic strength. Thus the same numerical scheme is used, where the input data are $\{pH, n_k^b\}$ and the only unknown is ζ.

Results and discussion

In this section, before the model is used to predict the zeta potential of organic polymer capillaries, we firstly discuss some examples involving classical materials, as an illustration of the model application. In this sense, Fig. 2 shows data of ζ as a function of pH, which were obtained with a vitreous silica capillary at 25 °C. The experimental conditions are described in Ref. [21]. Also in Fig. 2, lines represent the model predictions with the parameter values reported in this figure legend. In this sense, one may observe that typical values of C_{in} are in the range

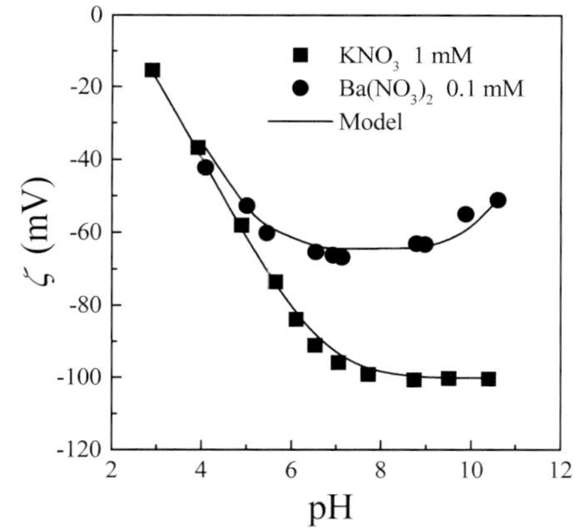

Fig. 2 Zeta potential of a vitreous silica capillary as a function of pH, for different salt solutions at 25 °C. The *symbols* are experimental data from Ref. [21]. The *lines* are the prediction of Eq. (5) with the following parameter values: for KNO_3, $pK_s = 3.5$, $C_{in} = 0.17$ F m^{-2} and $n_s = 0.8 \times 10^{17}$ m^{-2}; for $Ba(NO_3)_2$, $pK_s = 3.7$, $C_{in} = 0.5$ F m^{-2} and $n_s = 0.45 \times 10^{17}$ m^{-2}

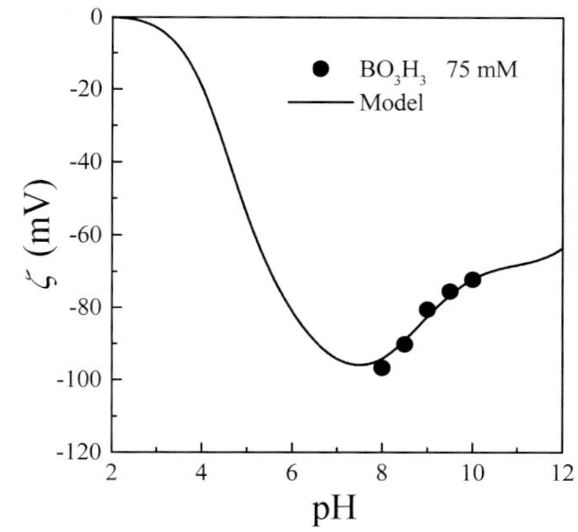

Fig. 3 Zeta potential of a fused silica capillary as a function of pH, when a buffer solution is used as the background electrolyte at 25 °C. The *symbols* are experimental data from Ref. [7]. The *lines* are the prediction of Eq. (5) with the following parameter values: $pK_s = 6$, $C_{in} = 1$ F m^{-2} and $n_s = 3.5 \times 10^{17}$ m^{-2}

0.1–1 F m^{-2} [4] and the number density n_s reported in the literature is around 10^{17} m^{-2} [3, 9]. On the other hand, in the case of solutions of KNO_3, the ionic strength remains almost constant since the concentration of salt (1 mM) exceeds greatly the concentration of acid or base added to adjust the pH. In contrast, the ionic strength of $Ba(NO_3)_2$ solutions changes significantly for pH > 7, owing to the contribution of $Ba(OH)_2$ added to increase the pH. In fact, we found that the minimum of the curve ζ(pH) in Fig. 2 (an experimental feature discussed in the literature [21]) is predicted well by the model when the equivalents of $Ba(OH)_2$ are considered in the evaluation of $\sigma_d\left(n_k^b, \zeta\right)$ [7].

Another example concerning a buffer solution (no added salt) is analyzed in Fig. 3. The data correspond to a fused silica capillary with a 75 mM solution of boric acid, the pK_a value of which is 9.2 at 25 °C. The experimental conditions are described in Ref. [7]. In this figure, the line represents the model prediction with the parameter values reported in the figure legend. In particular, the value of pK_s found had been previously reported elsewhere [27]. The experimental data are in the zone where boric acid has an effective buffering action. The model prediction is also shown out of this zone to observe the response of ζ as a function of pH. It is worth noting that the model predicts the point of zero charge at a pH of around 2.5, in agreement with results reported in the literature [4, 28].

In the following we apply the model to zeta potential data obtained from EOF experiments carried out with synthetic organic polymer capillaries: polyfluorocarbon (PFC), polyethylene (PE) and poly(vinyl chloride) (PVC) [29]. The experimental details are well described in Ref. [29]. Here it is relevant to mention that the BGE, which is a mixture of buffer and salt aqueous solutions, was formulated to kept the ionic strength relatively constant at 10 mM for the whole range of pH. The experimental values from PFC and PVC capillaries are presented in Fig. 4. Data from the PE capillary are not shown for the sake of clarity, as they rather superpose the PFC data. Also in Fig. 4, lines refer to the model prediction with the parameter values reported in Table 1. A remarkable agreement between theory and experiments is observed.

The values of pK_s found coincide with those reported for carboxylate groups [4, 22], being slightly higher in the case of PVC (see also Ref. [29] and references therein). In particular, the model predicts $C_{in} \rightarrow \infty$ for all these capillaries, which means that $\zeta \approx \psi_0$. This is also an

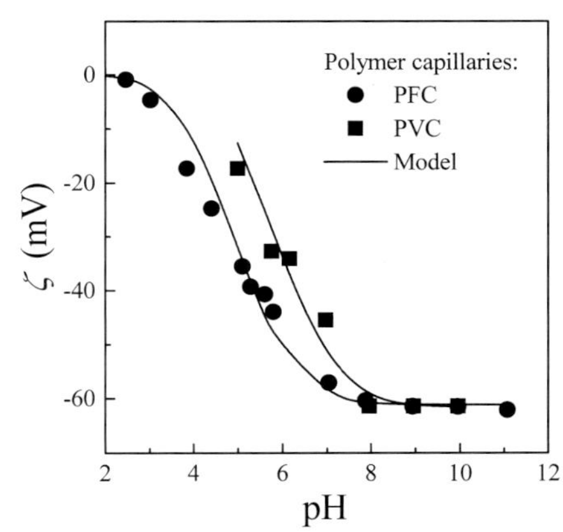

Fig. 4 Zeta potential of synthetic organic polymer capillaries [polyfluorocarbon, *PFC*, poly(vinyl chloride), *PVC*] as a function of pH at 24 °C. The *symbols* are experimental data from Ref. [29]. The *lines* are the prediction of Eq. (5) with the parameter values reported in Table 1

Table 1 Parameter values of the model applied to synthetic organic polymer capillaries, as obtained from electroosmotic flow experimental data reported in Ref. [29]

Capillary	pK_s	n_s (10^{17} m^{-2})
Polyfluorocarbon	4.55	1.09
Polyethylene	4.95	1.14
Poly(vinyl chloride)	5.50	1.10

expected result for carboxylate groups, which allow counterions to approach closely the surface and hence the compact layer thickness d is very thin [4]. For instance, values of C_{in} higher than 10^4 F m^{-2} were required in Ref. [22] to fit experimental data involving carboxylate groups attached to latex particles (see also Ref. [30] for results involving interfaces containing both carboxylate and sulfate groups). Therefore, as a result of the model, it is clear that the value of the zeta potential developed in polymeric capillaries (PFC, PVC and PE) is practically the same as that of the surface potential, which is accessible from titration experiments.

On the basis of the results discussed here, one may conclude that the model appears robust enough, in the sense that it is able to describe ζ satisfactorily for different capillary materials and BGEs within a wide range of pH and ionic strength. These calculations are thus useful to quantify and control the zeta potential in experimental programs of capillary electrophoresis.

Acknowledgements This research was supported by SEPCYT-FONCyT, PICT 09–09752, and SeCyT-UNL, CAI+D 2002, Argentina. C.L.A.B. and J.A.D. gratefully thank SECyT, CONICET, AAIFQ and the organizing committee of the 11th ICSCS for financial aid.

References

1. Dukhin SS, Derjaguin BV (1974) In: Matijevic E (ed) Surface and colloid science: electrokinetic phenomena, vol 7. Wiley, New York
2. Probstein RF (1989) Physicochemical hydrodynamics. Butterworths, New York
3. Russel WB, Saville DA, Schowalter WR (1989) Colloidal dispersions. Cambridge University Press, Cambridge
4. Hunter RJ (1992) Foundations of colloid science, vol II. Clarendon, Oxford
5. Gaš B, Štědrý M, Kenndler E (1995) J Chromatogr A 709:63
6. Griffiths SK, Nilson RH (2000) Anal Chem 72:4767
7. Berli CLA, Piaggio MV, Deiber JA (2003) Electrophoresis 24:1587
8. Grossman PD, Colburn JC (1992) Capillary electrophoresis: theory and practice. Academic, New York
9. Jandik P, Bonn G (1993) Capillary electrophoresis of small molecules and ions. VCH, New York
10. Khun R, Hoffstetter-Khun S (1993) Capillary electrophoresis: principles and practice. Springer, Berlin Heidelberg New York
11. Kenndler E (1998) Chem Anal 146:25
12. Kenndler E, Gassner B (1990) Anal Chem 62:431
13. Schwer Ch, Kenndler E (1991) Anal Chem 63:1801
14. Kenndler E, Friedl W (1992) J Chromatogr 608:161
15. Tavares MFM, McGuffin VL (1995) Anal Chem 67:3687
16. Kenndler E (1998) J Microcolumn Sep 10:273
17. Saville DA, Palusinski OA (1986) AIChE J 32:207
18. Palusinski OA, Graham A, Mosher RA, Bier M, Saville DA (1986) AIChE J 32:215
19. Reijenga JC, Kenndler E (1994) J Chromatogr A 659:403
20. Piaggio MV, Deiber JA (2003) Lat Am Appl Res 33:171
21. Wiese GR, James RO, Healy TW (1971) Discuss Faraday Soc 52:302
22. Healy TW, White LR (1978) Adv Colloid Interface Sci 9:303
23. Salomon K, Burgi DS, Helmer JC (1991) J Chromatogr 559:69
24. Dukhin SS (1995) Adv. Colloid Interface Sci 61:17
25. Dukhin SS, Zimmermann R, Werner C (2001) Colloids Surf A 195:103
26. Attard P, Antelmi D, Larson I (2000) Langmuir 16:1542
27. Hewlett-Packard (2000) Hewlett-Packard instrument guide CD-ROM, plate 010–023, CE basics. Hewlett-Packard
28. Kosmulski M (2002) J Colloid Interface Sci 253:77
29. Schützner W, Kenndler E (1992) Anal Chem 64:1991
30. Gisler T, Schultz SF, Borkovec M, Sticher H, Schurtenberger P, D'Aguanno B, Klein R (1994) J Chem Phys 101:9924

Progr Colloid Polym Sci (2004) 128: 25–29
DOI 10.1007/b97076

Jorge A. Bertolotto
Graciela B. Roston
María E. Ascheri

Electro-optical properties of DNA

J.A. Bertolotto (✉) · G.B. Roston
M.E. Ascheri
Facultad de Ciencias Exactas y Naturales, Universidad Nacional de La Pampa, Av. Uruguay 151, 6300 Santa Rosa (LP), Argentina
e-mail: jbertolotto@exactas.unlpam.edu.ar
Tel.: +54-2954-425166
Fax: +54-2954-432535

Abstract We calculate the steady-state electric birefringence for a flexible bent rod molecule due to free electrophoresis and the interaction of its induced dipole moment with an externally applied electric field. We consider the solution of flexible bent rod as a mixture of rigid bent rod molecules with different angles of conformation. We use the Fokker–Planck equation in order to find the orientational and conformational distribution function for the steady state. We first calculate the contribution of the particles having a fixed angle of conformation to the electric birefringence and then, averaging all these values, we obtain the final birefringence value. Using convenient parameters to characterize the DNA restriction fragments we explain the existence of a positive value of the electric birefringence, for low electric fields, and the sign reversal of it when we increase the applied electric field. We also corroborate the order of the inversion field value.

Keywords Electric birefringence · DNA · Free electrophoresis · Flexibility

Introduction

The steady-state electric birefringence of macromolecules and colloids in dilute solutions is theoretically interpreted considering the interaction of its permanent and induced dipole moments with an externally applied electric field. $\vec{E}$ O'Konski et al. [1] and Holcomb and Tinoco [2], among others, calculated the dependence of the steady-state electric birefringence on field strength under the assumptions of rigid molecular structures. Yoshioka [3] extended these calculations to nonrigid molecules. A comprehensive review of works on the electro-optical properties of DNA is given by Houssier [4].

Porschke [5] found that DNA restriction fragments with chain lengths between 870 and 3,000 Å show a positive electric dichroism and birefringence at low electric field. The electric birefringence Δn and dichroism ΔA changed to the usual negative values at higher electric fields. In previous theories, this phenomenon was explained by considering an arc model [5] or a bent rod model [6]. In both models the molecules have an electric polarizability tensor and a permanent dipole moment owing to the asymmetric charge distribution with respect to its centre of diffusion. Nevertheless, DNA has only a pure induced dipole moment and this apparent dipole is introduced to explain the phenomenon but it is not originated from one dynamic equation of the system. In an electro-optical experiment the charged molecules, in the presence of an applied electric field, change not only their orientation but also their position. This migration is the electrophoresis phenomenon. Δn and ΔA due to free-solution electrophoresis can be applied to interpret electro-optical properties of DNA. In a previous work [7], we showed that the coupling of the translational–rotational movement and a spin-free terminal motion of charged particles, in the presence of an electric field $\vec{E}$, give rise to an electric birefringence and dichroism. In this case, the model we applied to represent the DNA restriction fragments was a rigid bent rod molecule (BRM) one. In order to obtain the steady-state birefringence due to free-solution electrophoresis we found the orientational distribution function of the system by solving the Fokker–Planck equation for the steady state at low electric fields. This solution was given in powers of $\vec{E}$ until the second order. The calculated electric birefringence value was positive, in concordance with the experimental results [5]. Although this model explained the positive value of the electric birefringence and dichroism at low electric fields, it is unable to adjust the experimental curves of the electric birefringence and dichroism as a function of $\vec{E}$. This is due to the fact that the model, as an electrical characteristic of the system, only takes into account the electric charge without considering the electrical polarizability tensor of the BRM. There are preliminary studies considering the connection between electro-optics and electrophoresis. Stoylov [8] has done a review on the subject and basically analysed how the orientational distribution function of the particles influences the electrophoretic mobility. He also showed the relevance of the electrophoretic methods to obtain the necessary parameters for the analysis and interpretation of the electro-optical data. Antosiewicz et al. [9] studied a special case of hydrodynamical coupling for the case of tRNA molecules. The comparison of the data with and without this coupling effect indicated a positive dichroism contribution from the coupling effect.

In this work we develop an example of the relation between electro-optics and electrophoresis. We demonstrate that there is a contribution to the birefringence of the solution from particles having an electrophoretic mobility and such a shape to give rise to a coupling between the rotational and translational movement.

The orientation mechanism introduced here differs completely from the one originating from steric interactions during electrophoretic migration through a gel [10]. We calculate the dependence of Δn on the field strength for the case of a hinged BRM with an electric charge q, not polar but polarizable.

This model acceptably adjusts the experimental results of DNA restriction enzyme fragments at low electric fields.

Theory

The system is considered to be a dilute solution composed of hinged bent rodlike charged molecules; it is formed by two rodlike arms of length L, having an angle $\mathcal{X}$ between them with a bending potential at the hinge $V = \frac{1}{2}\varepsilon(\chi - \pi)^2$, where ε is the spring stiffness constant and π is the bending angle with the minimum energy configuration.

Each BRM has an optical polarizability tensor $\overleftrightarrow{\alpha}^{O}$ and an electric polarizability tensor $\overleftrightarrow{\alpha}^{E}$. This model is similar to the one presented in Ref, [6]. The BRM is placed in a molecule coordinate system $OXYZ$ with its origin fixed to the centre of mass of the molecule. The bent rod system lies on the ZX-plane.

The position and orientation of the particle with respect to the laboratory system $O'X'Y'Z'$ is described in terms of x', y', z' coordinates and the three Euler angles ϕ, θ and ψ, respectively. As the BRM is flexible at the hinge, the angle χ between the rods determines its conformation coordinate. The electric field $\vec{E}$ is applied along the Z' positive axis of the laboratory coordinate system. The components in the instantaneous directions XYZ of the displacement vectors and the infinitesimal rotations of the BRM with respect to the laboratory coordinate system are dx, dy, dz and dα, dβ, dγ, respectively. In this way, the system coordinates (x, y, z, α, β, γ) are defined (equivalent to the "local coordinate system" described by Perrin [11]).

In order to calculate the electric birefringence it is necessary to have the orientational distribution function for hinged BRMs in dilute solution, $f(x', y', z', \theta, \phi, \psi, \chi, t)$. Just for this, we have to obtain the Fokker–Planck equation associated with the function $f(x, y, z, \alpha, \beta, \gamma, \chi, t)$ which is written as follows [7, 12]:

$$\frac{\partial f}{\partial t} = -\vec{\nabla}[(\vec{M}_1)f] + \frac{1}{2}\vec{\nabla}[[M_2]\vec{\nabla}f],$$

where $\vec{\nabla} = \left(\frac{\partial}{\partial x}, \frac{\partial}{\partial y}, \frac{\partial}{\partial z}, \frac{\partial}{\partial \alpha}, \frac{\partial}{\partial \beta}, \frac{\partial}{\partial \gamma}, \frac{\partial}{\partial \chi}\right)$ is the differential operator and the statistical averages which appear in the components of the moment vector $\vec{M}_1$ and the moment matrix $[M_2]$ are made over all conformations as well as over all spatial orientations and positions. They could have been calculated from the seven Langevin equations of the system; three equations resulting from the translational movement of the molecule considered as a whole, three equations resulting from the rotational movement and one equation for the BRM deformational movement. To solve this nonlinear system of stochastic equations is too complicated because the χ coordinate is in all of them. Then, we should have to write the resulting Fokker–Planck equation related to the laboratory coordinate system.

We simplify the problem as follows: we assume the solution of hinged BRMs as a mixture of rigid BRMs with different angles of conformation. This assumption implies that, in order to calculate Δn, we use two steady-state distribution functions instead of using $f(x, y, z, \alpha, \beta, \gamma, \chi)$. One of them, $f_\chi(x, y, z, \alpha, \beta, \gamma)$, describes the rigid BRM position and orientation for a fixed angle χ at a given value of the applied electric field $\vec{E}$, and the other one, $G(\chi)$, describes the deformational movement of the hinged BRM for the same value of $\vec{E}$. Thus, we have to obtain two Fokker–Planck equations associated with these distribution functions.

This approximation is similar to the one used by Hagerman and Zimm [13] to investigate the relationship between the longest relaxation time and the flexibility of a linear polymer. Roitman and Zimm [14] justified this approximation, the so-called "rigid-body ensemble", by solving the diffusion equation for a trumbell model (three beads connected by two rigid rods with a bending potential at the hinge in the middle bead) and calculating the longest relaxation time. They found that this value is in good agreement with the one calculated by Hagerman and Zimm [13]. This type of "rigid-body" ensemble average has been shown to be a reasonable approximation for the calculation of steady-state properties of solutions [13–15].

The Fokker–Planck equation associated with the function $f_\chi(x, y, z, \alpha, \beta, \gamma)$ is obtained in a similar manner to the one developed in Ref. [7], the difference being the addition of a torque originating from interaction of the polarizable hinged BRM with $\vec{E}$ in the Langevin equations corresponding to the rotational movement and the BRM deformational movement. We assume that the solution of BRMs is sufficiently extensive so that, in the period of time the molecules attain an orientational equilibrium, the effect of electrophoresis does not change the spatial concentration of them. In this way, we may consider the orientational distribution function as $f_\chi(\alpha, \beta, \gamma)$. We write the Fokker–Planck equation related to the laboratory coordinate system taking into account that, for the case of particles oriented in the presence of an applied electric field E, the function f_χ becomes independent of the ϕ coordinate because the electric field introduces a cylindrical symmetry in the system. The resulting differential equation for the function

$$
\begin{aligned}
&\left(R_{11}\cos^2\psi + R_{22}\sin^2\psi\right)\frac{\partial^2 f_\chi}{\partial\theta^2} + \frac{\partial^2 f_\chi}{\partial\psi^2}\left(R_{11}\sin^2\psi\cot^2\theta + R_{22}\cos^2\psi\cot^2\theta + R_{33}\right)\\
&\quad + 2\frac{\partial^2 f_\chi}{\partial\psi\partial\theta}(R_{22}-R_{11})\sin\psi\cos\psi\cot\theta + \frac{\partial f_\chi}{\partial\theta}\Big\{\cot\theta\left(R_{11}\sin^2\psi + R_{22}\cos^2\psi\right).\\
&\quad - E^2\cos\theta\sin\theta\left[(a_2-a_3)R_{11}\cos^2\psi - (a_3-a_1)R_{22}\sin^2\psi\right] + \frac{Eq}{kT}P_{23}\cos\theta\sin\psi\Big\}\\
&\quad + \frac{\partial f_\chi}{\partial\psi}\frac{\cos\psi}{\sin^2\theta}\Big\{\frac{E^2}{kT}\sin^2\theta\sin\psi\left[R_{33}(a_2-a_1)\sin^2\theta + R_{22}(a_3-a_1)\cos^2\theta + R_{11}(a_2-a_3)\cos^2\theta\right]\\
&\quad + \frac{Eq}{kT}\sin\theta\left(P_{23}\cos^2\theta - P_{32}\sin^2\theta\right) + (R_{11}-R_{22})\sin\psi\left(\cos^2\theta+1\right)\Big\}\\
&\quad - f_\chi\frac{1}{\sin\theta}\Big[\frac{Eq}{kT}(P_{23}-P_{32})\sin\psi\sin^2\theta + \frac{E^2}{kT}R_{33}(a_1-a_2)\sin^3\theta\left(2\cos^2\psi - 1\right)\\
&\quad + \frac{E^2}{kT}R_{11}(a_2-a_3)\sin\theta\left(\cos^2\theta - \sin^2\theta\cos^2\psi\right) + \frac{E^2}{kT}R_{22}(a_3-a_1)\sin\theta\left(\sin^2\theta\sin^2\psi - \cos^2\theta\right)\Big] = 0,
\end{aligned}
$$

$f_\chi = f_\chi(\theta, \psi)$, is where k is the Boltzmann constant, T is the Kelvin temperature, R_{ii} for $i = 1,2,3$ are the terms of the rotational diffusion tensor and P_{23} and P_{32} are the nonzero terms of the translational–rotational coupling diffusion tensor [16, 17]. These terms were calculated by means of the Hydrox program by García de La Torre and Bloomfield.

The principal terms of the electric polarizability tensor, a_i for $i = 1,2,3$, were calculated in Ref. [6]. In this previous work, in order to obtain these quantities we considered the contribution of the parallel and perpendicular electric polarizabilities, $\alpha_{\|}$ and $\alpha_{\perp}$, of each arm to the electric polarizability tensor of the BRM. For the purpose of introducing the characteristic quadratic dependence of the electric polarizability of DNA fragments on the length [17], we proposed a simplified mechanism for the electric polarizability for a rodlike macroion:

1. The electric polarizability, in the direction of the macroion longitudinal axis, only originates from the displacement of counterions under the action of an applied electric field which gives, as a result, the appearance of two point charges opposite in sign in the extremes of the BRM.
2. The magnitude of this charge is proportional to the macroion length, L, and to the applied electric field, E.

With these assumptions the parallel electric polarizability for a rodlike macroion resulted in $\alpha_{\|} = K_{\|}L^2$. Considering that the charge which is moved transversally a length a by an electric field transverse to the rod is also proportional to L and E, we obtained the perpendicular electric polarizability for a rodlike macroion, $\alpha_\perp = K_\perp La$, with a being the transversal distance to the symmetry axis of each rod where the counterions move. We considered $K_{\|} = K_\perp = K$, where K is the electric polarizability constant.

Applying this calculation method to the BRM, we obtained the principal terms of the electric polarizability tensor:

$$a_1 = 2KL^2\cos^2(\chi/2) + 2KLa\sin^2(\chi/2),$$

$$a_2 = 2KLa$$

and

$$a_3 = 4KL^2\sin^2(\chi/2) + 2KLa\cos^2(\chi/2).$$

To solve Eq. (2) we apply a perturbation method, which gives us the solution expanded in powers of $\vec{E}$ [19]:

$$
\begin{aligned}
f_\chi(\theta,\psi) = {}& A + D_1 E\sin\psi\sin\theta + \frac{E^2}{16\pi^2 kT}\\
&\left(a_3\cos^2\theta + a_2\cos^2\psi\sin^2\theta + a_1\sin^2\psi\sin^2\theta\right)\\
&-\frac{1}{64\pi^2}\left(\frac{Eq}{kT}\right)^2\left[N_1\cos(2\psi) - N_2\cos(2\theta) - N_1\cos(2\theta)\cos(2\psi)\right]\\
&+\frac{E^4}{32\pi^2k^2T^2}\Big(\frac{1}{2}a_3^2\cos^4\theta + \frac{1}{2}a_2^2\sin^4\theta\cos^4\psi\\
&+\frac{1}{2}a_1^2\sin^4\theta\sin^4\psi + a_1a_2\sin^4\theta\sin^2\psi\cos^2\psi\\
&+a_2a_3\sin^2\theta\cos^2\psi\cos^2\theta + a_1a_3\sin^2\theta\sin^2\psi\cos^2\theta\Big),
\end{aligned}
$$

where

$$
\begin{aligned}
A = {}& \frac{1}{8\pi^2} - \frac{E^2}{960\pi^2k^2T^2}\Big[3E^2\left(a_1^2 + a_2^2 + a_3^2\right) + 20kT(a_1 + a_2 + a_3)\\
&+ 2E^2(a_3a_1 + a_2a_3 + a_1a_2) - 5q^2N_2\Big],
\end{aligned}
$$

$$D_1 = \frac{1}{8\pi^2}\frac{q}{kT}\frac{P_{32} - P_{23}}{R_{22} + R_{33}},$$

$$N_1 = N_3R_{22}P_{32} + N_3R_{11}P_{32} - N_3R_{11}P_{23},$$

$$N_2 = N_3R_{22}P_{32} + 2N_3R_{33}P_{23} + N_3R_{11}P_{23} - N_3R_{11}P_{32}$$

and

$$N_3 = \frac{P_{32} - P_{23}}{(R_{22}R_{11} + R_{22}R_{33} + R_{11}R_{33})(R_{22} + R_{33})}.$$

Then, we write the Fokker- + Planck equation for the deformational movement of the hinged BRM to find $G(\chi, t)$. In this equation the inertial terms are very small compared with the frictional ones; therefore, we can neglect them. We also assume $3\cos^2\left(\frac{1}{2}\chi(t)\right) << 1$ (this relation is valid for $\chi > 150$). With these assumptions we have

$$\frac{1}{\Theta}\frac{\partial G(\chi,t)}{\partial t} = \frac{\partial^2 G(\chi,t)}{\partial\chi^2} + \frac{\varepsilon + \Psi E^2}{kT}\frac{\partial}{\partial\chi}\left[(\chi - \pi)G(\chi,t)\right],$$

where

$$\Psi = -\frac{1}{2}LK\left(L\sin^2\theta\sin^2\psi - a\sin^2\theta\sin^2\psi - 2L\cos^2\theta + a\cos^2\theta\right)$$

and Θ is the diffusion coefficient of the deformational movement of the BRM.

Equation (9) is similar to the one described by Pathria [20] and its solution is the following:

$$G(\chi,t) = \sqrt{\frac{(\varepsilon+\Psi E^2)/2\pi kT}{1-\exp\left[-2\left(\frac{\varepsilon+\Psi E^2}{kT}\right)\Theta t\right]}} \times \exp\left(-\frac{(\varepsilon+\Psi E^2)(\chi-\pi)^2}{2kT\left\{1-\exp\left[-2\left(\frac{\varepsilon+\Psi E^2}{kT}\right)\Theta t\right]\right\}}\right).$$

In the steady state ($t\rightarrow\infty$) it results in

$$G(\chi) = \sqrt{\frac{(\varepsilon+\Psi E^2)}{2\pi kT}}\exp\left(-\frac{(\varepsilon+\Psi E^2)(\chi-\pi)^2}{2kT}\right).$$

The function distribution described in Eq. (12) is a Gaussian distribution with its peak at $\chi=\pi$. For a very high electric field strength we have a distribution close to Dirac's δ function, with an almost zero standard deviation σ. When we decrease the electric field strength, σ increases until a maximum value $\sigma=\sqrt{\frac{kT}{\varepsilon}}$ for $E=0$.

The application of the BRM model for studying the dependence of Δn on E^2 for the case of DNA dilute solutions yields values of ε which are of the order of 2×10^{-13} g cm^2 s^{-1}[6], if $E<2$kV cm^{-1}, then, the term ΨE^2 can be neglected in Eq. (12), and the hinged BRMs in solution do not change significantly their conformation distribution. Besides, $G(\chi)$ results in a function which is independent of the orientation of the BRMs with respect to $\vec{E}$, i.e., independent of the angles θ and ψ. Therefore, the fact that $G(\chi)$ only depends on the χ angle of conformation validates the assumption of the solution of hinged BRMs as a mixture of rigid BRMs with different angles of conformation.

Having this distribution function we can calculate the contribution of the particles with a fixed χ conformation angle to the electric birefringence of the solution from the following expression:

$$\Delta n_\chi = \frac{4\pi^2 c_1}{n}\int_0^{2\pi}\int_0^{\pi} f_\chi(\theta,\psi)\left[\frac{1}{2}A_0\left(3\cos^2\theta-1\right) - \frac{3}{4}B_0\left(1-\cos^2\theta\right)\left(2\sin^2\psi-1\right)\right]\sin\theta\, d\theta\, d\psi,$$

where c_1 is the number of molecules per volume unit and n the refractive index of the solution,

$$A_0 = \left(\alpha_\parallel^O - \alpha_\perp^O\right)\left[3\sin^2(\chi/2)-1\right],$$

$$B_0 = -2\left(\alpha_\parallel^O - \alpha_\perp^O\right)\cos^2(\chi/2)$$

and $\alpha_\parallel^O$ and $\alpha_\perp^O$ are the optical polarizabilities in the directions parallel and perpendicular to the symmetry axis of each rod, respectively.

Solving the integral in Eq. (13), we obtain

$$\Delta n_\chi = -\frac{\pi c_1}{420n}\left(\frac{E}{kT}\right)^2\left(E^2\left[2A_0(-a_3a_2 - a_3a_1 + 2a_1a_2 - 6a_3^2 + 3a_2^2 + 3a_1^2) + 3B_0(a_1-a_2)(3a_1+3a_2+a_3)\right] + \left\{28A_0KT(a_1 - 2a_3 + a_2 - \pi^2N_2q^2kT). + 42B_0\left[(P_{32}-P_{23})N_1q^2 + kT(a_1-a_2)\right]\right\}\right).$$

The birefringence Δn of the mixture is obtained by averaging Δn_χ over all χ angles, using the conformational distribution function of the ensemble of BRMs, $G(\chi)$ (from Eq. 12):

$$\Delta n = \int_0^\pi \Delta n_\chi G(\chi)d\chi \Big/ \int_0^\pi G(\chi)d\chi$$

After solving Eq. (17) using numerical methods, we finally obtain the Δn value of the hinged BRM in solution.

Discussion

The expression obtained for the electric birefringence of hinged BRMs due to free-solution electrophoresis and the interaction between the induced electric dipole moment and the applied electric field can be used to interpret electro-optical properties of DNA. Porschke [5] found that DNA restriction fragments exhibit a positive value of the electric dichroism and birefringence for low electric fields. These values change to their usual negative ones for higher electric fields.

The curve of Δn values versus E obtained through Eq. (17) is shown in Fig. 1 for the following parameters to characterize the DNA restriction fragments: $L=996$ Å (corresponding to a chain length of 1,992 Å or 586 base pairs), $K=1.7\times10^{-5}$ cm (this value corresponds to an electric polarizability anisotropy close to $\alpha_\parallel^E - \alpha_\perp^E = 6.8\times10^{-15}$ cm^3, $\alpha_\parallel^O - \alpha_\perp^O = -7.3\times10^{-21}$ cm^3, $q=-1.55\times10^{-7}$ statC, $c_1=1.68\times10^{13}$ cm^{-3} (corresponding to a concentration of 28 nM in units of double helices) and $\varepsilon=2.5\times10^{-13}$ g cm^2 s^{-2} [6].

The value of the electrophoretic charge used in the fitting is $q=0.28e$ per phosphate group, where e is the electronic charge. In order to compare the theoretical curves of Δn versus E with the experimental ones we used Porschke's data from Fig. 2 in Ref. [5]. In this

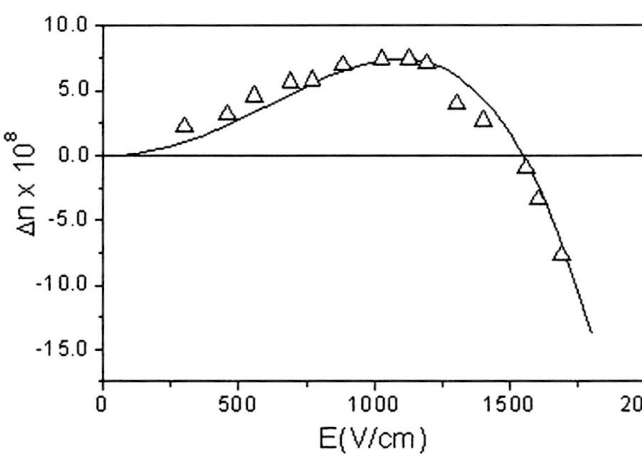

Fig. 1 Curve of Δn values versus E (*continuous line*) for a solution of hinged bent rod molecule with the characteristic parameters for DNA restriction fragments described in the text. The *triangles* represent the experimental data from Porschke [5]

figure, he showed the stationary values of the electric dichroism for DNA fragments with 587 base pairs (ionic strength $I \approx 12.4$ mM). The two quantities compared showed the same type of dependence on the applied electric field but their values were different. So, to obtain a fit of the experimental curve we have to multiply the corresponding values with a scale factor. From Fig. 1 we observe good agreement between the theoretical values and the experimental ones. A positive value of the electric birefringence (or dichroism) for low electric fields and a sign reversal of it when we increase the applied electric field are also shown. In addition, the order of the inversion field value is corroborated.

Conclusions

In this work we demonstrated the importance of considering the effect of the free-solution electrophoresis in the analysis of the electro-optical properties of macromolecules and colloids in dilute solutions. If the charged particles have a particular shape to originate a translational-rotational coupling and attain free-spin terminal motion, they contribute to the electric birefringence of the solution. This quite simple charged and polarizable BRM model shows that the effect of free-solution electrophoresis can also, for low electric fields, reverse the birefringence sign compared with the value resulting from the interaction of the permanent and induced dipole moments with an externally applied electric field.

Summarizing, this model adequately describes the steady-state electro-optical behaviour of DNA restriction fragments in dilute solutions. In our opinion, the theory of the electro-optical properties of macromolecules and colloids should be extended to consider the effect of translational-rotational coupling that takes place in the free-solution electrophoresis of particles with a particular shape.

References

1. O'Konski CT, Yoshioka K, Orttung H (1959) J Phys Chem 63:1558
2. Holcomb DN, Tinoco I (1963) J Phys Chem 67:2691
3. Yoshioka K (1987) J Chem Phys 86:491
4. Houssier C (1981) In: Krause S (ed), Molecular electro-optics. Electro-optics of macromolecules and colloids in solution. Plenum, New York, pp 363–398
5. Porschke D (1994) Biophys Chem 49:127
6. Bertolotto JA, Campo MG, Roston GB, Ascheri ME (2002) J Colloid Surf A 203:167
7. Bertolotto JA, Roston GB, Ascheri ME, Campo MG (2003) Physica A 327:185
8. Stoylov S P (1991) Colloid electro-optics: theory, techniques, applications. Academic, London
9. Antosiewicz J, Grycuk T, Porschke D (1991) J Chem Phys 95:1354
10. Åkerman B, Jonsson M, Moore D, Schellman J (1990) In: Birren B, Lai E (eds) Electrophoresis of large DNA molecules: theory and applications. Current communications in cell and molecular biology, vol 1. Cold Spring Harbor Laboratory, Plainview, NY, pp 23–41
11. Perrin F (1934) J Phys Radium 5:497
12. Risken H (1996) The Fokker–Planck equation. Methods of solution and applications. Springer, Berlin Heidelberg New York
13. Hagerman P, Zimm BH (1981) Biopolymers 20:1481
14. Roitman DB, Zimm BH (1984) J Chem Phys 81:6348
15. Fixman M (1983) J Chem Phys 78:1588
16. García de la Torre J, Bloomfield VA (1981) Q Rev Biophys 14:81
17. Brenner H (1961) Chem Eng 19:599
18. Stellwagen NC (1981) Biopolymers 20:399
19. Collatz L (1966) The numerical treatment of differential equations. Springer, Berlin Heidelberg New York
20. Pathria RK (1996) Statistical mechanics, Butterworth–Heinemann, Oxford

Progr Colloid Polym Sci (2004) 128: 30–35
DOI 10.1007/b97078

Graciela E. Magaz
Luis García Rodenas
Pedro J. Morando
Miguel A. Blesa

Chromium oxides in sediments: comparison of surface characterization by electrophoretic mobilities and Fourier transform IR–attenuated total reflection measurements

G.E. Magaz · L. García Rodenas
P.J. Morando · M.A. Blesa (✉)
Unidad de Actividad Química,
Centro Atómico Constituyentes,
Comisión Nacional de Energía Atómica,
Avenida General Paz 1499,
1450 San Martín,
Provincia de Buenos Aires, Argentina
e-mail: miblesa@cnea.gov.ar
Tel.: +54-11-67727153
Fax: +54-11-67727886

Abstract Adsorption of oxalate and chromate onto hydrous chromium(III) oxide and onto α-Cr_2O_3 was studied by spectroscopic methods (Fourier transform IR) and by measurement of electrophoretic mobilities. The Fourier transform IR spectra demonstrate the formation of surface complexes, and also the surface dimerization of chromate to give surface-bound dichromate. The mobility profiles are interpreted in terms of two successive surface complexation equilibria. The first stage yields uncharged species, and is responsible for a sharp decrease in mobilities in the acidic range. The second stage gives rise to negatively charged species, and is responsible for the charge reversal observed at higher concentrations. The postulated equilibria are given by equations (6)–(10). The dependence of the isoelectric point (pH_{piep}) on the total oxalate concentration is modeled on the assumption that the surface can be described as a mixture of oxalated and non-oxalated oxide, with two distinct extreme pH_{piep} values. The linear dependence of pH_{piep} on log[Cr(VI)], with slope 1.5, was explained in terms of the contribution of the stoichiometry of adsorption by the two complexation modes.

Keywords Oxalate · Chromate · Competitive adsorption · Surface complex

Introduction

The dynamics of chromium in aquatic systems, a priority metal pollutant [1] and potential carcinogen [2], is determined by the reductive precipitation/oxidative dissolution of the couple Cr(VI)/Cr(III) and by the complexation of Cr(III), especially by organic ligands. Precipitation involves Cr(VI) reactions with organic reductants, such as oxalic acid, to obtain Cr(III) that precipitates as $Cr_2O_3\ xH_2O$. The particles thus formed may adsorb inorganic and organic ions, and eventually may redissolve.

Oxalic acid is a natural metabolite and a component of industrial wastes that adsorbs significantly onto metal oxides, and has the ability to dissolve them; it can be considered as a very simple model of organic carboxylic acids present in natural environments. Chromate, in turn, is always associated with Cr-containing waters in oxidic conditions; it is also representative of the behavior of inorganic oxoanions as sulfate, arsenate, etc.

The present paper reports a study of the adsorption of oxalate and chromate on α-Cr_2O_3 and hydrous chromium(III) by electrophoretic and Fourier transform (FT) IR–attenuated total reflection (ATR) measurements. Surface complexation models that account for the results are proposed.

Experimental

Amorphous $Cr_2O_3\ 14H_2O$ (Brunauer–Emmett–Teller surface area 240 $m^2\ g^{-1}$) was synthesized as follows. A 2×10^{-2} mol dm^{-3} $Cr(NO_3)_3$ solution was quickly neutralized with NaOH until the molar ratio $[OH^-]_o/Cr(III)_o$ was 2.5. The suspension of the "active" $Cr(OH)_3$ thus obtained [3] was heated at 80 °C for 48 h; this step leads to polymeric soluble species [4]. The polymeric solution was neutralized until the molar ratio $[OH^-]_o/Cr(III)_o$ was 3.0, and the resulting suspension was aged at 80 °C for 48 h. The solid was filtered off on a cellulose nitrate membrane of 0.2-μm pore size, washed repeatedly with doubly distilled water and dried in a vacuum oven at 50 °C. The oxide was dialyzed until the conductance of the supernatant was 0.1 mS cm^{-1}.

The crystalline oxide, α-Cr_2O_3 (1.70 m^2 g^{-1}), was a commercial sample labeled 99.999% pure.

Electrophoretic measurements

For the pure (hydr)oxides, stock solutions containing 1 g L^{-1} of each solid were poured into 20 cm^3 10^{-2} mol dm^{-3} KCl, and the pH was adjusted to the desired value (in the range 3–11); the mobilities were measured after 24-h equilibration at 25.0±0.1 °C using PenKem S3000 equipment. Precautions to avoid contact with atmospheric CO_2 were taken. Analogous experiments were carried out with stock suspensions treated for 2 h with a solution containing chromate or oxalate, both in the concentration range 5×10^{-3}–5×10^{-6} mol dm^{-3} at pH 4.0, 25.0±0.1 °C, and ionic strength I=0.01 mol dm^{-3} (KCl). Mobilities were measured as before, in a dilute suspension equilibrated for 24 h. Other experiments were carried out with stock suspensions of α-Cr_2O_3 treated for 2 h with mixed chromate-/oxalate-containing solutions (total concentration 5×10^{-4} and 10^{-2} mol dm^{-3} and [Cr(VI)]/[oxalate]=1) at pH 4.0, 25.0±0.1 °C, and ionic strength 0.01 mol dm^{-3} (KCl).

FTIR–ATR spectra

For solutions, a single drop of each solution was placed on the horizontal surface of a 45° single bounce ZnSe ATR prism (Spectra Tech) and the spectra were recorded using pure water as a blank. For suspensions, an oxide layer was deposited on the prism by evaporating 100 μL of an aqueous suspension containing 0.1 g of the (hydr)oxide in 100 mL of solution at room temperature The spectra of the treated (hydr)oxide were recorded for samples exposed to chromate, oxalate and mixed (chromate-/oxalate-containing) solutions for 40 min, after repeated rinsing with pure water to eliminate nonadsorbed ligand. Untreated (hydr)oxide was used as a blank. All the spectra were recorded with a Nicolet 560P spectrometer fitted with a liquid-N_2-cooled microcrystalline cellulose triacetate detector.

Results and discussion

FTIR analysis

The spectrum of α-Cr_2O_3 treated with oxalic acid is shown in Fig. 1. A comparison with the spectrum of dissolved $Cr(C_2O_4)_3^{3-}$ (not shown) demonstrates that surface complexes with chelated oxalate are formed. The main peaks [1,710 (shoulder), 1,680 and 1,420 cm^{-1}) can be attributed to the =Cr–(C_2O_4) moiety with both carboxylate groups bound, either to the same surface Cr(III) ion or to two adjacent ones. In the solids $K_3[Cr(C_2O_4)_3]$ and $K_3[Cr(C_2O_4)_2(H_2O)_2]$, the same bands are observed [5, 6]. The peaks at 1,620 and 1,310 cm^{-1} are due to residual, noncomplexed oxalate in solution.

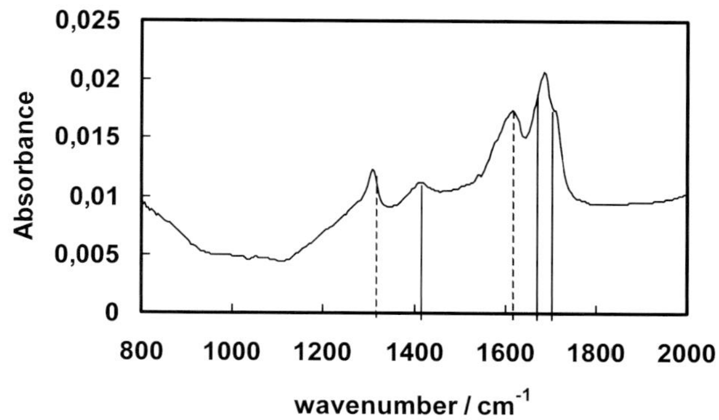

Fig. 1 Fourier transform (*FT*) IR–attenuated total reflection (*ATR*) spectra of α-Cr_2O_3 in contact with aqueous oxalate solution. [oxalate]=0.01 mol dm^{-3}, pH 3.6

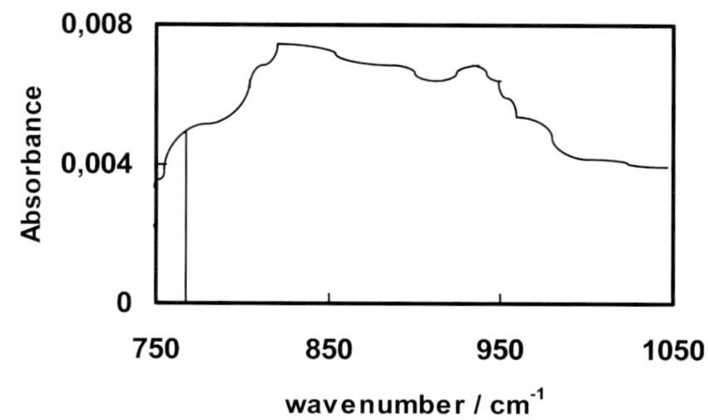

Fig. 2 FTIR–ATR spectra of α-Cr_2O_3 in contact with aqueous chromate solution. [Cr(VI)]=0.001 mol dm^{-3}, pH 4.0

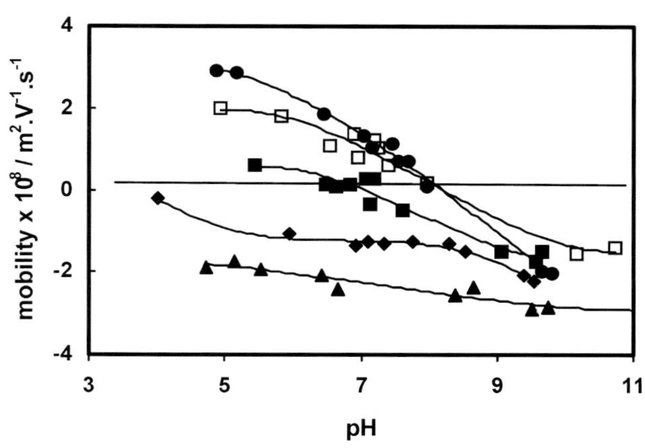

Fig. 3 Mobility/pH profile of α-Cr_2O_3 in oxalate solution. $[oxalate]_T$/mol dm^{-3}: 0 (*circles*), 5×10^{-6} (*open squares*), 5×10^{-5} (*closed squares*), 5×10^{-4} (*diamonds*) and 5×10^{-3} (*triangles*)

The spectrum of α-Cr_2O_3 treated with chromate is shown in Fig. 2. The bands in the range 790–970 cm^{-1} are assigned to an α-Cr_2O_3–chromate surface complex; they are broader than those of the aquoion and

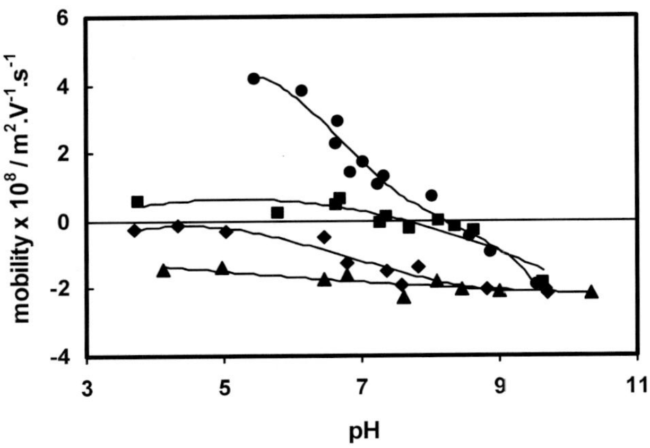

Fig. 4 Mobility/pH profile of Cr_2O_3 14 H_2O in oxalate solution. $[oxalate]_T$/mol dm^{-3}: 0 (*circles*), 5×10^{-5} (*squares*), 5×10^{-4} (*diamonds*) and 5×10^{-3} (*triangles*)

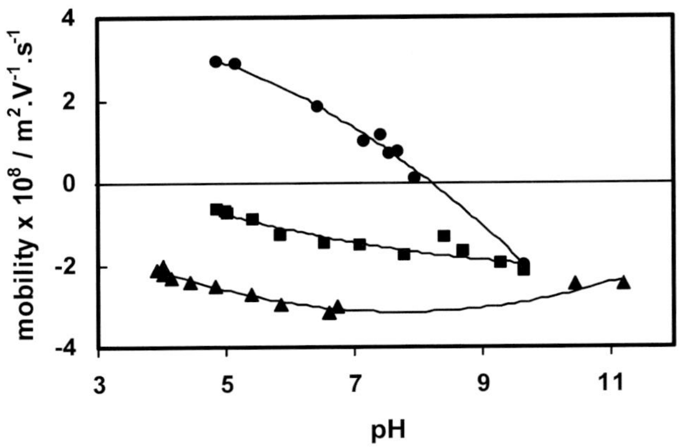

Fig. 7 Mobility/pH profiles of α-Cr_2O_3 in oxalate/chromate solution. $[oxalate]_T + [Cr(VI)]$/mol dm^{-3}: 0 (*circles*), 5×10^{-4} (*squares*) and 10^{-2} (*triangles*); $[Cr(VI)]/[oxalate] = 1$

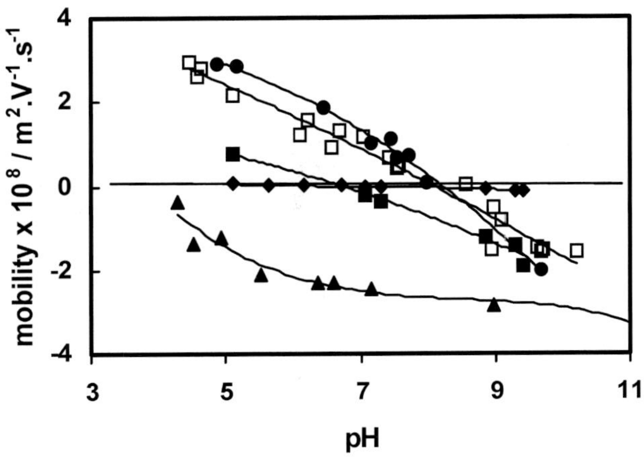

Fig. 5 Mobility/pH profile of α-Cr_2O_3 in chromate solution. [Cr(VI)]/mol dm^{-3}: 0 (*circles*), 5×10^{-6} (*open squares*), 5×10^{-5} (*closed squares*), 5×10^{-4} (*diamonds*) and 5×10^{-3} (*triangles*)

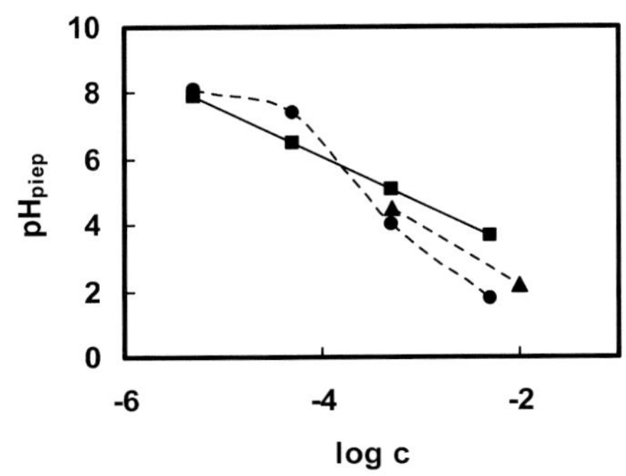

Fig. 8 Concentration dependence of pH_{piep} of α-Cr_2O_3 in aqueous solution containing oxalate (*circles*), chromate (*squares*) and mixed oxalate/chromate (*triangles*). The *x*-axis represents total ligand concentration on a logarithmic scale

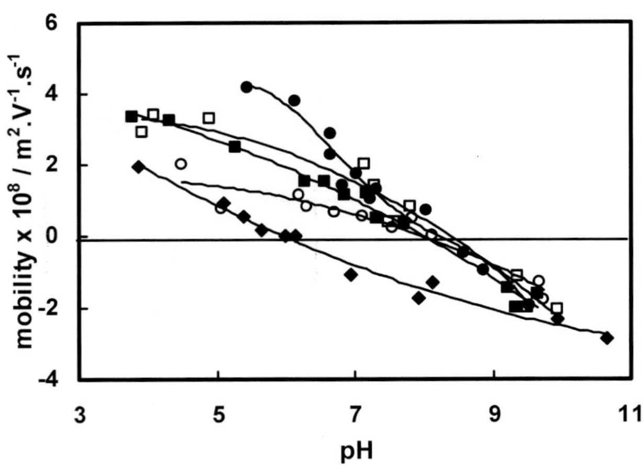

Fig. 6 Mobility/pH profile of Cr_2O_3 14H_2O in chromate solution. [Cr(VI)]/mol dm^{-3}: 0 (*closed circles*), 5×10^{-6} (*open squares*), 5×10^{-5} (*closed squares*), 2.5×10^{-4} (*open circles*) and 5×10^{-4} (*diamonds*)

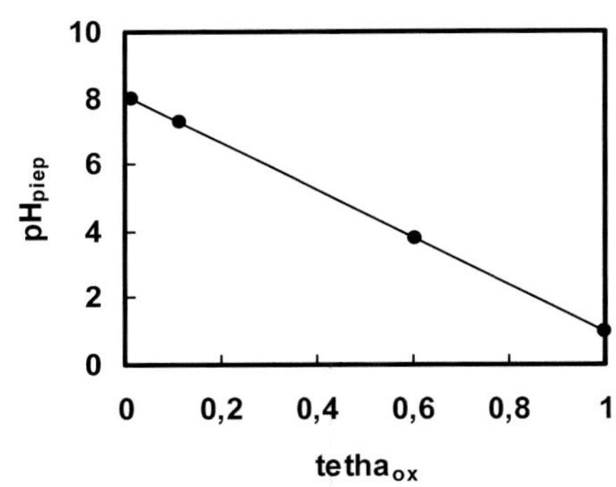

Fig. 9 ϑ_{ox} values required to fit the pH_{piep} values in oxalate media (see Fig. 8), according to Eq. (11)

correspond to the $(\nu_1 + \nu_3)$ CrO_4 bands in $NaCr(CrO_4)_2$ [7]. The shoulder at 770 cm^{-1} coincides with the ν_{as} (Cr–O–Cr) in dissolved $Cr_2O_7^{2-}$ [8, 9] and suggest that surface-induced dimerization takes place to a certain extent. Equation (10) (see later) provides a possible description of the formation of dichromate on the surface at a high degree of coverage.

Mobility analysis

The mobility/pH profiles for the two solids measured at various oxalate concentrations are shown in Figs. 3 and 4; Figs. 5 and 6 show the profiles of the solids in the presence of chromate, and Fig. 7 shows the data for equimolar Cr(VI)–oxalate mixtures. The isoelectric points of the crystalline and amorphous oxides in mineral media are 8.1 and 8.3, respectively, as reported before [9].

Oxalate and chromate bring about a decrease in the mobility values, the pH of the isoelectric point (pH_{piep}) shifts to lower values and charge reversal takes place at higher ligand concentrations and acidic conditions. This last effect demonstrates that oxalate and chromate are chemisorbed on chromium oxides. Amorphous and crystalline oxides differ in the magnitude of the mobility in the acidic range at low ligand concentration.

To explain the curves of Figs. 3–6, we propose the adsorption modes described in the following subsections for each ligand. These models are the simplest possible ones.

Adsorption modes for oxalate

Two modes of adsorption must be considered:

1. A charge-neutralization mode, implying electroneutral adsorption in the pH range close to the isoelectric point in mineral acids (Eq. 1) and neutralization of positive charge in more acidic media (Eq. 2). The equilibrium constants for Eqs. (1) and (2) are linked through the surface first acidity constant $K_{a1}{}^{s}$

$$\equiv Cr - OH + H^+ + HC_2O_4^- = \equiv Cr - OH_2^{+\cdots}C_2O_4H^-, \quad (1)$$

$$\equiv Cr - OH_2^+ + HC_2O_4^- = \equiv Cr - OH_2^{+\cdots}C_2O_4H^-. \quad (2)$$

Although Eqs. (1) and (2) are written as involving ion pairs, the nature of these species is discussed later.

2. A charge-reversal mode (Eqs. 3, 4, 5):

$$\equiv Cr - OH + H^+ + HC_2O_4^- = \equiv Cr - C_2O_4^- + H_3O^+ \quad (3)$$

$$\equiv Cr - OH_2^+ + HC_2O_4^- = \equiv Cr - C_2O_4^- + H_3O^+, \quad (4)$$

$$\equiv Cr - OH_2^+ + C_2O_4^{2-} = \equiv Cr - C_2O_4^- + H_2O. \quad (5)$$

Again, Eqs. (3) and (4) are linked through the surface first protolytic equilibrium; Eqs. (4) and (5) are linked through the second protolysis of oxalate.

The first mode operates at low oxalate concentration, and therefore it does not involve ion-pair formation. Chemisorption is required, and it is therefore proposed that the two modes correspond to two successive surface complexation reactions, as described in Eqs. (6) and (7):

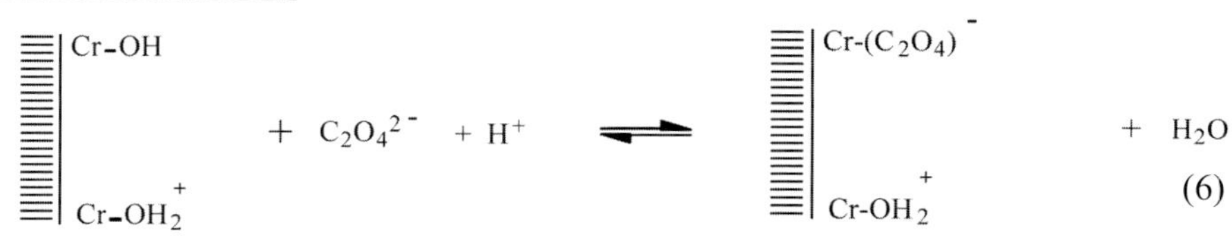

(6)

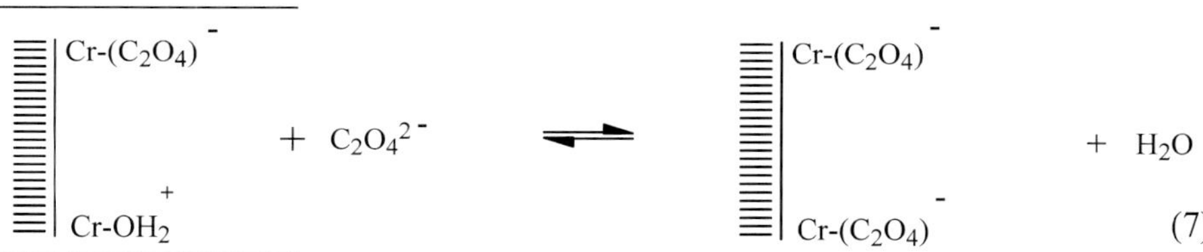

(7)

The main reaction at low oxalate concentration is (Eq. 6) conducive to a decrease of positive charge, but without the charge reversal brought about by Eq. (7). The involvement of two vicinal sites makes these modes sensitive to surface topography, thus explaining why charge neutralization, Eq. (6), is more important on the amorphous oxide than on α-Cr_2O_3 [10].

Adsorption modes for chromate

Figures 5 and 6 are accounted for by considering equilibria as shown in Eqs. (8) and (9), equivalent to Eqs. (6) and (7), with CrO_4^{2-} as the surface-bound ligand:

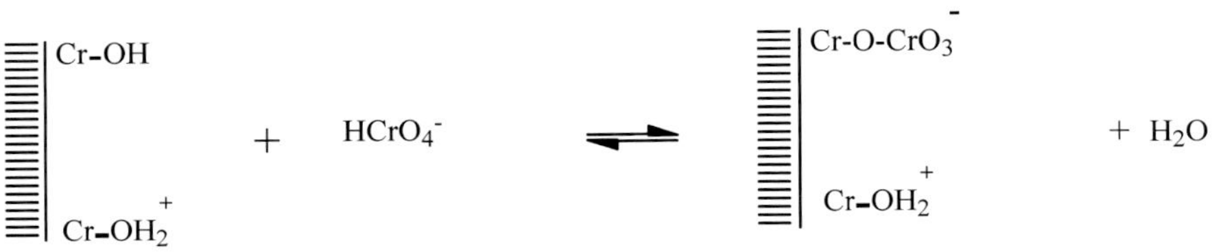

(8)

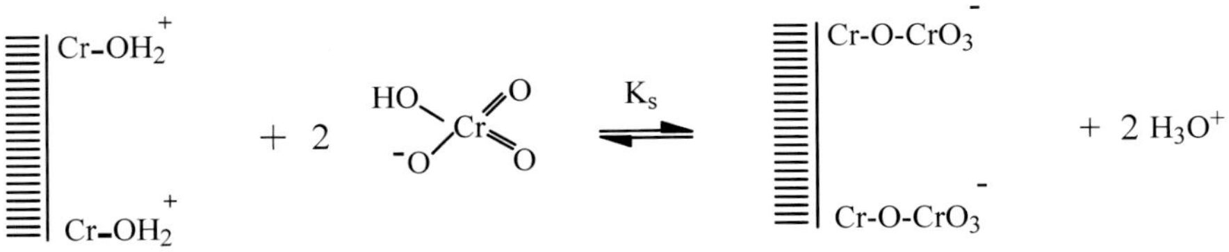

(9)

The FTIR band at 770 cm^{-1} suggests that a third reaction, the surface dimerization of chromate, should also be taken into account:

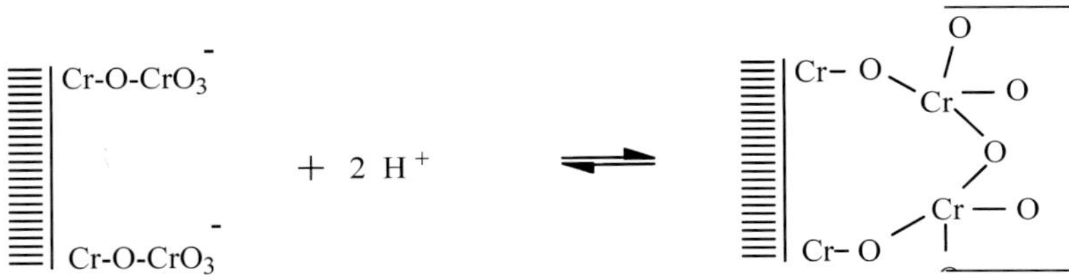

(10)

The high surface concentration favors, as expected, the dimerization of chromate.

Dependence of pH_{piep} on total ligand concentration

The dependence of pH_{piep} of α-Cr_2O_3 on log[oxalate]$_T$, log[Cr(VI)], and log([oxalate]$_T$ + [Cr(VI)]) is shown in Fig. 8.

Conceptually, the surface of the oxide at the various isoelectric points obtained as the total oxalate concentration varies can be envisaged as resulting from a mixture of oxalated sites (mole fraction ϑ_{ox}) and nonoxalated sites (mole fraction ϑ_{free}), each part being characterized by a different value of pH_{piep}. The pH_{piep} of the partially covered surface is calculated simply as the weighted average of both extreme pH_{piep} values:

$$pH_{piep} = 1 \times \vartheta_{ox} + 8.1 \times \vartheta_{free} = 8.1 - 7.1 \times \vartheta_{ox}. \quad (11)$$

This empirical equation is simpler than those used to describe the pH_{piep} of mixed oxides [3]; more elaborated equations, taking into account the ΔpK_a values of the two types of sites are not warranted by the limited set of data. Even the use of Eq. (11) is impaired by the lack of a good estimate of the pH_{piep} of the totally oxalated surface. Arbitrarily assuming it to be about 1 and using the value of 8.1 for the pH_{piep} of the pure oxide, the ϑ_{ox} values required to describe the changes in pH_{piep} can be calculated; the results, shown in the ϑ_{ox} values that lead to the linear relationship proposed by Eq. (11), as shown in Fig. 9, are reasonable taking into account the changes in oxalate concentration and pH. Further validation of the model would require the measurement of adsorption isotherms at various pH values, to derive conditional stability constants of the surface complexes.

The linear dependence of pH_{piep} on the total chromate concentration shown in Fig. 8 can be explained assuming that Eqs. (8)–(9) make contributions in an adequate ratio to the adsorption density. In the limiting case that Eq. (9) dominates adsorption, and the neutral species makes no contribution to the adsorption density, a slope of 1 is predicted for the straight line in Fig. 8. Additional surface coverage by electroneutral species formed in Eqs. (8) and (10) increases the slope by a factor that depends on the stoichiometric ratio. The observed slope (1.5) suggests important contributions from both adsorption modes at the pH_{piep} values.

Figure 8 shows that the effect of adsorption on pH_{piep} at low concentrations is more important for chromate than for oxalate, whereas the reverse is true for concentrations above roughly 1×10^{-4} mol dm^{-3}. This behaviour is simply due to a higher affinity for $HC_2O_4^-$ than for $HCrO_4^-$ and to the shifts in the protolytic equilibria in oxalate; the deprotonation of $HCrO_4^-$ takes place only above about pH 6, and does not affect the linearity observed in Fig. 8.

The pH_{piep} values of solutions containing mixtures of both ligands adsorbed onto α-Cr_2O_3 fall between the pH_{piep} values of each one adsorbed in separate experiments; therefore, it is reasonable to assume competitive chemisorption, i.e. the sites involved in oxalate adsorption are also involved in chromate adsorption.

Acknowledgements This work was supported by CNEA (program 5) and ANPCyT (grant PICT 06–00000–00306). P.J.M. and M.A.B. are members of CONICET.

References

1. US Environmental Protection Agency (1987) Fed Regist 52:12866
2. Venitt S, Levy LS (1974) Nature 250:493
3. Blesa MA, Magaz GE, Salfity JA, Weisz AD (1997) Solid State Ionics 101–103:1235
4. Spiccia L, Marty W (1986) Inorg Chem 25:266
5. Schmelz MJ, Miyazawa T, Mizushima S, Lane TJ, Quagliano JV (1957) Spectrochim Acta 9:51
6. Hug S, Sulzberger B (1994) Langmuir 10:3587
7. Saavedra MJ, Parada C, Baran EJ (1996) J Phys Chem Solids 57:1929
8. Brown RG, Ross SD (1972) Spectrochim Acta 28:1263
9. García Rodenas LA, Weisz AD, Magaz GE, Blesa MA (2000) J Colloid Interface Sci 230:181
10. Magaz GE, García Rodenas LA, Morando PJ, Blesa MA (1998) Croat Chem Acta 71:917

Progr Colloid Polym Sci (2004) 128: 36–43
DOI 10.1007/b97083

Nina M. Kovalchuk
Dieter Vollhardt

Nonlinear oscillations at liquid interfaces by surfactant transfer

D. Vollhardt (✉)
Max Planck Institute of Colloids and Interfaces, 14424 Potsdam/Golm, Germany
e-mail: vollh@mpikg-golm.mpg.de

N.M. Kovalchuk
Institute for Problems of Material Science, 03142 Kiev, Ukraine

Abstract Systems are reviewed in which spontaneous nonlinear oscillations occur at the interface by surfactant transfer inside the system. Oscillations of the electrical potential through the interface and of the interfacial tension were observed by transfer of ionic species through an oil/water interface and through a liquid oil membrane interposed between two aqueous solutions. Auto-oscillations of the surface tension occur at the air/water interface by dissolution of a surfactant droplet in the bulk of water. The last system is the simplest from the chemical point of view, and the mechanism of this phenomenon can be explained on the basis of direct numerical simulations of the dynamics of the model system.

Keywords Nonequilibrium systems · Liquid interfaces · Nonlinear oscillations · Marangoni instability

Introduction

Processes induced by the mass transfer through liquid interfaces have attracted attention for a long time because of their importance for fundamental science and technology. Studies of these processes can be helpful for a better understanding of the behavior of living cells, liquid–liquid extraction, stability of foams and emulsions, processes in alveolar layers, etc. Often such systems far from equilibrium demonstrate significant nonlinear dynamics, in particular, spontaneous nonlinear oscillations, a subject of special interest in modern science.

The main objectives for such studies of nonlinear oscillations are (1) to find systems experimentally, different in chemical composition, that demonstrate such behavior, (2) to determine the driving forces and to investigate the effect of each of them under various conditions, (3) to determine the mechanisms that can lead to spontaneous oscillations by mass transfer through the liquid interface, and (4) to derive the general mechanisms of nonlinear oscillations in different systems.

Three types of systems developing spontaneous nonlinear oscillation at liquid interfaces are being studied intensively at present. We consider these systems, focusing attention mainly on the experimental aspects of the problem.

Oscillations at the oil/water interface

Systematical experimental investigations of spontaneous nonlinear oscillations at liquid/liquid interfaces were started by Dupeyrat and Michel [1]. They observed movement at the interface between two immiscible liquids (water and oil) and variations of the interfacial tension. Both liquids contained ionic solutes of equilibrium concentrations. The movement began after a current was passed through the solutions for a short time and then switched off. The current brought the system into a nonequilibrium state owing to electrolysis. The system transition to the equilibrium state was accompanied by the appearance of instability, which revealed itself in the convective motion at the interface. The phenomenon occurred only if one of the solutes was a surfactant. Therefore, it was suggested that the driving force of instability is a variation of the interfacial tension (Marangoni effect) [2].

A system similar to the systems considered in Ref. [1] can be obtained under conditions far from equilibrium also without using current, if each of the ionic solutes is dissolved in the phase in which it is least soluble. In such a system nonlinear oscillations were discovered and investigated by Nakache and Dupeyrat [2–4]. They observed spontaneous movements of liquids accompanied by oscillations of the interfacial tension and the interfacial potential difference by two-directional solute transfer through an oil/water interface (Fig. 1). The experiments were performed, for example, as follows [4]. The organic solution (35 cm^3) was poured in a beaker (4.5-cm diameter). Then using a pipette 35 cm^3 water solution was very carefully poured over the organic solution. Oscillations began after an induction period and could be related to the relaxation type according to the shape. A sharp decrease of the surface tension was followed by a rather gradual increase.

Nakache and Dupeyrat performed comprehensive studies of the phenomenon by using various oil substances and soluble compounds. The oscillations were observed with different organics, such as nitroethane and nitrobenzene. The surface-active compounds used were alkyltrimethylammonium halogenides ($C_{12}Br$, $C_{16}Cl$, $C_{18}Cl$) that were dissolved in the aqueous phase. Hydrophilic compounds such as potassium iodide or picric acid (HPi) were dissolved in the organic phase. The main conditions required for the appearance of oscillations were established in these studies. Two chemical compounds, one being surface active, have to be dissolved in such a manner to generate a system far from equilibrium. For instance, the long-chain compound $C_{12}Br$, which is very soluble in the organic phase, was dissolved in water, whereas HPi, which is very soluble in water, was dissolved in nitroethane. Oscillations were not observed when one single compound, $C_{12}Br$ or HPi, was dissolved in one of the phases, or when both compounds were dissolved in the same phase, or when each of them was dissolved in the phase where it is most soluble. Oscillations occurred in a rather narrow concentration range of the solutes. Although the oscillations were rather irregular, it was proved that the characteristics of the oscillations depend strongly on the concentrations of the compounds and temperature.

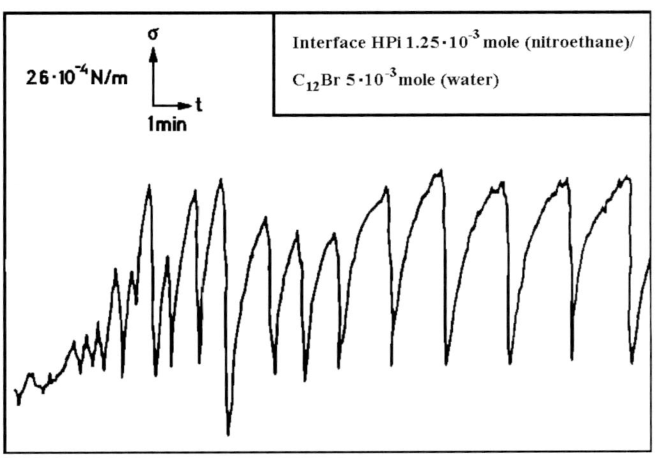

Fig. 1 Temporal oscillations of the interfacial tension in the system of picric acid (*HPi*) solution in nitroethane/dodecyltrimethylammonium bromide (*C_{12}*Br) solution in water [2]

Yoshikawa and Matsubara [5] extended the list of organic solvents suitable for the reproduction of the oscillations. They observed regular synchronized oscillations of pH in the aqueous phase and of the electrical potential between the two phases in the system where HPi was dissolved in 2-nitropropane (or 1-nitropropane) and $C_{16}Br$ in the aqueous phase (Fig. 2). The oscillations occurred when the $C_{16}Br$ concentration in the aqueous solution exceeded 2 mM. The presence of ethanol in the aqueous solution reduced this threshold concentration. Oscillations were also observed by using $C_{12}Br$ and 1-laurylpyrydinium chloride as surfactants. Later Magome and Yoshikawa [6] performed experiments with systems containing $C_{18}Cl$ in the aqueous phase that is in contact with iodine solution in nitrobenzene saturated with potassium iodide.

The last system was investigated in detail by Kai and coworkers [7–9]. About 120 combinations of solutions ranging from 10^{-4} to 10^{-1} M were prepared in this study. Cylindrical and annular containers were used with various diameters and heights. The measurements of the electrical potential between the interface and the bulk of aqueous solution showed that strictly periodic, periodic with superimposed noise or highly nonperiodic (chaotic) temporal behavior depending on the initial solute concentration can occur. It was found that the instabilities occur successively from one type of motion to another, depending on the initial concentration of the solution as well as the shape and the aspect ratio (ratio of diameter to the height) of the containers. Solitary waves, periodic oscillations with and without propagation in space, steady wave propagation between a source and a

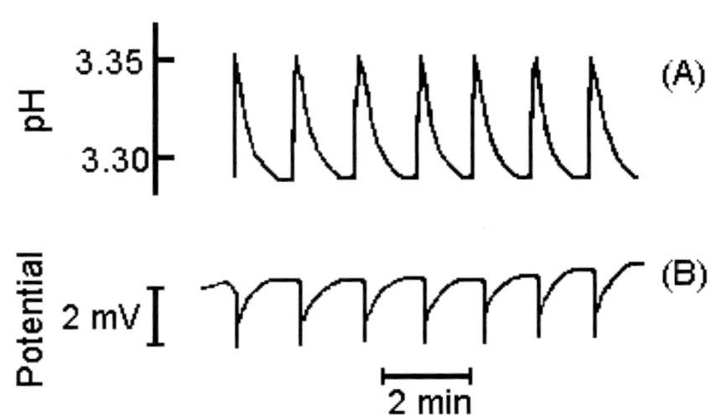

Fig. 2 **a** Temporal oscillation of pH in the aqueous phase and **b** electrical potential between aqueous and organic phases by the mass transfer through the interface HPi solution in 2-nitropropane/hexadecyltrimethylammonium bromide (*C_{16}*Br) solution in water [5]

sink, isotropic stretching motion from the center of the cylindrical container to the boundary and vice versa, and periodic rotation and chaotic motions of waves were observed. As a result of the studies a phase diagram of the interfacial motions in the concentration plane was proposed. The diagram distinguishes between five regions for which different types of motion are characteristic.

To characterize another surfactant system which gives nonlinear oscillations Shioi and coworkers [10, 11] investigated a system having *n*-heptane surfactant solution in contact with water or aqueous solution. The main distinction of this system from those systems investigated earlier is the presence of a surfactant in the oil phase instead of the water phase. Di(2-ethylhexyl) phosphoric acid (DEHPA) and sodium bis(2-ethylhexyl) sulfosuccinate (AOT) were used as surface-active compounds. Using AOT as a surfactant did not lead to the appearance of oscillation for all the aqueous solutions studied. No oscillation was observed when the organic phase containing DEHPA was in contact with pure water or an aqueous solution of barium chloride and sodium chloride. Oscillations of interfacial tension were found only in the three systems DEHPA/$CaCl_2$, DEHPA plus $Ca(OH)_2$/$CaCl_2$ and DEHPA plus $Ca(OH)_2$/KH_2PO_4. They were classified into two groups: type A oscillations were always weak and random, whereas type B oscillations were strong and included sometimes quasi-periodic parts. The amplitude of the oscillation decreased as the container diameter increased. Experiments with simultaneous measurements of the interfacial tension and surface tension of the oil phase were also carried out but no changes in the surface tension were noticed when the interfacial tension oscillated [11].

The characteristics of interfacial flow in the system considered were studied and the kinetic energy of the interfacial flow was evaluated, which proved to decay exponentially with time after the contact of the oil and aqueous phases. Flow visualization by means of silica powder has shown that after the organic and aqueous phases were brought in contact, radial propagating spontaneous flows appeared in many points of the interface. These flows formed quasi-circular convective patterns. The patterns look very like those observed by Orell and Westwater [12] by extraction of acetic acid out of ethylene glycol with ethyl acetate. Unfortunately, the data concerning the temporal dependence of the interfacial tension are absent in Ref. [12].

Another system with nonlinear behavior was investigated by Takahashi et al. [13]. They also measured the interfacial tension and the interfacial potential difference at the water/nitrobenzene interface. But they used pure nitrobenzene in their experiments, whereas two compounds (NaCl and sodium dodecyl sulfate, SDS) were dissolved in the water phase. The surface-active compound (SDS solution) was injected with a microsyringe inserted below the air/water interface at a point far from the water/nitrobenzene interface. Therefore, in contradiction to the systems described earlier, a nonuniform surfactant distribution in the bulk of water occurs in the system considered at the very beginning. The system displays a single well-shaped oscillation of the interfacial tension and the interfacial potential difference after a certain induction period. Successive oscillations were observed by continuous SDS injection using a peristaltic pump.

Oscillations in systems containing a liquid membrane

To bring the study of nonlinear oscillation nearer to real biological objects, Yoshikawa and Matsubara [14, 15] extended the experiments with the oil/water system described earlier to a liquid oil membrane (nitrobenzene containing HPi) interposed between two aqueous solutions, one of them containing a surfactant ($C_{16}Br$) and an aliphatic alcohol. The experiments were performed using a U-shaped glass tube (Fig. 3). Well-shaped regular spontaneous oscillations of the electrical potential difference between the two aqueous solutions were obtained using the homologous series of aliphatic alcohols from ethanol to pentanol added to the $C_{16}Br$ solution. The oscillations started after a certain induction period and continued for about 1 h. No oscillations were observed in the absence of alcohol or $C_{16}Br$. The oscillation frequency increased as the alcohol concentration increased (Fig. 4). The threshold alcohol concentration needed to induce the oscillations decreased as the alcohol hydrophobicity increased. The oscillation amplitude decreased with time. The decrease of the amplitude was found to be smaller for the system in the presence of sucrose in another aqueous phase. Use of aromatic alcohol instead

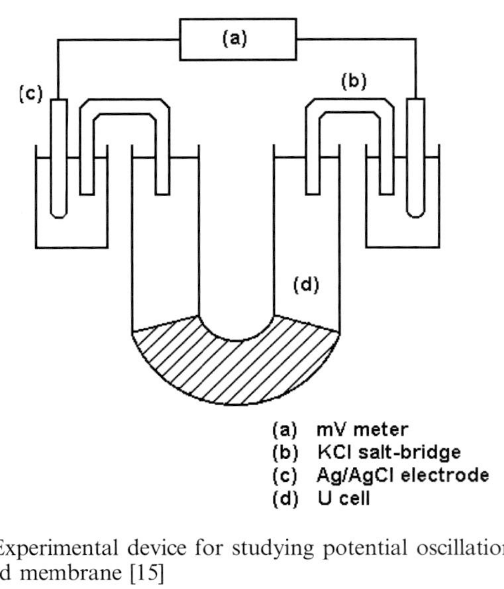

Fig. 3 Experimental device for studying potential oscillations across the liquid membrane [15]

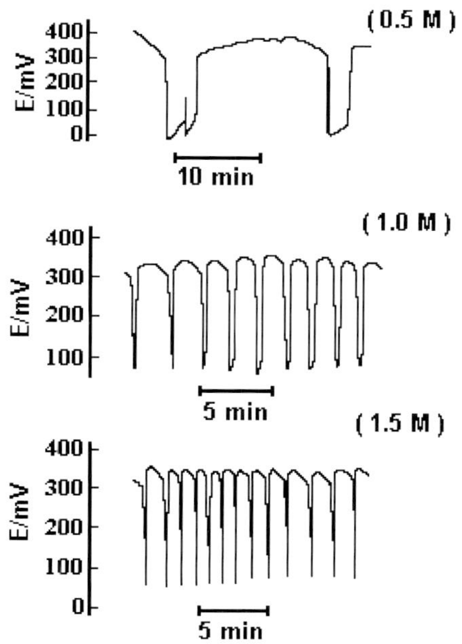

Fig. 4 Oscillation of electrical potential across the liquid membrane with different concentrations of ethanol in the aqueous phase. The concentration of $C_{16}Br$ in the same aqueous phase was 5 mM; the concentration of sucrose in another aqueous phase was 0.1 M [15]

of aliphatic alcohol inverted the oscillation mode (Fig. 5).

Arai and coworkers [16–18] showed that oscillations also occur in water/1-octanol/water systems (octanol membrane). They performed simultaneous measurements of the electrical potential across the two oil/water interfaces to determine the particular interface at which the potential oscillations occur. It was proved that, as a rule, the potential oscillation is generated at the interface between the organic phase and the aqueous phase, which initially does not contain surfactant (denoted as a1, whereas the other aqueous phase with added surfactant and alcohol was denoted as a2). However in a certain range of surfactant and alcohol concentrations, the oscillations are generated at both interfaces or only at

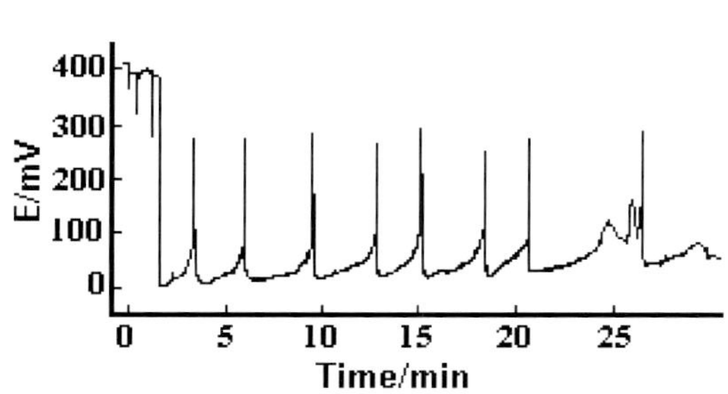

Fig. 5 Oscillation of electrical potential across the liquid membrane with 0.1 M benzyl alcohol in the aqueous phase. The concentration of $C_{16}Br$ in the same aqueous phase was 5 mM; the concentration of sucrose in another aqueous phase was 0.1 M [15]

a2. When oscillation is generated at a1 it seems to be more regular and well shaped.

Two types of surfactant were studied in Ref. [16]: cationic $C_{16}Br$ and anionic SDS. Both surfactants show similar behavior, except the oscillation modes were opposite to each other (such as, in Figs. 4, 5). Later the number of the surfactants used was considerably extended by Arai et al. and other researchers (see, e.g., Ref. [17] and references therein). The transfer of charged surfactant ions across the interface can be controlled by an external electrical potential difference. Application of constant current to interface a1 facilitates or inhibits the surfactant transfer and thus changes the mode of the potential oscillations [17]. The oscillation characteristics (potential just after the start of measurement, potential at the induction period, duration of the induction period, oscillation period and amplitude) also depend on the surfactant and alcohol added to a2 as well as on their concentrations, on the material and on the geometry of the measuring cell [16, 17]. They can also be strongly affected by a substance added to aqueous phase a1. The effect of the presence of an electrolyte in a1 on the oscillation of the electrical potential across the octanol membrane was investigated in Ref. [16]. It was then supposed that the system using the liquid membrane could serve as a sensor of a new type. To clarify the possibility of a liquid membrane system for taste evaluation, the response of the system to the presence of primary taste substances (bitterness, sweetness, sourness, saltiness, umami), artificial sweetness and two test substance combinations were examined [18]. It was found that the oscillation characteristics differ noticeably for the various substances studied. The liquid membrane was also found useful for evaluation of drug features, for example, pharmacological activity [17].

The nonlinear oscillations just considered have obviously much in common. They are induced if mass transfer of an ionic surfactant through a liquid/liquid interface occurs in the presence of an electrolyte. Convective motion because of the Marangoni effect accompanies the oscillations. A number of models were proposed to explain the oscillation mechanisms [4, 5, 9, 11, 14, 16, 19–21]. All the available models are mainly qualitative and are based on the consideration of chemical, namely ionic, interactions together with adsorption/desorption at the interface; however, there is no generally accepted model of these phenomena. At the same time the role of hydrodynamics in the oscillation mechanism is largely ignored although the rate and direction of the reactions at the interface are to a great extent determined by the rate of supply of the compounds to the interface and by taking away the reaction products. It is obvious that hydrodynamics can play a significant role in the development of the oscillation because it influences the mass transfer in the system and therefore it should be taken into account.

A quantitative description of the systems considered here from first principles, including chemical interactions and hydrodynamics, is too difficult because of their complicated chemical and ionic compositions. Therefore, a simpler system revealing nonlinear oscillations could be helpful for improving the existing models and for a comprehensive understanding of the mechanisms causing the nonlinear oscillations at the liquid/liquid interfaces. Auto-oscillations of the surface tension discovered recently [22] represent such model systems.

Surface tension auto-oscillation at the air/water interface

Auto-oscillations of the surface tension occur at the air/water interface by dissolution of a surfactant droplet in the bulk water. A scheme of a measuring cell having a diameter of about 4–6 cm is presented in Fig. 6. A droplet (diameter 2–4 mm) is formed at the capillary tip by gradual addition of a surfactant, having a limited solubility in water, into the capillary. The dynamic surface tension is measured with a Wilhelmy plate. The measurement begins immediately after the drop formation. During a certain time (induction period) the surface tension remains unchanged. Then, the oscillation of the surface tension begins (Fig. 7). The oscillation has an asymmetrical shape similar to that discussed in the preceding sections, with an abrupt decrease of the surface tension followed by a gradual increase. This suggests the nonlinear nature of the phenomenon considered.

Auto-oscillations were observed for such surfactants as diethyl phthalate [22], fatty acids [23] and aliphatic alcohols [24]; they last for some hours. The maximum fixed duration was 72 h. The oscillation period increases with time and depends on the surfactant used. For example, in the experiment with octanol the period was about 7 min at the beginning of the process and 22 min after 72 h. In the series of aliphatic alcohols the initial period increases from about 40 s for pentanol to about 7 min for octanol. It was established in Ref. [24] that the oscillations appear only at a large enough capillary immersion depth. At small depths only a single pulse occurs and then the surface tension decreases monotonously. The critical immersion depth of the capillary, dividing an oscillatory and a nonoscillatory regime, depends on the surfactant used. For example, it decreases as the chain length of the aliphatic alcohol increases. Comprehensive experimental studies of the effect of the system geometry on the appearance and characteristics of the auto-oscillations [25] show that the critical depth (for a certain surfactant) increases as the cell diameter increases. Thus, the ratio of the cell diameter to the capillary immersion depth (aspect ratio) determines the regime of the system dynamics.

Talcum powder spread over the surface reveals fast surface motion directed from the capillary to the wall during the abrupt decrease of the surface tension. Thus, it has been supposed that the phenomenon is related to the convective instability driven by the surface tension gradient (Marangoni instability) [26]. Linear stability analysis [27, 28] shows that oscillatory instability (sometimes called overstability) at the air/liquid interface of infinite extension occurs only by the surfactant transfer from the gaseous phase to the liquid. Therefore the nonlinear effects and confined system geometry should be taken into account to explain the auto-oscillation phenomenon. To do this, numerical simulation of the

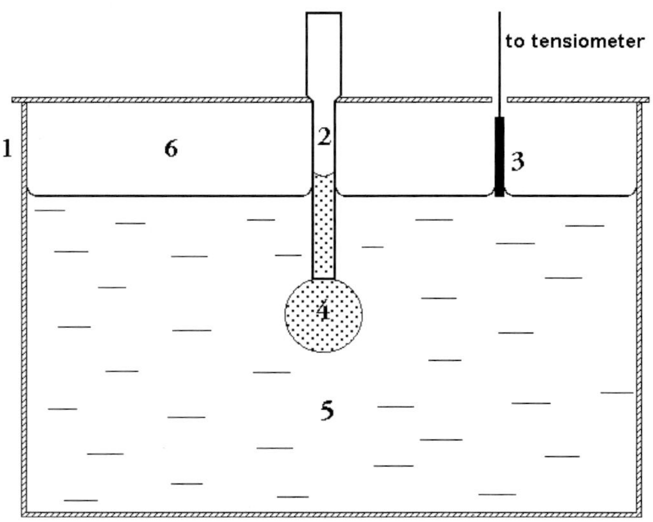

Fig. 6 Measuring cell for the investigations of the surface tension auto-oscillation: cylindrical glass vessel with covering plate (*1*), capillary (*2*), Wilhelmy plate (*3*), surfactant droplet (*4*), water (*5*), air (*6*)

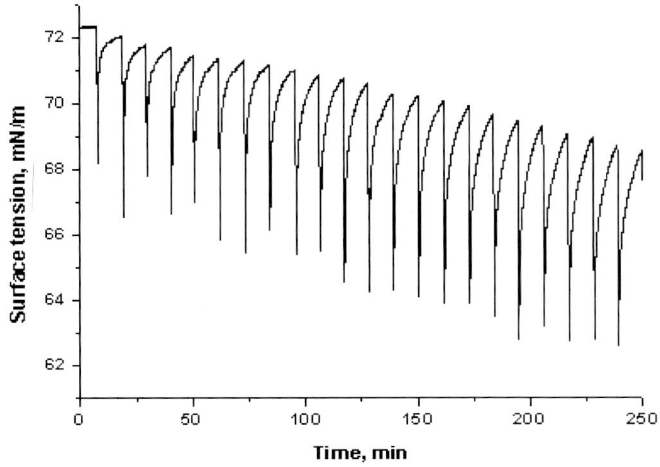

Fig. 7 Auto-oscillations of the surface tension in the octanol/water system [23]

evolution of the model system was performed on the basis of the full nonstationary nonlinear equations of physicochemical hydrodynamics, namely, the Navier–Stokes equation, a continuity equation and a convective diffusion equation with appropriate boundary conditions.

First the dissolution of a surfactant droplet under the free surface of a semi-infinite liquid layer was studied [29]. The calculations have shown that after a certain induction period instability arises in the system, resulting in an abrupt increase of the surface velocity accompanied by an abrupt decrease of the surface tension. The duration of the induction period as well as the values of the surface velocity and surface tension reached depend on the surfactant properties [30]. However, repeated oscillations were not obtained in the frame of this model. Only a single oscillation with a consequent monotonous decrease of the surface tension was obtained. In agreement with experimental results, this displays that the system dimensions are crucial for the onset of auto-oscillations.

An improved model taking into account finite system dimensions was proposed in Ref. [31]. The simulations displayed that, depending on the capillary immersion depth and the cell diameter, the evolution of such a model system to the final equilibrium surfactant distribution can occur in two different ways, namely, in an oscillatory or in a nonoscillatory regime [32]. This is in complete agreement with the experimental results. The oscillation shape is also very similar to that observed in experiments (Fig. 8). The comparison of Figs. 7 and 8 shows that the oscillation period calculated is several times longer than that observed experimentally. One of the main reasons for this is that the model used in Refs. [31, 32] assumes that the solution density does not depend on the concentration and that it is equal to the water density. Our recent calculations show that taking into account of the buoyancy force results in a decrease of the oscillation period to values which agree better with those experimentally observed.

The analysis of the numerical results allows an understanding of the mechanism of auto-oscillations at the air/liquid interface, which is discussed in detail in Ref. [31]. Initially (in the absence of buoyancy), only the diffusion mass transfer occurs in the system. It is seen from Fig. 6 that the surfactant distribution in the bulk water should be nonuniform in both vertical and radial directions. When the solute reaches the surface, the nonuniform surfactant distribution in the radial direction causes convective motion because of the Marangoni effect. The convection intensifies the solute supply to the surface, especially in the vicinity of the capillary, where the concentration gradients and, therefore, the velocity of convection are largest. This causes an increase of the surfactant concentration and the concentration gradient at the surface near the capillary, leading to a further increase of the surface velocity. The convection in the system and the concentration gradients in the radial and vertical directions increase gradually until the vertical concentration gradient reaches a threshold value. Then, instability develops in the system according to the known mechanism [26]. Initially, instability arises in the vicinity of the capillary where the concentration gradient has a maximum and then propagates over the whole surface. The convection spreads the solute very quickly over the surface, causing an abrupt decrease of the surface tension.

The temporal dependence of the surface velocity corresponding to the surface tension oscillation presented in Fig. 8 is shown in Fig. 9. It is seen that the abrupt decrease of the surface tension correlates with an abrupt increase of the surface velocity, i.e., with the

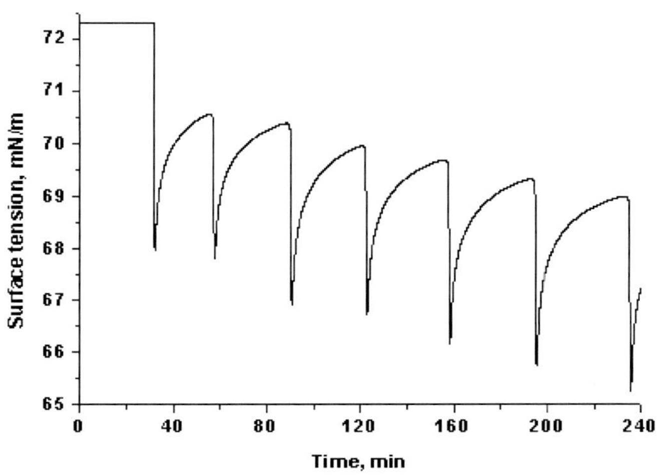

Fig. 8 Auto-oscillation of the surface tension in the octanol/water system (numerical simulation) [32]

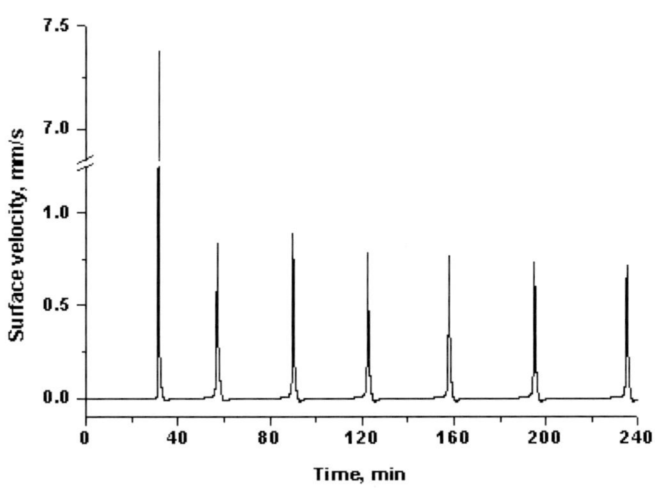

Fig. 9 Time dependency of the surface velocity for the octanol/water system (numerical simulation) [32]

increase of convective instability. The regime with high velocity (the fast stage of the oscillation period) continues for only a very short time. When the intensive convection fades, the system returns to the slow stage, during which the relaxation of the surface tension occurs. The main cause for the fading of instability is the accumulation of the solute near the wall of the cell. A radial concentration gradient appears here which slows down the convective motion directed from the capillary to the wall. The decrease of the velocity causes a decrease of the surfactant supply in the capillary region, and therefore a decrease of the radial concentration gradient here, and so on. The liquid in most parts of the surface even begins to flow in the opposite direction (from the wall to the capillary). This reverse motion corresponds to the negative values of the surface velocity in Fig. 9. It prevents the supply of the solute from the droplet to the surface in the capillary region. In this slow stage the velocity of the convective motion becomes some orders of magnitude lower than during the fast stage. The surface tension increases gradually because of desorption of surfactant from the interface to the adjacent bulk phase in most parts of the system. A calming of the convective motion is the precondition for the appearance of the next oscillation. With time the diffusion of the solute from the droplet restores the vertical concentration gradient at the surface in the capillary region. When the gradient becomes large enough the instability arises again and the following oscillations develop by the same scenario.

Conclusions

Three types of systems which show nonlinear interfacial dynamics were considered: (1) superimposed layers of oil and water in which two ionic compounds, one being surface active, are dissolved in such a way that solute transfer between the layers begins after they are brought into contact; (2) systems similar to system 1 but an oil layer is interposed between two aqueous solutions; (3) a surfactant droplet is dissolved under the air/water interface. All these systems are far from equilibrium and are characterized by the surfactant transfer to or across the interface. In these systems, oscillations of the interfacial tension and/or electrical potential through the interface are accompanied by the appearance of the convective motion at the interface. The convection affects the rate and direction of the solute transfer in the interfacial region and, therefore, the rate and direction of adsorption/desorption processes as well as of chemical reactions at the interface. So far, a generally accepted oscillation mechanism, taking into account nonlinear hydrodynamics coupled with the interaction of the compounds at the surface, does not exist for the first two systems considered because of the complexity of the electrochemical processes within them. The simplest system, in which the auto-oscillations of the surface tension are generated by dissolving a surfactant droplet under the surface, represents the simplest nonlinear oscillatory system at liquid interfaces. The time evolution of this system can be described by numerical simulations on the basis of rigorous equations of physicochemical hydrodynamics. The proposed model, explaining the auto-oscillation mechanism, should be useful to understand more complicated nonlinear oscillations in similar systems.

References

1. Dupeyrat M, Michel J (1971) Experientia Suppl 18:269
2. Nakache E, Dupeyrat M (1982) In: Zieper J, Oertel H (eds) Convective transport and instability phenomena. Braun, Karlsruhe, pp 319–337
3. Dupeyrat M, Nakache E (1978) Bioelectrochem Bioenerg 5:134
4. Nakache E, Dupeyrat M, Vignes-Adler M (1983) J Colloid Interface Sci 94:187
5. Yoshikawa K, Matsubara Y (1983) J Am Chem Soc 105:5967
6. Magome N, Yoshikawa K (1996) J Phys Chem 100:19102
7. Kai S, Ooishi E, Imasaki M (1985) J Phys Soc Jpn 54:1274
8. Kai S, Muller SC, Mori T, Miki M (1991) Physica D 50:412
9. Kai S, Mori T, Miki M (1992) In: Kai S (ed) Pattern formation in complex dissipative systems, World Scientific, Singapore, pp 144–170
10. Shioi A, Sugiura Y, Nagaoka R (2000) Langmuir 16:8383
11. Shioi A, Kumagai H, Sugiura Y, Kitayama Y (2002) Langmuir 18:5516
12. Orell A, Westwater JW (1962) AIChE J 8:350
13. Takahashi T, Yui H, Sawada T (2002) J Phys Chem 106:2314
14. Yoshikawa K, Matsubara Y (1983) J Biophys Chem 17:181
15. Yoshikawa K, Matsubara Y (1984) J Am Chem Soc 106:4423
16. Arai K, Fukuyama S, Kusu F, wTakamura K (1995) Electrochim Acta 40:2913
17. Arai K, Kusu F (2001) In: Volkov AG (ed) Liquid interfaces in chemical, biological and pharmaceutical applications. Surfactant science series 95 Dekker, New York, pp 699–727
18. Arai K, Kusu F, Takamura K (1998) Anal Chim Acta 365:279
19. Maeda K, Nagami Sh, Yoshida Y, Ohde H, Kihara S (2001) J Electroanal Chem 496:124
20. Pimienta V, Etchenique R, Buhse T (2001) J Phys Chem A 105:10037
21. Shioi A, Abe T (2002) Chem Phys Lett 365:197
22. Kovalchuk VI, Kamusewitz H, Vollhardt D, Kovalchuk NM (1999) Phys Rev E 60:2029
23. Kovalchuk NM, Vollhardt D (2002) Mater Sci Eng C 22:147

24. Kovalchuk NM, Vollhardt D (2000) J Phys Chem B 104:7987
25. Grigorieva OV, Kovalchuk NM, Grigoriev DO, Vollhardt D (2003) J Colloid Interface Sci 261:490
26. Sternling CV, Scriven LE (1959) AIChE J 5:514
27. Hennenberg M, Bish PM, Vignes-Adler M, Sanfeld A (1979) J Colloid Interface Sci 69:128
28. Chu XL, Velarde MG (1989) J Colloid Interface Sci 131:471
29. Kovalchuk NM, Vollhardt D (2001) Phys Rev E 63:031604
30. Kovalchuk NM, Vollhardt D (2001) J Phys Chem B 105:4709
31. Kovalchuk NM, Vollhardt D (2002) Phys Rev E 66:026302
32. Kovalchuk NM, Vollhardt D (2003) J Phys Chem B 107:8439

Progr Colloid Polym Sci (2004) 128: 44–51
DOI 10.1007/b97094

Ludmila Boinovich

Molecular structure of liquids and surface forces

Abstract The purpose of this contribution is to analyse the surface forces and the stability of thin-layer systems falling outside the Derjaguin–Landau– Verwey–Overbeek theory. It is shown that peculiarities of fluid behaviour in a confined geometry allow one to understand the origin of the previously mentioned phenomena. Emphasis is placed on the liquid interlayers constituted from molecules possessing internal rotational degrees of freedom.

Keywords Surface forces · Static and dynamic structure · Surface-induced phase transitions

L. Boinovich
Institute of Physical Chemistry,
Russian Academy of Sciences,
Leninsky prospect 31, 119991 Moscow,
Russia
e-mail: ame@phyche.ac.ru
Fax: +7-095-9525308

Introduction

The investigations of the stability of colloid and disperse systems have a very long history. The basis for the quantitative analysis of the problem was put down by the Gouy–Chapman theory of diffuse ionic atmospheres [1–3]. This theory, together with London's theory of molecular forces [4], constituted a background for the development of the Derjaguin–Landau–Verwey–Overbeek (DLVO) theory of stability of liophobic colloids [5, 6]. The theory, in which the total interaction energy is represented by two terms, namely an electrostatic and a van der Waals (dispersion) contribution, succeeded in describing quantitatively the equilibrium in many colloid and disperse systems and in revealing the role of electrolytes in their destabilisation. Further improvement of the theory included the development of a macroscopic, continuum approach for dispersion interactions by Lifshitz and coworkers [7, 8] and generalisation for the heterosystems by Derjaguin [9].

Despite the evident success of the DLVO theory in the quantitative description of various systems containing thin liquid films, there are many reports in the literature where experimental results of forces between surfaces do not appear to conform to the classical DLVO theory [10–13]. These data have given rise to the development of new theoretical approaches, as well as new experimental techniques, providing better insight into the structure and properties of liquids in a confined geometry.

At present, it is evident that the main shortcomings of the DLVO theory directly followed from its advantages. In particular, just owing to a continual approach the classical theory allows us to calculate the interactions between two semi-infinite bodies through the liquid interlayer on the basis of macroscopic properties only, such as the frequency dependence of dielectric permittivities of contacting phases, ion charges, bulk electrolyte concentrations, surface charges or potentials. All structure deviations, related to the confinement of the liquid, as well as to the interaction of liquid molecules (not electric charges) with the interfaces, appear to be outside the frame of classical theory. Attempts to take the previously mentioned peculiarities into consideration have been made in a series of papers [14–32]. On one hand, they allowed the description of some characteristic features of particular experimental observations; on the other hand, they required the introduction of the number of parameters, the magnitude or the physical nature of which is not always clearly defined.

Thus, the nonuniformity of liquid interlayers due to adsorption or variation of liquid structure in the vicinity of confining surfaces has been accounted for in the calculations of the van der Waals interactions in the frames of the model of a stratified system, with constant dielectric properties within every layer [14–17]. Alternatively, the smoothly varying dielectric permittivity across the whole interlayer according to different relations was considered in Ref. [18].

As regards the electrostatic interaction across the interlayer of electrolyte solutions, the development of the theory consists in more and more detailing of the system under consideration. Thus, the consideration of the discrete character of the charge [19] or the dipole [20–23] distribution at the interfaces allowed one to elucidate the origin of correlation and image charge forces. The extra (to DLVO) repulsion in the films of electrolyte solutions can also be attributed to the effect of the finite size of the ions [24, 25] or to a variation of the static dielectric permittivity of the solvent in the field of the charged interface [26]. The effect of the presumed preferential alignment of the solvent dipoles near the surface, manifested in the continuum model as dielectric saturation, was analysed in Ref. [27]. It was shown that the dielectric saturation, usually represented by a reduced dielectric constant in the vicinity of the surface, gives rise to a repulsive hydration force.

The first attempt to consider the processes induced in the solvent itself due to the confinement of the liquid interlayer was made by Marcelja and Radic [28]. They used a Landau expansion of the free-energy density over the order parameter, which was later interpreted as the local water polarisation [29]. It was shown that for small fluctuations of the order parameter, the accounting for the quadratic order parameter and its gradient terms of expansion only allowed one to describe exponentially decaying repulsion between confining surfaces. Later this idea of surface-induced water polarisation as the origin of hydration forces was exploited in many different approaches [30–32]. Leikin and Kornyshev [30, 31] considered two different mechanisms of creation of water polarisation at the interfaces. The nonlocal electrostatics was used to take into account spatial correlations of polarisation fluctuations in water under the influence of a charge distribution in neutral lipid membranes [30]. The chemical mechanism of water ordering considered by means of a free-field model with a phenomenological order parameter allowed the analysis of the difference in the water states near the polar surfaces and in the bulk [31]. Both free-field and nonlocal electrostatic approaches predicted the exponentially decaying forces, determined by the lateral ordering of polar groups along the interacting surfaces.

In all of the previously mentioned approaches, the solvent (or the liquid constituting the film) was considered as a structureless medium. However numerous experimental data obtained in last decade (which will be discussed later) indicate the strong deviation of the static and the dynamic molecular structure of the confined liquid from that in the bulk.

The theoretical approach based on the analysis of the alterations of the vibrational spectrum at the transition from bulk liquid to confined liquid film, due to both intermolecular interactions and interactions with the confining phases, was developed in Refs. [33, 34] and we discuss it in the following section.

Phonon mechanism of long-range surface forces

Here we consider briefly the theoretical approach exploiting the peculiarities of molecular structure in liquids. In this approach, the dynamics of molecules, which includes intermolecular and intramolecular rotations and vibrations, is considered, during some characteristic time, as a harmonic process, described by a set of normal modes. Each normal mode is determined by the parameters of oscillators involved in the vibrational process, as well as by their interaction with each other, both in the same molecule and between the neighbouring molecules. The entire spectrum of normal modes is characterised by the distribution of the density of vibrational states over the frequencies of the vibrations, ω. The spectra of the normal modes for the bulk liquid and the confined interlayer differ from each other for the following reasons. The thermal motion in liquids leads to the damping of collective vibrations, which is reflected in the localisation of normal modes. In the theory, the localisation is characterised by the free path length, $L(\omega)$, of corresponding phonon. In the bulk liquid the localisation is very strong, and the vibrations of different oscillators (fragments of molecules) can be considered as localised within one molecule, located in the strong nonuniform field created by surrounding molecules. In the confined geometry the thermal motions and hence the localisation of normal modes will change owing to surface-induced energy barriers for translations and gyrations of molecules in the liquid [35]. Owing to the interaction of molecules with the confining phases, the set of energetically preferential mutual orientations of molecules, as well as intermolecular distances, will also change in comparison with those in the bulk liquid [36–39]. Finally, the surface-induced energy barriers for intramolecular rotations give rise to the redistribution of the conformational composition in the confined liquid layer.

All these factors, affecting the density of vibrational states, will lead to the thickness-dependent contributions to the excess free energy of the liquid interlayer, and thus to the appearance of long-range interactions, additional to the conventional DLVO forces. The magnitude of these interactions, attributed to the phonon component of the disjoining pressure, can be described by the following equation [33]:

$$\Pi(H) = \sum_j \int_0^\infty \left(-\frac{dK_j}{dH} - \frac{K_j H}{L_j^2(\omega)} \frac{dL_j(\omega)}{dH} + \frac{K_j}{L_j(\omega)} \right) \times \exp\left[-H/L_j(\omega)\right] d\omega. \quad (1)$$

Here we use the substitution

$$K_j = Z_j(\omega) kT \ln \frac{\mathrm{sh}(\hbar\omega/2kT)}{\mathrm{sh}(\hbar\omega_{j0}/2kT)},$$

where $Z_j(\omega)$ is the configuration average density of vibrational states for oscillators of the *j*th kind, obtained by the diagonalisation of dynamic matrices for the liquid film with thickness H and frozen thermal jumps of molecules, ω_{j0} is the eigenfrequency of the corresponding vibration in the bulk liquid, $\hbar$ k and T are the Plank constant, the Boltzmann constant and absolute temperature, respectively, and $L_j(\omega)$ is the free path of an *j*-type phonon with frequency ω. The product $Z_j(\omega)\exp(-H/L_j)$ plays the role of the density of phonon states in the liquid film, taking into account the finite time of mode relaxation. As shown in Ref. [34], for quasi-one-dimensional waves, $L_j(\omega)$ can be determined by solving the transcendent equation

$$L_j(\omega) = \upsilon_j(\omega)\tau_0 \left[\int_{L_j(\omega)} dz \rho(z) \exp\left(\frac{U(z)}{kT}\right) \right]^{-1}. \quad (2)$$

Here the coordinate axis z is introduced along the direction of phonon propagation, $\rho(z)$ is the linear density of oscillators, τ_0 is the lifetime between the thermal jumps in the bulk liquid, $U(z)$ is the distance-dependent (owing to the interaction with phases confining the film) energetic barrier for thermal jumps of oscillators and $\upsilon_j(\omega)$ is the group velocity of the *j*th phonon. So the increasing interaction with the confining phases will cause the enhancement of the phonon mechanism in the film stability through the increasing density and free path length of the phonons, according to Eq. (2). In contrast, near lyophobic surfaces the intermolecular interaction is strongly enhanced as was shown, for example, for a hydrogen-bond network in aqueous media [40]. This is immediately reflected in the increase of the group velocity of the correspondent phonons. Thus, being very sensitive to the energetic barrier $U(z)$, the magnitude of $L_j(\omega)$ might vary with even minor chemical changes at the interfaces, as was observed for different lipid membrane systems [41, 42]. In the case of inhomogeneous boundary conditions, it will fluctuate along the surface.

Equation (1) contains the summation over *j* different types of phonons. The signs of these summands depend on the relation between the mode frequencies in a film and the corresponding frequencies in the bulk liquid, whereas the magnitudes are determined not only by the frequencies themselves, but also to a higher degree by the density of phonon states in the liquid film. Typically the summands differ from each other by orders of magnitude, and only the less localised optical modes essentially contribute to the total disjoining pressure [33].

The character of dependencies $\Pi(H)$ provided by the three terms in parentheses in Eq. (1) can be tentatively associated with three regimes on the disjoining pressure–distance curve. The final term plays the dominant role for films thick enough when the values of L and K become almost independent of thickness. In general, this term describes the dependence of the disjoining pressure on the film thickness by the sum of several exponentially decaying contributions. This type of behaviour correlates well with numerous experimental data [43–48].

The other two terms in Eq. (1) will play a major role at lower thicknesses where the discreteness of the liquid structure has to be taken into account. This discreteness leads to a nonuniform distribution of molecules across the interlayer, due to both the exclusion-volume effect and the interaction of the liquid with interfaces confining the film. The former effect causes the oscillations of the local density inside the film with the film thickness. The corresponding oscillations of the force constants in the dynamic matrix lead, in turn, to the oscillations of the density of vibrational states, $Z(\omega)$, and, as a result, to the oscillations of the excess free energy and the disjoining pressure in the film.

At intermediate film thicknesses the first term vanishes, while the second one still contributes essentially, leading to the deviation from the exponential dependence provided by the final term. The magnitude and the sign of the derivative dL/dH, contained in the second term, are determined, according to Eq. (2), by the character of the interaction between the molecules (oscillators) and the phases confining the film, i.e. by surface philicity/phobicity.

The role of the phonon mechanism of the disjoining pressure in the force balance in thin liquid films is strongly affected by the real deviation in the structure of liquids in a confined geometry from that in the bulk. In the following sections we consider some experimental data convincing one of considerable changes in the parameters of liquids in the vicinity of an interface as well as in the confined interlayer.

Static structure and confinement-induced phase transitions

The confinement of a liquid either from one side, as happens at the interface between two bulk phases, or from both sides, as in the interlayer, generates the inhomogeneity in the static liquid properties in the vicinity of interfaces. First of all it causes strong density variations normal to the wall for systems both of small molecules and of long polymers [49, 50], liquid metals [51] and liquid crystals [52]. The origin of wall-induced layering is related to exclusion-volume effects and takes place even near inert walls. The strength of the interaction between the fluid molecules and confining phases can significantly affect the density profile in the thin interlayer [36–38, 53], leading to the an increase of density in the layers adjacent to the lyophilic surface and to the opposite effects for the lyophobic wall [37].

The orientation and conformational ordering, obtained from numerous experimental data, pointed to confinement-induced phase transitions of different types. Since such transitiosn can be considered as a direct indication of the structural reorganisation, we summarise briefly the main peculiarities of the phenomena.

It has recently been reported for long-chain *n*-alkanes and *n*-alcohols that confinement on one side leads to surface freezing at temperatures different from that for the bulk phase. Such a transition was detected for an alkane–air interface by ellipsometry [54], X-ray reflectometry [55] and surface tension measurements [56, 57]. X-ray studies have revealed the formation of a hexagonally ordered monolayer in which hydrocarbon chains are oriented normal to the surface of the liquid. Sum-frequency vibrational spectroscopy [58] permits a better understanding of the nature of this transition. It was established that either before or after surface melting the chains are oriented more or less along the surface normal, but the higher temperature spectrum pointed out the significant number of trans–gauche defects in the chains, whereas after surface freezing subsurface molecules mainly attain all-trans conformations. It is important to note that the role of subsurface molecule interaction with the confining phase is very significant for the type of ordering. Sum-frequency-generation (SFG) spectroscopy revealed that the hexadecane molecules more or less stand up at the hexadecane/vapour interface and lie flat at the hexadecane/quartz interface. Similar confinement-induced solidlike static ordering with a small portion of trans–gauche defects and very effective chain–chain interaction has also been detected by means of sum-frequency spectroscopy for some short and medium-length alcohols [37] and acetone [70]. For alcohols the influence of the nature of the confining phases on the surface freezing temperature is very impressive. For example, the undecanol–air interface is characterised by a freezing temperature of 13 °C, for the undecanol–water interface the correspondent temperature was reported to be 21.1 °C [59], whereas for an undecanol monolayer on a water surface the ellipsometry measurement as well as surface diffraction studies gave 28.5 °C for 2D surface melting [60]. Similar confinement-induced first-order transitions were observed for adsorbed monolayers of medium-chain alcohols on water–air, water–oil, water–solid and mica–air interfaces [61–67]. It is interesting to underline the influence of the interaction between an alcohol molecule and the confining phase in establishing molecular orientations for doubly confined systems. At room temperatures the orientation of, for example, undecanol in the monolayer appears to be normal to the confining interfaces of water and air [63]. The undecanol bounded by highly oriented pyrolytic graphite (HOPG) demonstrates the extended conformation, planar to the interfaces [68]. At the same time undecanol between mica surfaces reveals the tendency to associate into planar oriented bilayers [67–69]. At elevated temperatures the molecular axes of undecanol in the monolayer acquire the planar orientation. In contrast the alcohol molecules between mica plates reorient into an orientation normal to the interface. In the HOPG confined interlayer molecules persist in the planar orientation over the whole temperature range investigated (until 60 °C). Thus the double confinement of a liquid may significantly change the ordering in the liquid film and shift the temperatures of the phase transitions. Besides the properties of the interlayer reveal the strong sensitivity to the film thickness.

It was found that cyclohexane confined within 50- and 20-nm silica gel pores demonstrates the strongly reduced (not increased, as happens in the case of one-side confinement) freezing point, dropping with decreasing mean pore diameter [35]. The depression of the freezing point of nitrobenzene and alcohols in porous glass, with a linear relationship between the shift of the melting temperature and the inverse pore radius, was reported in Refs. [71, 72].

The orientation order and mobility of a nematic liquid crystal in random nanometre pores, studied by combined NMR and broadband dielectric spectroscopy, demonstrated a similar tendency. It was found that 4-pentyl-4′-cyanobiphenyl (5CB) molecules in the boundary layer covering the pore walls reveal the reduced mobility, whereas the isotropic–nematic transition as well as the crystallisation of confined nematogen were considerably suppressed [73]. A surface probe balance with extremely high resolution in measuring forces allowed the transition between liquidlike and solidlike behaviour of films of cyclohexane, octamethylcyclotetrasiloxane, toluene [74], octanol and undecanol [69] to be probed in detail in the course of film thinning. The results revealed that these films, excluding alcohols, undergo an abrupt transition from liquidlike to solidlike behaviour below a critical thickness corresponding to several molecular layers of the liquid. The solidlike behaviour was manifested in extremely high creep viscosity and the ability to sustain a finite shear stress for macroscopic times. In contrast, the alcohols, manifesting contact angles on mica of nearly 50° [67], retained their bulklike lubrication properties down to a thickness of a single bilayer [69]. These features indicated rather the solidlike deviation of the dynamic structure and the thermal motion from those in the bulk than static ordering. This point has strong support from the results of a molecular dynamic simulation of the structure of *n*-octane films bounded by atomically smooth 10–4 Lennard-Jones planes [38]. It was concluded there that the solidification of nanoscopically thin films of linear alkanes is an energetically driven phenomenon and is accompanied by the freezing of molecular migrations and rotations. Thus we will now address the consideration of the dynamic structure or deviations of intermolecular and intramolecular motions appearing owing to the confinement of the liquid.

Confinement-induced dynamic structuring

Some examples reported in the previous section undoubtedly indicated that intermolecular interactions as well as intermolecular distances in the confined geometry differ from those in the bulk. These should change the force constants and eigenfrequencies of the dynamic matrix in the film and, as a result, the set of vibrational, rotational and translational frequencies. Here we present some experimental data validating this point. The direct observation of the dynamic structure of the confined liquids is now available from very different techniques, like NMR [35, 73] and broadband dielectric spectroscopy [73, 75], nonlinear optical spectroscopic techniques such as second harmonic generation [76] and SFG spectroscopy [58], and transmittance Fourier transform IR spectroscopy [77]. The last two methods provide vast information about IR active vibrational modes. The analysis of the spectra of liquids either in the vicinity of one interface or in a confined interlayer indicates the essential distinctions from the spectra of the bulk liquid. The SFG spectra of the water–air interface (air here plays the role of a phobic confining medium) exhibited three prominent peaks, at 3,680, 3,400 and 3,200 cm^{-1}, in the OH stretch region [58] instead of a wide peak at nearly 3,400 cm^{-1}, typical for bulk water. They were associated with a free OH bond, OH hydrogen-bonded to neighbours in a relatively disordered structure and OH hydrogen-bonded in a well-ordered icelike structure, respectively. The spectra revealed the strong dependence on the character of the interaction with the confining phase as was demonstrated by SFG spectra of water confined by different hydrophobic phases like quartz coated by a monolayer of octadecyltrichlorsilane, air, hexane [40] and carbon tetrachloride [78] (Fig. 1). The analysis of the OH stretching vibrations of doubly confined water was performed by a molecular dynamic method [37]. In this paper a simple point charge water model, which contained both intramolecular bond flexibility and instantaneous responsive electronic

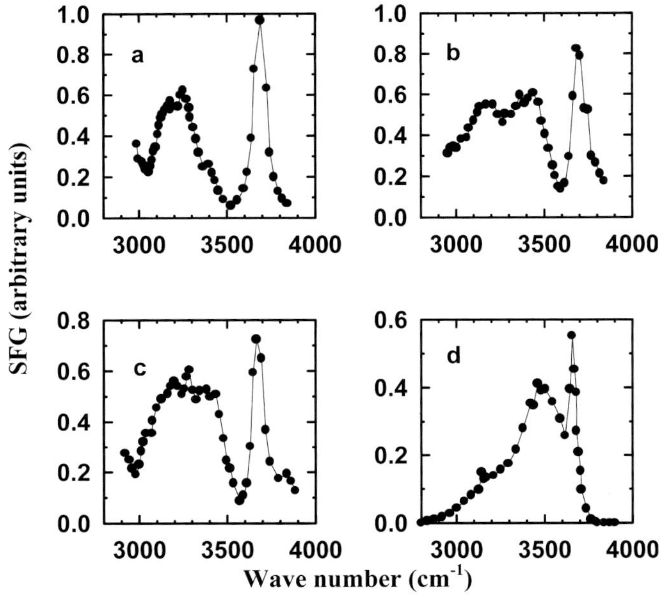

Fig. 1 Sum-frequency-generation (*SFG*) spectra from (**a**) the quartz–octadecyltrichlorosilane–water interface, (**b**) the air–water interface, (**c**) the hexane–water interface [39] and (**d**) the carbon tetrachloride–water interface [77]

polarisation, was employed to model the structure and dynamic properties of water between two rigid hydrophobic plates. The result of the computer simulations indicated that the O–H bond length is closer to the gas-phase value very near the wall, but is lengthened (even more than in bulk phase liquid water) in the central region of the 26-Å-thick liquid film. The H–O–H angle is also less distorted relative to the gaseous water than for bulk liquid molecules and tends smoothly to the bulk value with the plateau in the central part. The distortion of the OH bond length is a measure of the hydrogen-bond strength in the interlayer. So, a greater hydrogen-bond strength gives rise to a longer OH distance and as a result to a lower OH stretching vibrational frequency. Having in mind that a redshift of the OH stretching vibrational frequency should provide a negative value of *K* and respectively a negative contribution from the third term in the parentheses of Eq. (1), one should expect a negative disjoining pressure or attraction between hydrophobic plates divided by a water film. This conclusion corresponds to the observations in real experiments [48].

The SFG spectra of the surface layer of short-chain alcohols confined by air as well as the IR transmittance spectra of the same alcohols doubly confined by poorly wetted calcium fluorite plates also demonstrate a set of prominent peaks in the OH stretching region [58, 79] different from the OH vibrational band in bulk alcohol. The spectra, for example, for an ethanol–air interface and a 25-nm-thick ethanol film in Fig. 2 appear to resemble the essential part of the bulk solid and clearly

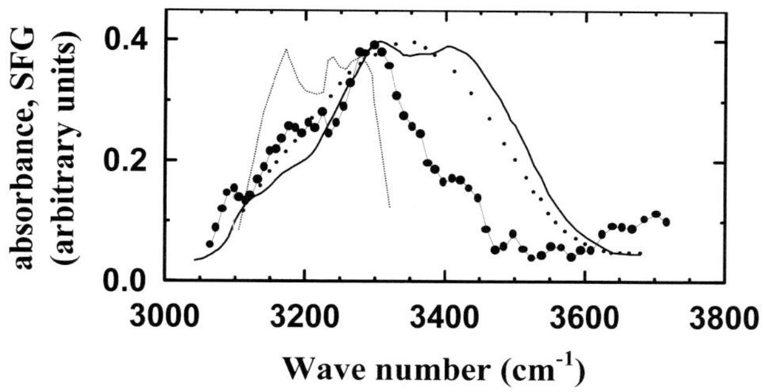

Fig. 2 IR spectra for ethanol in the OH stretch region. The *dotted* and *dashed curves* are the bulk liquid and the bulk crystalline phases, respectively; the *solid curve* corresponds to the 25-nm film of liquid ethanol, confined by fluorite plates [78]. The *circles* represent SFG spectra for the liquid surface [58]

indicate more strong hydrogen bonding in the confined geometry than in the bulk liquid. The analysis of the CH stretching vibrations of terminal methyl and methylene groups for the surface molecules [58] proves this conclusion, demonstrating a relatively small number of trans–gauche defects and strong chain–chain interaction.

The confinement of cyclohexane in silica pores of 4–50-nm diameter gives rise to substantial changes in molecular dynamics [35]. It was found that decreasing the pore diameter from 50 to 4 nm causes an increase in the rotational activation energy from 7.3 to 12.6 kJ mol^{-1}, indicating a significant restriction of rotational motion in small pores.

The dynamic behaviour of nematic liquid crystal 5CB is strongly affected by the type of interaction with the confining phases. For 5CB bounded by poorly wetted calcium fluorite plates the deviation of the IR transmittance spectrum of the film from the that of the bulk demonstrates the significant shift of the conformational equilibrium of the liquid crystal molecules towards enrichment with coiled conformations of hydrocarbon tails for films thinner than 1,000 nm [80].

The confinement in randomly oriented pores with two different mean sizes (1,000 and 100 Å) investigated by means of broadband dielectric spectroscopy [75] resulted in the following. It was shown that (1) a new, not present in the bulk, low-frequency relaxation process with a frequency of about 10 kHz appears, (2) the relaxation processes in the confined liquid crystal are not frozen even at temperatures about 20°C below the bulk crystallisation temperature and (3) the "isotropic" phase of confined 5CB is not a bulklike isotropic phase and some degree of orientational order still persists.

The acceleration of orientational relaxation processes was also found in water, confined by hydrophobic walls by molecular dynamics methods on the basis of the simple point charge–flexible/polarizable model [37]. By employing the TIP4P model, in Ref. [81] a decreased diffusion rate perpendicular to the surface was found. On the other hand, lateral diffusion was decreased near a hydrophilic surface and enhanced close to a hydrophobic wall.

The features of the molecular dynamics discussed clearly demonstrate that both the intermolecular and the intramolecular structures in thin films deviate from those in the bulk, leading to changes in the thermal motion as well as in the rotational–vibrational spectra of the interlayer and as a result to additional long-range forces according to the phonon mechanism. In the concluding section we discuss a few examples of surface forces, the physical mechanism of which should be related to the peculiarities of the dynamic and static molecular structure.

Conclusions

Direct indication of the influence of the molecular structure of liquids on the character and magnitude of surface forces may be found from the oscillating behaviour of the force–distance curve obtained at short distances by means of surface force apparatus or atomic force microscopy [11, 12] as well as from the stepwise behaviour of adsorption isotherms for some liquids [82]. According to the phonon mechanism of long-range forces [83] these oscillations arise owing to spatial dispersion and result in oscillations of collective vibrational state density varying with film thickness. This, in turn, causes the variation of the force of interaction between confining phases according to the first term in the parentheses of Eq. (1). The period of oscillations and the steps on the isotherms characterise the periodicity of the static structure and are associated with the characteristic dimensions of molecules in the direction normal to the interface. Since for *n*-alkanes, fatty alcohols and cyanobiphenyls with an extended planar orientation this characteristic dimension has the sense of the radius of gyration of the hydrocarbon tail and has a magnitude of nearly 0.5 nm, the phonon mechanism, in agreement with known experimental data [13, 52, 68], provides the same period of oscillations for all the previously mentioned liquids. As for molecules in the liquid state the extended form of the hydrocarbon tail in the planar orientation appears to be surface-induced [38]. The shift of the conformational equilibrium with an increase in the portion of coiled conformations should take place when moving out of interfaces to the middle plane of the liquid film. Hence the period of layering is expected to increase slightly [84] in the course of film thickening. This is exactly what has been obtained in experiments [52]. The modification of surfaces, confining a liquid film, results on one hand, in the establishing another preferable orientation of molecules in the subsurface layer. On the other hand, it will affect the energetic barriers for intermolecular and intramolecular

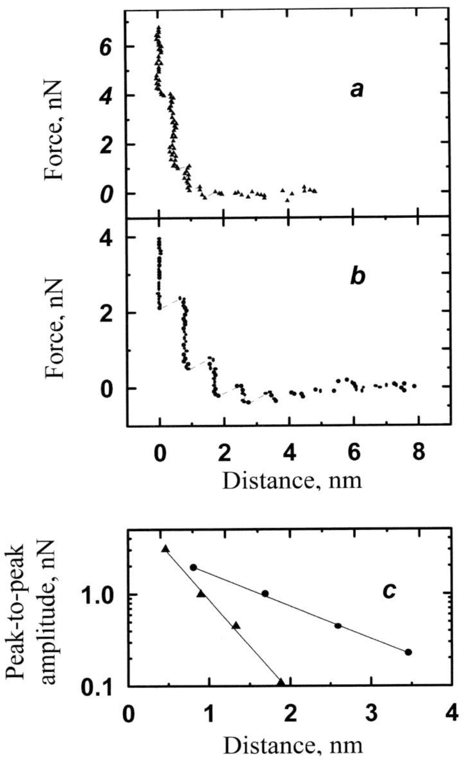

Fig. 3 The force–distance curves for **a** highly oriented pyrolytic graphite and **b** mica substrates immersed in *n*-dodecanol [68] and **c** peak-to-peak amplitudes versus distance on a logarithmic scale. The exponential decay length is 1.23 nm for mica (*circles*) and 0.43 nm for highly oriented pyrolytic graphite (*triangles*)

rotations. So in the case of a high enough potential one will again have layering but with another period, corresponding to new characteristic dimensions of molecules in the direction normal to the interface. The numerous data on alcohols between different confining surfaces demonstrate clearly this tendency of reordering [13, 60, 63, 68]. Moreover, the enhancement of the interaction with the interfaces leads to the restriction of molecular rotations and as a result to the manifestation of bilayer ordering [67, 68]. From the theoretical point of view this should cause the essential enhancement of the free path length of phonons and, as the result, more oscillations should be pronounced on the force curve. This theoretically predicted conclusion correlates well with experimental observations for dodecanol films, confined by HOPG and mica (Fig. 3). In contrast the decrease of surface–liquid molecule interaction facilitates both intermolecular and intramolecular rotations. The former circumstance causes oscillation smearing, whereas the latter leads to the shift of the conformational equilibrium towards coiled conformations. It seems plausible that the force–distance data of *n*-alcohols confined between mica surfaces coated with a monolayer of didodecyldimethylammonium bromide [85] are in the good conformity with the previous consideration. Namely, the monolayer-induced facilitating of intermolecular rotations leads to the higher spatial disorder in the molecular arrangement and to a decrease in the force constants in the dynamic matrix. Hence the first term in Eq. (1), providing the oscillations, fades away. At the same time the facilitating of intramolecular rotations provides the essential enlargement of the portion of high-energy coiled conformations and as the result the appearance of repulsion between didodecyldimethylammonium bromide-coated mica surfaces.

The combined findings of real and numeric experiments presented here have revealed rich alterations in the static and the dynamic behaviour of liquids at interfaces and in the interlayer. These alterations appear to be essentially affected by the details of the molecular structure of liquids, the character of their interaction with confining phases and also significantly by the interlayer thickness. The few examples presented here clearly demonstrate that a theoretical approach, based on accounting for details of molecular structure and their variations with film thickness, can be fruitfully applied for the description of many experimental observations attributed to non-DLVO forces.

Acknowledgement This work was performed with financial support from the Russian Academy of Sciences (grant no. 132–99).

References

1. Gouy G (1910) J Phys 9:457
2. Gouy G (1917) Ann Phys 7:129
3. Chapman DL (1913) Philos Mag 25:475
4. London F (1930) Z Phys 63:245
5. Derjaguin BV, Landau LD (1941) Acta Physicochim URSS 14:633
6. Verwey EJW, Overbeek JTG (1948) Theory of the stability of liophobic colloids. Elsevier, Amsterdam
7. Lifshitz EM (1955) Zh Eksp Teor Fiz 29:94
8. Dzyaloshinsky IE, Lifshitz EM, Pitaevsky LP (1961) Soviet Phys Usp 73:381
9. Derjaguin BV (1954) Discuss Faraday Soc 18:85
10. Derjaguin BV, Churaev NV, Muller VM (1987) Surface forces. Plenum, New York.
11. Israelachvili J (1992) Intermolecular and surface forces. Academic, London.
12. Ninham BW (1999) Adv Colloid Interface Sci 83:1
13. Christenson HK (1988) J Dispersion Sci Technol 9:171
14. Ninham BW, Parsegian VA (1970) J Chem Phys 53:3398
15. Churaev NV (1975) Colloid Polym Sci 253:120
16. Langbain D (1971) J Phys Chem Solids 32:133
17. Langbain D (1974) J Adhes 6:1
18. Barash YS, Ginzburg VL (1975) Soviet Phys Usp 116:5

19. Derjaguin BV, Muller VM (1975) Dokl Akad Nauk USSR 225:601
20. Attard P, Mitchell DJ (1988) J Phys Chem 88:4391
21. Jonsson B, Wennerstrom H (1983) J Chem Soc Faraday Trans II 79:19
22. Martynov GA Smilga VP (1965) Colloid J 27:250
23. Boinovich LB, Emelyanenko AM (2003) Adv Colloid Interface Sci 104:93
24. Paunov VN, Dimova RI, Kralchevsky PA, Broze G, Mehreteab A (1996) J Colloid Interface Sci 182:239
25. Paunov VN, Binks BP (1999) Langmuir 15:2015
26. Basu S, Sharma MM (1994) J Colloid Interface Sci 165:355
27. Henderson D, Lozada-Cassou M (1994) J Colloid Interface Sci 162:508
28. Marcelja S, Radic N (1976) Chem Phys Lett 42:129
29. Gruen DWR, Marcelja S, Parsegian VA (1984) In: Perelson AS, Delisi C, Wiegel FW (eds) Cell surface dynamics. Dekker, New York, p 59
30. Leikin S, Kornyshev AA (1990) J Chem Phys 92:6890
31. Kornyshev AA, Leikin S (1989) Phys Rev A 40:6431
32. Gruen DWR, Marcelja S (1983) J Chem Soc Faraday Trans II 79:225
33. Boinovich LB, Emelyanenko AM (2002) Adv Colloid Interface Sci 96:37
34. Boinovich LB, Emelyanenko AM (1992) Z Phys Chem 178:229
35. Aksnes DW, Gjerdaker L (1999) J Mol Struct 475:27
36. Anastasiou N, Fincham D, Singer K (1983) J Chem Soc Faraday Trans II 79:1639
37. Zhu SB, Robinson GW (1991) J Chem Phys 94:1403
38. Ballamudi RK, Bitsanis IA (1996) J Chem Phys 105:7774
39. Yu CJ, Evmenenko G, Richter AG, Datta A, Kmetko J, Dutta P (2001) Appl Surf Sci 182:231
40. Quan Du, Freysz E, Shen YR (1994) Science 264:826
41. Lis LJ, McAlister M, Fuller N, Rand RP, Parsegian VA (1982) Biophys J 37:657
42. Rand RP, Fuller N, Parsegian VA, Rau DC (1988) Biochem 27:7711
43. Le Neveu DM, Rand RP, ParsegianVA (1976) Nature 259:601
44. Pashley RM, McGuiggan PM, Ninham BW, Evans DF (1985) Science 229:1088
45. Pashley RM (1982) Colloid Polym Sci 16:57
46. Tsao Y, Yang SX, Evans DF, Wennerstrom H (1991) Langmuir 7:3154
47. Rabinovich YI, Derjaguin BV (1987) Colloid J 49:682
48. Christenson HK (1992) In: Schrader ME, Loeb G (eds) Modern approach to wettability: theory and applications. Plenum, New York, p 29
49. Christenson HK (1988) J Dispersion Sci Techn 9:171, and references therein
50. Horn RG, Hirz S J, Hadziioannou G, Frank CW, Catala JM (1989) J Chem Phys 90:6767
51. Hiusman WJ, Peters JF, Zwanenburg MJ, de Vries SA, Derry TE, Abernathy D, van der Veen JF (1997) Nature 390:379
52. Horn RJ, Israelachvili JN, Perez E (1981) J Phys 42:39
53. Abraham FF (1978) J Chem Phys 68:3713
54. Pfohl T, Beaglehole D, Riegler H (1996) Chem Phys Lett 260:82
55. Ocko BM, Wu XZ, Sirota EB, Sinha SK, Gang O, Deutsch M (1997) Phys Rev E 55:3164
56. Earnshaw JC, Hughes CJ (1992) Phys Rev A 46:4494
57. Maeda N, Yaminsky VV (2000) Phys Rev Lett 84:698
58. Miranda PB, Shen YR (1999) J Phys Chem 103:3292
59. Aratono M, Takiue T, Ikeda N, Nakamura A, Motomura K (1993) J Phys Chem 97:5141
60. Renault A, Legrand JF, Goldmann M, Berge B (1993) J Phys II 3:761
61. Casson BD, Braun R, Bain CD (1996) Faraday Discuss 104:209
62. Lin M, Firpo J-L, Mansoura P, Barret JF (1979) J Chem Phys 71:2202
63. Berge B, Renault A (1993) Europhys Lett 21:773
64. Motomura K, Matubayasi N, Aratono M, Maura R (1978) J Colloid Interface Sci 64:356
65. Salajan M, Glinski J, Chavepeyer G, Platten JK (1994) J Colloid Interface Sci 164:387
66. Johal MS, Usadi EW, Davies PB (1996) J Chem Soc Faraday Trans 92:573
67. Mugele F, Baldelli S, Somorajai GA, Salmeron M (2000) J Phys Chem B 104:3140
68. Nakada T, Miyashita S, Sazaki G, Komatsu H, Chernov AA (1996) Jpn J Appl Phys 35:52
69. Mugele F, Salmeron M (2001) J Chem Phys 114:1831
70. Yeh YL, Zhang Ch, Held H, Wei X, Lin SH, Shen YR (2001) J Chem Phys 114:1837
71. Ritter MB, Awschalom DD, Schafer MW (1988) Phys Rev Lett 61:966
72. Liu G, Li Y-Z, Jonas J (1989) J Chem Phys 90:5881
73. Cramer C, Cramer T, Kremer F, Stannarius R (1997) J Chem Phys 106:3730
74. Klein J, Kumacheva E (1998) J Chem Phys 108:6996
75. Sinha GP, Aliev FM (1998) Phys Rev E 58:2001
76. Eisenthal KB (1996) J Phys Chem 100:12997
77. Boinovich LB, EmelyanenkoAM (1991) Surf Interface Anal 17:764
78. Scatena LF, Brown MG, Richmond GL (2001) Science 292:908
79. Boinovich LB, Gagina IA, Emelyanenko AM (1995) Colloid J 57:851
80. Boinovich LB, Emelyanenko AM (2001) Izv Ross Akad Nauk Ser Khim 305 (in Russian)
81. Rossky PJ, Lee SH (1989) Chem Scr A 29:93
82. Beaglehole D, Christenson HK (1992) J Phys Chem 96:3395
83. Boinovich LB, Emelyanenko AM (1999) Prog Colloid Polym Sci 112:64
84. Zerbi G, Magni R, Gussoni M, Moritz KH, Bigotto A, Dirlikov S (1981) J Chem Phys 75:3175
85. Herder CE, Ninham BW, Christenson HK (1989) J Phys Chem 90:5801

Progr Colloid Polym Sci (2004) 128: 52–56
DOI 10.1007/b97080

Alexandre Emelyanenko

The application of digital image processing to study surface phenomena

A.M. Emelyanenko
Institute of Physical Chemistry,
Russian Academy of Sciences,
Leninsky prospect 31, 119991 Moscow,
Russia
e-mail: ame@phyche.ac.ru
Fax: +7-095-9525308

Abstract The investigation of two- or three-phase boundaries is an essential part in the study of different surface phenomena like surface tension, contact angles, wetting hysteresis, swelling, and flotation. At present, the most accurate methods to investigate these parameters are based on the digital image processing of interfacial profiles. In this paper we describe several methods employing the digital processing of video images of fluid–liquid interfaces in different geometries. The errors in determining the interface parameters arising in the methods based on the digital image processing are analysed. Some experimental data obtained by the methods of pendant, sessile drops and drops on the filament are presented and discussed.

Keywords Surface tension · Contact angles · Wetting · Experimental methods · Digital image processing

Introduction

The investigation of surface phenomena in two- or three-phase systems is of great fundamental and practical importance. Many methods of study of these phenomena are based on the information about the profile of the interfaces in the systems under investigation. During the last 20 years there increased popularity was gained by the methods of studying the phenomena based on the digital processing of video images of the interfacial profiles. The theoretical and methodological bases as well as the computational procedures for this group of methods were intensively developed by several research groups [1–14].

To date, the methods based on digital processing of pendant drop images are the most accurate ones to determine the interfacial tension and those based on sessile drop image processing are believed to be the most accurate for measurements of contact angles greater than 20°. The aim of this paper is to briefly characterize the most popular implementations of methods with respect to their accuracy and ranges of applicability. Particular emphasis is given to the new experimental possibilities of interfacial tensiometry, which arose owing to advantages of digital image processing.

Factors influencing the accuracy of methods based on digital image processing

The first applications of digital image processing for measurements of interfacial tension and contact angles were aimed at the automation of classical methods based on measurements of several characteristic dimensions of pendant or sessile drops and comparison with the appropriate characteristics of theoretical profiles, either tabulated or computed within the optimization program [6, 8]. However, owing to fast improvement of computer capability it was very soon appreciated that the processing of complete interfacial profiles (usually 500–2,000 data points along the profile) provides a qualitatively new level of interfacial measurements. The application of numerical solution of the Laplace equation, determining the profile of the interface, together with the appropriate choice of optimization scheme, allows one to extract not only the interfacial tension, but other characteristics of the interface as well, like interfacial area and contact radius.

While the reproducibility of methods based on digital processing of interfacial profiles is usually excellent, several points should be noted regarding their accuracy in general. The following factors affecting the accuracy of surface tension and contact angle measurement have been discussed in the literature: (1) the inclination of the video camera's vertical axis with respect to direction of gravity and optical malalignment of the experimental device [11]; (2) the nonsquare shape of the pixels of the camera and/or the digitizer [11]; (3) random errors of interface segmentation [4, 7, 10, 11, 15]; (4) improper choice of illumination and the settings of the video camera and/or the digitizer [10, 14]; (5) vibrations of the interface [4, 10]; (6) the discrete character of the capture and the representation of the optical image [16].

The bottleneck determining the achievable accuracy of the methods is the quality of the set of data points determining the coordinates of interfacial profile. One of the most important stages in digital image processing is the analysis of the digital image field with subsequent segmentation and recognition of the object boundaries. To conciliate the requirements of the speed and accuracy, many approaches were developed to solve this problem, from simple global thresholding [5, 6] to dynamic gradient thresholding, based on the comparison of intensity gradients in fields of the working image and the background image [17]. The latter segmentation technique provides correct accounting for light scattering and diffraction at the interfaces, as well as for nonuniformity of illumination of the image field.

Another factor influencing the quality of the representation of the actual interfacial profile is the data discretization. Discretization is inevitable during digital image processing. First of all, the photosensitive zone of a video camera represents a matrix of CCD picture elements that results in discretization of the optical image at its capture. Besides, the digitizing video processor performs discretization of input analogue video signals, both on the coordinates of a pixel position, and on the intensity, representing an output video signal as a set of integer values of a gray level. Therefore the distribution of gray levels in the digitized field of the image will depend on the arrangement of the optical image in relation to the CCD picture elements of the sensitive zone of the camera. That, in turn, will have an influence on the results of segmentation. Such variations of the mutual arrangement are inevitable, for example, on refocusing the microscope on the object, owing to a finite focus depth and imperfections of the system adjustment in real experimental conditions. Finally, the coordinates of the points at the interfacial profile are also determined as discrete pixel coordinates. The subpixel resolution, as frequently used by many workers, is a result of different interpolation procedures between these discrete coordinates. Thus, the final results are still affected by discretization.

The effect of discretization on the accuracy of pendant and sessile drop methods has been investigated numerically using the method of model profiles [16]. The results of this study show that while the method of pendant drops almost always provides high accuracy of surface tension measurements (typically 0.05–0.5%), the sessile drop method can be validly used for determination of surface tension only for the case of contact angles greater than 80–90° (Fig. 1). At the same time, image discretization does not affect significantly the accuracy of contact angle measurements. For contact angles larger than 20° the accuracy was reported to be in the range 0.05–0.5° [3, 4, 19, 22].

It is necessary to note that since the previously mentioned estimations are related to the image discretizations during digital data processing, the limits of the applicability of the methods will change with the further development of video capturing and digitizing devices.

Thus, the proper accounting for the previously mentioned factors and the correct choice of the experimental methods and computational procedure allow one to obtain reliable results.

New experimental techniques based on digital image processing

It is necessary to note that the new developments do not just improve the speed and the accuracy of interfacial measurements. In this part, several examples will be given describing the new experimental possibilities which hardly would be attainable without the advances in digital image processing for interfacial tensiometry.

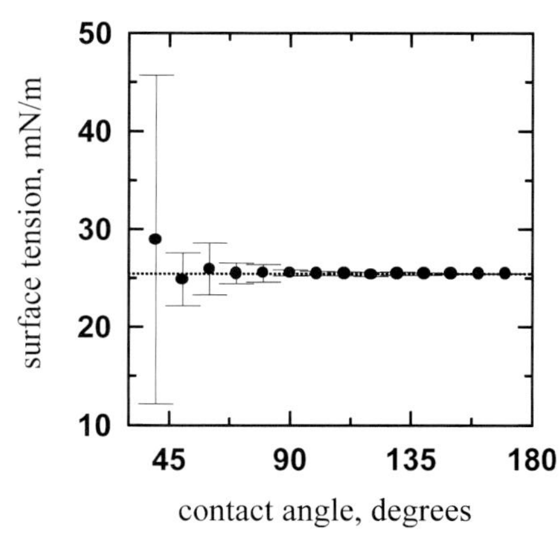

Fig. 1 Computed values of surface tension (*points*) and their standard deviations (*error bars*) owing to discretization of ideal profiles of sessile drops versus contact angle. The *solid line* denotes the prescribed value of the surface tension

One very interesting possibility arises owing to the fact that while the Laplace equation cannot be solved analytically for the interfacial profile, it might be solved for the volume of liquid included between the drop apex and any horizontal cross section of the drop profile. The minimization of the mean square difference between the experimentally measured and the theoretically calculated volumes for cross sections drawn through all points of the experimental drop profile, with respect to the radius of the drop apex and the capillary constant, considered as fit parameters, can be done without any iterative procedure. Therefore it becomes possible to organize a feedback control for changing the interfacial area or the interfacial tension in the real time of the experiment according to a given program. This idea was realized by Faour et al. [12] for the study of the kinetics of surfactant adsorption at a decane–water interface.

As an example of the application of this method, the kinetics of bovine serum albumin adsorption on a decane–water interface at a constant interfacial area is presented in Fig. 2. The changes in the interfacial area at constant interfacial tension, which under these conditions are directly proportional to the number of adsorbed molecules, are illustrated in Fig. 2. Thus, this technique provides a new possibility to follow the desorption (or dissolution) kinetics from monolayers at the interfaces. The energy barrier of adsorption can be more easily determined with such experiments.

Another interesting application of digital image processing arises when one considers a drop on a vertical filament. In distinction to the cases of sessile and pendant drops, in this geometry the interface does not contain a point where two principal curvatures coincide. Thus, the Laplace equation has to be solved with a different boundary condition, namely, with the condition of a vertical tangent at the position of the maximal drop diameter.

The analysis of the condition of the mechanical equilibrium of a drop on a vertical filament [21] has shown that the stability of such a drop is determined mainly by the wetting hysteresis. It was also shown that the specific peculiarity of this geometry allows, on one hand, both the advancing and the receding contact angles to be measured simultaneously and, on the other hand, small contact angles to be measured with high accuracy. As an example of the application of this method we present in Fig. 3 the time evolution of contact angles formed by a pentanol droplet on a glass filament of 185-μm diameter, in the course of evaporation under conditions of extra-low undersaturation.

This method allows one to investigate in the same experiment the dependence of contact angles as well as wetting hysteresis on the diameter of fibers, with direct applications to many practical and technological problems. Besides, as shown in Ref. [21], the comparison of the theoretical and experimentally measured profiles allows one to investigate the influence of wetting defects on the shape of the interface, as well as to study the transition zone between the liquid meniscus and the equilibrium wetting film.

Exploiting the advantages of digital processing to improve the possibilities of classical sessile and pendant drop methods

As already mentioned, the digital image processing of fluid–fluid interfaces provides a convenient and powerful

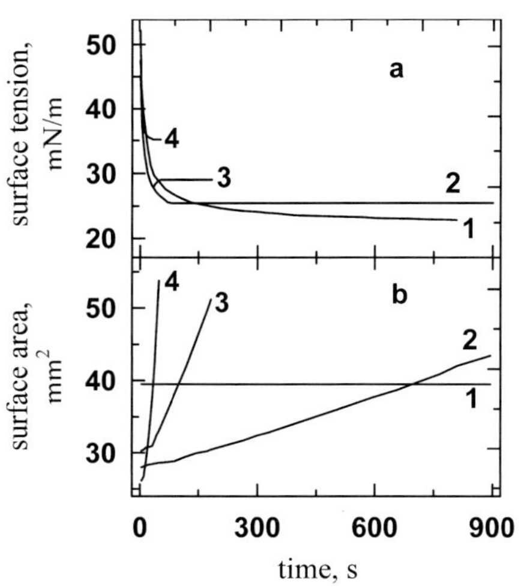

Fig. 2 The time evolution of *a* the surface tension and *b* the surface area of pendant drops of *n*-decane in water under conditions of bovine serum albumin adsorption at the interface [12]. Regulation of the interfacial area (*1*); regulation of the interfacial tension (*2–4*)

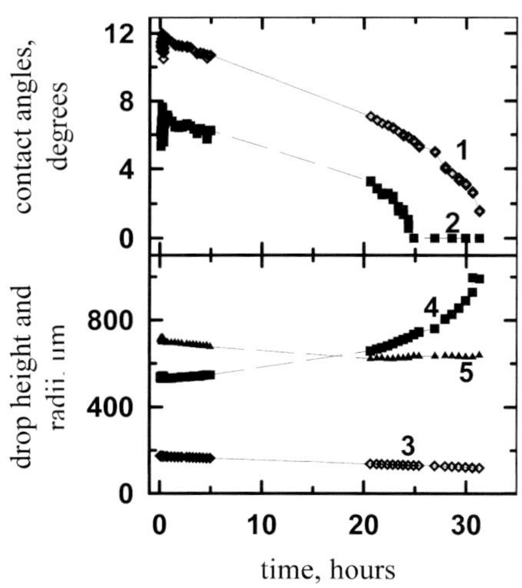

Fig. 3 The time dependencies of advancing (*1*) and receding (*2*) contact angles, equatorial (*3*) and meridional (*4*) radii and drop height (*5*), i.e., the distance between the upper and lower lines of three-phase contact, of a pentanol drop on a glass filament with diameter 185 μm

technique to determine interfacial tensions and contact angles from the shape of axisymmetric menisci. From the theoretical profile that best matches the drop profile from the image of a real drop, many additional characteristics, like drop volume, contact radius and surface area, are readily available. In combination with the capability to record a sequence of images of the drop evolution, it provides the possibility for different dynamic experiments.

Measurements of contact angles as a function of the three-phase contact radius at different rates of advancing by pumping the liquid steadily into the sessile drop from below the solid surface [22] were used to study surface energetics of a copolymer with polar headgroup deposited on silicon wafers. A sequence of pictures of the growing drop was recorded at a rate of about one picture every 2 s. The actual rate of advancing was determined by linear regression when plotting the three-phase contact radius versus time. It was shown that even carefully putting an initial liquid drop from above on a solid surface can result in a contact angle somewhere between advancing and receding, in spite of good homogeneity of the surface. Thus, it takes time for the initial drop front to start advancing when the liquid is added from below the surface. Further increase in the drop volume causes the three-phase contact line to advance, with two possible behaviors of the contact angle θ versus contact radius R. For some liquids (e.g., water) θ remained essentially constant as R increased (Fig. 4a), while for some others (like diiodomethane) well-pronounced stick–slip behaviour was registered (Fig. 4b). The latter manifested itself in continuous increasing of θ with drop volume at substantially constant R. This is accompanied by increasing of the local curvature of the vapour–liquid interface, and hence a pressure difference through the interface, at the zone of three-phase contact. At some moment the liquid profile in the transition zone between the drop and precursor wetting film loses its mechanical stability, resulting in the overflow of liquid from the drop onto the precursor film, which causes θ to decrease suddenly. Apparently it looks like liquid front slipping. The drop front jumps from one location to the next, and, as more liquid is supplied into the drop, the process is repeated. This mechanism of slipping was first proposed by Frenkel [23] and was later proved theoretically by Martynov et al. [24].

Another experimental problem for which the application of digital image processing has allowed progress to be made is the investigation of the surface tension of liquid-crystalline substances. Many available experimental results, obtained by different techniques, were inconsistent either in view of the temperature-dependence behavior of the surface tension near the phase-transition temperature or with respect to the absolute values of the surface tension. The detailed studies performed by Song and Springer [25] by the pendant drop method showed that the surface tension of freshly formed drops of many liquid crystals depended on the time of drop ageing. The authors suggested that such unusual time behavior was caused by the presence of certain gas molecules, either as a result of the interaction between the gas molecules and the molecules at the liquid surface or as a consequence of gas-induced reorganization of liquid-crystal molecules at the interface. Further experiments with liquid drops of *p*-methoxybenzylidene-*p*-*n*-butylaniline in the atmosphere of different gases and under different pressures [26] confirmed this suggestion. It was shown that the type of surrounding gas strongly influences both the magnitude of the surface tension (Fig. 5) and the equilibration time, during which the drop exposed in a gas atmosphere acquires its steady-state value of surface tension. The change of surface tension of several liquid crystals during the time of drop formation was proven to be nonnegligible in certain gas atmospheres, especially at high gas pressure.

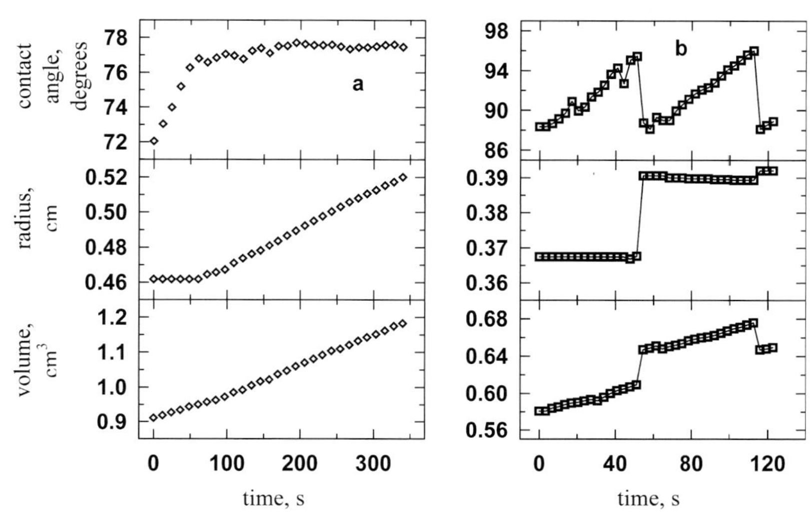

Fig. 4 Different behavior of the contact angle, formed by a liquid drop in the course of increasing the drop volume [22]: **a** increasing the contact radius at a constant value of the (advancing) contact angle; **b** slip–stick behavior

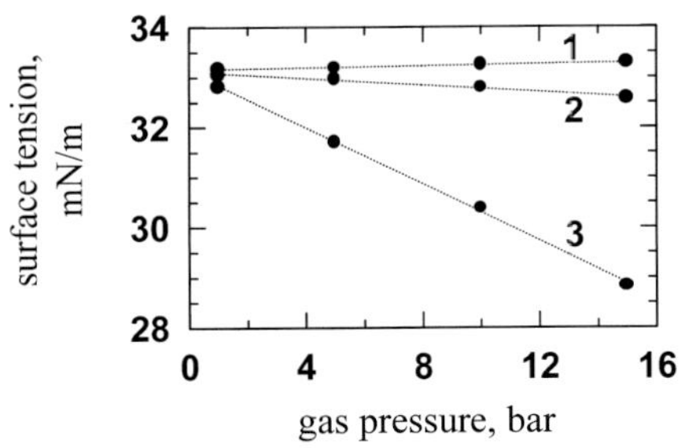

Fig. 5 The influence of the surrounding atmosphere on the surface tension of a freshly formed drop of *p*-methoxybenzylidene-*p*-*n*-butylaniline [26]: He (*1*); N_2 (*2*); CO_2 (*3*)

Conclusions

The examples presented in this paper demonstrate that the application of digital image processing of interfacial profiles for the study of surface/interfacial phenomena provides a qualitatively new level of investigation. The technical advantages of computer and video techniques, based on the facilities for automatic storage and processing of large amounts of data, averaging and statistical analysis, made it possible to outstrip the accuracy attainable with conventional methods, such as the Wilhelmy technique, capillary rise and maximum bubble pressure, by almost an order of magnitude. It is equally satisfactory that these methods might be applied at low and high temperatures, at high pressures and under vacuum conditions. Since the profiles of the drops can be rapidly recorded, these methods are used to determine the surface tension and the contact angle of aging systems, i.e., systems where the properties are changing with time, and thus to study the kinetics of processes affecting the surface properties of substances. The fast acquisition and calculation times allow one to provide a feedback control of the parameters of the system under investigation. The group of methods based on digital processing of interfacial profiles have been successfully used to study the temperature dependence of surface and interfacial tensions in a series of interesting systems, including both low molecular and polymeric substances, and both simple and liquid-crystal substances in various environments.

Acknowledgement This work was performed with financial support from the Russian Academy of Sciences (grant no. 133–99).

References

1. Rotenberg Y, Boruvka L, Neumann AW (1983) J Colloid Interface Sci 93:169
2. Skinner FK, Rotenberg Y, Neumann AW (1989) J Colloid Interface Sci 130:25
3. Li D, Cheng P, Neumann AW(1992) Adv Colloid Interface Sci 39:347
4. Cheng P, Li D, Boruvka L, Rotenberg Y, Neumann AW (1990) Colloids Surf A 43:151
5. Girault HH, Schiffrin DJ, Smith BDV (1982) J Electroanal Chem 137:201
6. Girault HH, Schiffrin DJ, Smith BDV (1984) J Colloid Interface Sci 101:257
7. Anastasiadis SH, Chen JK, Koberstein JT, Siegel AF, Sjhn JE, Emerson JA (1987) J Colloid Interface Sci 119:55
8. Hansen FK, Rodsrud G (1991) J Colloid Interface Sci 141:1
9. Thiessen DB, Chione DJ, McCreary CB, Krantz WB (1996) J Colloid Interface Sci 177:658
10. Song B, Springer J (1996) J Colloid Interface Sci 184:64
11. Song B, Springer J (1996) J Colloid Interface Sci 184:77
12. Faour G, Grimaldi M, Richou J, Bois A (1996) J Colloid Interface Sci 181: 385
13. Mat F, De Coninck J, Emelyanenko AM, Boinovich LB (1996) In: Mathieu E, Reihl B, Briggs D (eds) ECASIA'95. Wiley, Chichester, pp 561–564
14. Emelyanenko AM, Boinovich LB (2001) Colloid J 63:159
15. Jennings JW Jr, Pallas NR (1988) Langmuir 4:959
16. Emelyanenko AM, Boinovich LB (2001) Colloids Surf A 189:197
17. Emelyanenko AM, Boinovich LB (2002) Prib Tekh Eksp 1:52
18. del Rio OI, Neumann AW (1997) J Colloid Interface Sci 196:136
19. del Rio OI, Kwok DY, Wu R, Alvarez JV, Neumann AW (1998) Colloids Surf A 143:197
20. Alvarez JM, Amirfazli A, Neumann AW (1999) Colloids Surf A 156:163
21. Emelyanenko AM, Ermolenko NV, Boinovich LB (2004) Colloids Surf A 239: 106
22. Kwok DY, Gietzelt T, Grundke K, Jacobasch H-J, Neumann AW (1997) Langmuir 13:2880
23. Frenkel YI (1948) J Exp Theor Phys 18:659
24. Martynov GA, Starov VM, Churaev NV (1977) Colloid J 39:472
25. Song B, Springer J (1997) Mol Cryst Liq Cryst 293:39
26. Chen GH, Springer J (1997) Mol Cryst Liq Cryst 307:89

Progr Colloid Polym Sci (2004) 128: 57–62
DOI 10.1007/b97081

Abdelrahman I. ElSherbini
Anthony M. Jacobi

Critical contact angles for liquid drops on inclined surfaces

Abstract The contact angles of liquid drops on inclined plane surfaces have been investigated at critical conditions of sliding. Drops of two liquids were tested on eight surfaces covering a wide range of advancing and receding contact angles. Contact angles and contact line dimensions were observed for various drop sizes and surface inclination angles. The contact-angle variation along the circumference of the drop was best fit by a third-degree polynomial of the azimuthal angle. The findings lead to the development of a general relation between the advancing contact angle, the receding contact angle, and the maximum Bond number for a liquid–surface combination. This relation is well supported by data from the literature. The new relation allows the characterization of a surface by its advancing contact angle for a given liquid, and prediction of the corresponding receding angle and maximum drop size.

Keywords Contact angle · Surface tension · Drop size · Retention

A.I. ElSherbini · A.M. Jacobi (✉)
Department of Mechanical and Industrial Engineering,
University of Illinois at Urbana-Champaign,
1206 W. Green St.,
Urbana, IL 61801, USA
e-mail: a-jacobi@uiuc.edu
Tel.: +1-217-3334108
Fax: +1-217-2446534

Introduction

The profile of a liquid drop resting on a horizontal surface usually shows one value for the contact angle between the surface and the vapor–liquid interface. When the surface is tilted, however, two different values are observed for the contact angles. If the contact angles were equal, the surface-tension force holding the drop would vanish, and the drop would slide on the surface owing to gravity. Therefore, the contact angles need to be accurately determined when studying the retention of liquid drops–such as condensate, paint, or pesticide – on surfaces. In this work, we investigate the contact angles of critical drops which are on the verge of sliding on surfaces.

Earlier investigations of drops on vertical and inclined surfaces have disagreed on contact angles. Numerical studies of drops on plane surfaces [1, 2] suggested that the advancing and receding contact angles of a drop at incipient motion depend on the size of the drop and the inclination angle of the surface. In contrast, the experiments of MacDougall and Ockrent [3] showed these angles to be constant. In addition to differences regarding advancing and receding contact angles, previous studies disagreed on the variation of contact angles within a three-dimensional drop along the circumference of its base. Several different forms were predicted [1, 2, 4] or assumed [5–7] for the variation of the contact angle, θ, with azimuthal angle, ϕ, in a drop. The goal of this paper is to determine the contact angles in drops at incipient motion, and to study the relationship between these angles at various critical conditions.

Setup and procedures

The contact-angle experiments involved two liquids, several hydrophilic and hydrophobic surfaces, different drop sizes, and various surface inclination angles, in order to cover a wide range of conditions. The liquids used were water and ethylene glycol. The surfaces tested had eight commercially available coatings on rolled aluminum sheets used as heat exchanger fins. The surfaces were designated by letters A–H and identified by their characteristic contact angles. Advancing and receding contact angles were measured by feeding and withdrawing liquid, respectively, through a microsyringe onto a horizontal surface. A contact-angle goniometer (Rame-Hart 100-00) was used to take the measurements. Three specimens of each surface were cut, wiped once with alcohol for cleaning, and dried. The advancing and receding angles were measured on three different spots of each specimen, totaling 18 measurements for each liquid–surface combination.

An apparatus, shown schematically in Fig. 1, was designed and constructed to test drops on vertical and inclined surfaces. The drop to be examined rests on the test surface, which is attached to a fixed plate. A camera, microscope, and light source are connected to an arm that rotates around the drop. The fixed plate and rotating arm are attached to a larger plate which can be tilted to different angles of inclination. The digital camera was connected to the microscope, which had ×20 magnification, with a special adaptor that could hold the camera against gravity when needed. Xenon light was used at an angle of 180° from the camera, illuminating around the drop profile. The light source was covered by black velvet to improve contrast. When testing water drops, the fixed plate was replaced by another plate inside a vapor-tight, transparent box that was saturated with water vapor to mitigate evaporation. Tests were conducted to verify that refraction through the box did not affect the recorded images.

A rectangular piece of the test surface was cut and attached to the fixed plate. It was then cleaned by wiping it once using alcohol. The surface was tilted to the angle of inclination, α, to be tested. A drop was injected onto the test surface from a microsyringe. Microsyringes with capacities of 1, 5, 10, and 25 μl were used to cover a range of drop sizes with good accuracy. The maximum uncertainty in the volume measured from syringes was 2.5%.

Profile images of drops were recorded with the camera positioned at various azimuthal angles. The camera angle was changed by rotating the arm carrying the camera and light source from $\phi = 0°$ to 165° or from 180° to 345°, where ϕ is measured as in Fig. 1. At least two rounds of pictures were recorded so that average values of contact angles were obtained. Images of drop contours were also recorded and the dimensions of the contact line were calculated. The drops tested ranged in volume from 0.7 to 25.0 mm^3, with diameters from 1.18 to 5.80 mm. Measurements of contour dimensions resulted in maximum uncertainties of 2.3% in equivalent diameter, D, and 4.6% in aspect ratio, β. The

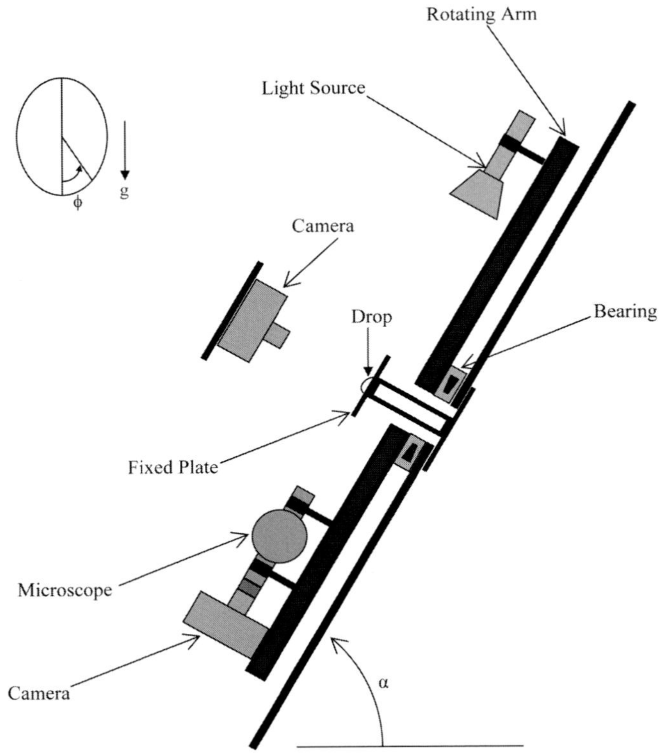

Fig. 1 Cross section of the apparatus used to study drops on inclined surfaces

propagated uncertainty in the Bond number reached 5.8%. The surfaces tested had advancing angles from 49° to 112° with receding angles from 12° to 78°. Surface inclination angles were varied from 0° (horizontal) to 90° (vertical).

Three software packages were used to analyze the profile images and obtain contact angles. The profile image was rotated, when needed, so that the base of the drop coincided with the horizontal. The first software was then used to adjust the contrast and find edges of the portions of the profile close to each end. Another software package was utilized to obtain $x - y$ coordinates of the edges. Each data file with edge coordinates was analyzed by a third package to set an origin where the drop base meets the profile, fit a curve to the edge, and find the slope of that curve at the origin. The contact angle is the arctangent of the slope. A fourth-degree polynomial was found to best fit the edge coordinates, with coefficients of determination, r^2, always exceeding 0.98.

Although this procedure for measuring contact angles is lengthy, it is highly repeatable. Different individuals conducting the experiments obtained very close angle measurements. The measurement uncertainty due to both the experimental procedure and data analysis was determined to be 3° from repeated tests. Since the surfaces tested were real engineering surfaces, the results showed higher scatter due to inhomogeneity. A quicker and easier method to find contact angles from images is by fitting two circles to the profile and calculating the angles. The method is described elsewhere [8]. For drops with noncircular contours, the line of sight is not always perpendicular to the cross section in which the contact angles lie, which leads to errors in the azimuthal angle of that cross section. Also, when the contact angle is greater than 90°, the "apparent" contour dimensions may be larger than those at the drop base. Such errors have been accounted for, as explained by ElSherbini [8].

Results and discussion

Advancing and receding contact angles

The characteristic contact angles of both liquids on the surfaces tested are listed in Table 1. The contact angles were measured on horizontal surfaces, and compared with angles obtained from images of drops on inclined surfaces. The advancing contact angles, θ_A, of some tilted surfaces were slightly higher than those for horizontal surfaces, confirming the results of MacDougall and Ockrent [3]. The angles reported in the table have been adjusted whenever a slight increase in θ_A was observed. The advancing angles ranged from 49° to 112°, and the receding angles ranged from 12° to 78°. The contact-angle hysteresis spanned a range from 230° to 51°.

Contact-angle measurements for drops at noncritical conditions lead to interesting conclusions about the contact angles at the critical condition of sliding. Drops of various sizes were tested on surfaces at different inclination angles. The maximum contact angle was observed to be almost constant and equal to the characteristic θ_A of the surface for all conditions tested. Thus, for a drop at incipient motion on an inclined surface, the maximum contact angle is equal to θ_A of that surface, and is independent of the drop size or the surface inclination angle.

The minimum contact angle, θ_{min}, was also tested for drops at various conditions. The Bond number is used to present the data in dimensionless form,

$$Bo = \frac{\rho g D^2 \sin \alpha}{\gamma}, \tag{1}$$

where ρ is the liquid density, g is the acceleration of gravity, D is the equivalent base diameter, α is the surface inclination angle, and γ is the surface tension of the liquid–vapor interface. It was found that as the Bond number increases, the value of θ_{min} decreases almost linearly and approaches θ_R. The results suggest that the minimum contact angle in a drop at a critical condition is equal to the characteristic θ_R of the surface, regardless of the size of the drop. Contrary to earlier predictions [1, 2], this conclusion agrees with other measurements [3] and is confirmed by further analysis, as discussed later.

Another interesting finding regarding minimum contact angles is noticed when results from the various surfaces and liquids are combined, as shown in Fig. 2. The $\theta_{min}/\theta_{max}$ data are plotted for all surfaces tested, liquids, drop sizes, and inclination angles and are well fitted by a single curve, with $r^2 = 0.90$. The figure indicates that the rate of decrease in θ_{min} relative to θ_{max} is the same for different liquid–surface combinations and is, therefore, independent of θ_R. The equation for θ_{min} which provides the best fit to the overall model presented in this work is

Table 1 Contact-angle measurements for the liquids and surfaces used

Surface	Ethylene glycol		Water	
	θ_A	θ_R	θ_A	θ_R
A	50	15	72	29
B	50	15	49	12
C	59	25	75	52
D	59	27	83	56
E	69	19	97	46
F	65	30	88	58
G	69	28	80	55
H	89	59	112	78

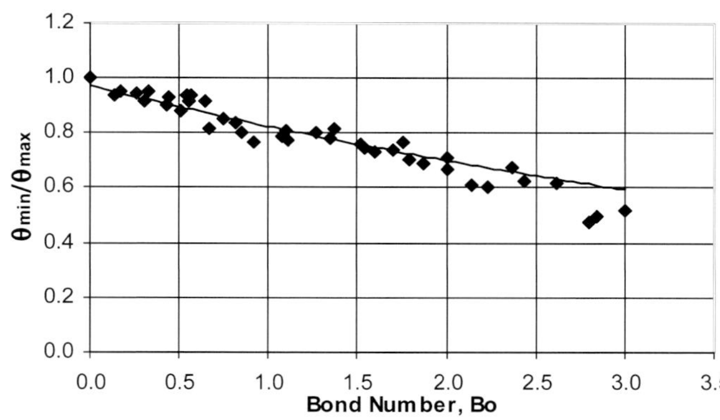

Fig. 2 Minimum contact angle, normalized by the maximum angle, as it varies with the Bond number for all liquids and surfaces tested. A single curve fits the data with $r^2 = 0.90$

$$\frac{\theta_{min}}{\theta_A} = 0.01Bo^2 - 0.155Bo + 0.97, \tag{2}$$

where $\theta_A \approx \theta_{max}$ as found previously. The minimum contact angle will follow the curve of Fig. 2 until the drop reaches a critical condition and slides. The relation between the contact angles and the Bond number at critical conditions will be discussed later.

Contact angles were also measured along the circumference of the drop contour. A typical variation of the contact angle with the azimuthal angle in a drop on a vertical surface, where the orientation of θ is as in Fig. 1, is shown in Fig. 3. The contact angle can be fit by a third-degree polynomial of θ, with a coefficient of determination of 0.98. The contact-angle function can be described by [9]

$$\cos\theta = 2\frac{\cos\theta_{max} - \cos\theta_{min}}{\pi^3}\phi^3 - 3\frac{\cos\theta_{max} - \cos\theta_{min}}{\pi^2}\phi^2 + \cos\theta_{max}, \tag{3}$$

where θ_{max} and θ_{min} are the maximum and minimum contact angles of the drop, respectively. At critical conditions, these angles are equal to θ_A and θ_R, respectively.

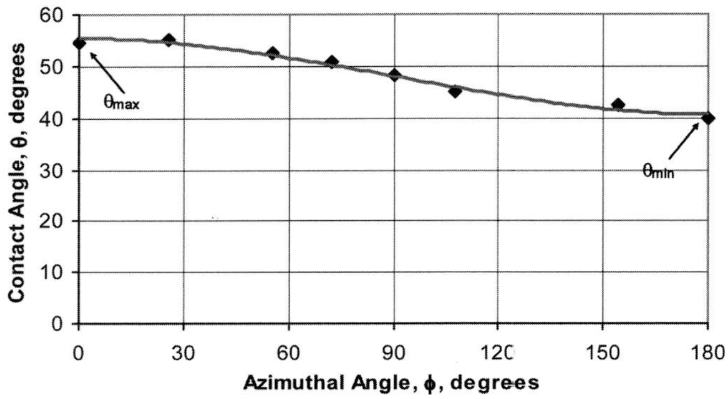

Fig. 3 Typical results for the variation of contact angle with azimuthal angle (drop volume 0.75 mm^3, surface C, $\alpha = 90°$). A third-degree polynomial fits the data with $r^2 = 0.98$

Relation between critical contact angles and Bond number

It was shown in Fig. 2 that the minimum contact angle of a drop decreases as the Bond number is increased, until a critical condition is reached. Consider a balance of forces acting on the drop, as shown in Fig. 4. For the x-direction,

$$F_{sx} + F_{gx} = 0, \tag{4}$$

where F_{sx} and F_{gx} are the x-components of the surface tension and gravitational forces, respectively. The surface-tension force can be found by integrating the differential forces over the drop contour,

$$F_{sx} = -2\gamma \int_0^\pi \zeta \cos\theta \cos\phi \, d\phi, \tag{5}$$

where γ is the surface tension, and ζ is the contour radius. Using an equivalent diameter to represent the drop contour, and substituting for $\cos\theta$ from Eq. (3), the surface-tension force becomes

$$F_{sx} = \frac{24}{\pi^3}\gamma D(\cos\theta_{min} - \cos\theta_A). \tag{6}$$

The gravitational force can be expressed in terms of the volume, V, as

$$F_{gx} = -\rho g V \sin\alpha. \tag{7}$$

The volume of a drop can be expressed as a function of its equivalent diameter and contact angles, if the drop shape is known. In the interest of obtaining closed-form equations for drop parameters at critical conditions, we first consider a simplified approximation of the drop as a spherical cap. A more accurate numerical approximation is then used.

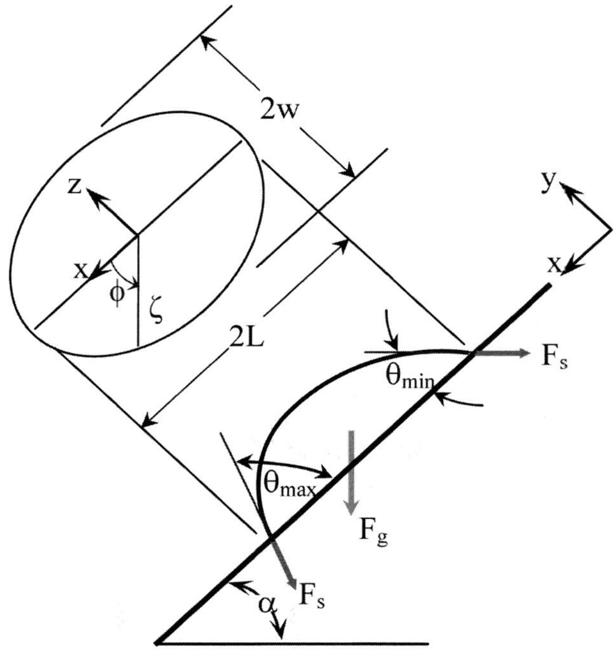

Fig. 4 Force balance on a drop on an inclined plane

Analysis based on the spherical-cap approximation

The volume of a drop approximated as a spherical cap is

$$V = \frac{\pi D^3}{24}\left(\frac{2 - 3\cos\theta_{av} + \cos^3\theta_{av}}{\sin^3\theta_{av}}\right), \tag{8}$$

where $\theta_{av} = \frac{\theta_{min}+\theta_A}{2}$. The gravitational force may be rearranged to

$$F_{gx} = -\gamma D Bo \frac{V}{D^3}. \tag{9}$$

Substituting Eqs. (6), (8), and (9) into Eq. (4) and simplifying, we obtain an expression for the Bond number:

$$Bo = \left(\frac{576}{\pi^4}\right)\left[\frac{\sin^3\theta_{av}(\cos\theta_{min} - \cos\theta_A)}{2 - 3\cos\theta_{av} + \cos^3\theta_{av}}\right]. \tag{10}$$

This equation can be rearranged to describe the minimum angle as

$$\cos\theta_{min} = \cos\theta_A + \frac{\pi^4}{576}\left(\frac{2 - 3\cos\theta_{av} + \cos^3\theta_{av}}{\sin^3\theta_{av}}\right)Bo. \tag{11}$$

Notice that the Bond number can also be defined in terms of the volume of a drop, instead of the diameter, as

$$Bo_V = \frac{\rho g}{\gamma} V^{2/3} \sin\alpha, \tag{12}$$

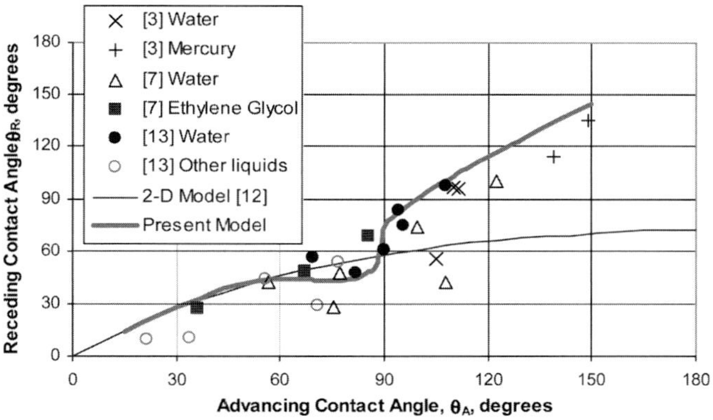

Fig. 5 Prediction of receding contact angles for given advancing angles, compared with measurements of other researchers for different liquids and surfaces ($r^2 = 0.95$)

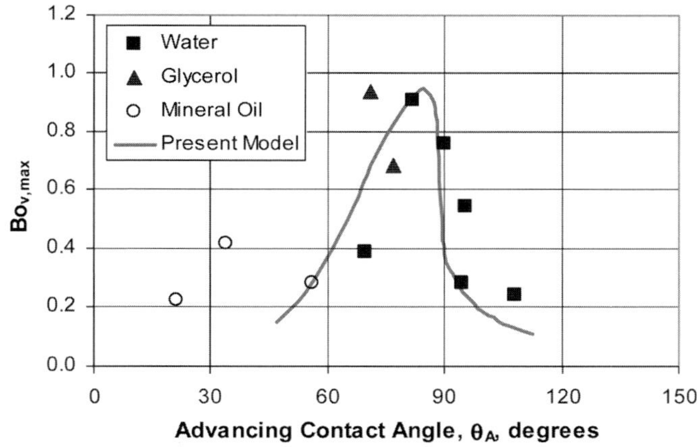

Fig. 6 Prediction of the maximum Bond number based on volume for various advancing angles, compared with measurements for several liquids and surfaces by Goodwin et al. [13]

and an expression similar to Eq. (10) can be obtained for Bo_V.

The analysis presented here leads to a very interesting result. Equation (2) relates the variables θ_{min}, θ_A, and Bo, and Eq. (11) sets limits for these variables when applied at critical conditions. For a given θ_A, the two equations can be solved simultaneously to get the critical Bond number and minimum contact angle. This result indicates that at critical conditions, θ_{min} and Bo_{max} depend only on θ_A, regardless of the specific liquid–surface combination used or the inclination angle. Although θ_A itself depends on the surface, various liquid–surface combinations can give the same advancing angle. Consider a liquid drop on a surface with advancing angle θ_A. The minimum angle at incipient motion will be the same for different surface inclinations; therefore, the minimum angle at critical conditions is characteristic of the surface, and is the receding contact angle of the surface, θ_R. When the surface inclination angle is increased, the maximum diameter, D_{max}, decreases to keep Bo_{max} constant.

This analysis confirms earlier observations [3] that θ_A and θ_R are independent of the surface inclination angle. In addition, it generalizes that finding by pointing out that θ_R depends only on θ_A when the behavior of the liquid–surface combination can be captured by Eq. (2). Therefore, given an advancing angle θ_A, the receding angle and the maximum Bond number are expected to be constant for other liquid–surface combinations having the same advancing angle.

Analysis based on the two-circle approximation

The foregoing analysis assumed the volume of a drop to be approximated by a spherical cap. Moreover, the calculated surface-tension force was based on an equivalent circular contour. ElSherbini and Jacobi [10] developed a method to reliably calculate the volumes of drops on inclined surfaces. In this method, drop profiles are approximated as parts of two circles sharing a common tangent, and the volume is obtained from a numerical integration that takes the form

$$V \approx \int f[\zeta(\phi), \theta(\phi)] \mathrm{d}\phi, \tag{13}$$

where $\zeta(\phi)$ describes the contour shape, and $\theta(\phi)$ is the contact-angle function for the drop. Using the two-circle approximation, Eq. (4) is solved numerically.

With the numerical calculations, the same behavior of parameters at critical conditions is observed. The contact angles and maximum Bond number remain constant. The equations provide a first approximation, but the numerical calculations are needed to get accurate values. Depending on the range of θ_A, the spherical-cap analysis can underestimate Bo_{max} by up to 66%.

Experiments reported in the literature support the proposed relation between the advancing and receding contact angles. The numerically calculated receding angles are compared in Fig. 5 are compared in Fig. 5 with measurements of several researchers using various liquids and surfaces. The data from three sources present measurements involving five liquids and 15 surfaces. Each point represents a liquid–surface combination. Good agreement between the curve and data is observed, with an r^2 of 0.95, confirming the existence of a general relation between the characteristic advancing and receding contact angles. Merte and coworkers [11, 12] studied drop profiles in two dimensions and suggested a relation between their maximum and minimum angles. Although their relation agrees with the data for small advancing angles, it deviates at higher angles, as shown in Fig. 5.

The prediction of a maximum Bond number for given contact angles is compared with data by Goodwin et al. [13], who measured volumes of drops of different liquids on vertical surfaces. Since Goodwin et al. did not provide diameter data, the Bond number based on volume, defined in Eq. (12), is used. The variation of $Bo_{v,max}$ with the advancing angle for water, glycerol, and mineral oil on several surfaces is shown in Fig. 6. The calculated maximum Bond numbers agree reasonably with measurements. For mineral oil, two measurements showed a larger deviation from the prediction at low advancing angles (21° and 34°), which may be due to higher measurement errors at such low values.

Since $Bo_{v,max}$ is constant for a given liquid–surface combination, Eq. (12) shows that $V_{max}^{-2/3}$ is proportional to $\sin\alpha$. This proportionality was empirically observed by Briscoe and Galvin [14] for sessile drops. Thus, the present analysis provides an explanation for their observations. Measurements for pendant drops showed a different exponent for V_{max}, which needs more investigation.

Conclusions

The contact angles of critical-size drops resting on inclined surfaces have been studied. Experiments indicated that the maximum and minimum contact angles in a critical drop are approximately equal to the characteristic advancing and receding angles of the surface, respectively. The contact-angle variation along the circumference of the drop was best fit by a third-degree polynomial of the azimuthal angle. A general relation was found between the advancing and receding contact angles, for liquid–surface combinations showing common minimum-angle behavior. A first approximation was given by solving Eqs. (2) and (11), but a more accurate prediction involves numerical integrations based on the two-circle method. This relation is well supported by data from the literature. It was also found that the maximum Bond number is constant for a given θ_A, which allows the prediction of the maximum drop diameters and volumes at various surface inclinations and for different liquid–surface combinations having the same θ_A.

Acknowledgement The authors acknowledge the support of the Air Conditioning and Refrigeration Center of the University of Illinois at Urbana-Champaign.

References

1. Brown RA, Orr FM Jr, Scriven LE (1980) J Colloid Interface Sci 73:76
2. Milinazzo F, Shinbrot M (1988) J Colloid Interface Sci 121:254
3. MacDougall G, Ockrent C (1942) Proc R Soc Lond Ser A 180:151
4. Dimitrakopoulos P, Higdon JJL (1999) J Fluid Mech 395:181
5. Rotenberg Y, Boruvka L, Neumann AW (1984) J Colloid Interface Sci 102:424
6. Korte CM, Jacobi AM (2001) J Heat Transfer 123:926
7. Extrand CW, Kumagai YJ (1995) J Colloid Interface Sci 170:515
8. ElSherbini AI (2003) Modeling Condensate drops retained on the air-side of heat exchangers. PhD thesis. University of Illinois at Urbana-Champaign, Urbana
9. ElSherbini AI, Jacobi AM (2004) J Colloid Interface Sci 273:556
10. ElSherbini AI, Jacobi AM (2004) J Colloid Interface Sci 273:566
11. Merte H, Yamali C (1983) Waerme Stoffubertrag 17:171
12. Merte H, Son S (1987) Waerme Stoffubertrag 21:163
13. Goodwin R, Rice D, Middleman SJ (1988) J Colloid Interface Sci 125:162
14. Briscoe BJ, Galvin KP (1991) J Colloid Surf 52:219

Progr Colloid Polym Sci (2004) 128: 63–67
DOI 10.1007/b97082

Arlete Tavares Almeida
Hartmut Gliemann
Thomas Schimmel
Denise Freitas Siqueira Petri

Atomic force microscopy in a liquid and *in situ* ellipsometry as complementary techniques for the study of protein adsorption

A.T. Almeida · D.F.S. Petri (✉)
Instituto de Química,
Universidade de São Paulo,
São Paulo, Brazil
e-mail: dfsp@iq.usp.br
Tel.: +55-11-30913831
Fax: +55-11-3815-5579

H. Gliemann · T. Schimmel
Institut für Nanotechnologie,
Forschungszentrum Karlsruhe,
Karlsruhe, Germany

T. Schimmel
Institut für Angewandte Physik,
Universität Karlsruhe (TH),
Karlsruhe, Germany

Abstract The combination of two *in situ* techniques for the study of immobilization of biomolecules is seldom performed. Therefore, the combination of *in situ* ellipsometry with atomic force microscopy (AFM) in a liquid was used to follow the adsorption process of enolase, an enzyme of the glycolytic pathway. The adsorbed amount and topographic information of the adsorbed layers were obtained in the liquid environment. AFM images showed that surfaces with comparable adsorbed amounts of protein presented immobilized entities arranged in strings or in spherical clumps, evidencing the importance of combining *in situ* ellipsometry and AFM in a liquid.

Keywords Adsorption kinetics · Enolase · *In situ* ellipsometry · atomic force microscopy

Introduction

The importance of the study of adsorption process of proteins can be evaluated by several applications of proteins in industrial [1–3] and biomedical [4–6] fields. The advance of instrumental techniques has allowed one to go deeper into the study of adsorption behavior of proteins in liquids. While *ex situ* measurements techniques yield information about the thickness, the adsorbed amount, the chemical composition and topographic images, *in situ* measurements provide the adsorption kinetics with the corresponding three-dimensional structural arrangement [7]. Specially in the case of proteins and enzymes, the conformational structures are intimately correlated with their function. Most reports found in the literature combine *in situ* with *ex situ* techniques for the study of protein adsorption [8–21]. The main disadvantage of such a combination is that upon drying the characteristics of the adsorbed layer might no longer correspond to those in the liquid environment, where the biomolecules are active. Surface plasmon ressonance [11], atomic force microscopy (AFM) [13, 15, 19], ellipsometry [14, 16, 18, 20, 21] and reflectometry [17] are some examples of *in situ* techniques used in the study of the adsorption process of proteins. However, combining two *in situ* techniques in the same study is seldom performed. Therefore, in this work *in situ* ellipsometry and tapping mode AFM in a liquid are used for the study of the adsorption behavior of enolase on silicon surfaces in a liquid environment. Enolase (2-phospho-D-glycerayte hydrolyase) was chosen owing to its key role in the glycolytic pathway, and it acts as plasminogen receptor on the surface of a variety of hematopoetic, epithelial and endothelial cells [22].

Materials and methods

Enolase from *Saccharomyces cerevisiae* (EC 4.2.1.11, 46.7 kDa, Sigma, St. Louis, USA) was dissolved in NaCl solution (0.001 and 0.1 mol/L). Freshly cleaned silicon (100) wafers (1 cm^2) (Crystec, Berlin, Germany) with a native layer of SiO_x were used as substrates.

Ellipsometry

Ellipsometric measurements were performed in a vertical computer-controlled DRE-EL02 Ellipsometer (Ratzeburg, Germany), as described elsewhere [20]. The angle of incidence ϕ was set to 70.0° and the wavelength λ of the laser was 632.8 nm. For the data interpretation, a multilayer model composed by the substrate, the unknown layer and the surrounding medium should be used. Then the thickness (d_x) and refractive index (n_x) of the unknown layer can be calculated from the ellipsometric angles, Δ and Ψ, using the fundamental ellipsometric equation and iterative calculations with Jones matrices [23]:

$$e^{i\Delta}\tan\Psi = R_p/R_s = f(n_k, d_k, \lambda, \phi), \quad (1)$$

where R_p and R_s are the overall reflection coefficients for the parallel and perpendicular waves. They are a function of the angle of incidence ϕ, the wavelength λ of the radiation and of the indices of refraction and the thickness of each layer of the model, n_k, d_k.

First of all, the thickness of the SiO_2 layers was determined in air, considering the index of refraction for Si as $\tilde{n} = 3.88 - i0.018$ [25] and its thickness as an infinite one; for the surrounding medium, the index of refraction was measured at 23 °C with an Abbé refractometer. The indices of refraction amounted to 1.3335 and 1.3345 for 0.001 mol/L NaCl and 0.1 mol/L NaCl, respectively. Because the native SiO_2 layer is very thin, its index of refraction was set as 1.462 and just the thickness was calculated. The mean SiO_2 thickness measured for 50 samples amounted to 1.9 ± 0.1 nm. The adsorption from solution was monitored *in situ* with the help of a cell made of poly(methyl methacrylate). This cell has two quartz windows, one for the incident beam and the other for the reflected beam, both with an inclination angle of 70.0°. The measurements were done in a conditioned room at 23 °C ± 1 °C.

The small differences in the indices of refraction of the substrate, enolase and solution make an independent determination of $n_{enolase}$ and $d_{enolase}$ impossible; therefore, $n_{enolase}$ was kept constant at 1.50 and $d_{enolase}$ was calculated. The hydrodynamic thickness could be calculated if the optical contrast in the system were higher. Nevertheless, it is important to remember here that, if the index of refraction assumed for the adsorbing layer lies in a reasonable range (between 1.40 and 1.60), the product $n_{enolase}d_{enolase}$ should be a constant value [24]. This product is the parameter necessary to calculate the adsorbed amount of enolase Γ [26]:

$$\Gamma = \frac{d_{enolase}(n_{enolase} - n_0)}{dn/dc}, \quad (2)$$

where n_0 is the index of refraction of the solution, and dn/dc is the increment of refractive index determined with a differential refractometer. For our system, dn/dc amounted to 0.16 mL/g at 23 °C.

Atomic force microscopy

Tapping mode AFM images of silicon surfaces after contact with enolase solution at different adsorption times were obtained in a liquid at room temperature with a Nanoscope IIIa/Multimode (Veeco, USA). Silicon cantilevers covered with a film of W_2C (MikroMasch, USA) with a spring constant of 0.35 N/m were oscillated at a cantilever resonance of 8 kHz. Scan areas of 5 × 5 μm^2 with a resolution of 512 × 512 measured points (pixels) were obtained. Image processing and the mean roughness (root mean square) were performed by means of the Nanoscope III software.

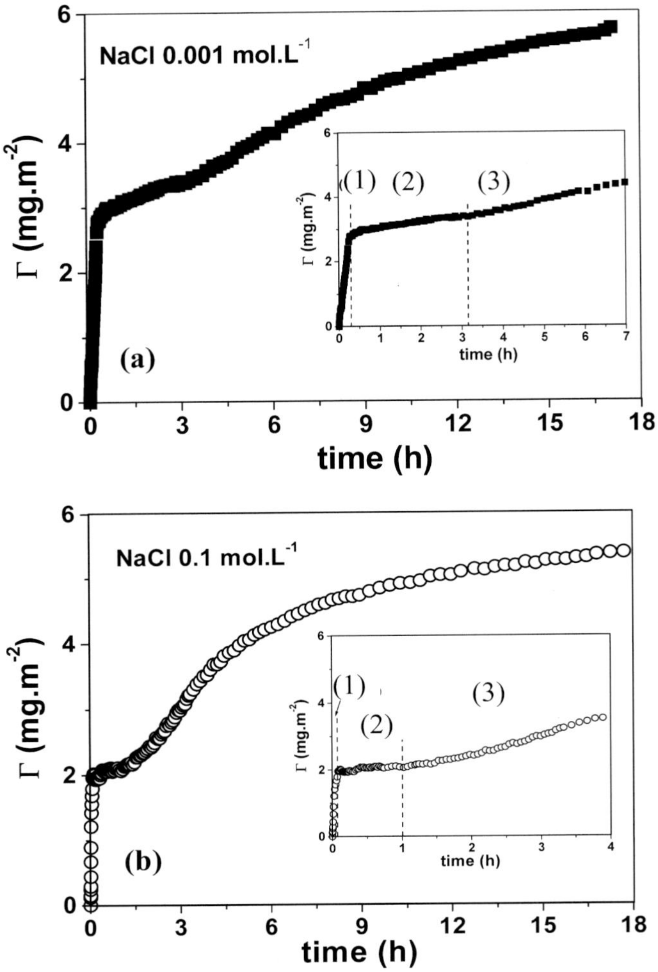

Fig. 1 Adsorption kinetics for enolase ($c = 0.01$ g/L) in **a** 0.001 mol/L NaCl and **b** 0.1 mol/L NaCl solution

Results and discussion

In previous work [20], it was demonstrated by ellipsometry that the adsorption process of enolase onto a silicon wafer in the dilute regime (0.01 g/L) presents three steps. Step 1 is a diffusion-controlled process, since it involves transport of molecules towards the interface. Initially, adsorption is controlled by diffusion of single or small aggregates of enolase molecules. During the second step, the adsorbed amount of enolase (Γ) reaches a saturation

Fig. 2 Atomic force microscopy images in a liquid environment corresponding to the three steps of the adsorption process of enolase under two ionic strengths: **a** 0.001 mol/L NaCl and **b** 0.1 mol/L NaCl. The roughness of the enolase film is evidenced as root-mean-square (*RMS*) values. Scan area 5 μm × 5 μm, *Z* scale 7 nm (steps 1 and 2) and 10 nm (step 3)

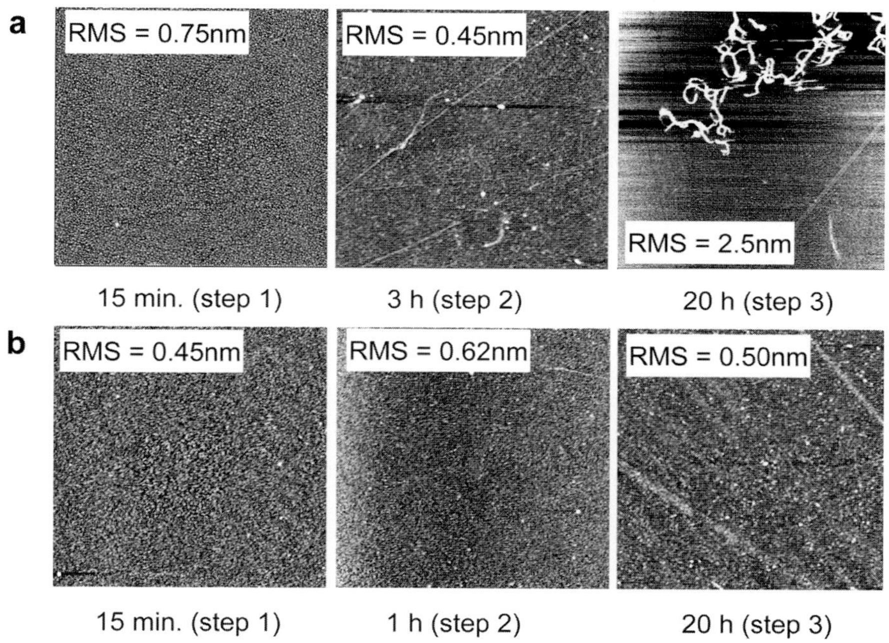

value, which might indicate the formation of the first adsorbed layer. Arriving molecules adsorb on free sites on silicon surfaces until a uniform monolayer is formed. This is a slow process because the number of adsorption sites decreases continuously with time. The increase of ionic strength leads to a small decrease of the plateau value, which was attributed to a contraction of the double electric layer on the silicon and enolase surfaces. All experiments were performed at pH 6.7, which is close to the experimental value of the isoeletric point of enolase (6.1) [27]. According to Loidl-Stahlhofen et al. [28], the silicon surface is negatively charged at pH 6.8; therefore, under such conditions, both enolase molecules and the silicon wafer might be charged. From Fig. 1, one can also observe that the formation of the first adsorbed layer (Γ plateau) is faster upon increasing the ionic strength. During the adsorption of a charged protein onto a charged surface, the electrical double layers overlap, resulting in a barrier for deposition. Increasing the ionic strength, charge screening favors the barrier reduction. Therefore, the adsorbed molecule spends less time rearranging its structure before new arriving molecules occupy the empty adsorption sites on the substrate surface. The third step is characterized by a continuous increase of Γ with time, which can be attributed to the deposition of enolase on the already adsorbed monolayer caused by cooperative binding. This is a common process in biological systems and can lead to multilayer adsorption at long adsorption times. After 18 h of adsorption, the adsorbed amount of enolase was 5.5 ± 0.2 mg/m^2, regardless of the ionic strength, indicating that the deposition of upper layers is driven not only by electrostatic interaction but also by hydrogen bonding. Lateral diffusion of the already adsorbed enolase molecules, which leads to clustering on the surface and opens up additional room for the arriving molecules, could also be considered as another reason for the observed effect.

AFM images obtained in a liquid environment allowed a direct comparison with kinetics obtained from ellipsometry. Figure 2 shows that proteins are already present at short adsorption times for both ionic strengths. At low ionic strength (Fig. 2a), the protein film becomes smoother after 3 h of contact. This adsorption time corresponds to step 2 of the ellipsometric adsorption kinetics, which is characterized by the saturation of the first adsorbed layer of enolase. In the case of adsorption at high ionic strength (Fig. 2b), after 1 h of adsorption the roughness of the adsorbed layer is higher because the saturation of the first adsorbed layer is faster under such conditions. These findings are in good agreement with ellipsometric measurements. The AFM image corresponding to the third step of the adsorption process shows an increase of the roughness of the enolase film after long adsorption times, which supports the idea of multilayer formation suggested by the kinetic curve at low ionic strength (Fig. 1a). Although ellipsometric measurements have shown that ionic strength has no effect on the Γ values obtained after long adsorption times (18 h), AFM images show that enolase aggregates might adopt different sizes and shapes according to the ionic strength (Fig. 2). These results show the importance of combining these two techniques in order to study the adsorption behavior of a protein. Ellipsometry

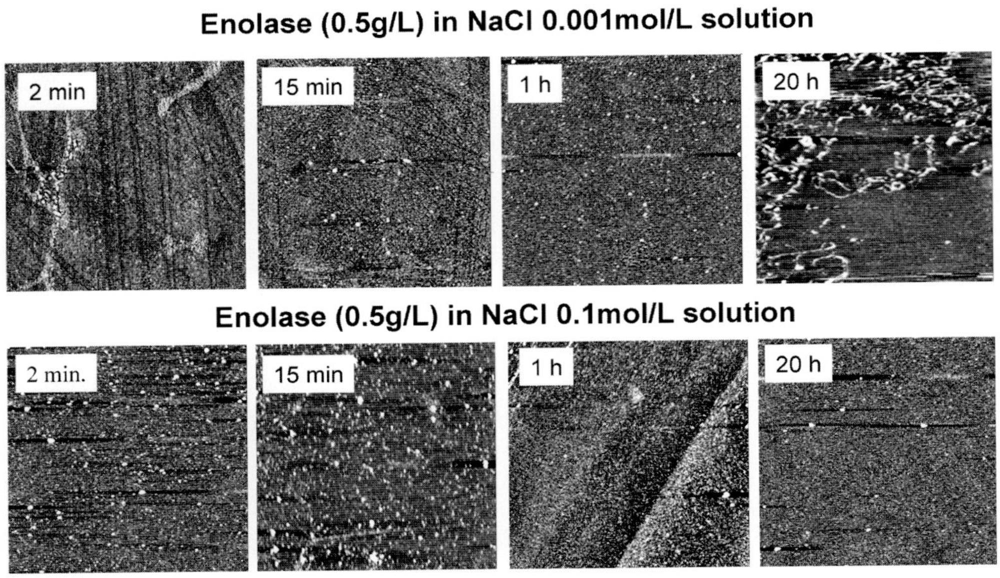

Fig. 3 Atomic force microscopy images in a liquid environment of silicon surfaces after contact with enolase (0.50 g/L) in NaCl solution: **a** 0.001 mol/L and **b** 0.1 mol/L. Scan area 5 μm × 5 μm, *Z* scale = 7 nm (steps 1 and 2) and 15 nm (step 3)

Table 1 Adsorbed amount (Γ) and root-mean-square (*RMS*) values of silicon surfaces after contact with enolase (0.50 g/L) in 0.001 and 0.1 mol/L NaCl solutions at different adsorption times

Adsorption time	0.001 mol/L NaCl		0.1 mol/L NaCl	
	Γ (mg/m^2)	RMS (nm)	Γ (mg/m^2)	RMS (nm)
2 min	1.1	0.62	1.8	0.52
15 min	1.6	0.89	3.2	0.96
1 h	2.9	0.85	4.5	0.90
20 h	4.6	1.7	6.3	1.1

yields mean thickness values resulting from analysis over spot areas of 3 mm^2, while the AFM images correspond to areas of 25 μm^2. Under low ionic strength, strip-shaped (Fig. 2a) enolase aggregates are observed, while at high ionic strength these aggregates appeare as spherical clumps (Fig. 2b). Similar behavior was observed on silicon surfaces after contact with concentrated enolase solution (0.50 g/L), as shown in the AFM images in Fig. 3. The corresponding adsorbed amount and roughness of the enolase films obtained at different adsorption times and ionic strength are shown in Table 1. Increasing the ionic strength, the adsorbed amounted increased, regardless of the adsorption time. In contrast to the adsorption behavior in the dilute range, screening effects favored the enolase adsorption in the concentrated regime. Such behavior might be explained by the screening-enhanced adsorption model proposed by van de Steeg et al. [29], where the adsorption process is driven only by electrostatic interactions.

Small-angle X-ray scattering measurements of enolase ($c = 6.0$ g/L) in 0.001 mol/L NaCl yielded a radius of gyration of 29 Å, confirming that only single molecules [27] are present in solution and that the aggregation is probably induced by the surface of the enolase film previously formed. The factor that triggers this aggregation process is still unknown. Probably the aggregates are induced by the enolase film surface and their size and shape can result from the balance between contributions of ionic strength and the characteristics of this surface.

Conclusion

The present results on the adsorption of enolase onto silicon wafers showed that the techniques AFM in a liquid and *in situ* ellipsometry complement each other. The combination of two *in situ* methods allowed us to obtain the adsorbed amount and the topography of the immobilized layer in a liquid environment. Protein films with comparable adsorbed amounts presented immobilized entities arranged in strings or in spherical clumps. Such structural differences are observed on a nanometric scale and cannot be determined by ellipsometry alone, since it yields only a mean value of the adsorbed amount in a spot of a few square millimeters.

Acknowledgements The authors acknowledge FAPESP (grants 97/13070-2 and 00/08051-3) and CNPq for financial support, Sabrina M. Pancera (IQ/USP) and Rosangela Itri (IF/USP) for the small-angle X-ray scattering measurements and discussion of the results, and the National Laboratory of Synchrotron Light for the use of their facilities (SAS no. 1247/2).

References

1. Fernandez-Lafuente R, Armisén P, Sabuquillo P, Fernández-Lorente G, Guisán JM (1998) Chem Phys Lipids 93:185
2. Mattedi A, Filho RM (2000) Comput Chem Eng 24:1111
3. Alonso J, Barredo JL, Armisén P, Díez B, Salto F, Guisán JM, Garcia JL, Cortés E (1999) Enzyme Microbial Technol 25:88
4. Werner C, König U, Augsburg A, Arnhold C, Körber H, Ralf Zimmermann HJ (1999) Colloids Surf A 159:519
5. Dahint R, Seigel RR, Harder P, Grunze M, Josse F (1996) Sens Actuators B 36:497
6. Slomkowski S (1998) Prog Polym Sci 23:815
7. Nakanishi K, Sakiyama T, Imamura K (2001) J Biosci Bioeng 91:233
8. Kim K, Kim C, Byun Y (2001) Langmuir 17:5066
9. Ortega-Vinuesa JL, Tengvall P, Lundstrom I (1997) J Colloid Interface Sci 186:129
10. Coen MC, Lehmann R, Gröning P, Bielmann M, Galli C, Schlapbach L (2001) J Colloid Interface Sci: 233, 180
11. Green RJ, Davies J, Davies MC, Roberts CJ, Tendler SJB (1997) Biomaterials 18:405
12. Vansteenkiste SO, Corneillie SI, Schacht EH, Chen X, Davies MC, Moens M, Van Vaeck L (2000) Langmuir 16:3330
13. Johnson CA, Yuan Y, Lenhoff AM (2000) J Colloid Interface Sci 223:261
14. Kull T, Nylander T, Tiberg F, Magnus Wahlgren N (1997) Langmuir 13:5141
15. Kurrat R, Walivaara B, Marti A, Textor M, Tengvall O, Ramsden JJ, Spencer ND (1998) Colloids Surf B 11:187
16. Malmsten M. (1999) Colloids Surf A 159:77
17. Gage RA, Norde W (1997) Colloids Surf B 9:139
18. Arwin H (2001) Sens Actuators A 92:43
19. Zlatanova J, Lindsay SM, Leuba SH (2000) Prog Biophys Mol Biol 74:37
20. Almeida AT, Salvadori MC, Petri DFS (2002) Langmuir 18:6914–6920
21. Tengvall P, Lundström I, Liedberg B (1998) Biomaterials 19:407
22. Pancholi V (2001) Cell Mol Life Sci 58:902
23. Azzam RMA, Bashara NM (1987) Ellipsometry and polarized light. North Holland, Amsterdam
24. Motschmann H, Stamm M, Toprakcioglu C (1991) Macromolecules 24:3681
25. Edward DP (ed) (1985) Handbook of optical constants of solids. Academic, London
26. De Feijter JA, Benjamins F, Veer FA (1978) Biopolymers 17:1759
27. Chin CCQ, Brewer JM, Wold F (1981) J Biol Chem 2563:1377
28. Loidl-Stahlhofen A, Schmitt J, Nöller J, Hartmann T, Brodowsky H, Schmitt W, Keldenich J (2001) Adv Mater 13:1829
29. van de Steeg HGM, Cohen Stuart MA, de Kaizer A, Bijsterbosch BH (1992) Langmuir 8:2538

Progr Colloid Polym Sci (2004) 128: 68–73
DOI 10.1007/b97084

Makiko Takahashi
Shinichi Hamaguchi
Hiroshi Ito
Toyoko Imae
Tetsuya Nakamura

Association of block copolymers with dendritic and perfluorinated side chains in solution and at an interface

M. Takahashi · S. Hamaguchi
H. Ito · T. Imae
Graduate School of Science,
Nagoya University, Chikusa,
Nagoya 464-8602, Japan

T. Imae (✉)
Research Center for Materials Science,
Nagoya University, Chikusa,
Nagoya 464-8602, Japan
e-mail: imae@nano.chem.nagoya-u.ac.jp
Tel.: +81-52-7895911
Fax: +81-52-7895912

T. Nakamura
NOF Corporation, Taketoyo, Chita,
Aichi 470-2345, Japan

Abstract Block copolymers, poly(3,5-bis(3,5-bis(benzyloxy)benzyloxy)benzyl methacrylate-*random*-methacrylic acid)-*block*-poly(2-perfluorooctylethyl acrylate) [P(D_3MA,MAA)-*b*-PFA], which contain benzyloxy dendron and perfluorinated alkyl chains in their side chains, were synthesized. The products were characterized by differential scanning calorimetry and IR absorption spectroscopy, and their degrees of polymerization were determined with gel permeation chromatography, elemental analysis, and UV–vis spectroscopy. It was confirmed from transmission electron microscopic and atomic force microscopic observation that P(D_3MA,MAA)-*b*-PFAs were associated into polydisperse, spherical particles in a chloroform solution. The surface pressure–area isotherms of Langmuir monolayers at the air–water interface displayed hysteresis, which was remarkable for a P(D_3MA,MAA)-*b*-PFA with a medium substitution fraction of the dendron block. The molecular orientation at the air–water interface estimated from atomic force microscope images of Langmuir–Blodgett films depended on the substitution fraction: Whereas the atomic force microscope image of a Langmuir–Blodgett film of a copolymer with a medium substitution fraction showed a flat surface, that with the high fraction showed a zone texture.

Keywords Block copolymer · Dendron · Perfluorinated alkyl chain · Surface pressure–molecular area isotherm · Atomic force microscopy

Introduction

Fluorinated polymers are useful materials owing to their low surface energy and thermal stability [1–18]. Especially, block copolymers which have a fluorocarbon side chain in one block and a solvophilic side chain in the other block exhibit great potential owing to their fluorocarbon character and amphiphilicity. Then, the copolymers can form self-assembled, ordered structures at the interface and micellar structures in the solution. Their self-assembling properties can be controlled by the changes of the chemical structure, i.e., the length and the ratio of hydrophobic and hydrophilic segments [19–21].

The fluorinated diblock copolymers poly(methyl methacrylate)-*block*-poly(2-perfluorooctylethyl methacrylate) (PMMA-*b*-PFMA) and poly(methacrylic acid)-*block*-poly(2-perfluorooctylethyl methacrylate) (PMAA-*b*-PFMA) consist of a randomly coiled solvophilic block and a rigid perfluorinated block [22, 23]. PMMA-*b*-PFMA forms small micelles in organic solvents, whereas PMAA-*b*-PFMA in methanol is associated into larger micelles than PMMA-*b*-PFMA and the size of its micelles depends cooperatively on the pH in the aqueous medium.

It is expected that when side chains of the PMAA block in PMAA-*b*-PFMA are replaced by bulky dendrons, the whole block copolymer should become rigid and have different properties from the original block copolymer. In the present work, we report the characterization of block copolymers with dendritic and perfluorinated side chains, poly(3,5-bis(3,5-bis(benzyloxy)benzyloxy)benzyl methacrylate-*random*-methacrylic acid)-*block*-poly(2-perfluorooctylethyl acrylate) [P(D_3MA,MAA)-*b*-PFA] and their association behavior in solution and at the interface.

Polymers that have dendritic side chains have been reported by some researchers [24–37]. Förster et al. [29] have investigated how dendron side chains stiffen the polymer backbone by means of small-angle neutron scattering. Bo et al. [30] synthesized amphiphilic polymers with dendrons and hydrophobic alkyl chains. With the development of science, multifunctional polymer materials are demanded. Copolymers consisting of a perfluorinated block and a dendron block, which are investigated in the present work will be attractive as novel multifunctional materials. However, such block copolymers have never been reported, as far as we know.

Experimental

Dendron, 3,5-bis(3,5-bis(benzyloxy)benzyloxy)benzyl bromide, was purchased from Tokyo Kasei Kogyo Co. and other chemicals were of commercial grade. Diblock copolymer, poly(methacrylic acid)-*block*-poly(2-perfluorooctylethyl acrylate) (PMAA-*b*-PFA, lot no. M2), was synthesized by eliminating, with concentrated hydrochloric acid, the *t*-butyl methacrylate (*t*-BMA) side chains of poly(*t*-butyl methacrylate)-*block*-poly(2-perfluorooctylethyl acrylate), which was synthesized by radical polymerization from *t*-BMA and 2-perfluorooctylethyl acrylate as monomeric units and polyperoxide as initiator.

Thermal properties were analyzed using differential scanning calorimetry (DSC). First heating was scanned from 30 to 150 °C at an elevation rate of 10 °C min^{-1}, and then the polymers were cooled down to 30 °C at a rate of 5 °C min^{-1}. Subsequently, a second heating, scanned from 30 to 200 °C at a rate of 10 °C min^{-1}, was carried out. Fourier transform IR (FT-IR) absorption spectra were recorded at room temperature (about 25 °C) using a Bio-Rad FTS 575C FT-IR spectrometer. KBr pellets of the compounds were prepared.

The molecular weight was determined using a Shimadzu SCL-10A gel permeation chromatogram (GPC). The eluent for the GPC was tetrahydrofuran (THF), and polystyrene was used as a standard of molecular weight. During the measurement, the temperature was kept at 40 °C. UV–vis spectra of THF solutions of the polymers and a dendron were measured at 25 °C with a Shimadzu UV-2200 spectrometer. Elemental analysis was carried out with an LCO CHN-900 elemental analyzer.

A Hitachi H-800 transmission electron microscope (TEM) was used to take photographs of the cast films on carbon-coated copper TEM grids from a methanol solution (1.0 mg cm^{-3}) of M2 and chloroform solutions (1.0 mg cm^{-3}) of M2den (Table 1). A Digital Instruments NanoScope III microscope was used to get the atomic force microscope (AFM) images of cast films and Langmuir–Blodgett (LB) films on mica. Microscopic observations were done at room temperature (about 25 °C).

A LB film deposition apparatus (Nippon Laser & Electronics Laboratory) was used to measure surface pressure–surface area (π–A) isotherms. A small amount of a chloroform solution (1.0 mg cm^{-3}) of the polymer was spread on the water subphase in the trough, and 30 min was allowed for solvent evaporation. Isotherms were measured at a compression rate (barrier speed) of 5 mm min^{-1} at a subphase temperature of 25 °C. Optical imaging of the monolayer on the water subphase was observed by a Brewster-angle microscope (Nippon Laser & Electronics Laboratory NL-EMM63) placed over a Langmuir film balance.

LB films were prepared at a surface pressure of 20 mN m^{-1}. The Langmuir film was transferred onto a mica substrate for the AFM observation. The Z-type and ZX-type LB films were prepared by one upstroke transfer and additional downstroke transfer to the Z-type film, respectively, at a rate (dipping speed) of 2 mm min^{-1}. After the transfer, the LB films were dried in vacuo for 8 h.

Results and discussion

Synthesis and characterization of block copolymers

PMAA-*b*-PFA (1.0 g for lot no. M2den1 or 0.19 g for lot nos. M2den2 and M2den3), 3,5-bis(3,5-bis(benzyloxy)benzyloxy)benzyl bromide (1.0 g for M2Den1 and M2den2 or 3.0 g for M2Den3) and 1,8-diazabicyclo[5.4.0]undec-7-ene (0.20 g) were dissolved in 10 cm^3 dimethyl sulfoxide, and the solution was heated at 50 °C for 5–16 h (Scheme 1). Then the solution was acidified with three drops of hydrochloric acid (12 N), and poured into excess methanol (200 cm^3). The resultant

Table 1 Characterization of block copolymers

Lot no.	$M_n/10^3$ [a]	M_w/M_n [a]	Number of dendrons [b]	C:H ratio [c]	Composition				Molecular length/nm			Collapse pressure/ mN m^{-1}	Occupied molecular area/nm^2 $molecule^{-1}$
					l	m–l	n	m+n	a	b	a+b		
M2	25.9	1.29	0	3.39:3.58	0	123	29	152				30–40	
M2 den1	8.3	1.19	2.5	3.99:4.81	3	55	3	61	12.6	0.7	13.3		
M2 den2	25.9	1.70	23.7	6.15:5.47	24	12	11	47	7.8	2.4	10.2	20–25	12 (solid state), 18 (liquid state)
M2 den3	27.1	1.72	28.0	6.16:5.45	28	2	8	38	6.5	1.7	8.2	13–20	17 (solid state)

[a]From gel permeation chromatography
[b]From UV–vis spectra
[c]From elemental analysis

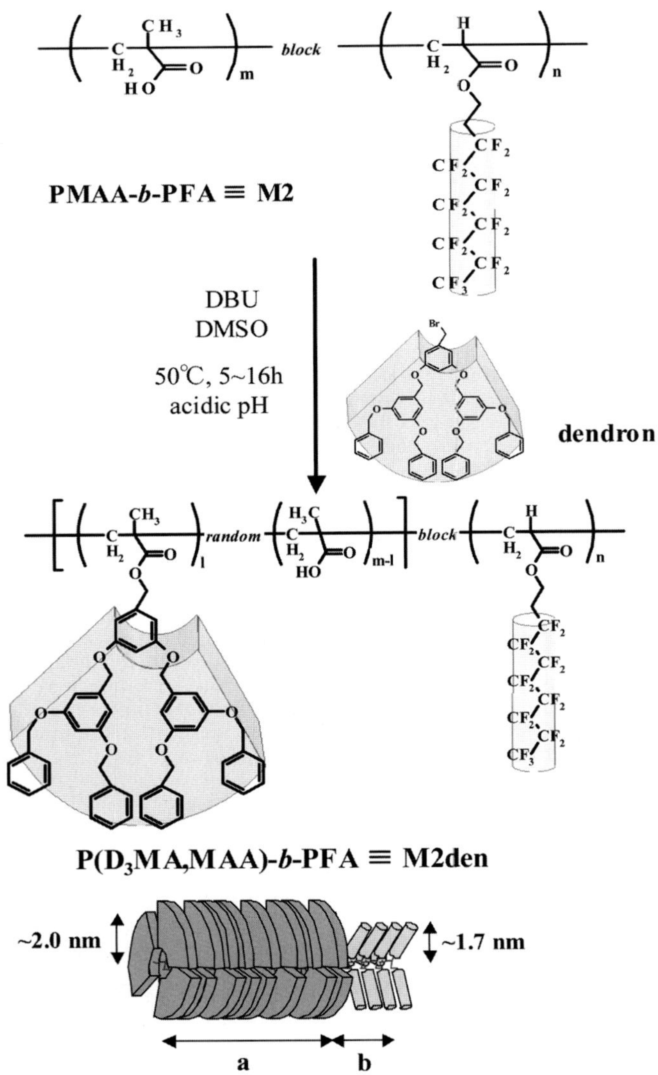

Scheme 1

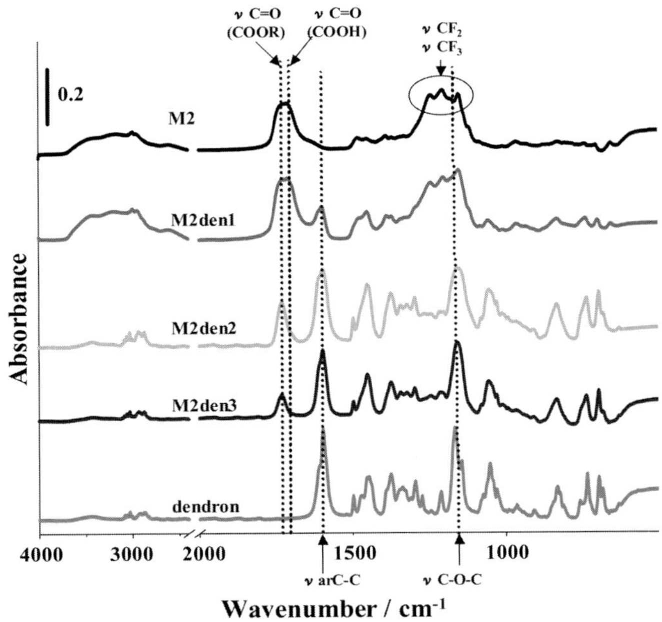

Fig. 1 IR absorption spectra of KBr pellets of block copolymers and dendrons

precipitates were separated from the solution, washed with water (800 cm^3), and dried in vacuo. Since the degree of substitution of the PMAA block should depend on the ratio of the 3,5-bis(3,5-bis(benzyloxy)benzyloxy)benzyl bromide dendron to PMMA-*b*-PFA, the products are random copolymers including the PMAA block and the dendron block and must be described as poly(3,5-bis(3,5-bis(benzyloxy)benzyloxy)benzyl methacrylate-*random*-methacrylic acid)-*block*-poly(2-perfluorooctylethyl acrylate) (M2den), where the degrees of polymerization of the PD_3MA, PMAA, and PFA blocks are denoted l, m–l, and n, respectively (Scheme 1).

In the DSC thermogram of the second heating process on the M2den copolymer, the melting temperature, T_m of the perfluorinated block was observed at 67.5 °C (M2den1), 68.5 °C (M2den2), or 67.5 °C (M2den3) as an endothermic peak. Additionally, the glass-transition temperature, T_g, of the dendron block was observed at 46.2 °C (M2den2) or 41.8 °C (M2den3) as the midpoint of the transition. This indicates the replacement of methacrylic acid by the dendron.

The replacement was confirmed even from IR spectra. As seen in Fig. 1, the spectrum of PMAA-*b*-PFA displayed ester C=O and carboxylic acid C=O stretching vibration bands at 1,730 and 1,713 cm^{-1}, respectively. CF_3 asymmetric, CF_2 antisymmetric, and CF_2 symmetric stretching bands appeared at 1,243, 1,204, and 1,152 cm^{-1}, respectively. The bands originating from PMAA-*b*-PFA weakened relatively with the replacement of methacrylic acid by the dendron, and absorption bands characteristic of the dendron strengthened. The bands at 1,596 and 1,156 cm^{-1} are assigned to the benzene ring C–C and ether C–O–C antisymmetric stretching vibration modes, respectively.

The number-average molecular weight, M_n, and the polydispersity, M_n/M_w, of the block copolymers were evaluated with the GPC, as listed in Table 1. The copolymers in THF have a dendron aromatic ring absorption band at 212 nm. The molar extinction coefficient (182,000 M^{-1} cm^{-1}) of the 212 nm band of 3,5-bis(3,5-bis(benzyloxy)benzyloxy)benzyl bromide in dilute THF solution was determined using the Lambert–Beer law and was utilized to calculate the number of dendrons per copolymer, i.e., the degree of polymerization, l, of a PD_3MA block (Table 1). From the elemental analysis, the C-to-H molar fraction of the

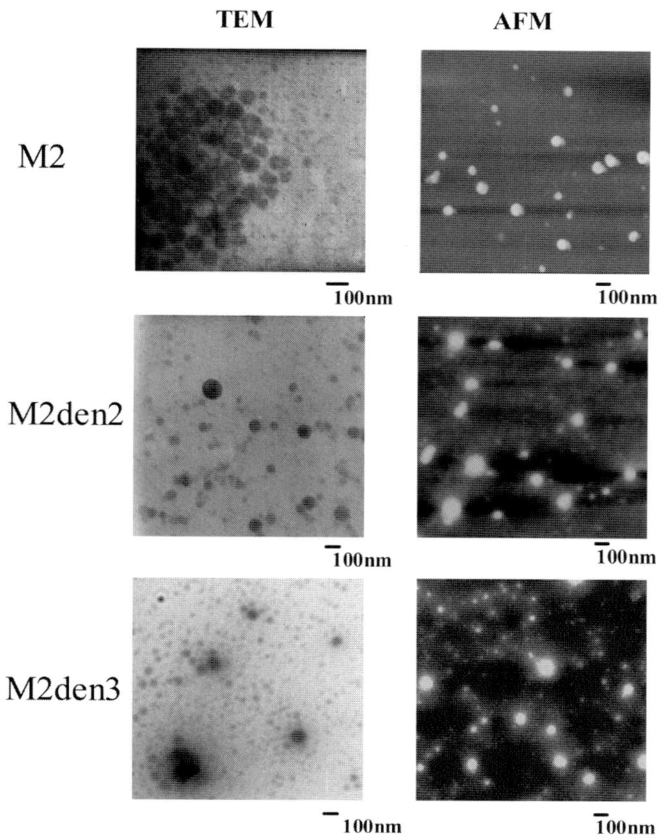

Fig. 2 Transmission electron microscope (*TEM*) and atomic force microscope (*AFM*) images of films cast from chloroform solutions of block copolymers

copolymers was obtained (Table 1). From the values of M_n and the C-to-H molar ratio, the degrees of polymerization of PMAA and PFA were calculated, as listed in Table 1. Consequently, the dendron was replaced in 3/58 portions of the PMAA block in M2den1, 2/3 in M2den2 and 28/30 in M2den3. The rich replacement to the dendron made us expect an increase in the rigidity of the copolymer structure because of the compactness of the dendron side chain in addition to the rigidity of the fluorocarbon side chain. Estimated molecular lengths are included in Table 1. The backbone main chains of the copolymers cannot take on a trans-zigzag configuration because of steric hindrance of the side chains, i.e., bulky dendron and rigid PFA, but in a trans–gauche-type configuration, the distance between side chains is wider than the thickness or the diameter of the side chains. Therefore, when the chain structures of the copolymers were approximated by the trans–gauche-type configuration, the molecular length [P(D_3MA,MAA) block length a plus PFA block length b] was estimated by multiplying the degree of polymerization by 2.17 Å, which is one pitch length of a repeating unit. Such a configuration of the main chain was already estimated for PMMA-*b*-PFMA [22]. As seen in Table 1, the evaluated lengths of the copolymers are only 8–13 nm.

Association in solution and at the interface

The intensity of the UV–vis band (212 nm) of THF solutions of P(D_3MA,MAA)-*b*-PFA deviates from the Lambert–Beer law at higher concentrations, indicating the formation of aggregates. The association of copolymers was examined in chloroform. The TEM photographs of cast films prepared from chloroform solutions of PMMA-*b*-PFA (M2) and P(D_3MA,MAA)-*b*-PFA (M2den2, M2den3) are shown in Fig. 2. The three copolymers all formed polydisperse, spherical particles. Similar textures were also obtained from AFM images (Fig. 2). This suggests that the association behavior of the copolymers in chloroform is similar and is not affected by the substitution with the dendron. Since the three

Fig. 3 Surface pressure–molecular area isotherms of Langmuir films of block copolymers

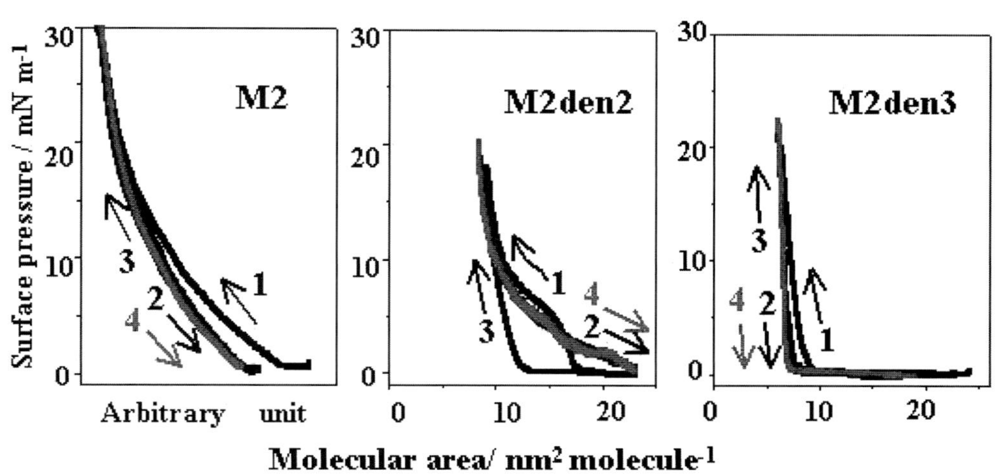

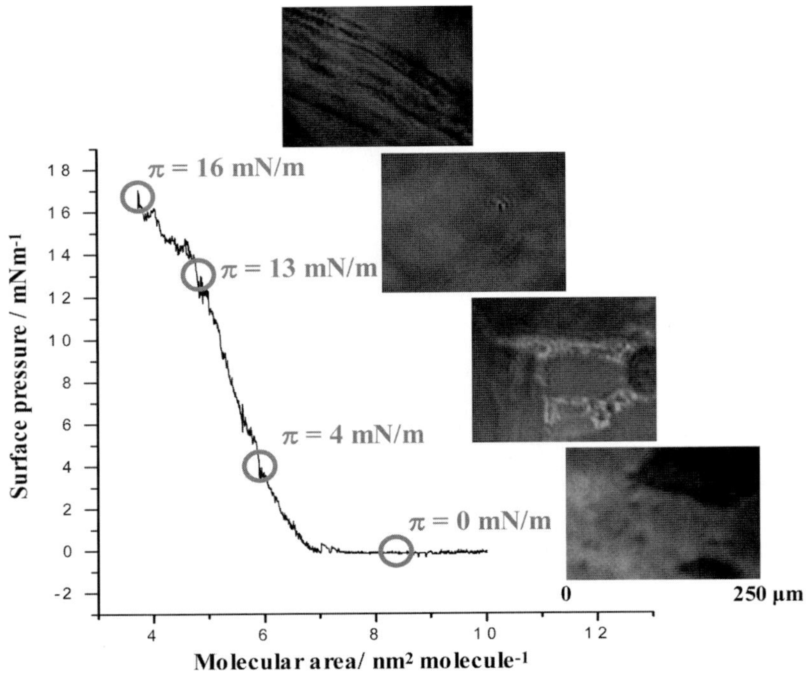

Fig. 4 Brewster-angle microscope images of an M2den3 Langmuir film at different surface pressures

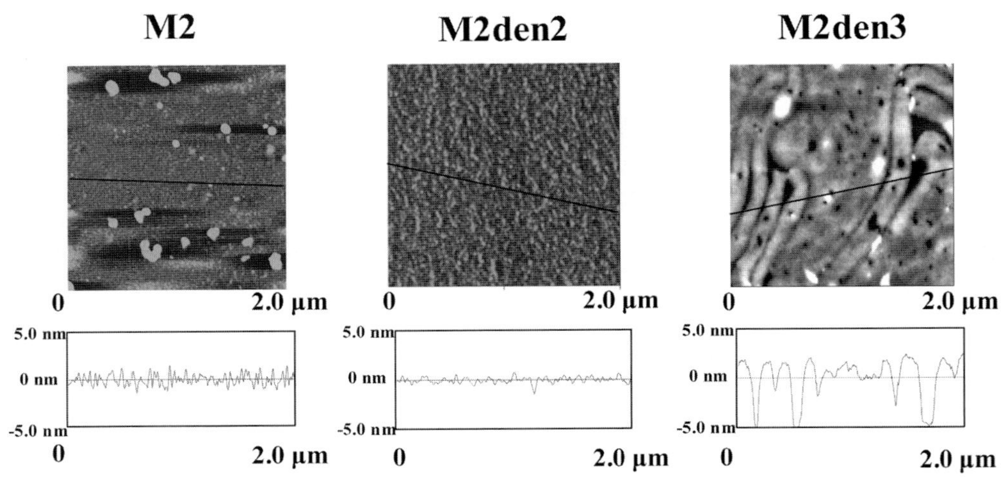

Fig. 5 AFM images of Z-type Langmuir–Blodgett films of block copolymers and their section analyses

copolymers consisted of solvophobic and solvophilic blocks, they all can easily form amphiphilic aggregates like micelles.

The surface area of the Langmuir film of the copolymer at the air/water interface was constantly compressed at a rate of 5 mm min^{-1}. After the liquid and solid states had been observed, the Langmuir films collapsed at a certain surface pressure, which changes with the degree of substitution of the dendron, although the collapse pressure for each copolymer changes with the width owing to the worse reproducibility of the surface pressure–area isotherms, as listed in Table 1. The hysteresis curves of the surface pressure–area isotherms of the copolymers are shown in Fig. 3. On the Langmuir films of M2 and M2den3, the isotherms after the first expansion curve passed on the same curves. This indicates that the molecular orientation at the first compression remains in the following processes. Remarkable hysteresis was observed on the isotherms of an M2den2 film. They were different not only for the compression and expansion processes of the first turn but also for the compression processes of the first and second rounds. This hysteresis behavior suggests that the molecular orientation in the film arranged at the first compression is somewhat released at the expansion and that the more compact molecular orientation at the following compression is also released in the next expansion process. At the higher turn, the area for the liquid state is diminished, although the surface area for the solid state remains constant. This behavior should be a characteristic of a moderately dendron substituted block copolymer.

The molecular areas occupied for liquid and solid states on the second compression scan are listed in Table 1. The occupied molecular area of M2 was not determined, because M2 slightly dissolved in chloroform and the molecular area was not accurately determined. Since the occupied area of the solid state on the M2den2 Langmuir film is close to the expected area [the cross-sectional area (about 12.6 nm^2) of the $P(D_3MA,MAA)$ block], it seems to be a reasonable value for the copolymer orientation, where the molecular axes orient with almost vertical close packing at the air/water interface. The occupied area of the M2den3 film is smaller than the expected area, if the axes of the molecules are almost vertically arranged at the interface. This suggests the overlap of molecules or the interdigitation of side chains at the interface.

Brewster-angle microscope images of an M2den3 Langmuir film at the first compression scan are shown in Fig. 4. At a surface pressure of 0 mN m^{-1}, many domains of different sizes spread on the water subphase. As the surface pressure was increased, the separated domains gathered and combined. The film had some defects at 4 mN m^{-1} but grew to be homogenous at 13 mN m^{-1}, and then collapsed at 16 mN m^{-1}. AFM images of Z-type LB films are shown in Fig. 5. The height roughness of the LB film was about 2 nm for M2 and 1 nm for M2den2, which are only 15% and 10% of a molecular length, respectively, indicating the flat surface of the Z-type LB film. The AFM image of the M2den3 film was very different from that of the other two copolymers. M2den3 formed the zone structure, where there were some cracks on and between the zones. The height of the zone was about 7 or 8 nm, which corresponds to the molecular length.

References

1. Miyamoto M, Aoi K, Saegusa T (1989) Macromolecules 22:3540–3543
2. Iyengar DR, Perutz SM, Dai C-A, Ober CK, Kramer EJ (1996) Macromolecules 29:1229–1234
3. Zhang Y, Wu C, Fang Q, Zhang Y-X (1996) Macromolecules 29:2494–2497
4. Kassis CM, Steehler JK, Betts DE, Guan Z, Romack TJ, DeSimone JM, Linton RW (1996) Macromolecules 29:3247–3254
5. Krupers MJ, Möller M (1997) Macromol Chem Phys 198:2163–2179
6. Thomas RR, Anton DR, Graham WF, Darmon MJ, Sauer BB, Stika KM, Swartzfager DG (1997) Macromolecules 30:2883–2890
7. Petit F, Iliopoulos I, Audebert R, Szönyi S (1997) Langmuir 13:4229–4233
8. Krupers MJ, Sheiko SS, Möller M (1998) Polym Bull 40:211–217
9. Akhremitchev BB, Mohney BK, Marra KG, Chapman TM, Walker GC (1998) Langmuir 14:3976–3982
10. Thomas RR, Anton DR, Graham WF, Darmon MJ, Stika KM (1998) Macromolecules 31:4595–4604
11. Park IJ, Lee S-B, Choi CK (1998) Macromolecules 31:7555–7558
12. Ishizone T, Sugiyama K, Sakano Y, Mori H, Hirao A, Nakahama S (1998) Polym J 31:983–988
13. Tanaka Y, Hasegawa H, Hashimoto T, Ribbe A, Sugiyama K, Hirao A, Nakahama S (1999) Polym J 31:989–994
14. Matsumoto K, Mazaki H, Nishimura R, Matsuoka H, Yamaoka H (2000) Macromolecules 33:8295–8300
15. Busse K, Kressler J, van Eck D, Höring S (2002) Macromolecules 35:178–184
16. Hillmyer MA, Lodge TP (2002) J Polym Sci Part A Polym Chem 40:1–8
17. Gavelin P, Jannasch P, Furó I, Perersson E, Stilbs P, Topgaard D, Södermann O (2002) Macromolecules 35:5097–5104
18. Imae T (2003) Curr Opin Colloid Interface Sci 8:307–314
19. Matsumoto K, Kubota M, Matsuoka H, Yamaoka H (1999) Macromolecules 32:7122–7127
20. Böker A, Reihs K, Wang J, Stadler R, Ober CK (2000) Macromolecules 33:1310
21. Arnold ME, Nagai K, Spontak RJ, Rreeman BD, Leroux D, Betts DE, DeSimone FM, DiGiano FA, Stebbins CK, Linton RW (2002) Macromolecules 35:3697–3707
22. Imae T, Tabuchi H, Funayama K, Sato A, Nakamura T, Amaya N (2000) Colloids Surf A 167:73–81
23. Ito H, Imae T (2004) J Colloid Interface Sci (in press)
24. Kaneko T, Horie T, Asano M, Aoki T, Oikawa E (1997) Macromolecules 30:3118–3121
25. Birol K, Wilhelm C, Katrin G, Wolfram S, Schlüter AD (1997) J Am Chem Soc 119:3296–3301
26. Bao Z, Amundson KR, Lovinger AJ (1998) Macromolecules 31:8647–8649
27. Bo Z, Rabe JP, Schlüter AD (1999) Angew Chem Int Ed Engl 38:2370–2372
28. Prokhorova SA, Sheiko SS, Ahn CH, Percec V, Möller M (1999) Macromolecules 32:2653–2660
29. Förster S, Neubert I, Schlüter AD, Lindner P (1999) Macromolecules 32:4043–4049
30. Bo Z, Zhang C, Severin N, Rabe JP, Schlüter AD (2000) Macromolecules 33:2688–2694
31. Malenfant PRL, Fréchet JMJ (2000) Macromolecules 33:3634–3640
32. Shu L, Schäfer A, Schlüter AD (2000) Macromolecules 33:4321–4328
33. Ouali N, Méry S, Skoulios A (2000) Macromolecules 33:6185–6193
34. Prokhorova SA, Sheiko SS, Mourran A, Azumi R, Beginn U, Zipp G, Ahn CH, Holerca MN, Percec V, Möller M (2000) Langmuir 16:6862–6867
35. (a) Sashiwa H, Shigemasa Y, Roy R (2000) Macromolecules 33:6913–6915; (b) Sashiwa H, Shigemasa Y, Roy R (2001) Macromolecules 34:3211–3214; (c) Sashiwa H, Shigemasa Y, Roy R (2001) Macromolecules 34:3905–3909
36. Marsitzky D, Vestberg R, Blainey P, Tang BT, Hawker CJ, Carter KR (2001) J Am Chem Soc 123:6965–6972
37. Andreopoulou AK, Kallitsis JK (2002) Macromolecules 35:5808—5815

Progr Colloid Polym Sci (2004) 128: 74–80
DOI 10.1007/b97153

Carlos A.R. Costa
Carlos A.P. Leite
Lay-Theng Lee
Fernando Galembeck

Control of nanoparticle organization in two dimensions by adsorbed polymer layers

C.A.R. Costa · C.A.P. Leite
L.-T. Lee (✉) · F. Galembeck
Instituto de Química, Instituto do Milênio de Materiais Complexos, Universidade Estadual de Campinas, CP 6154, 13083-970 Campinas, SP, Brazil
e-mail: laytheng@iqm.unicamp.br
Tel.: +55-19-37883014
Fax: +55-19-37883023

Abstract When a dilute dispersion of relatively monodisperse silica particles is dried on a solid substrate, close-packed arrays are formed. The ordering of the particles results from attractive capillary forces between neighboring particles, and convective transport of particles to the ordered domain during drying. In the presence of adsorbed polymer layers on the particles, the close-packed arrays are disrupted and the particles are disperse upon drying. This is due to steric repulsions of adsorbed polymer layers that prevent close contact of particles during drying, decreasing the probability for formation of the nucleus phase for close-packed ordering. Using electron spectroscopy imaging in a transmission electron microscope, we directly observed the adsorbed polymer layers on the dry dispersion. The thickness of the adsorbed layer and its effect on particle dispersion are investigated as a function of polymer concentration, chain length and solvent quality.

Keywords Stöber silica · Poly(*N*-isopropylacrylamide) · Adsorbed polymer layers · Electron spectroscopy imaging in a transmission electron microscope

Introduction

Uniform fine particles are often used as model systems in the study of adsorption, catalysis and size-dependent solid-state properties such as quantum confinement [1], film formation and particle self-arraying [2, 3]. Many years ago, Stöber et al. [4] reported an elegant method for the preparation of monodisperse spherical silica particles with sizes covering almost the entire colloidal range, by tetraethylorthosilicate (TEOS) hydrolysis, in an ethanolic medium in the presence of ammonia. Since this pioneering work, Stöber silica particles have been used as model colloids in a large number of experimental investigations [5–8].

When a dilute dispersion of silica particles is dried on a solid substrate, two-dimensional close-packed arrays are formed. These close-packed domains result from self-assembly of particles, a process that has been described by a two-stage mechanism [9]: (1) formation of a nucleus of the ordered phase, and (2) convective transport of particles toward the ordered domain. The formation of a nucleus takes place when the liquid level falls below the particle height. If neighboring particles are close together, deformations of their menisci result in capillary attraction of the particles, forming a nucleus cluster. Once the nucleus is formed, further evaporation within the nucleus increases the local capillary suction, resulting in convective transport of particles to the organized phase. Therefore, if open arrays of particles are desired, some degree of control over the self-assembly process is required. In this study, we examine the effects of adsorbed polymer layers on the organization of a submonolayer of silica particles upon drying. A direct observation of the adsorbed polymer layer is carried out using bright-field and electron spectroscopy imaging in a transmission electron microscope. The influence of polymer concentration, chain length and solvent quality on the adsorbed layer thickness is studied.

Experimental

Preparation of silica particles

Silica nanoparticles were prepared by the method of Stöber [10, 11]. TEOS, absolute ethanol used as the solvent and ammonium hydroxide were of reagent grade. Glassware was cleaned with 10% hydrogen chloride and rinsed with distilled water and absolute ethanol. TEOS (4 mL) were added to ethanol (50 mL) in a screw-cap vial in the presence of NH_4OH (3.5 mL). The mixture was placed in a water bath at 36 ± 0.1 °C under ultrasonic vibration (25 kHz, 200 W) for 120 min. The solid content from this synthesis procedure was determined gravimetrically by drying a dispersion at 80 °C until a constant weight was obtained [12]. The solid content was determined to be 2.5% by weight.

Characterization of particle size

Photon correlation spectroscopy

The effective diameter of the silica particles was measured using a ZetaPlus instrument (Brookhaven Instruments) with Bi-MAS software and a solid-state laser (15 mW, $\lambda = 670$ nm) as the radiation source. Dilute dispersions of the sample in distilled water and in absolute ethanol (100 μL in 2.5 mL in 3-mL dust-free acrylic cuvettes) were used for the measurements. Zeta-potential measurements were carried out in 10^{-3} M KCl. The data reported are the averages of six individual runs on two sample aliquots.

Atomic force microscopy imaging

The diameter of the dry silica particles was also determined by microscopy techniques. About 15 μL diluted silica dispersion was deposited on freshly cleaved mica and allowed to dry under atmospheric conditions. Noncontact atomic force microscopy (AFM) images were acquired using a TopoMetrix Discoverer instrument. The cantilever tip was oscillated mechanically to match the natural frequency of the system. The tip was lowered to 10 nm from the sample surface. The mechanical oscillation of the tip was tracked by a four-quadrant photodetector and analyzed by a feedback loop to control the distance between the tip and the sample surface, while scanning the sample at constant oscillation amplitude. The topographic image was built pixel by pixel (300 × 300) using the voltage fed to the z piezoelectric ceramic by the loop.

Field-emission scanning electron microscopy imaging

The sample was prepared as for AFM imaging, but with the mica substrate glued to the flat surface of the brass sample holder. Dry samples were coated with a thin carbon layer. A JEOL JSM-6370F field-emission scanning electron microscope was used. The nominal resolution of this instrument is 3 nm, when operated at 15 kV using a 70-μm aperture.

Elemental mapping

Energy-filtered analytical transmission electron microscopy (TEM) imaging was employed. A Carl Zeiss CEM 902 transmission electron microscope (80 KeV) equipped with a Castaing-Henry energy filter spectrometer and a Proscan Slow Scan CCD camera were used. The setup was controlled by a microcomputer running the AnalySis 3.0 system. The spectrometer uses inelastically scattered electrons to form energy-loss and element-specific images [13, 14]. When an electron beam passes through a sample, interactions of different elements with the electrons result in characteristic energy losses. A prism–mirror system deflects electrons with different energies to different angles so that electrons with a well-defined energy can be selected to create the electron energy-loss spectra. If elastically scattered electrons are selected ($\Delta E = 0$ eV) a transmission image with reduced chromatic aberration is obtained. When monochromatic inelastically scattered electrons are selected, electron spectroscopic images are formed, where the contrast is dependent on the local energy-loss spectrum and thus on the concentration fluctuations of the selected element. Bright areas in the elemental distribution map correspond to element-rich domains. The following procedure was used to acquire spectral images: a set of 38–42 images was acquired around the absorption border for the element of interest. The energy window used was 6 eV, and the energy step between images was 2.5 eV. This image set was used to define the three energy windows used for elemental mapping. In this three-window technique [15], two images were recorded at energy windows below the absorption threshold; these images were used to form the background image by extrapolation with a chosen function. A third image was obtained using an energy window set at the absorption band. The elemental map was obtained by subtracting the background from this third image. The R-map macro from the AnalySis software was used to check for signal saturation. Each elemental map was validated by three independent checks: (1) contrast inversion in the plasmon region; (2) spectral verification, and (3) absence of signal saturation.

In this study, owing to the high concentration of carbon in the polymer molecule, the energy-loss imaging for carbon, acquired at 284 ± 8 eV was used to map the spatial distribution of the polymer. One drop (about 15 μL) of the silica dispersion was deposited on carbon-coated parlodion films supported on 400-mesh copper grids (Ted Pella) and allowed to dry under controlled temperature. To ensure that the particles were not excessively thick, they were first observed at $\Delta E = 0$ eV, and then at $\Delta E = 15$–50 eV. Image-contrast inversion was always obtained, showing significant transmission of electrons through the particles (W. Probst, LEO-Zeiss Elektronenmikroskopie GmbH, private communication). This observation is understood by considering that the mean free path for 80-keV electrons in silica particles is greater than 80 nm for elastic scattering [16], and an estimated few hundred nanometers for inelastic scattering [17]. Image processing was performed with an IBM PC using the Image-Pro Plus 4.0 image analyzer program (Media Cybernetics).

Polymer adsorption

The polymer used was poly(N-isopropylacrylamide) (PNIPAM); the chemical structure is shown in Fig. 1. PNIPAM is a water-soluble polymer that exhibits lower critical solubility temperature (LCST) behavior [18]. Above the LCST, PNIPAM collapses and phase-separates out of water. Owing to its near corporal temperature, many potential applications for PNIPAM have been proposed, including protein conjugates [19], controlled-released of drugs [20, 21] and immobilized-enzyme reactors [22]. The PNIPAM samples used in this study were synthesized using the method described by Schild and Tirrell [23] and described in previous studies [24].

The polymer adsorption procedure is shown in Fig. 2. The silica dispersion was diluted in water and sonicated to disperse the particles. Polymer was added and allowed to adsorb for 30 min at about 20 °C. After adsorption, the sample was further diluted and dried at 20 °C or 40 °C.

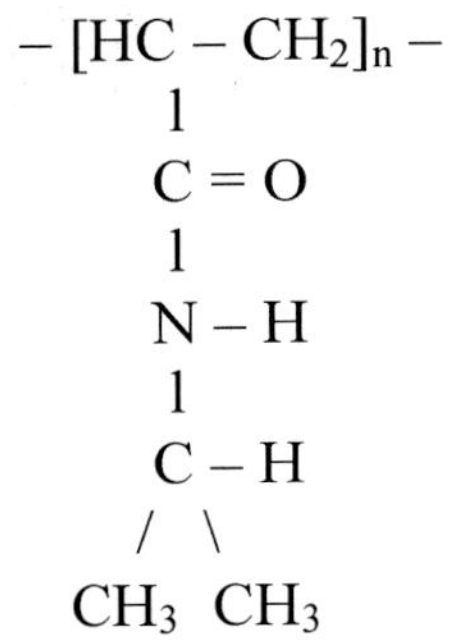

Fig. 1 Chemical structure of poly(*N*-isopropylacrylamide) (*PNIPAM*)

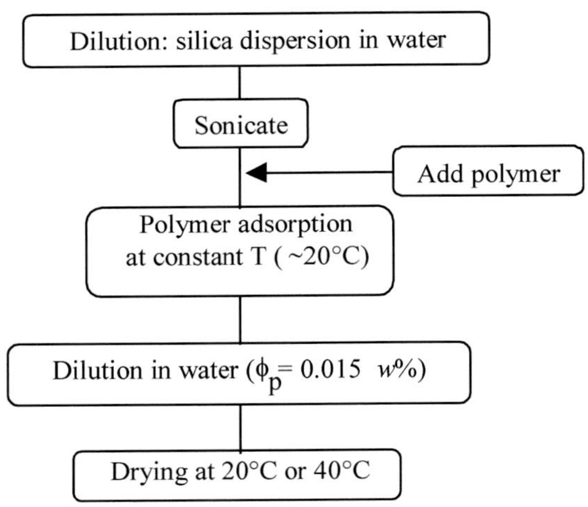

Fig. 2 PNIPAM adsorption procedure

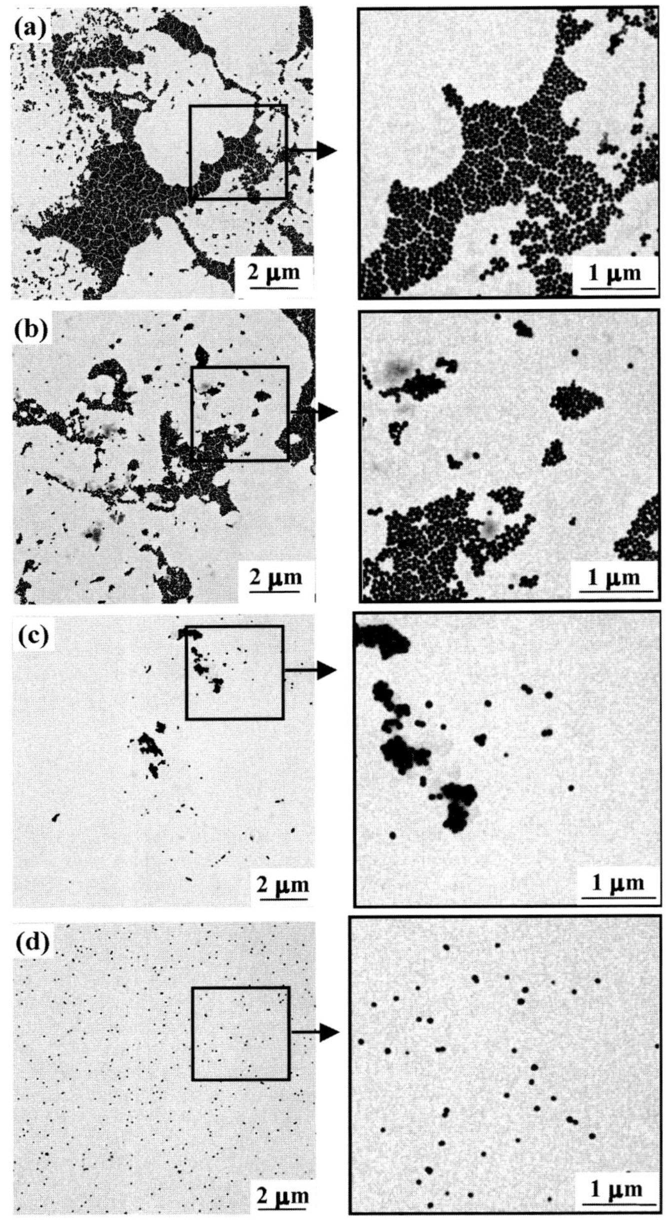

Fig. 3 Bright field images of submonolayers of silica particles dried at 20 °C in **a** water alone, and in PNIPAM at **b** 10^{-5} g mL^{-1}, **c** 10^{-4} g mL^{-1}, and **d** 10^{-3} g mL^{-1} ($M_w = 90{,}000$) The images on the *right* are enlargements of the areas indicated

Results and discussion

The silica particle diameter determined by various techniques and the zeta potential are given in Table 1. These results show that the particle diameter determined by photon correlation spectroscopy (PCS) is significantly larger than those determined by AFM, field-emission scanning electron microscopy (SEM) and TEM, suggesting considerable particle swelling. The effect of particle hydration under atmospheric conditions can also be seen from the difference in the results between AFM and SEM and TEM. The higher value obtained from SEM compared with that obtained from TEM can be attributed to the carbon coating.

Table 1 Particle size and zeta potential of silica particles

Sample environment during the measurement	Technique	Diameter (nm)	Zeta potential (mV)
Water, 25 °C	Photon correlation spectroscopy	103 ± 1	
Ethanol, 25 °C	Photon correlation spectroscopy	99 ± 1	
Air, 55% relative humidity, 25 °C	Atomic force microscopy	81 ± 8	
10^{-6} mbar	field-emission scanning electron micropscopy	79 ± 7	
10^{-6} mbar	Transmission electron microscopy	76 ± 7	
10^{-3} M KCl			–46

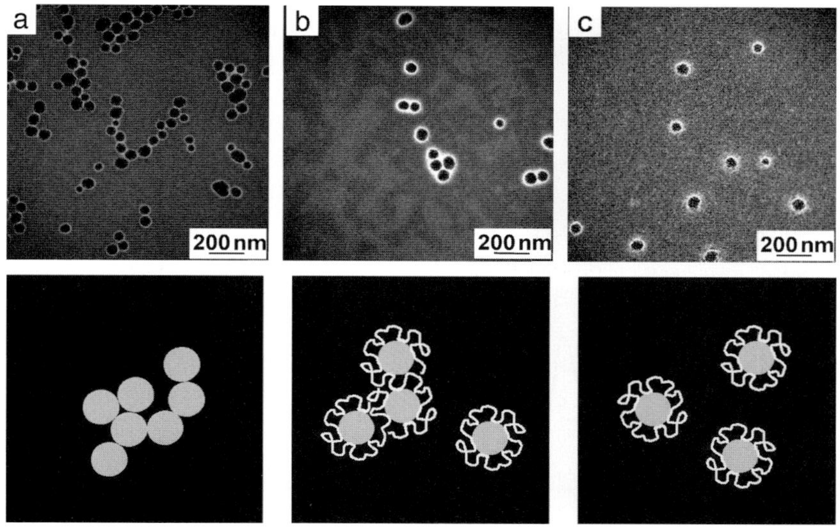

Fig. 4 Carbon maps of silica particles dried at 20 °C in **a** water, and in PNIPAM at **b** 10^{-4} g ml^{-1} and (**c**) 10^{-3} g mL^{-1} ($M_w = 90{,}000$)

TEM bright-field images of silica particles dried at 20 °C from a dispersion in pure water and in the presence of PNIPAM ($M_w = 90$ kg mol^{-1}) at different concentrations are shown in Fig. 3. The polymer concentration mentioned in the following discussion is the concentration during the adsorption step. In the absence of polymer, large regions of close-packed arrays of silica particles are formed. In the presence of low concentrations of polymer (10^{-5} g mL^{-1}), no significant effect on the particle ordering is seen. With increased polymer concentration however, a decrease in the size of the ordered domains and separation of particles are observed. At 10^{-3} g mL^{-1} polymer, an open array of dispersed particles is formed on the dry film.

Elemental maps for carbon acquired at 284 ± 8 eV energy loss are shown in Fig. 4. In these images, the brightness is proportional to the concentration. For all samples, a bright ring can be seen around the silica particles, indicating a higher concentration of carbon on the particle surface. In the absence of polymer, the bright ring is due to residual carbon from TEOS, used in the particle synthesis. In the presence of polymer, the rings are significantly thicker and more intense. Furthermore, the thickness of the rings is not uniform around the particles, and does not vary much with polymer concentration. We attribute these intense rings to adsorbed polymer layers. Accordingly, the thickness of this adsorbed layer increases with the molecular weight of the polymer, as seen from the enhanced thickness of the bright ring around the particle (Fig. 5). From the gray level intensity strip scan, the thickness of the adsorbed polymer layer can be evaluated, as shown in Fig. 6. These values are given in Table 2. Results obtained from PCS measurements are also shown for comparison and to confirm the presence of an adsorbed polymer layer on the particle.

In both cases, the adsorbed layer thickness increases with molecular weight. As expected, the values deduced from PCS are higher than those obtained from TEM since the former measures swollen layers in water, while the latter measures dry layers. Theoretically, their corresponding slopes (of thickness versus molecular weight) should also differ. Although the slope for the PCS results can be perceived to be higher qualitatively, a quantitative fit to the results is not given here owing to an insufficient number of points and to the imprecision of the results obtained at low molecular weight (90 kg mol^{-1}).

The results show that organization of silica particles upon drying is significantly modified in the presence of

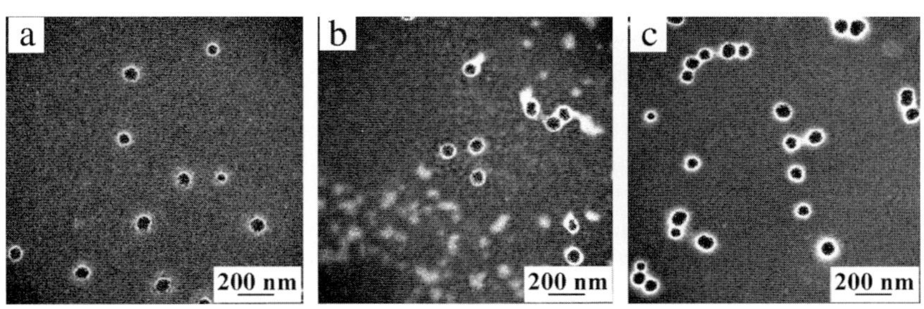

Fig. 5 Carbon maps of silica particles dried at 20 °C in the presence of PNIPAM with different molecular weights: **a** $M_w = 90{,}000$K; **b** $M_w = 190{,}000$; **c** $M_w = 1{,}050{,}000$. Polymer concentration 10^{-3} g mL^{-1}

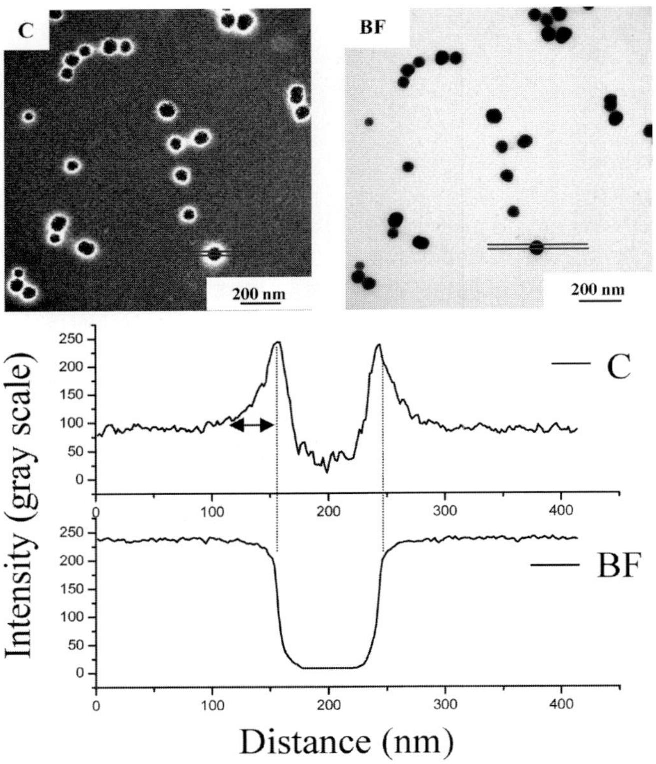

Fig. 6 Gray scale intensity strip scan from carbon and bright-field maps. The *arrow* on the line scan shows the width taken as the thickness of the adsorbed polymer layer

Table 2 Adsorbed polymer layer thickness determined from transmission electron microscopy and photon correlation spectroscopy measurements

M_w (kg mol^{-1})	Adsorbed polymer layer thickness (nm)		
	90	190	1,050
Transmission electron microscopy	10 ± 5	21 ± 5	38 ± 5
Photon correlation spectroscopy	10 ± 2	35 ± 2	129 ± 3

polymer; an increase in the dispersion of the dry particles is obtained with an increase in polymer concentration. The carbon elemental maps show clearly and directly that this effect is related to adsorbed polymer layers on the silica particles. These observations reflect the effects of adsorbed polymer layers on the stability of colloidal dispersions: above a critical polymer concentration when complete coverage of the particle surface by adsorbed polymer is obtained, steric repulsions of the adsorbed layers prevent attractive van der Waals interactions between particles [25]. Such stabilization is highly sensitive to solvent quality. Under good solvent condition, adsorbed polymer forms a swollen and extended layer and overlap of polymer segments from these adsorbed layers is energetically unfavorable. Interaction of adsorbed layers is therefore repulsive. This is the case for PNIPAM in water at 20 °C [24]. Under poor solvent conditions, the adsorbed layer is collapsed, and overlap of adsorbed layers reduces the net free energy. The resulting attractive interaction of adsorbed layers causes destabilization of the particles. As can be expected, organization of the dry particles in two dimensions is consequently affected. A comparison of silica particles dried under good (20 °C) and poor (40 °C) solvent conditions is shown in Fig. 7. At 40 °C (above the LCST), water is a poor solvent for PNIPAM, and the adsorbed layer is collapsed [24]. The results in Fig. 7 show a clear transition from disperse phase to an aggregated phase when the drying temperature is increased from 20 to 40 °C. The corresponding carbon distribution maps show a decrease in the thickness of the adsorbed polymer layer, confirming a collapse of the adsorbed polymer layer. Note in this case, where net interaction is attractive, open aggregates and clusters are formed. This is in contrast to close-packed domains that are formed when net interaction is repulsive in the absence of polymer (Fig. 8).

Thus, we have shown that in the presence of adsorbing polymer, nanoparticles of silica are dispersed upon drying. That this effect is due to the adsorbed polymer layer is demonstrated by the effects of polymer concentration, polymer molecular weight and solvent quality. However, although the adsorbed polymer layer clearly plays a critical role, an argument based on steric repulsion is necessary but insufficient. Recall that the bare silica particles carry a net negative charge and despite their repulsion produce close-packed structures upon drying. Furthermore, the net Coulombic repulsion of bare silica is more long range, with the double-layer thickness close to 1 μm in distilled water, compared with the range of steric repulsion, which is of the order of the adsorbed layer thickness, about 10–100 nm. Steric repulsion alone, therefore, cannot explain the dispersion of the particles.

A second effect of adsorbed polymer layers is attributed to modification of the silica surface heterogeneity. Indeed, it has been found that nanosize silica particles, in spite of their global negative charge, exhibit surface chemical heterogeneity as evidenced by heterogeneous elemental composition as well as electric potential [10]. The extent of heterogeneity increases with a decrease in the particle size. Such surface heterogeneity within the particles can lead to localized heterogeneous aggregation, especially in the late stage of drying when the particle concentration in the drying dispersion increases. This favors nucleus patch formation, the first stage in particle self-assembly. In the presence of an adsorbed polymer layer, the surface heterogeneity is masked, and heterogeneous aggregation is suppressed.

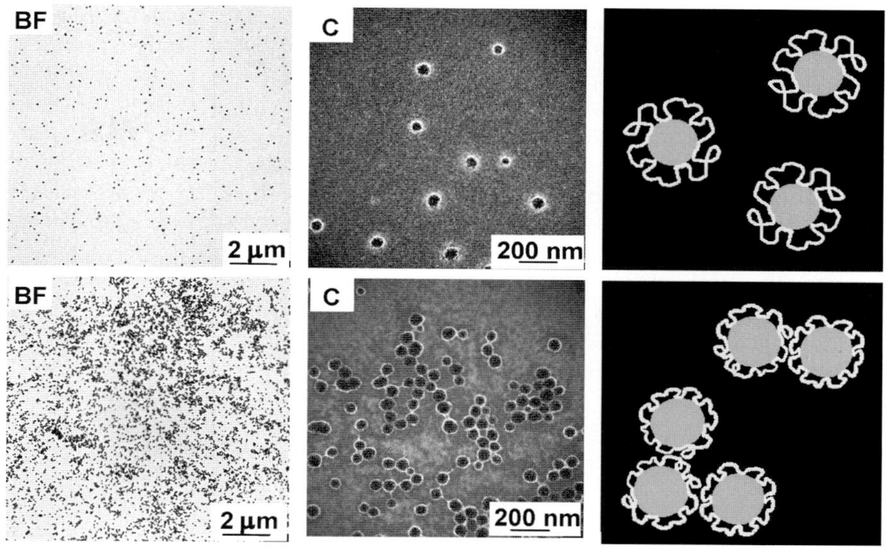

Fig. 7 Bright-field and carbon maps of silica particles dried in the presence of PNIPAM at 20 °C (*top*) and at 40 °C (*bottom*). Polymer concentration 10^{-3} g mL^{-1}, $M_w = 90{,}000$

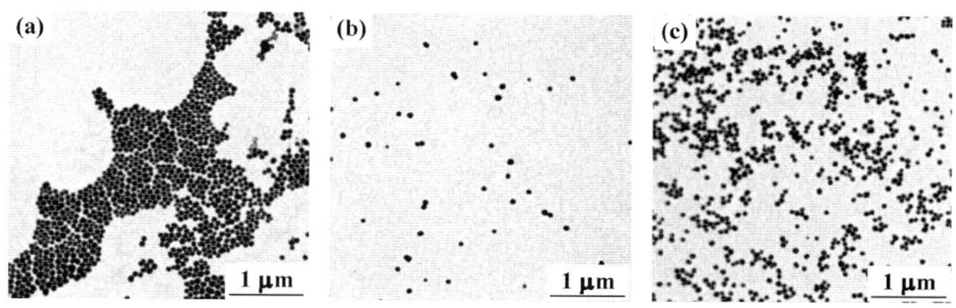

Fig. 8 Bright-field images of silica particles dried in **a** water alone, and in PNIPAM at **b** 20 °C and **c** 40 °C. Polymer concentration 10^{-3} g mL^{-1}, $M_w = 90{,}000$

Thus, the role of adsorbed polymer layers is twofold: masking surface chemical heterogeneity of the silica nanoparticles, and steric repulsion of the adsorbed layers preventing close contact of the particles. Work is currently underway in our laboratory to isolate these two effects, by using other types of adsorbates, such as short-chain molecules, on the silica particles.

The effect of polymer interaction with the solid substrate also needs to be considered. In the present system, hydrophobic interaction of the polymer with the substrate, in this case, carbon-coated parlodion film, is possible. Attractive interaction of the polymer layer and the substrate can lead to adhesion of the particle to the surface, resulting in reduction of particle mobility. Lateral movement of the particle in nucleus formation and in convective transport may thus be impeded. However, in the nucleus formation stage, the lateral capillary interaction energy is extremely high, calculated to be of the order of 10^4 kT for particles of about 100 nm [9]. Particle–substrate interaction therefore affects only the convective stage of particle organization, while the nucleus formation stage remains unaffected. Therefore, if particle adhesion to the substrate (via the adsorbed polymer layer) were the only mechanism responsible for dispersion of the particles upon drying, nucleus clusters with limited aggregation numbers should be obtained. This is not the case for a polymer in a good solvent (20 °C, Fig. 7, top) where the particles are completely disperse. Thus, the critical role of the adsorbed polymer layer on the nucleus formation stage, as described in the previous paragraph, is clear. For the case of a poor solvent, on the other hand, the effect of particle adhesion appears to be more significant. Here, open aggregates and clusters are formed owing to attraction of the collapsed polymer layers under poor solvent conditions (at 40 °C, Fig. 7, bottom). Is spite of the presence of clusters, the second stage of self-assembly – convective transport of particles to build up the cluster or aggregate size – does not appear to be operative. This may be explained by an enhanced particle-substrate adhesion due to increased hydrophobicity of the collapsed polymer, and thus an increased polymer–substrate interaction.

Conclusions

Analytical electron spectroscopy imaging has been shown to be a powerful tool for imaging the distribution of elements with high spatial resolution. This technique is therefore well adapted for the study of two-dimensional organization of nanosize particles and the influence of adsorbed polymer layers. In this study, clear and direct observations of adsorbed polymer layers on dry silica particles have been made and these results correlated with the organization of the particles. It was found that under high surface coverage and good solvent conditions, the silica particles are completely disperse upon drying, in contrast to close-packed arrays that are formed by self-assembly of particles when dried in the absence of polymer. This result is attributed to masking of microheterogeneity of the surface of the silica nanoparticles and steric repulsion of adsorbed polymer layers during the drying process. A decrease in solvent quality, induced by an increase in temperature for the thermosensitive polymer used in this study, results in collapse of the adsorbed layer. Under this condition, loose aggregates of particles are formed upon drying. Thus, close-packed arrays, disperse particles and loose aggregates can be formed, depending on the polymer concentration and the solvent quality.

Acknowledgements The authors thank FAPESP, Pronex/Finep/MCT and CNPq for financial support. This is a contribution from the Millennium Institute for Complex Materials.

References

1. Henglein A (1989) Chem Rev 89:1861
2. Averback RS, Höfler HJ, Tao R (1993) Mater Sci Eng 166:169
3. Galembeck F, Lima ECO, Masson NC, Monteiro VAR, Souza EF (1996) In: Pelizzetti E (ed) Fine particles science and technology: from micro to nanoparticles. Kluwer, Dordrecht, pp 267–279
4. Stöber W, Fink A, Bohn E (1968) J Colloid Interface Sci 26:62
5. Van Helden AK, Vrij A (1980) J Colloid Interface Sci 78:312
6. Kirkland JJ (1979) J Chromatogr 185:273
7. Kops-Werkhoven MM, Fijnaut HM (1981) J Chem Phys 74:1618
8. Tan CG, Bowen BD, Epstein N (1987) J Colloid Interface Sci 118:290
9. Denkov ND, Velev OD, Kralchevsky PA, Ivanov IB, Yoshimura H, Nagayama K (1992) Langmuir 8:3183
10. Costa CAR, Leite CAP, Souza EF, Galembeck F (2001) Langmuir 17:189
11. Leite CAP, Souza EF, Galembeck F (2001) J Braz Chem Soc 12:519
12. Costa CAR, Leite CAP, Galembeck F (2003) J Phys Chem B 107:4747
13. Egerton RF (1986) Electron energy-loss spectroscopy in the electron microscope. Plenum, New York
14. Joy DC (1986) Principles of analytical electron microscopy. Plenum, New York
15. Reimer L, Zepke U, Moesch J, Schulze-Hillert St, Ross-Messemer M, Probst W, Weimer E (1992) EELS spectroscopy: a reference handbook of standard data for identification and interpretation of electron energy loss spectra and for generation of electron spectroscopic images, Carl Zeiss, Oberkochen
16. Newbury DE (1986) In: Joy DC, Romig AD Jr, Goldstein JL (eds) Principles of analytical electron microscopy. Plenum, New York, p 8
17. Newbury DE (1986) In: Joy DC, Romig AD Jr, Goldstein JL (eds) Principles of analytical electron microscopy. Plenum, New York, p 20
18. Schild HG (1992) Prog Polym Sci 17 16:3
19. Chen GH, Hoffman AH (1993) Bioconjugate Chem 4:509
20. Yoshida R, Uchida K, Kaneko Y, Sakai K, Kikuchi A, Sakurai Y, Okano T (1995) Nature 374:240
21. Chu LY, Park SH, Yamaguchi T, Nakao SI (2001) J Membr Sci 192:27
22. Park TG, Hoffman AS (1988) Appl Biochem Biotech 19:1
23. Schild HG, Tirrell DA (1991) Langmuir 7:665
24. Lee LT, Jean B, Menelle A (1999) Langmuir 15:3267
25. Luckham PF (1987) In: Feast WJ, Munro HS (eds) Polymer surfaces and interfaces. Wiley, New York

Progr Colloid Polym Sci (2004) 128: 81–85
DOI 10.1007/b97086

Ney H. Moreira
Munir S. Skaf

Structural characterization of the H_2O/CCl_4 liquid interface using molecular dynamics simulations

N.H. Moreira · M.S. Skaf (✉)
Instituto de Química,
Universidade Estadual de Campinas,
Cx.P. 6154, Campinas,
SP 13083-970, Brazil
e-mail: skaf@iqm.unicamp.br
Fax: +55-19-37883023

Abstract We report a molecular dynamics computer simulation study of the liquid/liquid interface formed by water and carbon tetrachloride (H_2O/CCl_4) at 298 K. The primary focus of the work is the characterization of the structural features of the interface at a molecular level. The computed density profiles for the aqueous and organic phases and the distribution of surface widths indicate a fairly smooth interface approximately 5-Å thick. Analyses of the water structure near and far from the interface show a significant depletion in the average number of hydrogen bonds in the vicinity of the organic phase with respect to the bulk. The average orientation of the water molecules reveals that close to the interface the dipole moments tend to align parallel to the interfacial plane, while the vector joining the hydrogen atoms points slightly away from this plane. These results are discussed in the light of recently reported nonlinear spectroscopy measurements.

Keywords Liquid interfaces · Molecular dynamics simulations · Water/CCl_4 microstructure

Introduction

The heterogeneous nature of interfacial liquid/liquid systems has long been a subject of keen interest to scientists and has motivated widespread studies about microphysical behavior at interfaces [1–5]. At molecular scales, the system's broken symmetry imposes a nonconventional ambient in the immediate vicinity of the interface, resulting in molecular behavior that can be much different from that found in the bulk liquid phase [4]. Liquid interfaces also play an important role in biological systems, such as membranes and protein outer surfaces [6], and in colloid sciences, which very often involve systems of great practical and technological importance [7].

In recent years, new approaches based on nonlinear spectroscopic methods [3, 4] and neutron reflection techniques [5] have provided important insights and detailed information about the structural and dynamical behavior of molecules at interfaces. While a great deal of our current knowledge on interfacial systems has been obtained from measurements of thermodynamic macroscopic properties, nonlinear spectroscopic techniques such as second harmonic generation [3] and vibrational sum frequency (VSF) [4] have made it possible to investigate the behavior of molecules located in the vicinity of the interface without the interfering signal from molecules situated in the bulk.

Parallel to the experimental advances, theoretical approaches based on molecular dynamics (MD) computer simulations have contributed significantly towards our understanding of the microscopic nature of liquid interfaces and molecular interactions near surfaces. Several MD studies have been reported recently on interfacial systems, most notably in the presence of an aqueous phase including water/vapor interfaces [8], water/dichloroethane [9], water/carbon tetrachloride [10], water/supercritical CO_2 [11], and also reverse micellar systems [12]. Because of the very nature of the MD computational technique, these works provide a privileged look at the microscopic features of the interfacial systems, which are extremely rich in molecular details.

Of particular interest, the H_2O/CCl_4 interface emerges as an important prototype for hydrophobic interacting systems that grossly mimic biological membranes, and as a natural support media for surfactant systems in molecular simulation studies [13]. The solubility of CCl_4 in water at 298 K is only 2.21×10^{-2} g/dm^3. This liquid/liquid interface is particularly appealing to MD studies mainly because of the structural simplicity of CCl_4 molecules, which simplifies substantially structural analyses in computational studies. Moreover, the CCl_4 molecules are quite insensitive to IR radiation absorption, eliminating possible interferences in nonlinear spectroscopy of interfacial water molecules [4].

In this work, we present results of extensive MD simulations for the H_2O/CCl_4 liquid/liquid interface aimed at elucidating the general structural features of this interface at a microscopic level. The analyses reported here are focused on the density profiles of the liquids, the characterization of the interface roughness, and the intermolecular structural properties of water as functions of the distance from the interface. Our results indicate, for instance, that closest to the organic phase, the average orientation of the water molecules is such that the molecular plane nearly parallels the plane that defines the interface, with one of the hydrogens pointing slightly to the aqueous phase, while the other points preferentially towards the organic phase. These results are consistent with recent VSF spectroscopy data which indicate that near the interface there is a substantial fraction of water molecules with "free" OH bonds [4].

The remainder of this work organized as follows. First we describe the interaction potentials and computational details. Then we present and discuss our results. Finally, we present our concluding remarks.

Models and simulation details

The intermolecular interactions are described in terms of nonpolarizable, classical effective potentials given by the sum of short-range Lennard-Jones potentials plus Coulomb electrostatic terms between all pairs of atoms belonging to distinct molecules. In addition, molecular internal degrees of freedom are also taken into account by considering bond stretching and angle deformation potentials:

$$V = \sum_{i,j} 4\varepsilon_{ij}\left[(\sigma_{ij}/r_{ij})^{12} - (\sigma_{ij}/r_{ij})^{6}\right] + \sum_{i,j} \frac{q_i q_j}{4\pi\varepsilon_0 r_{ij}} + \sum_{\text{bonds}} K_b(b-b_0)^2 + \sum_{\text{angles}} K_\theta(\theta-\theta_0)^2,$$

where σ_{ij} and ε_{ij} are the Lennard-Jones distance and energy parameters between a pair of atoms i and j of distinct molecules separated by a distance r_{ij} and q_i is the partial atomic charge on atom i. The variables b and θ refer to bond length and angle, respectively. We chose the well-known simple point charge/flexible model [14] for water and the OPLS-AA model for CCl_4 [15]. The model parameters are given in Table 1.

The MD simulations were performed in the *NVE* ensemble on a system of 222 CCl_4 and 1,000 water molecules contained in a box of size $24 \times 24 \times 100$ Å^3, such that the density in each phase corresponded to the experimental data at room temperature. Periodic boundary conditions were applied in all directions and the dimension of the longer z-axis (normal to the interface) ensured that spurious artifacts from the presence of the multiple interfaces were kept negligibly small. No constraints were imposed on the molecules of each species; the interface appeared naturally as a result of the intermolecular forces. The simulations were performed using the TINKER program package [14] with a cutoff of 12 Å on the Lennard-Jones forces, while particle mesh Ewald sums were used to treat long-range Coulombic forces. The equations of motions were solved using the velocity Verlet algorithm [14] with a time step of 1 fs. The initial configuration was built using the molecular packing program PACKMOL [16]. After a long (200-ps) thermalization period, an ensemble of 60 MD trajectories of 10-ps each was collected for analysis. During each run the energy was conserved within 0.2% error.

Results and discussion

General interface structure

The first important characterization of the liquid interface resides in the knowledge of the average density profile of each species as one move across the coexisting phases. The density profiles along the z coordinate with respect to the laboratory frame for the aqueous and organic phases are shown in Fig. 1. Far from the interface, the density of both phases converges to experimental macroscopic values, with some fluctuations around the mean. These fluctuations reflect the fact that the density profile is essentially a local property. By collecting a larger ensemble of trajectories or simulating larger systems, these fluctu-

Table 1 Intramolecular and intermolecular interaction potential parameters for the simple point charge/flexible water [14. and OPLS-AA CCl_4 [14]

Vibrational parameters

Molecule	K_b (kcal/mol Å^2)	b_0 (Å)	K_θ (kcal/ mol degree^2)	θ (degree)
CCl_4	630.0	1.766	49.00	109.50
Water	527.2	1.000	37.95	109.47

Intermolecular parameters

Atom type	σ (Å)	ε (kcal/mol)	q (e)
C	3.410	0.100	0.248
Cl	3.450	0.285	−0.062
O	3.166	0.155	−0.820
H	0	0	0.410

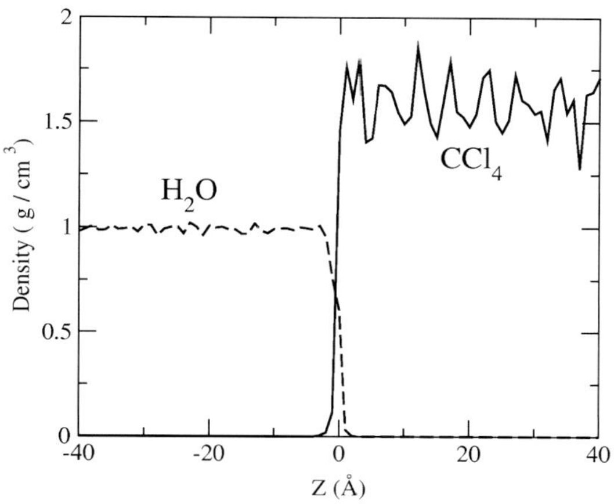

Fig. 1 Water and CCl_4 density profiles across the interface

ations can be made arbitrarily smaller. The interfacial region is characterized by a sharp change in the local densities that take place within a slab of about 5 Å on average centered around $z = 0$. The average width of the interfacial zone is thus a little larger than the molecular diameter of CCl_4. This result is consistent with the low miscibility of these solvents.

A detailed picture of the interface roughness can be obtained from the probability distribution of interface widths. The instantaneous interface width within a cross section A of the xy-plane can be defined as [11]

$$W(A,t) = \left|z_{H_2O}^{max}(A,t) - z_{CCl_4}^{min}(A,t)\right|, \qquad (1)$$

where $z_{H_2O}^{max}(A,t)$ and $z_{CCl_4}^{min}(A,t)$ correspond, respectively, to the largest center-of-mass z coordinate of a water molecule and the smaller z coordinate of a CCl_4 molecule found within a cross section A of the interface at time t. The quantities $z_{H_2O}^{max}(A,t)$ and $z_{CCl_4}^{min}(A,t)$ define the maximum penetration depths of one phase into the other in the parallelepiped of base A. From the MD trajectories, we compute the probability distribution of widths, $P_A(w)$, for different cross sections A, thus yielding valuable information on the interface roughness. Results for $P_A(w)$ are shown in Fig. 2a for different choices of cross-sectional areas, as indicated. For the smallest cross section considered (4.5 × 4.5 Å^2), the distribution peaks at $w \approx 4$ Å, which nearly equals the distance of closest proximity between a pair of water and CCl_4 molecules. For larger cross-sectional areas, $P_A(w)$ shifts towards the origin as expected on the basis of its own definition (Eq. 1). The important point to notice here is the very low fraction of widths larger than about 5 Å, thus suggesting that this interface is essentially sharp.

In order to provide further support, Fig. 2(b) shows the probability distribution, $P_A(\Delta)$, of the interface thickness, $\Delta(A,t) = z_{H_2O}^{max}(A,t) - z_{CCl_4}^{min}(A,t)$ which was computed for the entire interface cross section

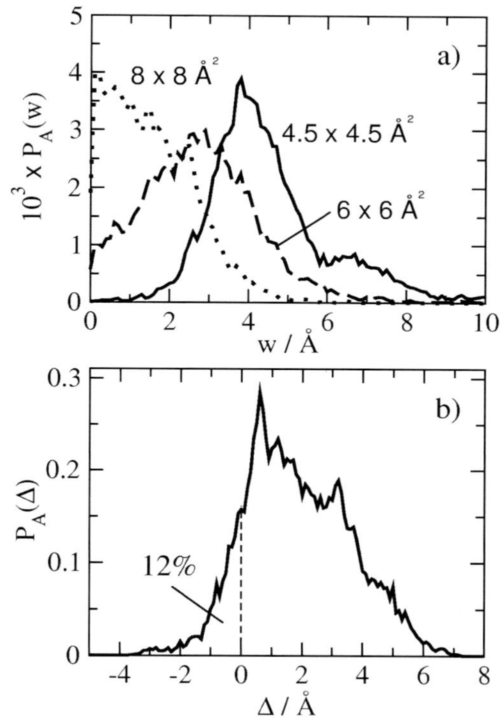

Fig. 2 Interface width probability distribution sampled over different areas of interface cross sections, as indicated. **b** Interface thickness probability distribution sampled over the entire xy-plane (24 × 24 Å^2). The area to the left of the *vertical dashed line* corresponds to the fraction of total configurations in which the surface is flat (i.e., no penetration of molecules of one phase into the other)

(24 × 24 Å^2). Negative values of Δ correspond to situations in which water molecules do not penetrate the organic phase and vice versa. Integrating $P_A(\Delta)$ for all values of $\Delta \leq 0$, we find that 12% of all configurations correspond to a noninterpenetrating, flat interface. Similarly, we also computed the distribution, $P_A(h)$, of the interface position, $h(A,t) = 1/2\left[z_{H_2O}^{max}(A,t) + z_{CCl_4}^{min}(A,t)\right]$ for several choices of A. We find (data not shown) that $P_A(h)$ is symmetrically centered at $z = 0$ and practically vanishes outside the interval $-2 < h(A,t) < 2$ Å for all choices of A, reinforcing the view of a sharp and fairly nonrough interface on molecular scales.

Water local structure

To study the local structure of water molecules near the interface, we must first establish a criterion to decide which molecules are in the interfacial region. Therefore, we invoked the results presented earlier and defined as the interfacial region a slab of the simulation box that goes from $z = -2$ Å to $z = 2$ Å. It is worth mentioning that in the cases of rough interfaces, for instance, in the cases where the interface position fluctuates substantially over the z-axis, a dynamic criterion based on the instantaneous position of interface would be necessary to actually define an interfacial region with respect to the laboratory frame.

The water structure is conveniently described in terms of the intermolecular radial distribution functions, $g(r)$, between the OO, HH, and OH atomic pairs. These functions are depicted in Fig. 3 for molecules located in the bulk phase, in the interfacial region $-2 \leq z \leq 2$, and for an intermediate region between the interface and the bulk, with $-6 \leq z \leq -2$. Figure 3 shows that the positions of the first maxima and minima of the $g(r)$'s remain practically unchanged as one moves from the bulk to the interface region, indicating that the molecular interactions between nearest-neighbor atoms are not significantly altered. The intensity of the first peaks decreases near the interface, pointing to a depletion of coordination numbers. This is particularly noticeable for the HH and OH $g(r)$ functions. The number of hydrogen bonds between water molecules can be estimated by properly integrating the radial distribution function for the OH pairs in spherical coordinates up to the first minimum near 2.5 Å. In the interface region, this estimate yields an average of 2.7 hydrogen bonds per water molecule, in contrast to 4.1 hydrogen bonds for molecules in the bulk. This implies that near the CCl_4 interface water molecules experience weaker restoring forces and, thus, may present dynamical behavior markedly different from that found in the bulk. If fact, recent VSF nonlinear spectroscopy measurements [4] have indicated that the vibrational dynamics of water molecules in the vicinity of a CCl_4 interface is markedly "freer" than in the bulk.

The average orientation of the water molecules with respect to the plane of the interface is an important structural feature that we have explored here. We computed the normalized probability distribution for the angle between the water dipole and the z-axis (normal to the interface), and also for the angle between the z-axis and the vector joining the hydrogen atoms of a given molecule. For clarity, we denote μ_z and HH_z the z components of the unitary dipole and HH vectors, respectively.

The results are depicted in Fig. 4 for bulk (dashed lines) and interfacial (solid lines) water molecules. For comparison, Fig. 4a also shows the distribution of the z components of the dipoles for an intermediate region

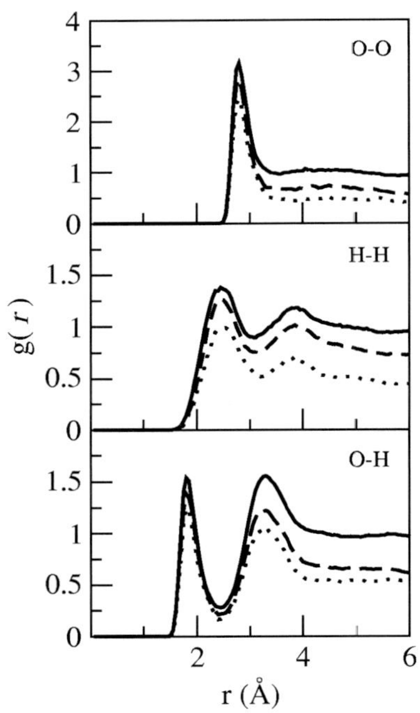

Fig. 3 Radial distribution functions for water molecules in the bulk (*solid lines*), at the interface $-2 \leq z \leq 2$ Å (*dotted lines*), and in an intermediate region $-6 \leq z \leq -2$ Å (*dashed lines*)

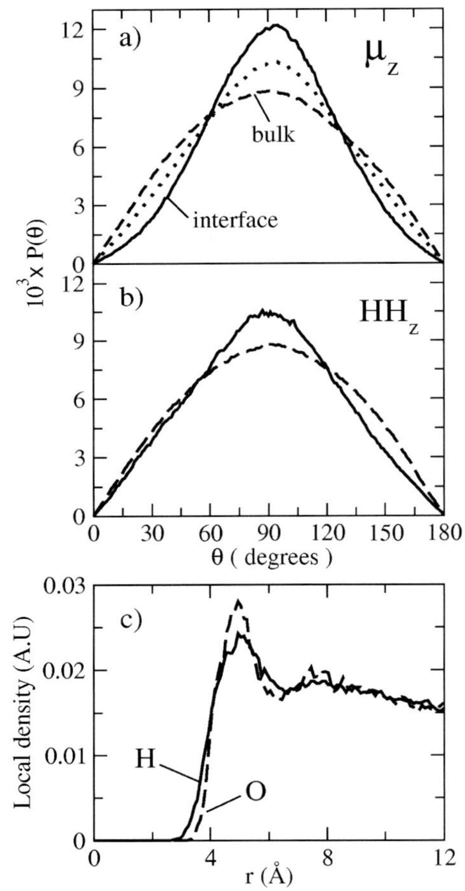

Fig. 4 Probability distributions for the angle between the z-axis (normal to interface) and **a** the unit dipole and **b** HH vectors for bulk and interfacial regions. **c** The local density of hydrogen and oxygen atoms measured from interfacial CCl_4 molecules

between the bulk and the interface (dotted line). For an isotropic environment such as a bulk liquid, the orientation of any molecular vector follows a uniform distribution, which in polar coordinates is simply given by $P(\theta) = (1/2)\sin\theta$. Indeed, the bulk distributions shown in Fig. 4a and b are sinusoidal. As one approaches the interface, there is a relative enhancement in the population of water dipoles oriented parallel to the interface with a concomitant decrease in the probability of finding a water molecule with the main symmetry axis pointing across the interface (Fig. 4a). Similar behavior is observed for the HH vectors (Fig. 4b). However, in this case one notices a slight asymmetry in the distribution, which implies that near the interface the HH vectors are, on average, slightly twisted at an angle from the interface plane. Similar conclusions have been reported for previous MD simulations of this interface [8, 10], although in less detailed molecular terms.

Further support for this observation is provided by the local density of hydrogen and oxygen atoms as measured from the center of masses of the CCl_4 molecules at the interface. The local density for the H (solid line) and O (dashed line) atoms is shown in Fig. 4c in arbitrary units. The local density of H atoms is divided by two, such that at larger distances it coincides with the local density of O atoms. Figure 4c clearly shows that in going from the organic to the aqueous phase one would encounter a water hydrogen first.

Concluding remarks

The water/CCl_4 interface is an important prototype system to study molecular level properties of liquid/liquid surfaces. In this work we performed a series of MD simulations to explore detailed structural features of this interface. We demonstrated, through the local density profiles and probability distributions of the surface width and position, that the interface is rather sharp and smooth on molecular scales.

The nonuniform charge distribution across the interface, reflected in the difference of dielectric constants of the aqueous and organic phases, promotes a preferential orientation of water molecules in the immediate vicinity of the interface, in which the dipoles tend to be parallel to the surface mean plane. The HH vectors of interfacial waters are approximately parallel to the surface, but are slightly twisted such that one of the hydrogens points into the organic phase while the other hydrogen points to neighboring water molecules. Near the interface, the average number of hydrogen bonds per water molecule is only 2.7, indicating the presence of free OH bonds.

Ongoing research is being conducted in order to investigate the intermolecular and intramolecular dynamical features on this interesting liquid/liquid interface.

Acknowledgements The authors thank FAPESP (01/09374–3) and CNPq for financial support of this work.

References

1. Benjamin I (1997) Annu Rev Phys Chem 48:407
2. Benjamin I (1995) Acc Chem Res 28:233
3. Farrer RA, Fourkas JT (2003) Acc Chem Res 36:605
4. (a) Richmond GL (2001) Annu Rev Phys Chem 52:357; (b) Richmond GL (2002) Chem Rev 102:2693
5. (a) Penfold J et al (1997) J Chem Soc Farad Trans 93:3899; (b) Krueger S (2001) Curr Opin Colloid Interface Sci 6:111
6. Tieleman DP, Marrink SJ, Berendsen HJC (1997) Biochim Biophys Acta 1331:235
7. Bourgeat-Lami E (2002) J Nanosci Nanotech 2:1
8. Schweighofer KJ, Essmann U, Berkowitz, ML (1997) J Phys Chem B 101:3793
9. Benjamin I (1992) J Chem Phys 97:1432
10. Chang TM, Dang LX (1998) J Chem Phys 108:818
11. da Rocha SRP, Johnston KP, Westacott RE, Rossky PJ (2001) J Phys Chem B 105:12092
12. Faeder J, Albert MV, Ladanyi BM (2003) Langmuir 19 (6):2514
13. Dominguez H, Smondyrev AM, Berkowitz ML (1999) J Phys Chem B 103:9582
14. Ponder JW TINKER software tools for molecular design
15. Forester TR, Smith W (1998) J Comput Chem 19:102
16. Martinez JM, Martinez L (2003) J Comput Chem 24:819

Progr Colloid Polym Sci (2004) 128: 86–91
DOI 10.1007/b97088

Mário Fernando Porto
Emerson Marcelo Girotto
Marcos Hiroiuqui Kunita
Maria do Carmo Gonçalves
Edvani Curti Muniz
Adley Forti Rubira
Eduardo Radovanovic

Atomic force microscopy, scanning electric potential microscopy and contact-angle surface analysis of low-density polyethylene grafted with maleic anhydride

M.F. Porto · E.M. Girotto
M.H. Kunita · E.C. Muniz
A.F. Rubira · E. Radovanovic (✉)
Grupo de Materiais Poliméricos e Compósitos, Departamento de Química, Universidade Estadual de Maringá, Av.Colombo 5790, 87020-900 Maringá, Paraná, Brazil
e-mail: eradovanovic@uem.br
Tel.: +55-44-2614332

M. do Carmo Gonçalves
Instituto de Química, UNICAMP, C. Postal 6154, 13083-970 Campinas, SP, Brazil

Abstract Polymers of great commercial use such as polyethylene, polypropylene and polystyrene have had their surface properties modified. Many researchers have recently altered the properties of polymers via grafting reactions. In this work, the surface of low-density polyethylene (LDPE) submitted to grafting with maleic anhydride followed by hydrolysis, chlorination, amination and protonation reactions was analyzed by atomic force microscopy, scanning electric potential microscopy, Fourier transform IR spectroscopy and water contact angle measurements. In different micrographs of the LDPE surface modified by four different processes it was possible to observe that the mountain-shaped structures with a global peak in the virgin polyethylene had become peaky in the grafting steps. It was possible to see that the LDPE surface becomes less rough with the hydrolysis process, going back to values close to those for the virgin polyethylene in the following steps. In scanning electric potential microscopy and topographic micrographs of a chlorinated sample it was possible to observe a correlation between the extent of surface modification and surface topography.

Keywords Polyethylene · Grafting · Atomic force microscopy · Scanning electric potential microscopy · Surface analysis

Introduction

Polymer materials are often surface-treated to change the chemical composition, increase the surface energy, or modify the surface topography while keeping the bulk properties unchanged [1–3]. Graft polymerization of hydrophilic monomers on an inert polymer such as polyethylene, polypropylene, and poly(tetrafluoroethylene) is one of the effective methods for improving surface properties [4]. In a grafted polymer one or more types of blocks are connected to the main polymer as side chains, which are constituted or configured differently from the main polymer, and these side chains could promote hydrophobic/hydrophilic changes in the polymer surface. For example, graft polymerization with vinyl monomers onto polyethylene substrates has attracted considerable interest because it imparts some desirable properties such as biocompatibility, ion exchange, dyeing ability and protein adsorption [5–10]. Polyethylene grafted with suitable monomers such as styrene, *N*-vinylpyrrolidone, acrylonitrile, acrylamide, and methacrylic acid has been reported [11–22]

The surface modification of low-density polyethylene (LDPE) has been made in our group by the grafting of maleic anhydride, in order to accomplish subsequent reactions, generating reactive functional groups on the LDPE surface [23]. The modifications generated in the polymer surface generally are studied by techniques of surface analysis and scanning probe microscopy techniques are among the main techniques used. These kinds of microscopy, first developed by Binnig and Rohrer [24], provided the ability to measure material surface properties with high spatial resolution. The principle of operation of atomic force microscopy (AFM) is based on detecting a diminutive force between the sample and the scanning tip attached to a small cantilever. With these apparatus it is possible to obtain a micrograph knowing the Z scale besides the X and Y scales and then it is possible to analyze an important characteristic: the surface roughness. Analyses of surface roughness have been used in a study of polishing of metallic surfaces [25], in a correlation study of surface roughness and gas permeation in polymeric membranes [26, 27], and in studies of the changes suffered by a polymer surface treated with a plasma beam [28]. Besides, using scanning electric potential microscopy (SEPM) with AFM it is possible to map the topography and surface potential distribution of the sample at the same time, making it possible to compare morphological structure and local electrical properties. In the polymer surface, differences in the electrical distribution could be related to differences in the chemical oxidation state of the polymer. Then, this technique that is used especially for the study of semiconductor devices [29, 30] also appears to be suited for biological samples [31] and to characterize the effect of electrochemical treatment on conducting polythiophene polymers [32, 33] as well as electrical charge distribution on dieletric polymers [34]. A description of the SEPM technique is found elsewhere [34]. In this work, the surface of LDPE submitted to grafting with maleic anhydride followed by hydrolysis, chlorination, amination and protonation reactions was analyzed by AFM, SEPM, Fourier transform IR (FTIR) spectroscopy and water contact angle measurements.

Experimental

LDPE in a pellet was pressed at 170 °C and 1.25 P_a to obtain films of average thickness of 110 μm. The sample was immersed in a solution of maleic anhydride (4.0 g) in acetic anhydride (80 mL), using benzoyl peroxide (0.22 g) as a radical initiator, and heated at 100 °C, under agitation, for 6 h. Afterwards, the sample was washed with acetone and hydrolyzed at 100 °C for 5 h. The next step was the immersion in a solution of 2 vol % of antimony pentachloride in carbon tetrachloride, for 24 h at 35 °C, under agitation. The next immersions were in triethanolamine and in phosphoric acid, also under agitation, both for 24 h at room temperature. Then, six samples were prepared and analyzed: neat LDPE; LDPE grafted with maleic anhydride; LDPE grafted with maleic anhydride followed by hydrolysis; LDPE chlorinated after the hydrolysis; LDPE reacted with triethanolamine after the chlorination; and LDPE reacted with a H_3PO_4 solution after the reaction with triethanolamine. AFM and SEPM images were acquired using a Topometrix Discoverer TMX-2010 instrument. A low resonance frequency pyramidal silicon cantilever, from Thermomicroscope, was used in the topographic AFM measurements. In SEPM mode, this instrument uses a Pt-coated conducting tip to detect electric potential gradients throughout the scanned area [34]. Image processing was performed by the Topometrix image analyser program. The contact angles of water drops (Milli-Q) deposited on the surface of neat and modified polymers were measured at 25 °C using a Tantec contact angle meter. IR spectra of LDPE, before and after the surface modifications, were obtained with a FTIR Bomem MB-100, with a resolution of 4 cm^{-1}, in the range 4,000–400 cm^{-1}.

Results and discussion

The IR analyses are especially suitable to accompany the grafting of LDPE with maleic anhydride and posterior reactions. The spectra of neat LDPE and LDPE submitted to four sequential reactions are shown in Fig. 1. In the spectrum of LDPE grafted with maleic

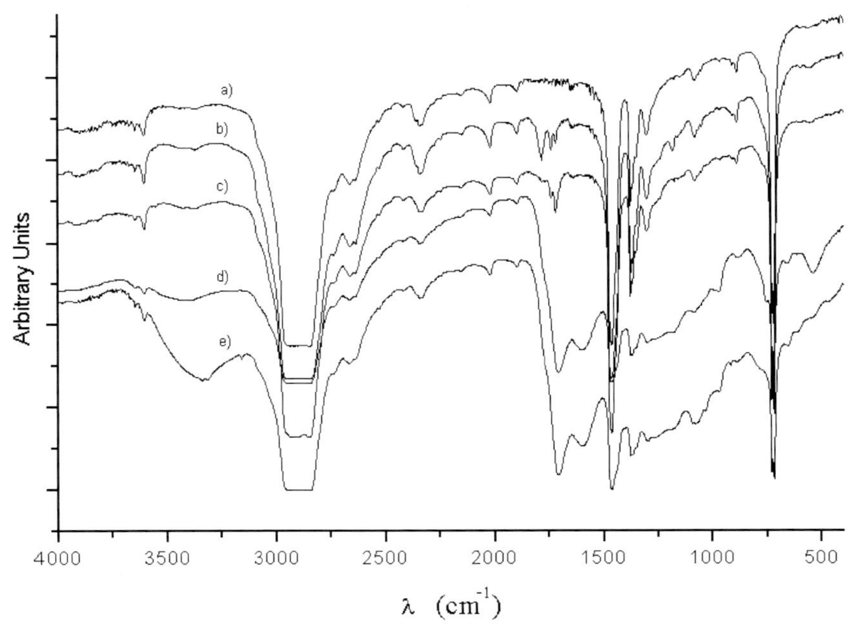

Fig. 1 Fourier transform IR spectra of neat low-density polyethylene (*LDPE*) (**a**), LDPE grafted with maleic anhydride (**b**), LDPE grafted with maleic anhydride followed by hydrolysis (**c**), LDPE chlorinated after the hydrolysis (**d**) and LDPE reacted with triethanolamine after the chlorination (**e**)

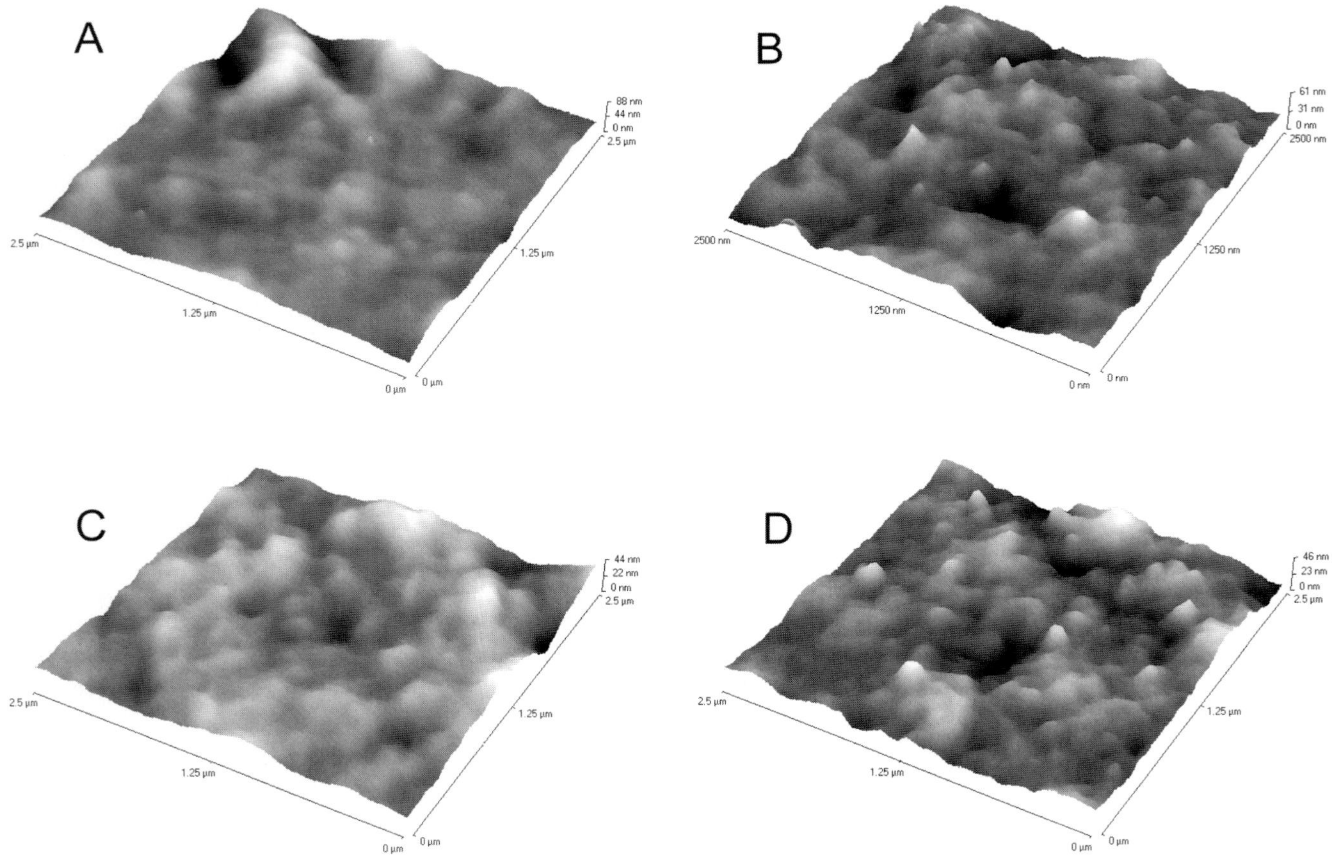

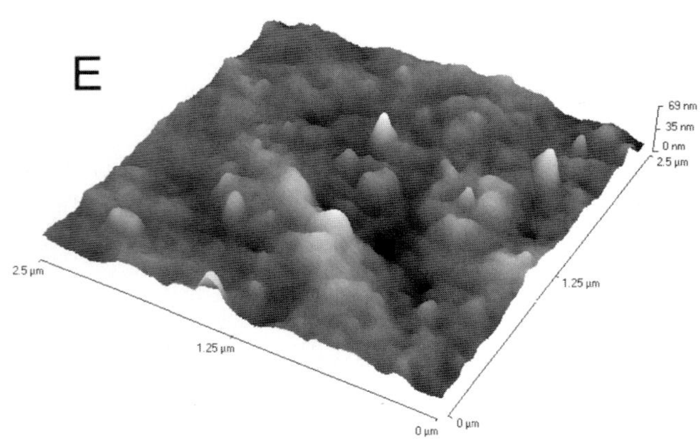

Fig. 2 Topographic atomic force microscope image of **a** neat LDPE, **b** LDPE grafted with maleic anhydride, **c** LDPE grafted with maleic anhydride followed by hydrolysis, **d** LDPE chlorinated after the hydrolysis and **e** LDPE reacted with triethanolamine after the chlorination

anhydride (Fig. 1, spectrum b) characteristic peaks of maleic anhydride were observed, i.e., C=O stretching peaks at 1,782, 1,741, and 1712 cm^{-1} and C–O stretching peak of cyclic anhydride at 1,184 cm^{-1} [20]. These peaks were not observed in the spectrum of neat LDPE (Fig. 1, spectrum a). In the hydrolyzed sample (Fig. 1, spectrum c) we only observed a peak attributed to C=O stretching, at 1,719 cm^{-1}, characteristic of carboxylic acid groups, and the peak at 1,184 cm^{-1} disappeared. After the chlorination reaction, the spectrum of LDPE (Fig. 1, spectrum d) showed characteristic peaks of C–Cl stretching at 753 and 540 cm^{-1}, besides the peak at 1,596 cm^{-1}, characteristic of C=C stretching. These double bonds were generated by the antimony pentachloride acting as a Lewis acid in the reaction [21–23]. After the reaction with triethanolamine, it was possible to observe in the LDPE spectrum a large characteristic peak of OH stretching at 3,332 cm^{-1} and the disappearance of the peaks assigned to C–Cl vibrations.

In different micrographs of the neat polymer surface, as well as in different micrographs of the LDPE surface modified by the four subsequent processes (representative AFM images of each step of the modification are shown in Fig. 2), it was possible to observe that the broad structures in the neat polyethylene had become peaky after the grafting steps. This change could be correlated with a higher degree of insertion of maleic anhydride and other polar groups in these polymer surface protrusions.

A roughness parameter directly calculated by the software of the microscope was used in these analyses: the average roughness (R_a), which corresponds to the arithmetic mean of the deviations (Z_i) in height from the image mean value ($\dot{z}$):

$$R_a = \frac{1}{N}\sum_{i=1}^{N}\left|Z_i - \dot{Z}\right|. \tag{1}$$

In the investigation of the modification of polyethylene via grafting, the surface roughness can be used to follow the surface modification process. The R_a values for the neat LDPE and modified LDPE films, calculated from at least six 10×10-μm-area images taken after each modification step, are presented in Fig. 3.

The R_a values from different scans of the same sample varied between 6% and 25%. In spite of these great deviations, it is possible to conclude that the polyethylene surface becomes less rough after the hydrolysis process, going back to values close to those of the neat polyethylene in the following steps. This fact can probably be associated with the formation of a less hydrophobic structure with the addition of acidic groups, in comparison with the other samples. This was confirmed by the water drop contact angle analysis shown in Fig. 4, which suggested that a good correlation between surface energy and roughness in this polymer exists.

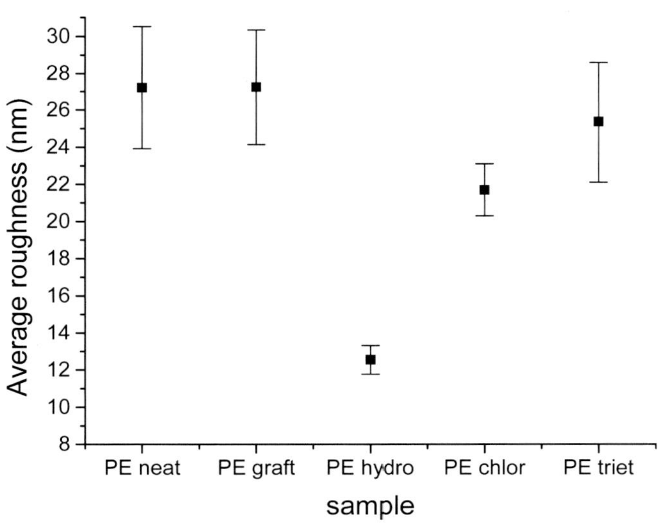

Fig. 3 R_a values for the neat LDPE and modified LDPE films, calculated from at least six images of each step of modification, in images of 10×10 μm

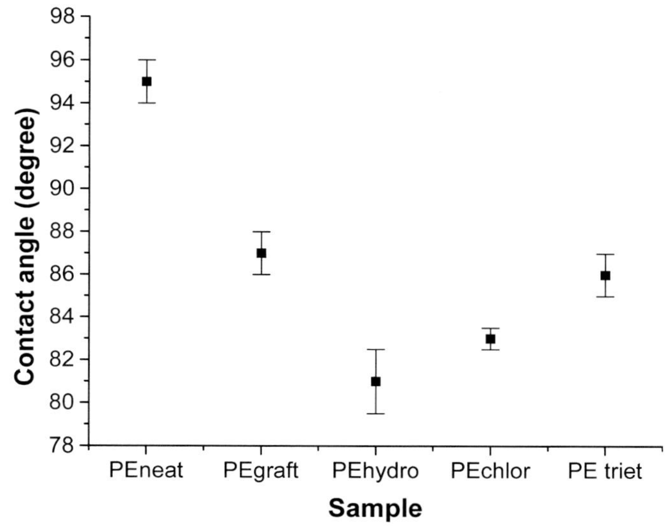

Fig. 4 Water contact angle for the neat LDPE and modified LDPE films. Drop volume 10 μL

Three pairs of topographic and electric potential images relative to neat LDPE and the four subsequent modified polymers are shown in Fig. 5. In the neat polyethylene (Fig. 5a, b), the two images (topographic and electric potential maps) show some similar features and most protrusions carry excess positive charge. The polyethylene grafted with maleic anhydride (Fig. 5c, d) presented similar behaviour. After the chlorination of grafted polyethylene (Fig. 5e, f) it is possible to see a good match between the topographic and electric potential images.

The bright spots in the electric potential image accumulated at the top of the elevations, suggesting that the modification reactions suffered by the polymer are concentrated at these points. The literature shows that the protruding regions in polymers can accumulate charges [34, 35], and probably these regions were more sensitive to the modification reactions performed in this work.

Conclusions

The surface of polyethylene changed from hydrophobic to acidic character following the maleation process, and to basic character following the amination process. It was possible to follow the surface modifications by FTIR and

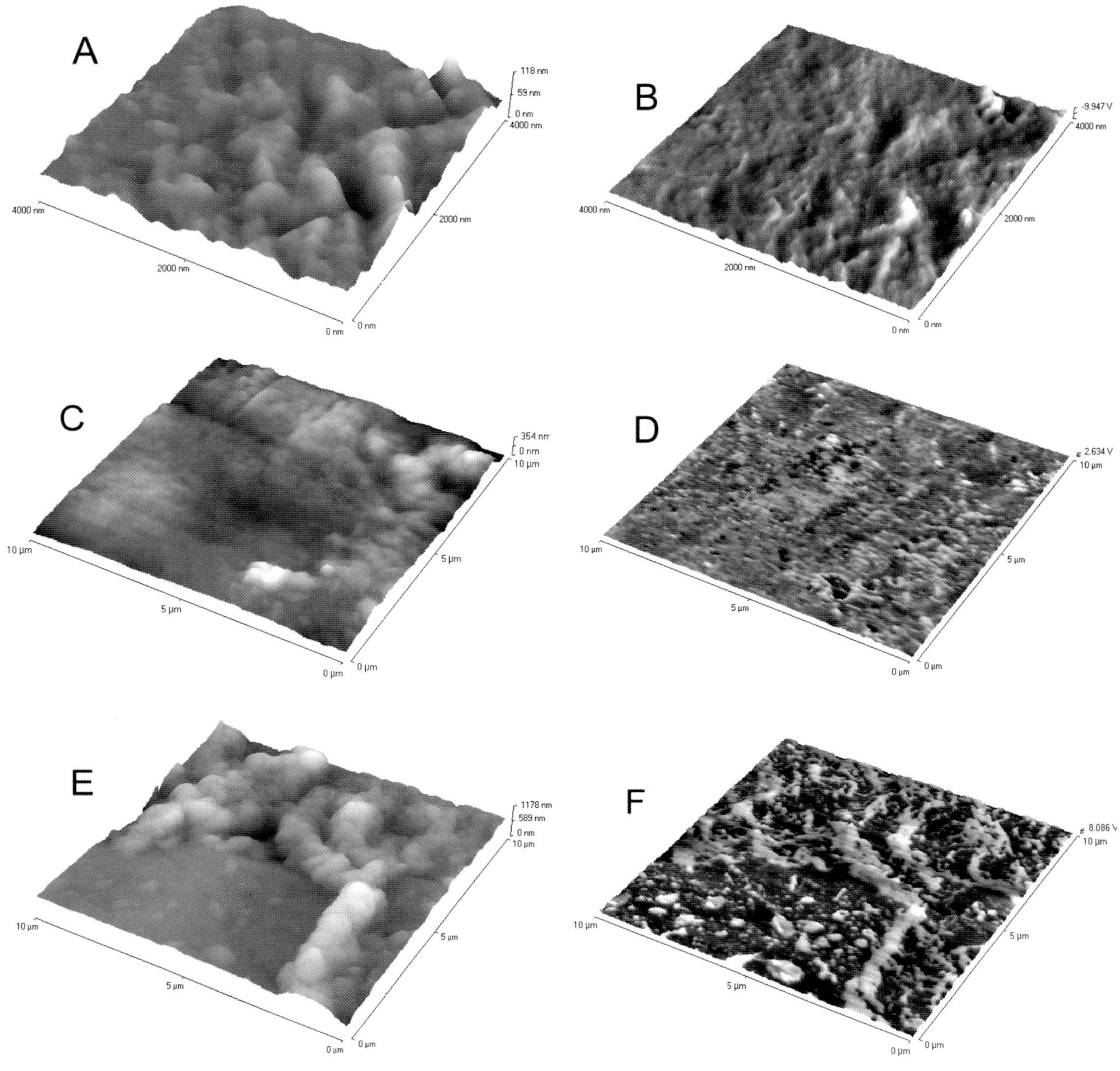

Fig. 5 Topographic (**a**, **c**, **e**) and scanning electric potential microscope (**b**, **d**, **f**) images of **a**, **b** neat LDPE, **c**, **d** LDPE grafted with maleic anhydride and **e**, **f** LDPE chlorinated after the hydrolysis

water contact angle measurements. AFM and SEPM images showed that reactions of surface modification produce changes in the topography, where the more hydrophylic surface is the smoother one. A correlation was observed between the extent of the surface reaction and the surface topography, in the case of chlorination.

Acknowledgements The authors thank CNPq, Capes, Fapesp, Fundação Araucária, Unicamp and UEM.

References

1. Wu S, (1982) Polymer interface and adhesion. Dekker, New York
2. Brewis DM, Briggs D (1981) Polymer 22:7
3. Fourche G (1995) Polym Eng Sci 35:968
4. Ishihara K, Iwasaki Y, Ebihara S, Shindo Y, Nakabayashi N (2000) Colloids Surf B 18:325
5. Nho YC, Sugo T, Tsuneda S, Makuuche K (1994) J Appl Polym Sci 51:1269
6. Cohn D, Hoffman AS, Ratner BD (1987) J Appl Polym Sci 33:1
7. Guthrie JT (1987) J Appl Polym Sci 37:39
8. Omichi H, Okamoto J (1985) J Appl Polym Sci 30:1277
9. Kim M, Saito K, Furusaki S, Sugo T, Okamoto J (1991) J Membr Sci 56:289
10. Zeng H, Li J, Wang X, Wen P (1999) Zhongguo Suliao 13:34
11. Mehta IK, Kumar S, Chauhan GS, Misra BN (1990) J Appl Polym Sci 41:1171
12. Mokhtar SM, Sabaa MW (1997) Polym Int 42:340
13. Gawish SM, Kantouch A, El-Naggar AM, Mosleh S (1992) J Appl Polym Sci 44:1671
14. Nitto K, Suzuki SI, Miyasaka K, Ishikawa K (1982) J Appl Polym Sci 27:6337
15. Kritskaya DA, Pomogailo AD, Ponomarev NA, Dyaschkovskii FS (1980) J Appl Polym Sci 25:349
16. Dilli S, Garnett JL, Kenyon SR, Martin EC, Phuoc DH, Yen Y, Leeder JD (1972) J Macromol Sci Chem A 6:729
17. Dilli S, Garnett JL (1967) J Appl Polym Sci 11:859
18. Wilson JE (1974) J Macromol Sci Chem A 8:733
19. Odian G, Rossi A, Trachtenberg EN (1960) J Polym Sci 42:575
20. Butich C, Yahiaoui A (1987) J Polym Sci Part A Polym Chem 25:3479
21. Nho YC, Jin JH (1997) J Appl Polym Sci 63:1101
22. Nho YC, Sugo T, Tsuneda S, Makuuchi K (1994) J Appl Polym Sci 51:1269
23. Kunita MH, Girotto EM, Radovanovic E, Goncalves MC, Ferreira OP, Muniz EC, Rubira AF (2002) Appl Surf Sci 202:223
24. Binnig G, Rohrer H (1982) Helv Phys Acta 55:726
25. Collins KE, Franchon AC, Jardim ICSF, Radovanovic E, Gonçalves MC (2000) LC-GC 18:106
26. Stamatialis DF, Dias CR, Pinho MN (1999) J Membr Sci 160:235
27. Bowen WR, Hilal N, Lovitt RW, Williams PM (1996) J Membr Sci 110:229
28. Zhang J, Cui CQ, Lim TB, Kang E-T, Neoh KG (1998) J Adhes Sci Technol 12:1205
29. Vatel O, Tanimoto M (1995) J Appl Phys 77:2358
30. Xu Q, Hsu JWP (1999) J Appl Phys 85:2465
31. Leng Y, Williams CC (1994) Colloids Surf A 93:335
32. Sememikhin OA, Jiang L, Iyoda T, Hashimoto K, Fujishima A (1996) J Phys Chem 100:18603
33. Sememikhin OA, Jiang L, Iyoda T, Hashimoto K, Fujishima A (1997) Electrochim Acta 42:3321
34. Galembeck A, Costa CAR, da Silva MCVM, Souza EF, Galembeck F (2001) Polymer 42:4845
35. Galembeck F, da Silva MCVM, Leite CAP, Costa CAR, Galembeck A (2002) Mol Cryst Liq Cryst 374:159

Progr Colloid Polym Sci (2004) 128: 92–97
DOI 10.1007/b97090

Önder Pekcan
Saziye Ugur

Dissolution of a percolation cluster on UV-cured polymeric films: a fluorescence study

Ö. Pekcan (✉) · S. Ugur
Department of Physics,
Istanbul Technical University,
34469 Maslak, Istanbul, Turkey
e-mail: pekcan@itu.edu.tr
Tel: +90-212-2853213
Fax: +90-212-2856386

Abstract The steady-state fluorescence technique was used for studying dissolution of UV-cured polymer films. These films were formed from poly(methyl methacrylate) chains labeled with pyrene which was introduced as a fluorescence probe to monitor the dissolution processes. Poly(methyl methacrylate) films with various ethylene glycol dimethacrylate contents, cured by UV radiation for 3 and 25 h, were dissolved in chloroform–heptane (20–80%) mixtures. The percolation threshold ($p_c = 0.35$) found for films irradiated for 3 h obeys the site percolation model, however the percolation threshold ($p_c = 0.25$) measured for films irradiated for 25 h obeys the bond percolation model.

Keywords Fluorescence · Cross-linking · Percolation · Dissolution

Introduction

Polymer network formation can be produced by chemical means, or by radiation [1–3], which has great advantages for quantitative analysis and the study of the resultant physical behavior. The simplest form of cross-linking is one where two radicals on adjacent molecules are linked together to form a direct cross-link [4]. This process is generally called gelation, which is the phase transition from a state without a gel to a state with a gel, i.e., gelation involves the formation of an infinite network. In the sol–gel phase transition, an infinitely large macromolecule is formed which is called a gel. The conversion factor p is the fraction of bonds which have been formed between the monomers of the systems, i.e., the ratio of the actual number of bonds at a given moment to the maximum possible number of such bonds [5, 6]. Thus for $p = 0$ no bonds have been formed and all monomers remain isolated. In the other extreme, $p = 1$, all possible bonds between monomers have been formed and thus all monomers in the system have clustered into one infinite network, with no sol phase left. Thus, for small p no gel is present, whereas for p close to unity one such network exists. Therefore, there is, in general, a sharp phase transition at some intermediate critical point $p = p_c$, where an infinite cluster starts to appear; a gel for p above p_c, a sol for p below p_c.

Fluorescence quenching and depolarization methods have been used for penetration and dissolution studies in solid polymers [7–9]. A steady-state fluorescence (SSF) study on dissolution of both annealed latex film [10] and poly(methyl methacrylate) (PMMA) discs using real-time monitoring of fluorescence probes [11] were reported. The fast transient fluorescence technique was recently applied to investigate polymer dissolution mechanisms [12]. Recently, the effect of γ [13] and UV radiation [14] on latex film dissolution were studied using the SSF technique.

In this work dissolution of PMMA films cured by UV radiation for two different irradiation times (3 and 25 h) were compared in chloroform–heptane mixtures using the SSF technique. A good–bad solvent mixture was preferred to initiate and control the partial dissolution processes. Various films with various cross-linker, ethylene glycol dimethacrylate (EGDM), contents were prepared and irradiated with UV light for 3 and 25 h to form a network before dissolution. In situ SSF experiments were performed to observe dissolution processes. The desorption rates, D_d, were measured using curves of intensity versus time. The maximum pyrene (P) intensity, $I_{P\infty}$, was measured after the dissolution process was complete for each film sample at various EGDM contents. D_d values were measured for films irradiated for 3 and 25 h.

Theoretical considerations

The theories of fluorescence quenching with and without diffusion are well established [15]. A chromophore P in its ground state is converted to an electronically excited state P^* by absorption of a photon. Then the excited state P^* decays to P by fluorescence, or the excited state P^* can be quenched by collision with a quencher Q.

In the language of percolation, one may think of monomers occupying the vertices of a periodic lattice, and chemical bonds as corresponding to the edges joining these vertices at any given moment with probability p. Only, the nearest neighbors which are occupied or connected by open bonds form clusters. At some intermediate critical point ($p=p_c$), an infinite cluster extending from one side of the system to the other starts to appear. This point is called the percolation threshold. At this point, the largest cluster spans the lattice connecting the left and right edges to the bottom edge and is called a "percolating cluster". Extensive simulations and theoretical work have shown that the percolation probability vanishes according to a power-law near p_c:

$$P_\infty(p) \approx (p - p_c)^\beta \quad (1)$$

for $p>p_c$, and $p \rightarrow p_c$. In a simple cubic lattice p_c is found to be 0.31 for the site percolation model and 0.25 for the bond percolation model [16]. The exponent β for a simple cubic lattice is 0.45. The Fickian diffusion model includes the solution of a unidirectional diffusion equation for a set of boundary conditions, which was cited by Crank and Parks [17]. For a constant diffusion coefficient, D, and fixed boundary conditions, the absorption and dissolution transport in and out of a thin slab is given by the following relation:

$$\frac{M_t}{M_\infty} = 1 - \frac{8}{\pi^2}\sum_{n=0}^{\infty}\frac{1}{(2n+1)^2}\exp\left(\frac{-(2n+1)^2 D\pi^2 t}{d^2}\right). \quad (2)$$

Here, M_t represents the amount of material absorbed or dissolved at time t, M_∞ is the equilibrium amount of material, and d is the thickness of the slab. n comes from the solution of the diffusion equation in the form of a trigonometrical series.

Experimental

The poly(isobutylene) (PIB) sterically stabilized PMMA latexes were prepared separately in two-steps [18]. In the first step, short PMMA sequences were grafted onto a PIB sample of nominal molecular weight 10^4. These PIB samples contained sites of unsaturation for reaction with either initiator radicals or growing PMMA chains. Polymerization was terminated before the PMMA sequences became sufficiently long as to render the polymer insoluble in the cyclohexane reaction mixture. This short graft copolymer, called the "dispersant", was precipitated from solution with ethanol and purified. In the second step of preparation, the dispersant methyl methacrylate and an initiator, 1-pyrenyl methyl methacrylate, were combined in cyclohexane solution and refluxed over-night. A white dispersion formed. This procedure produced spherical particles of relatively narrow size distribution. Using P–PMMA as the comonomer yielded latexes containing P groups covalently bound to the PMMA. These materials were freeze-dried from cyclohexane, and stored as a powder. The compositions of the materials were determined by NMR and UV absorption studies of their solutions in $CDCl_3$ and ethyl acetate, respectively. These analyses indicated that the particles contained 4 mol % PIB and 0.037 mmol P groups per gram of polymer.

The film preparation was carried out in the following manner. P-labeled PMMA chains were dissolved in chloroform and 5% of 2-benzyl-2-dimethylamino-1-(4-morpholinophenyl)-butanone-1(Irgacure 369) as photoinitiator was added to the solution. Then, using this stock solution 12 different films were prepared by adding different amounts of (EGDM) (0, 6, 17, 25, 35, 42, 46, 51, 56, 59, 65 and 70 wt %) to the 12 different test tubes. These solutions were placed onto 0.8x2.5-cm^2 glass plates and allowed to dry in air to produce film samples. We prepared two different sets of films in this way; the films in one set were cured by UV light for 3 h and the other set was irradiated for 25 h in a merry-go-round-type photoreactor equipped with 15 Philips lamps emitting light nominally at 350 nm at room temperature. The films were weighed before and after dissolution to measure the amount of dissolved PMMA material in both cases.

In situ fluorescence experiments were performed using a PerkinElmer LS-50 spectrofluorimeter at room temperature. Dissolution experiments were performed in a 1×1-cm quartz cell which was placed in the spectrofluorimeter. Samples were placed at one side of a quartz cell filled with a chloroform–heptane mixture (80–20 vol %) and the cell was then illuminated with 345-nm excitation light. The pyrene fluorescence intensity, I_P, was monitored during the dissolution process at 395 nm using the "time drive" mode of the spectrofluorimeter.

Results and discussion

Plots of P intensity versus dissolution time for the films cured for 3 h by UV radiation are shown in Fig. 1. It is seen that all curves increase at early times and by reaching a plateau at later times present exponential behavior. The maximum P intensity values for UV-cured films are plotted versus EGDM content in Fig. 2a, where it is seen that the increase in $I_{P\infty}$ for UV-cured samples starts above a content of 35 wt % of EGDM. The UV-cured film samples were weighed before and after the dissolution experiments and the amount of PMMA dissolved is plotted versus the EGDM content in Fig. 2b. As seen from the figure, below the critical point (35 wt % of EGDM), 65 wt % of PMMA dissolved from the UV-cured samples. On the other hand, 20 wt % of PMMA dissolved when the film samples were prepared by adding more than 35 wt % of EGDM and were cured by UV radiation. These findings together with intensity data predict that above 35 wt % of EGDM inclusion creates

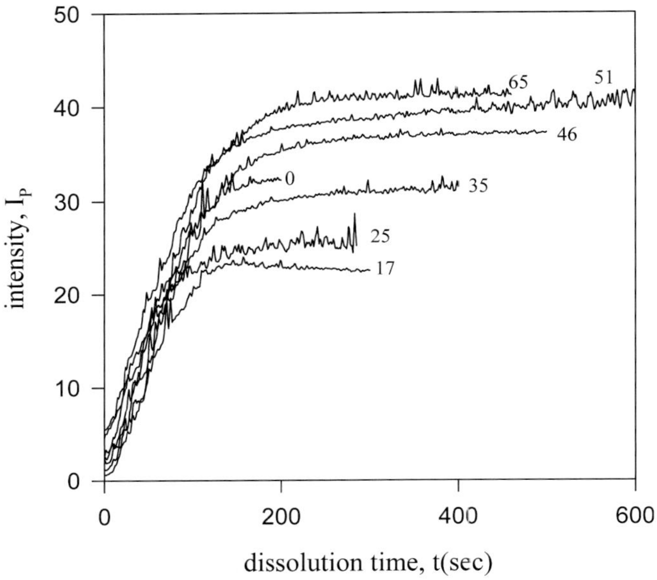

Fig. 1 Pyrene intensity, I_P, versus dissolution time for the 3-h UV-cured film samples dissolved in chloroform–heptane mixtures at low stirring speed. The *numbers* on each curve represent the amount of ethylene glycol dimethacrylate (*EGDM*) in weight percent

large PMMA clusters in the film samples which are difficult to dissolve. Below this critical point (35 wt %) the low $I_{P\infty}$ values most probably come from the high quenching rate of excited P molecules by solvent molecules. Owing to the low cross-linking of the films, P molecules are exposed to solvent molecules and are quenched immediately. High $I_{P\infty}$ values above the critical EGDM content (35 wt %) indicate that P molecules are not quenched owing to formation of large PMMA clusters in which P molecules are protected against solvent quenching.

In order to support these findings film samples with various EGDM contents which were not cured by UV radiation were dissolved in chloroform–heptane mixtures. The $I_{P\infty}$ values and amount of dissolved PMMA of these film samples are compared with the respective values and amounts for the UV-cured samples in Fig. 2, where it is seen that the $I_{P\infty}$ values for the noncured samples do not change as the EGDM content is increased and the noncured samples had much lower intensities compared with UV-cured samples. It was observed that samples that were not treated by UV radiation dissolved completely, indicating that no cross-linking had occurred. In comparing the noncured films with the UV-cured films which contain small and large PMMA clusters these results are explained in the following manner: since noncured samples have no cross-linked points, they dissolve completely and P-labeled PMMA chains are completely exposed to solvent molecules and are quenched totally, resulting in very low $I_{P\infty}$ intensities.

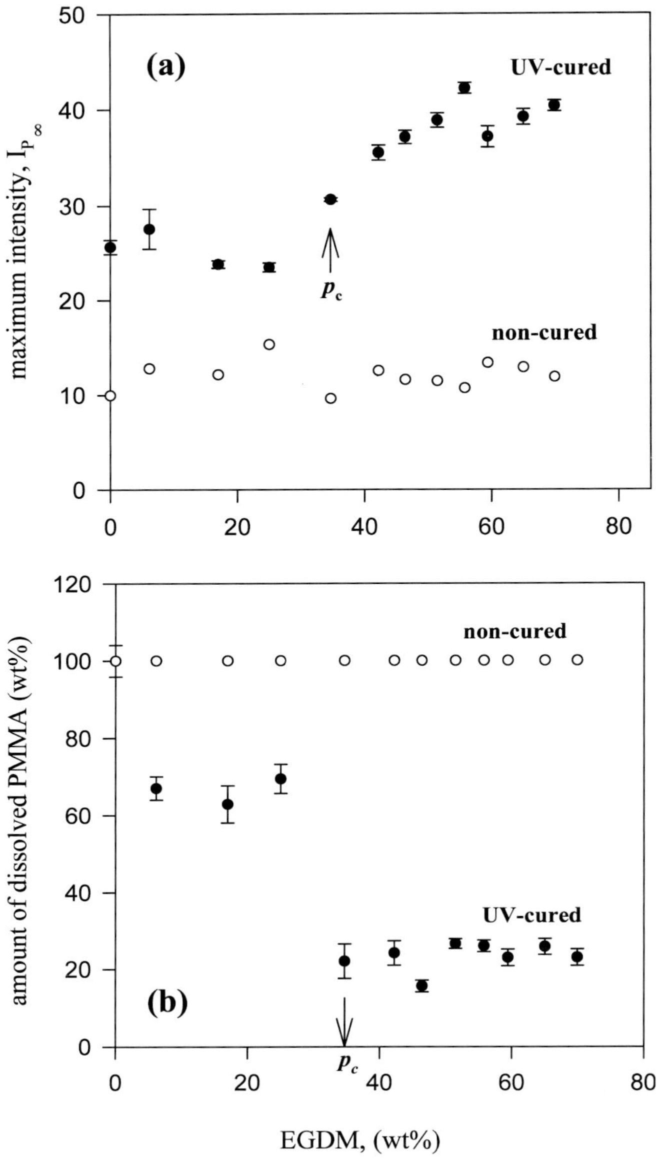

Fig. 2 **a** Maximum pyrene intensity, $I_{P\infty}$, and **b** the amount of dissolved poly(methyl methacrylate) (*PMMA*) material versus EGDM content for UV-cured (3 h) (*closed circles*) and noncured (*open circles*) film samples

The critical EGDM value, 35 wt %, most probably corresponds to the p_c for site percolation at which the largest cluster spans the lattice by connecting the left and right edges to the bottom edge [15]. The behavior of $I_{P\infty}$ for UV-cured samples in Fig. 2 also supports the formation of a percolation cluster at $p_c \approx 0.35$. The measured percolation threshold value predicts that the formation of a UV-cured film on a glass surface obeys the site percolation model for 3 h of irradiation. Above p_c, the fluorescence technique allows us to detect large lattice animals where $I_{P\infty}$ increased to its maximum value.

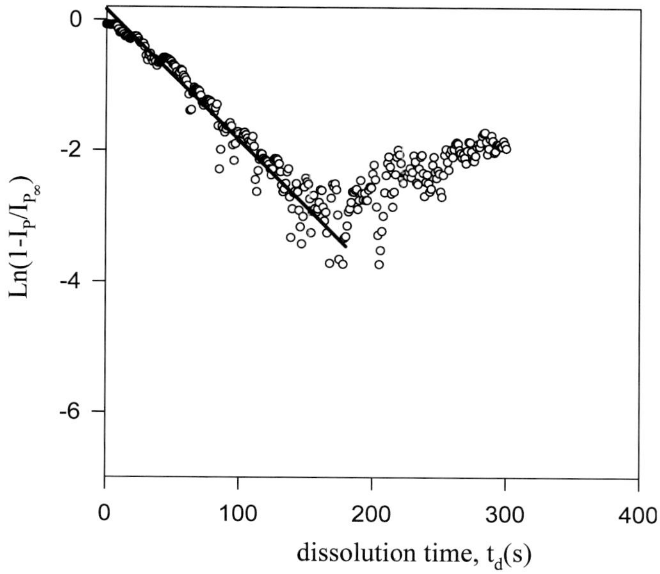

Fig. 3 Dissolution curves fitted to Eq. (3) for the film samples formed with 25 wt % of EGDM

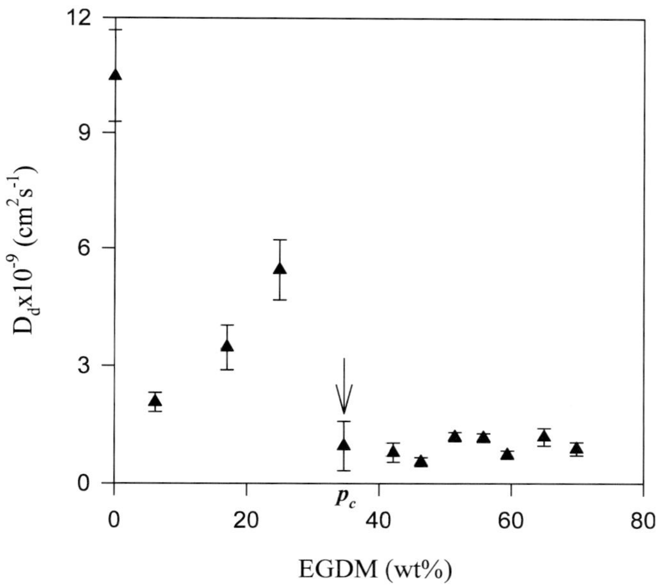

Fig. 4 Dissolution coefficient, D_d, versus EGDM content for short irradiation time (3 h)

The curves in Fig. 1 seem to follow a Fickian diffusion model. In processing the dissolution data, it is assumed that I_p is proportional to the number of P-labeled PMMA chains dissolving from the PMMA film. The logarithmic form of Eq. (2) is written for $n=0$, with $A_d = D_d\pi^2/d^2$ and $B_d = \ln(8/\pi^2)$ as follows:

$$\ln(1 - I_P/I_{P\infty}) = B_d - A_d t. \qquad (3)$$

Here, $I_{p\infty}$ represents the number of P molecules at equilibrium, D_d is the dissolution coefficient, and d is the thickness of the latex film. The dissolution curves fitted to Eq. (2) for films formed with 25 wt % of EGDM are presented in Fig. 3. The curve in Fig. 3 are quite linear and strictly obey the Fickian diffusion model. When the linear curves in Fig. 3 are compared with computations using Eq. (3), the dissolution coefficients, D_d, for P-labeled polymer molecules are obtained, and are plotted in Fig. 4 together with those for other samples against EGDM content. The results in Fig. 4 support our previous arguments, where highly cross linked films above p_c dissolve at least 4 times slower than films with low cross-linking.

In order to understand the percolation behavior of the films, the films were irradiated for a longer time (25 h). The maximum emission spectra of P at infinite time $I_{P\infty}$ (i.e., at the plateau) and the amount of dissolved PMMA are plotted versus EGDM content in Fig. 5. It is seen in Fig. 5a that the maximum P intensities are much lower for film samples with a high EGDM content than for film samples with a low EGDM content. In fact the $I_{P\infty}$ values and the amount of dissolved polymer drop dramatically above the critical, 25 wt %, EGDM content. Below the critical point, 45 wt % of PMMA dissolves from the UV-cured samples, indicating that some of the PMMA chains are slightly cross-linked and form small clusters in the film. On the other hand, only 20 wt% of PMMA dissolved when the film samples were prepared by adding more than 25 wt % of EGDM. $I_{P\infty}$ also had low values above this critical EGDM content, indicating that most of the PMMA chains are highly cross-linked by forming large clusters which prevent the film from undergoing dissolution. Most probably a percolation cluster is formed films with 25 wt % EGDM ($p_c = 0.25$) content that have been irradiated for 25 h.

If one assumes that the bond formation probability, p, is equal to the weight percent of EGDM content in the film sample then the data in Fig. 5b can be treated according to Eq. (1). The percolation probability now becomes the gel fraction [5, 6] where the P molecules are embedded. Since $I_{p\infty}$ is proportional to the amount of dissolved polymeric material, $(1-I_{p\infty}/I_{p\infty 0})$ gives information about the undissolved part of the polymeric material, i.e., $(1-I_{p\infty}/I_{p\infty 0})$ is proportional to the gel fraction formed on the partially dissolved film. Here $I_{p\infty 0}$ is the $I_{p\infty}$ from the sample with the lowest EGDM content. The slope of the log–log plot of $(1-I_{p\infty}/I_{p\infty 0})$ versus $(p\text{-}p_c)$ in Fig. 6 produces a critical exponent β of 0.55, which is somewhat larger than the value known for a simple cube ($\beta = 0.45$).

The logarithmic forms of the dissolution curves of the films irradiated for 25 h were fitted to Eq. (3). The desorption coefficient, D_d, for P-labeled PMMA materials were obtained, and are plotted in Fig. 7 against the EGDM content; the results support our previous

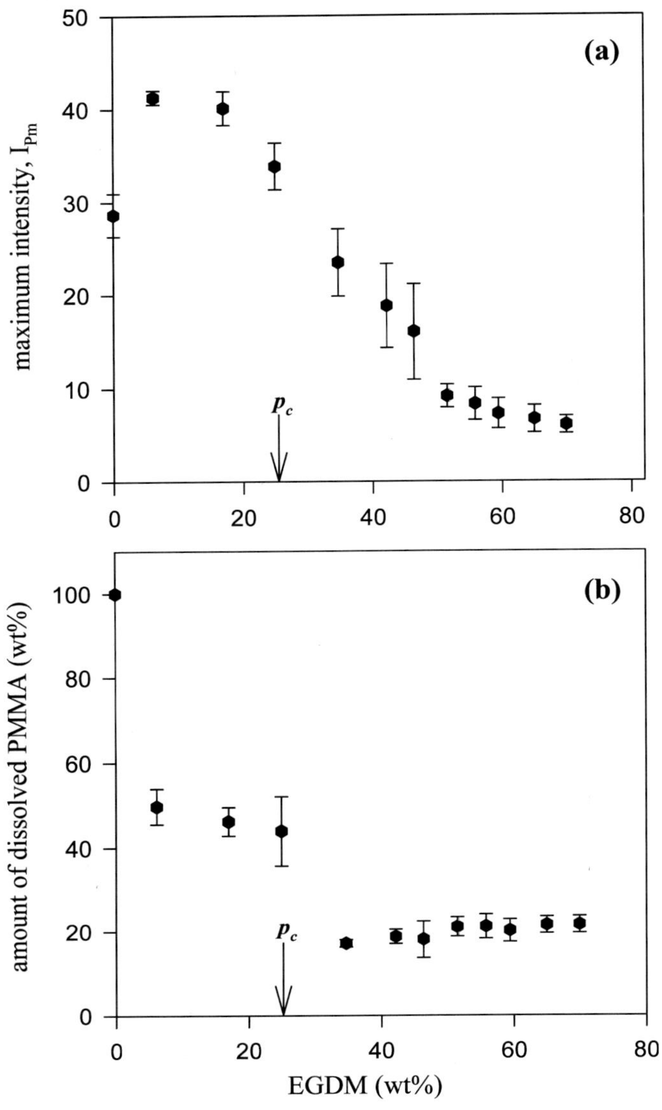

Fig. 5 **a** The maximum pyrene intensity, $I_{P\infty}$, and **b** the amount of dissolved PMMA material versus EGDM content for 25-h UV-cured films

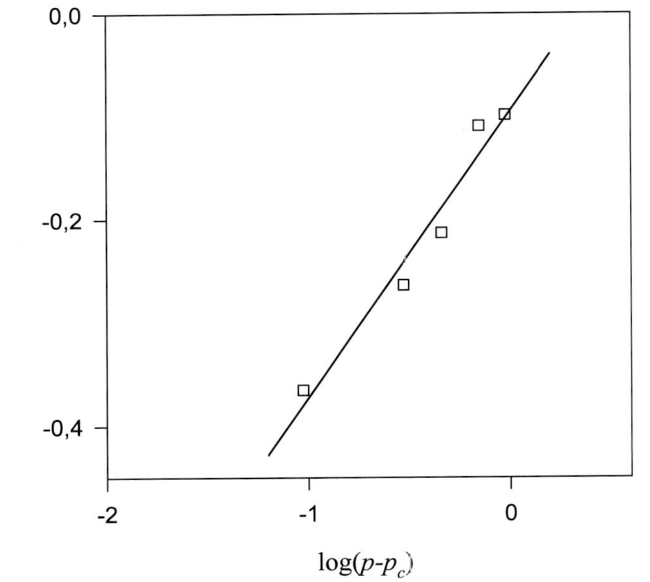

Fig. 6 The log–log plot of the percolation probability versus the bond formation probability

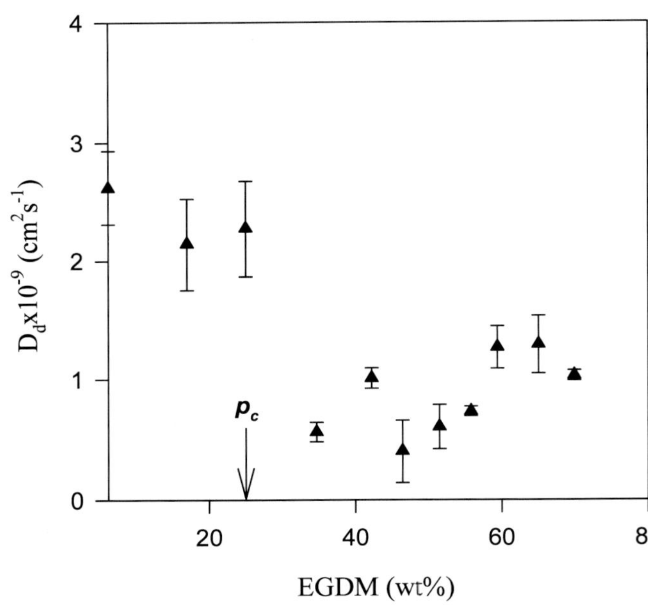

Fig. 7 Desorption coefficient, D_d, versus EGDM content for the films irradiated for 25 h

arguments, i.e., films with low cross-linking dissolve much faster than highly cross-linked films.

Conclusions

In conclusion, this work presents results obtained from the partial dissolution of UV-cured polymeric films. The films formed a percolating cluster during partial dissolution if they contained 35 and 25 wt % of cross-linker and obeyed the site and bond percolation models when exposed to UV radiation for 3 and 25 h, respectively. Here one may argue the effect of irradiation time on the percolation behavior of these films. In site percolation ($p_c = 0.31$) the sites may or may not be occupied, whereas bonds between sites always exist; however, in bond percolation ($p_c = 0.25$) the sites are always occupied and bonds between them may or may not be formed [19]. This picture fits our experimental observations where irradiation for 3 h creates few occupied sites but bond formation between the sites caused by EGDM is highly probable. On the other hand, longer irradiation (25 h) generates many occupied sites but few of them contribute to the bond formation using EGDM, which results in a low p_c value. These films dissolved much faster below p_c and dissolved partially above p_c, producing high and low desorption coefficients, D_d, for both irradiation times.

References

1. Makhlis A (1975) Radiation of physics and chemistry of polymers. Wiley, New York
2. Kloosterboer JG (1988) Adv Polym Sci 84:1
3. Vollenbroek FA (1988), Adv Polym Sci 84: 85
4. Florry PJ (1953) Principles of polymer chemistry. Cornell University Press, Ithaca
5. Stauffer D (1985) Introduction to percolation theory. Taylor and Francis, London
6. Stauffer D (1981) Pure Appl Chem 53, 1479
7. Guilet JE (1986) In: Winnik MA (ed) Phototphysical and photochmical tools in polymer science. Reidel, Dordrecht
8. Nivaggioli T, Wank F, Winnik MA (1992) J Phys Chem 96:7462
9. Pascal D, Duhamel J, Wank, J, Winnik MA, Napper DH, Gilbert R (1993) Polymer 34:1134
10. Pekcan Ö, Ugur S, Yilmaz Y (1997) Polymer 38:2183
11. Ugur S, Pekcan Ö (1997) Polymer 38:5579
12. S. Ugur, Ö. Pekcan (2000) Polymer 41:1571
13. Aydin K, Ugur S, Pekcan Ö (2001) J Colloid Interface Sci 233:91
14. Ugur S, Pekcan Ö (2003) Polymer 44:2041
15. Birks JB (1970) Photophysics of aromatic molecules. Wiley-Interscience, New York
16. Stauffer D, Aharony A (1992) Introduction to percolation theory. Taylor and Francis, London
17. Crank J, Park GS (1968) Diffusion in polymers. Academic, London
18. Pekcan Ö, Winnik MA, Egan L, Croucher MD (1983) Macromolecules 16:699
19. Stauffer D, Coniglio A, Adam M (1982) Gelation and critical phenomena. Advances in polymer science 44. Springer, Berlin Heidelberg New York

Progr Colloid Polym Sci (2004) 128: 98–104
DOI 10.1007/b97096

Christine C. Dupont-Gillain
Elzbieta Pamula
Frédéric A. Denis
Paul G. Rouxhet

Nanostructured layers of adsorbed collagen: conditions, mechanisms and applications

Abstract The nanometer-scale organization of collagen (type I) adsorbed on polystyrene (PS) and plasma-oxidized PS (PSox) was investigated using atomic force microscopy (AFM) as well as radioassays and X-ray photoelectron spectroscopy. The surface properties induced by plasma treatment of PS were examined: PSox was more hydrophilic than PS, owing to oxygen incorporation, and was covered by a layer of polyelectrolyte which could be swollen by water. Collagen adsorbed on PS formed elongated structures, the density and the height of which increased with the adsorption duration. On PSox, a smooth and homogeneous collagen layer was formed after up to 2 h of adsorption, but elongated structures appeared after 24 h. A similar contrast was found between CH_3- and OH-terminated self-assembled monolayers, indicating the role of surface hydrophilicity. The adsorption duration affected the nanostructure of adsorbed collagen layers, not only through the adsorbed amount but also owing to reorganizations of the adsorbed phase. Compared with collagen adsorbed on PSox, collagen adsorbed on PS was more easily picked up for attachment to the AFM probe and more easily displaced along the surface plane by scanning with the atomic force microscope or by drying. This suggests that collagen molecules adsorbed on PS leave segments protruding into the solution, while they form a felt of lying molecules on PSox. The observed elongated structures may be due to association of collagen segments, and/or to particular interactions with the atomic force microscope probe. The respective importance of these two mechanisms may depend on the chemical nature of the substratum. Nanostructured collagen layers offer attractive perspectives to control mammalian cell response or as a template to create nanostructured polymer surfaces.

Keywords Collagen · Protein adsorption · Atomic force microscopy · Polystyrene · Plasma treatment

C.C. Dupont-Gillain · E. Pamula
F.A. Denis · P.G. Rouxhet (✉)
Unité de chimie des interfaces,
Université catholique de Louvain,
Croix du Sud 2/18,
1348 Louvain-la-Neuve,
Belgium
e-mail: rouxhet@cifa.ucl.ac.be
Tel.: +32-10-473587
Fax: +32-10-472005

Present address: E. Pamula
Department of Biomaterials,
Faculty of Materials Science
and Ceramics, AGH–University of Science
and Technology, 30 Mickiewicza Ave.,
30-059 Kraków, Poland

Introduction

Understanding protein adsorption onto solid surfaces is a key step towards many applications, including biomaterial and biosensor design. While, in a first approach, evaluation of the adsorbed amount as a function of time and concentration is often performed, it is now challenging to control the spatial organization of adsorbed protein layers at the nanometer scale. The supramolecular organization of extracellular matrix proteins may affect the cell response [1]. The design of new nanostructured materials could also be inspired by the way biological molecules form organized layers [2, 3].

Collagen, an extracellular matrix protein, is the most abundant structural protein in the animal kingdom. The collagen molecule (length about 300 nm; diameter 1.4 nm) is a helix formed by three polypeptides. It may aggregate and form fibrils, in vivo as well as in vitro [4]. The fibrils formed in vivo [5, 6] and structures deposited in vitro by evaporation of solutions [7] were analyzed regarding their morphology, revealing ordered nanostructures. Collagen thus offers promising perspectives in view of creating nanostructured layers, owing to its shape and dimensions and to its self-assembling properties.

In this report, data obtained in our laboratory in the last few years [8–11] are put together with the aim of giving insight into the organization of adsorbed collagen layers. Polystyrene (PS) and plasma-oxidized PS (PSox), which is a known support for cell cultures, were chosen as substrates for adsorption. Self-assembled monolayers of alkanethiols on gold, terminated with CH_3 or OH groups, served as model surfaces, for the purpose of comparison. The surface of the substrates was characterized using multitechnique analysis, including the Wilhelmy plate method. The adsorbed collagen amounts were assessed using radioassays and X-ray photoelectron spectroscopy (XPS). Atomic force microscopy (AFM) was used, in different modes, to probe the properties of the adsorbed layers at the nanometer scale.

Substrate characterization

PS was spin-coated on glass or silicon supports, except for the Wilhelmy plate method, for which PS Petri dishes were used. PSox was obtained by submitting PS samples to a plasma discharge (13.56 MHz; 50 W; 30 s) in oxygen (7×10^{-4} bar) [8, 11, 12]. Self-assembled monolayers were produced by immersion of cleaned gold supports in 1 mM alkanethiol solutions in ethanol. The alkanethiols were $HS(CH_2)_{17}CH_3$ and $HS(CH_2)_{11}OH$ to obtain CH_3 and OH unifunctional surfaces, respectively [9].

Table 1 O-to-C ratio measured by X-ray photoelectron spectroscopy, water contact angle (θ), and root-mean-square roughness (R_{rms}) measured on 5×5 μm^2 atomic force microscope images for the different substrates used in this study for polystyrene (*PS*), plasma-oxidized PS (*PSox*), CH_3 and OH substrates

Substratum	O/C	θ (°)	R_{rms} (nm)
PS	0.01	88	0.3
PSox	0.26	59	0.3
CH_3	< 0.004	110	0.5
OH	0.15	24	0.5

The wettability of the substrates was evaluated using water contact angle (θ) measurement (sessile drop method), their surface chemical composition was determined with XPS, and their surface morphology was observed by AFM. The information obtained is summarized in Table 1. All the substrates were smooth and featureless, as indicated by their very low roughness. PS and CH_3 surfaces were hydrophobic (high θ), which can be linked to the very low amount or to the absence of oxygen as detected by XPS. In contrast, PSox and OH surfaces were also more hydrophilic (low θ). The ordering of the alkane chains in the self-assembled monolayers is at the origin of the more contrasted θ for CH_3 and OH surfaces compared with PS and PSox [13].

The changes induced at the PS surface by plasma treatment were investigated further [12]. The variation of cos θ measured using the Wilhelmy plate method on PS and PSox during repeated wetting cycles at pH 5.6 is shown in Fig. 1. For PS, the shape of the hysteresis remained constant for repeated cycles, with advancing (θ_{adv}) and receding (θ_{rec}) contact angles of 97° and 78°,

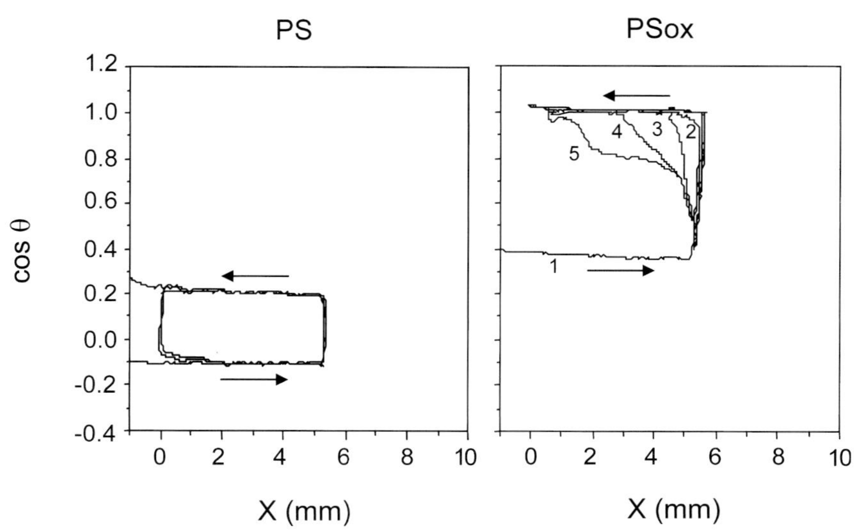

Fig. 1 Variation of cos θ as a function of the three-phase contact line position *X* during repeated immersion–emersion cycles for polystyrene (*PS*) and plasma-oxidized PS (*PSox*) at pH 5.6. On PS, three successive cycles were performed. On PSox, cycles 1 and 2 were successive, and a pause of 5, 10 and 30 min was used after cycles 2, 3 and 4, respectively. Reprinted with permission from *Langmuir* 16:8194. Copyright (2000) American Chemical Society

respectively. This behavior was not affected by pH. For PSox, at pH 5.6, θ_{rec} was always 0°, while θ_{adv} was 68° for the first cycle but 0° if a second cycle was performed immediately. If the sample was then submitted to pauses of increasing duration between successive immersion–emersion cycles, θ_{adv} increased progressively. The shape of the hysteresis was strongly affected by pH. At pH 3.0, θ_{adv} and θ_{rec} were 66° and 0° for all cycles. At pH 11.0, θ_{adv} and θ_{rec} were always 0°, even when pauses were performed between cycles, except for the first cycle, where $\theta_{adv} = 66°$.

Analysis of the C_{1s} peak obtained by XPS on PSox revealed the presence of a variety of oxygen-containing groups, including ionizable ones (carboxylic and carbonic acids, phenols). Moreover, the ζ potential estimated using streaming potential measurements was more negative for PSox compared with PS at all pH values. Compared with PS, PSox was more sensitive to atomic force microscope probe damage and showed lower friction in lateral force mode AFM. Taken together, these data indicate that the PSox surface may be viewed as covered by a polyelectrolyte. Deprotonation of oxygen-containing groups at high pH provokes a repulsion between adjacent polymer chains, allowing the surface to be swelled by aqueous solutions, depending on their pH. This close interaction with water explains why θ_{adv} is 0° at high pH. θ_{adv} of the first cycle is, however, higher because, in that case, the water meniscus defines a border between nonsolvated and solvated macromolecules. At pH 5.6, the swelling is more limited, and water evaporation during the pauses allows hydrophobicity to be recovered.

Supramolecular structure of adsorbed collagen as a function of time and of the substratum

Type I collagen (from calf skin; purchased either from Sigma-Aldrich in lyophilized form, or from Roche Diagnostics in solution at pH 3.0) was diluted to a concentration of 40 μg ml^{-1} in phosphate buffered saline (pH~7.2; ionic strength around 160 mM). It was checked that under these conditions collagen fibril formation in solution was very limited, the optimum pH for such fibril formation being around 5.0 [14]. Adsorption was then performed at 37 °C, using 5 ml of solution, on 1.3-cm^2 PS or PSox samples. The adsorption duration was varied, from 1 min to 24 h. Samples were rinsed with water and either examined in water using AFM, or dried and analyzed with XPS. The adsorbed amounts were determined using collagen labeled by reductive methylation with ^{14}C formaldehyde [15]. Collagen layers were also prepared on CH_3- and OH-terminated surfaces. The adsorption conditions were similar, but the collagen concentration was 30 μg ml^{-1}.

The evolution of the collagen adsorbed amount as a function of the adsorption duration, for PS and PSox, is presented in Fig. 2. The adsorbed amount increases sharply up to 2 h, after which a plateau is reached. The adsorbed amount is always smaller for PSox compared with PS, especially for the shorter adsorption times. The plateau level is about 0.9 and 1.1 μg cm^{-2} for PSox and PS, respectively. The adsorption kinetics was also followed by XPS. For both polymers, the evolution of the N-to-C ratio with the adsorption duration showed similar trends compared to those obtained from the radioassays.

AFM images, acquired in water, of collagen layers on PS and PSox after 2 and 24 h of adsorption, and on CH_3- and OH-terminated surfaces after 2 h of adsorption are presented in Fig. 3. On PS, a collagen layer showing elongated structures (height around 10 nm) is formed after 2 h. This was already the case after 1 min (not shown here; height around 5 nm). After 24 h, the density of the elongated structures increased. On PSox, the surface organization obtained up to 2 h is clearly different: the collagen layer is rather homogeneous and smooth. After 24 h, however, a pattern made of elongated structures (height around 5 nm) is observed. The morphologies obtained on CH_3- and OH-terminated surfaces after 2 h of collagen adsorption are remarkably similar to those observed on PS and PSox, respectively, for the same adsorption duration. It has been shown that the observed elongated structures are not due to fibril formation in solution but result from processes occurring in the adsorbed phase [14].

The adsorption duration is thus identified as a factor affecting the nanostructure of adsorbed collagen layers. This may partly be explained by the evolution of the adsorbed amount with time (Fig. 2). This is the case for the height and the density increase of the elongated structures from 1 min to 2 h of adsorption on PS. However, changes in the organization of the collagen

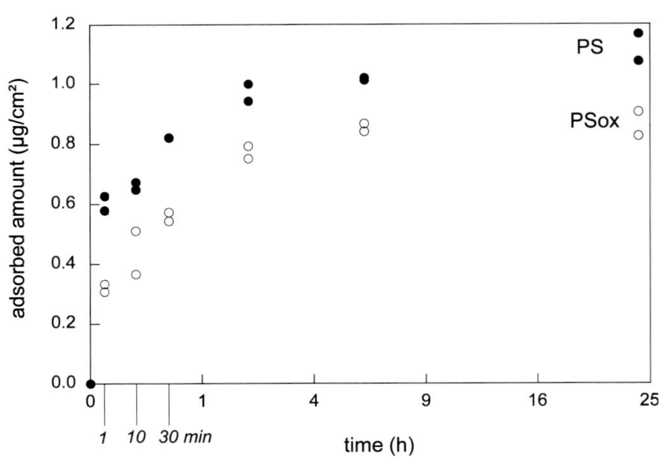

Fig. 2 Collagen adsorbed amount determined using radioassays on PS (•) and PSox (○) samples treated for different periods of time with a 40 μg ml^{-1} collagen solution. The abscissa scale was built based on the square root of the time, for the sake of clarity

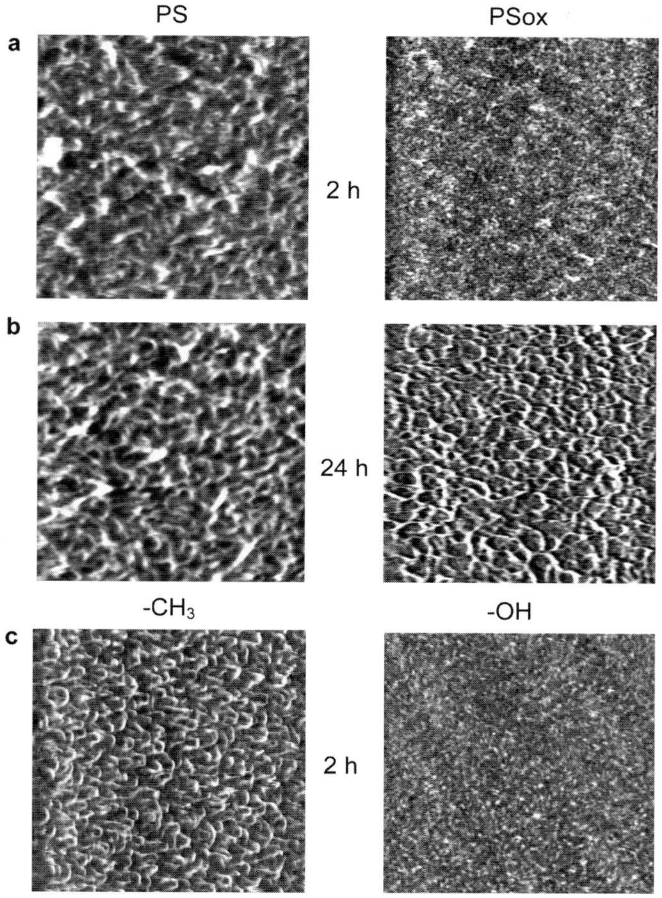

Fig. 3 Atomic force microscopy (*AFM*) images (5×5 μm^2) obtained in water on PS and PSox after **a** 2 h and **b** 24 h of adsorption from a 40 $\mu g\ ml^{-1}$ collagen solution ($z = 10$ nm on PS and $z = 5$ nm on PSox), and **c** on CH_3- and OH-terminated self-assembled monolayers after 2 h of adsorption from a 30 $\mu g\ ml^{-1}$ collagen solution ($z = 20$ nm on $-CH_3$ and $z = 10$ nm on –OH). **c** Reprinted with permission from *Langmuir* 18:819. Copyright (2002) American Chemical Society

layer are still observed between 2 and 24 h, at almost constant adsorbed amount. This is especially clear on PSox, where a smooth layer is obtained after 2 h, while the surface is covered with elongated features after 24 h. These reorganizations, which are not linked to evolution of the net adsorbed amount, may be explained either by exchanges between adsorbed collagen and collagen molecules in solution, or by reorganization of adsorbed collagen at the interface.

The chemical nature of the substrat also clearly influences the organization of the adsorbed collagen layers. The different organizations observed at short adsorption durations on PS and PSox may partly be attributed to the smaller adsorbed amount on PSox compared with PS. However, the amount adsorbed after 2 h is quite similar for PS and PSox, while the nanostructure of the layers remains very different. Given the particular surface properties of PSox, this behavior could tentatively be attributed to collagen interaction with the polyelectrolyte layer present at the PSox surface; however, similar behavior is observed when comparing CH_3 surfaces with OH surfaces on which no polyelectrolyte layer is present. Although it is difficult to establish to what extent the characteristic properties of the PSox surface play a role in the nanostructuration of adsorbed collagen, it seems that the effect of the surface hydrophilicity induced by the plasma treatment is predominant.

Understanding the formation and properties of adsorbed collagen layers

AFM is not only useful to observe the nanometer-scale morphology of adsorbed layers; it also allows the physicochemical properties of these layers to be probed. A first possibility is to record the tip–sample interaction as a function of separation, i.e., to acquire force–distance curves. The force–distance curves recorded for PS and PSox after collagen adsorption for 30 min from a 40 $\mu g\ ml^{-1}$ collagen solution are presented in Fig. 4a. For both polymers, the approach curve shows a curvature in the contact region; in certain sets of experiments, the curvature was marked at a larger distance for PS compared with PSox. For PS, in contrast with PSox, multiple adhesion events were recorded upon retraction [11].

By increasing the applied force, AFM can also be used to probe the mechanical resistance [8] or the thickness [9] of adsorbed layers. Figure 4b shows 10×10-μm^2 AFM images of collagen layers (adsorption for 2 h from a 60 $\mu g\ ml^{-1}$ solution) on PS and PSox, obtained in water after prescanning of the 5×5-μm^2 central area at a loading force of 7.1 nN. On PS, the adsorbed collagen layer is markedly modified upon scanning: collagen reorganizes into large strips running perpendicular to the scanning direction. In contrast, no damage is created on the collagen layer adsorbed on PSox.

Perturbation of the adsorbed layer may alternatively be provoked by drying [8]. AFM images of collagen layers formed by adsorption (7 $\mu g\ ml^{-1}$; 2 h) on PS and PSox followed by drying in 95% relative humidity (slow drying) are shown in Fig. 4c. While the collagen layer on PSox remains smooth after slow drying, it reorganizes strongly on PS, leading to the formation of a netlike structure. This behavior may be due to the rupture of the liquid film by dewetting, collagen being displaced by the water meniscus. Alternatively, it may be due to aggregation of collagen molecules as the water activity decreases upon evaporation. The insert in the image obtained for PS is an adhesion map, i.e., an image constructed using the adhesion force measured at each location by means of a force–distance curve. It shows in a direct way that the

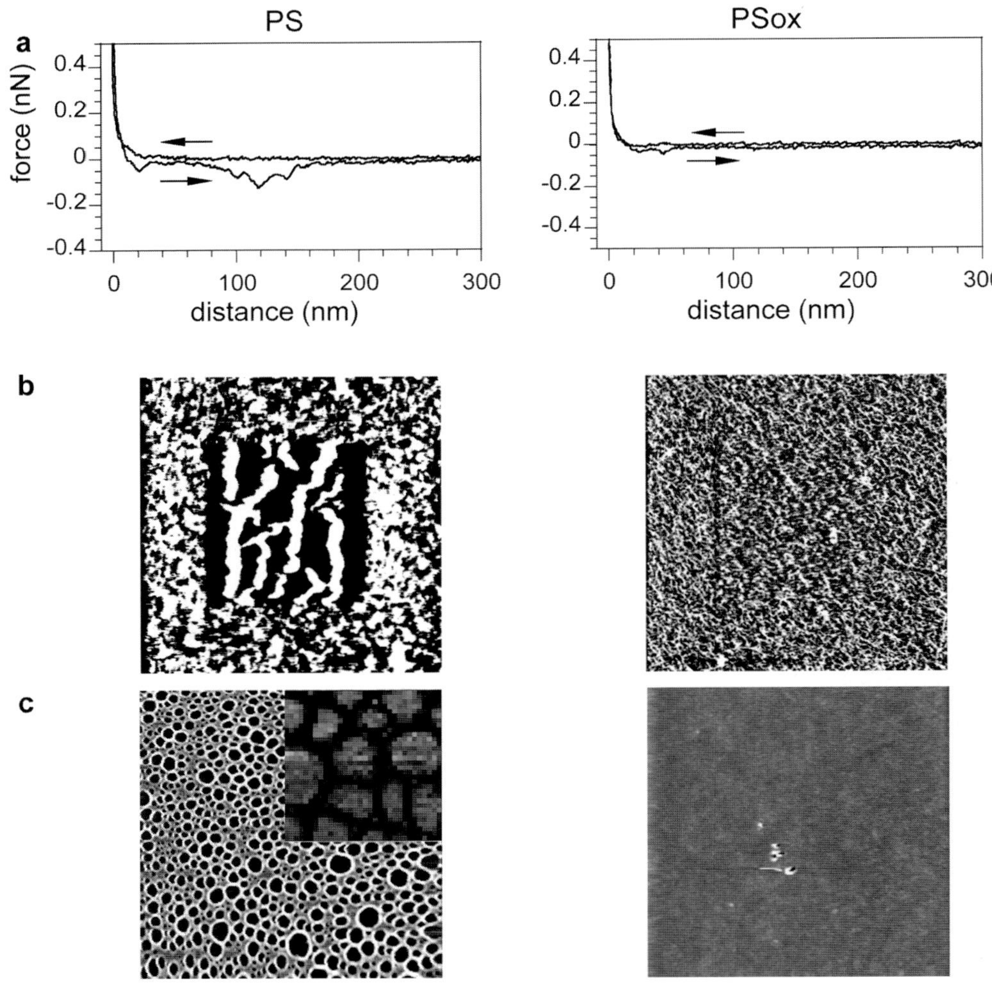

Fig. 4 Characterization of the properties of adsorbed collagen layers on PS and PSox: **a** typical force–distance curves recorded in water with AFM (adsorption from a 40 $\mu g\ ml^{-1}$ solution); **b** AFM images (10 × 10 μm^2; $z = 10$ nm) showing the perturbation induced by prescanning the central 5 × 5 μm^2 area (adsorption from a 60 $\mu g\ ml^{-1}$ solution); **c** AFM images (5 × 5 μm^2; $z = 10$ nm) obtained after slow drying (adsorption from a 7 $\mu g\ ml^{-1}$ solution). The *insert* in the image of PS after slow drying shows an adhesion map (2 × 2 μm^2; $z = 1.5$ nN) obtained using AFM in water. **b**, **c** Reprinted with permission from *Langmuir* 17:7261. Copyright (2001) American Chemical Society. Insert in **c** reprinted from *Surface Science* 539:145. Copyright (2003), with permission from Elsevier

contrast involved in the pattern is not only topographic but also of a chemical nature, with PS domains present at the outermost surface in the holes of a collagen net. The low adhesion measured on the net threads is indeed comparable to that measured on a continuous collagen coating, while the higher adhesion measured in the holes of the net is typical of PS [10].

The multiple adhesion events recorded in water for PS with adsorbed collagen may be attributed to the presence of dangling collagen segments, which can be picked up by the atomic force microscope probe. On PSox, adsorbed collagen molecules would in contrast lie on the surface, or would be entangled in such a way that no free segments may bridge with the atomic force microscope probe. The absence of perturbation induced upon scanning the adsorbed collagen layer on PSox may also be due to the stronger entanglement of the collagen molecules, which would confer high mechanical resistance to the layer, or to a strong collagen–PSox interaction. In contrast, the collagen–PS interaction would be weaker, or the organization of collagen on PS would not provide stability to the adsorbed layer. The stronger reorganization of the adsorbed collagen layer provoked by slow drying on PS compared with PSox may be explained by the same reasons, although in that case it might simply be due to the more hydrophobic character of PS, which would enhance the effect of dewetting.

Collagen adsorbed on PS may thus be viewed as interacting weakly with the substratum and leaving free segments dangling in solution, while collagen adsorbed on PSox would interact strongly with the substratum and/or would form a mechanically resistant layer of entangled molecules. This can be linked to the morphology of the adsorbed layers. The smooth collagen layers obtained on PSox after up to 2 h of adsorption would result from the formation of a felt of lying collagen molecules. This is supported by the higher thickness and compressibility of the collagen layer revealed by AFM on CH_3 surfaces compared with OH surfaces [9, 16]. It is also confirmed by the difference between the two systems regarding the viscoelastic properties of the adsorbed layer as deduced from quartz crystal microbalance measurements [16].

The origin of the elongated structures observed on PS whatever the adsorption duration, as well as on PSox after 24 h of adsorption, may be explained by different mechanisms, which are depicted in Fig. 5. In the first model, the elongated structures would be formed within the adsorbed phase, as a result of the balance of collagen–collagen and collagen–surface interactions. In the second model, collagen molecules leave free segments in solution, which can form bundles owing to the tendency of collagen to self-associate. These bundles of molecules are then pushed along the surface plane by the scanning atomic force microscope probe if a sufficient loading force is applied. Such a model was also proposed for collagen adsorbed on poly(ethylene terephthalate) [15]. Finally, it may not be excluded that the elongated structures result from a dynamic interaction between the atomic force microscope probe and a lawn of macromolecular segments (model 3).

Conclusions and perspectives

The nanometer-scale organization of collagen adsorbed on PS and PSox is shown to depend on both the adsorption duration and the chemical nature of the substratum. On PS, elongated structures, the density and the height of which increase with time, are observed. On PSox, a smooth and homogeneous layer is found after up to 2 h of adsorption, but elongated structures are also formed after 24 h. The effect of the adsorption duration may not only be explained by the increase of the adsorbed amount; reorganization within the adsorbed layer or exchanges with collagen in solution must also take place. The particular surface properties of PSox, i.e., the presence of a hydrated layer of polyelectrolyte, may contribute to the difference observed between PS and PSox; however, comparison with CH_3- and OH-terminated self-assembled monolayers suggests that surface hydrophobicity plays a predominant role.

Collagen adsorbed on PS is more easily picked up for attachment to the atomic force microscope probe and is more easily displaced along the surface plane by scanning with the atomic force microscope, compared with collagen adsorbed on PSox. This indicates that collagen molecules adsorbed on PS leave free segments protruding into the solution. On PSox, a felt of lying collagen molecules would be formed, thereby offering stronger mechanical resistance. The elongated structures observed by AFM may be due to association of collagen segments; however, particular interactions between the atomic force microscope probe and adsorbed molecules may influence the imaging process.

Being able to control the nanostructure of adsorbed collagen layers offers perspectives in different domains. Recently, it was shown that the morphology of smooth muscle cells could be modulated by the supramolecular

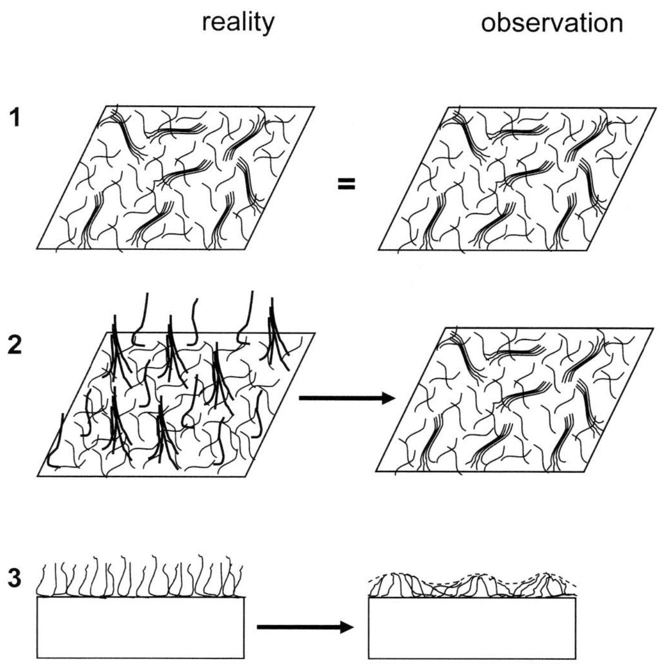

Fig. 5 Schematic representation of three possible models to account for the observation of the collagen elongated structures

organization (density of fibrils) of collagen present on the substratum [17]. We also showed that the adhesion of rat hepatocytes and of MCF-7/6 breast cancer cells was enhanced on nanopatterned compared with smooth collagen layers [18]. A better understanding of the mechanisms of organization of adsorbed collagen could thus lead to better control of cell-material interactions, which is a key step towards the design of biomaterials.

Collagen and, more generally, biological molecules are able to form nanostructures which can be used as templates to create other nanostructured materials. By spin-coating a good solvent of poly(methyl methacrylate) (PMMA) on top of a heterogeneous collagen–PMMA surface (obtained by adsorption and slow drying), we have shown that it was possible to create nanostructured PMMA surfaces, based on the collagen template [19]. The same should be possible using other polymers or substrates.

Acknowledgements The authors wish to thank V. De Cupere for the radiolabeling and for helpful discussions, I. Jacquemart for the adhesion map, as well as late Prof. P. Grange and P. Bertrand for the use of the atomic force microscope and of the spin-coater, respectively. C.D.-G. is a postdoctoral researcher of the Belgian National Foundation for Scientific Research (F.N.R.S.). The support of the F.N.R.S., of the Foundation for Training in Industrial and Agricultural Research (F.R.I.A.), of the Federal Office for Scientific, Technical and Cultural Affairs (Interuniversity Poles of Attraction Program) and of the Research Department of the Communauté Française de Belgique (Concerted Research Action) is gratefully acknowledged.

References

1. Brash JL, Horbett TA (1995) In: Horbett TA, Brash JL (eds) Proteins at interfaces II. ACS symposium series 602. American Chemical Society, Washington, DC, pp 1–23
2. Lowe CR (2000) Curr Opin Struct Biol 10:428
3. Dujardin E, Mann S (2002) Adv Mater 14:775
4. Kadler K (1994) Protein Profile 1:519
5. Baselt DR, Revel JP, Baldeschwieler JD (1993) Biophys J 65:2644
6. Baranauskas V, Vidal BC, Parizotto NA (1998) Appl Biochem Biotechnol 69:91
7. Maeda H (1999) Langmuir 15:8505
8. Dupont-Gillain CC, Rouxhet PG (2001) Langmuir 17:7261
9. Denis FA, Hanarp P, Sutherland DS, Gold J, Mustin C, Rouxhet PG, Dufrêne YF (2002) Langmuir 18:819
10. Dupont-Gillain CC, Jacquemart I (2003) Surf Sci 53:145
11. Pamula E, De Cupere VM, Dufrêne YF, Rouxhet P J Colloid Interface Sci (2004)
12. Dupont-Gillain CC, Adriaensen Y, Derclaye S, Rouxhet PG (2000) Langmuir 16:8194
13. Ulman A (1996) Chem Rev 96:1533
14. Dupont-Gillain CC, Jacquemart I, Rouxhet PG (submitted)
15. De Cupere VM, Van Wetter J, Rouxhet PG (2003) Langmuir 19:6957
16. Denis FA (to be published)
17. Elliott JT, Tona A, Woodward JT, Jones PL, Plant AL (2003) Langmuir 19:1506
18. Dupont-Gillain CC (2000) PhD thesis. Université catholique de Louvain, Louvain-la-Neuve, Belgium
19. Dupont-Gillain CC, Rouxhet PG (2001) Nanoletters 1:245

Progr Colloid Polym Sci (2004) 128: 105–108
DOI 10.1007/b97098

Masashi Mizukami
Kenichi Kusakabe
Kazue Kurihara

Shear resonance measurement on structuring of liquids confined between mica surfaces

Abstract The behavior of liquid molecules confined in nanometer-scale spaces is quite different from that in the bulk. The structuring of liquids confined between solid surfaces was investigated using a shear force resonance method, which was developed in order to examine in detail the ordering behavior of liquid molecules confined between two mica surfaces. The frequency and the amplitude of the resonance peak are highly sensitive to the long-range order and properties of the confined sample, affording information about the structuring, and frictional and lubrication properties of the liquids from the nanometer thickness to the bulk. This measurement was used to study the properties of octamethylcyclotetrasiloxane (OMCTS) confined between mica surfaces, and was compared with those of 4-cyano-4′-hexylbiphenyl, a liquid crystal molecule. Both liquids are known to exhibit an oscillation force. The resonance peak intensity measured in the presence of OMCTS showed no decrease down to the surface distance of 5 nm upon compression, and abruptly disappeared at 4.6 nm. This means that the resonance peak intensity was damped by the abrupt increase in the viscosity of OMCTS induced by the confinement. Broad and weak resonance peaks centered at angular frequencies of 250–305 s^{-1} were observed while further decreasing the distance, indicating the enhanced structuring of OMCTS between the surfaces. Unlike 4-cyano-4′-hexylbiphenyl, finally, OMCTS was completely removed from the gap, and a resonance peak similar to that of mica contact in air appeared.

Keywords Nanostructuring of liquids · Confined liquid · Shear resonance measurement · Nanotribology · Octamethylcyclotetrasiloxane

M. Mizukami · K. Kusakabe
K. Kurihara (✉)
Institute of Multidisciplinary Research for Advanced Materials,
Tohoku University, Katahira, Aoba-ku,
Sendai 980-8577, Japan
e-mail: kurihara@tagen.tohoku.ac.jp

Introduction

Liquid molecules confined in a nanometer-scale space, a pore of a porous crystal or a nanospace between two solid surfaces, behave quite differently from those in the bulk [1, 2]. They show peculiar properties, such as the formation of an ordered structure and a phase transition owing to the confinement effect and/or the interactions between liquid molecules and solid surfaces. Elucidation of their behavior has become increasingly important not only in basic surface science but in such applications as micromachines, liquid crystal displays and lubrication techniques.

The surface forces apparatus (SFA) is a unique tool for characterizing the properties of confined liquids as a function of the surface separation, i.e., liquid film thickness. The oscillatory forces were observed for liquid octamethylcyclotetrasiloxane (OMCTS) [3], and liquid crystal [4] thin films with a thickness of several molecular layers. The period of the oscillation corresponds to the thickness of a molecular layer, which is characteristic of the structuring of liquid molecules.

Taking advantage of the precise determination of the distance by the surface forces measurement, various types of shear force measurements were also developed and employed to study the structuring of confined liquids by monitoring the rheological and/or frictional responses [5–9]. Among them, the method we developed, the resonance measurement, was used to study liquid crystal thin films [7–12]. The frequency and the amplitude of the resonance peak are highly sensitive to the long-range order within the confined sample, affording information about the structuring, friction and lubrication properties of liquids from the nanometer thickness to the bulk. We studied the response of OMCTS, which previously showed an oscillation force, and compared it with the resonance response of 4-cyano-4′-hexylbiphenyl (6CB).

Method

A homebuilt shear force resonance unit, described earlier (Fig. 1), was used in combination with the SFA (Mark IV Anutech Co.) [7–9]. Here, the surfaces were hung by a pair of stiff leaf springs and moved laterally by a piezoelectric element, which was driven by the sinusoidal input voltage (U_{in}) of frequency f. In this case, the upper surface vibrates at a constant angular frequency $\omega(2\pi f)$. The deflection (Δx) of the leaf spring was detected using a static capacitance probe (U_{out}). The U_{in} and the U_{out} signals were recorded. The amplitude of the oscillation was then defined as the ratio of U_{out} to U_{in}, which was measured as a function of the frequency. The details of the setup can be found elsewhere [7–9].

Mica sheets used as the substrates were glued onto cylindrical quartz disks (the radius of curvature was about 2 cm, and the diameter was about 1 cm) with the silver-covered sides down on the disk, and were then subjected to the experiments.

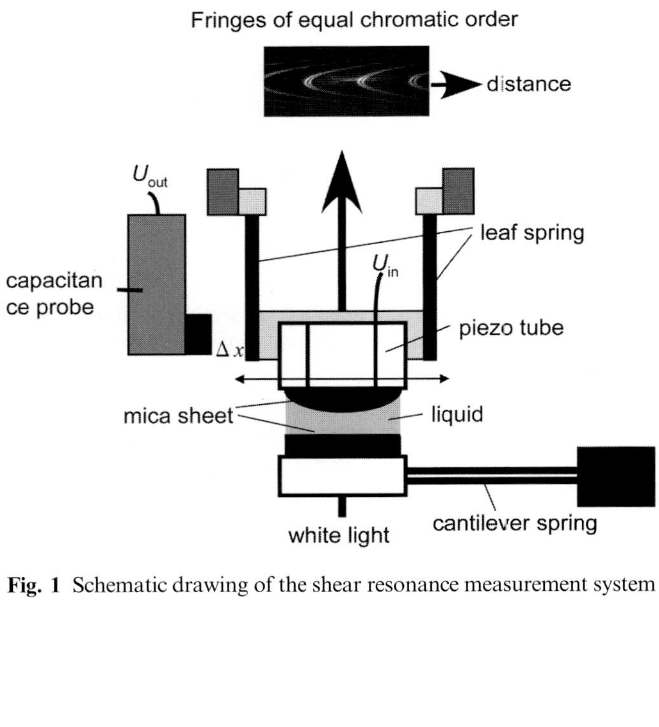

Fig. 1 Schematic drawing of the shear resonance measurement system

The measurements were first carried out with the mica surfaces separated (AS) in air, and the resonance frequency ($f_{0,AS}$) and the amplitude were obtained. In this case, the value f_0 was determined using the mass and the spring constant of the top component of the resonance unit. A similar set of measurements was performed for the surfaces in contact (MC), where the resonance frequency ($f_{0,MC}$) was determined using the mass and the spring constant of the top and the bottom unit together.

The liquid or liquid crystal sample (about 20 μl) was then injected into the gap between the mica surfaces. The distance between the mica surfaces (D) was determined using the fringes of equal chromatic order with a resolution of 0.1 nm [4]. The measurement was started at D=100–200 nm. The resonance curve was recorded at each separation as the bottom surface approached the top. 6CB (Merck) was used as received. OMCTS (Tokyo Kasei) was dried with a 4-Å molecular sieve (Nacalai Tesque) and was distilled prior to use.

Properties of liquid crystal thin films

First, the results for the 6CB confined between the mica surfaces are described to demonstrate how the resonance properties for the confined liquids change [10–12]. 6CB molecules are known to align parallel to the mica surface. Resonance curves obtained for 6CB in the nematic phase at room temperature (about 23 °C) are shown in Fig. 2.

At D = 104 nm, the resonance peak appeared at an angular frequency of 212 s^{-1} which was practically the same as the AS peak. The resonance amplitude was lower than that observed for air, which should be due to the higher viscosity of the liquid crystals than air. This resonance behavior did not significantly change when the distance was decreased to 18 nm (data not shown). The

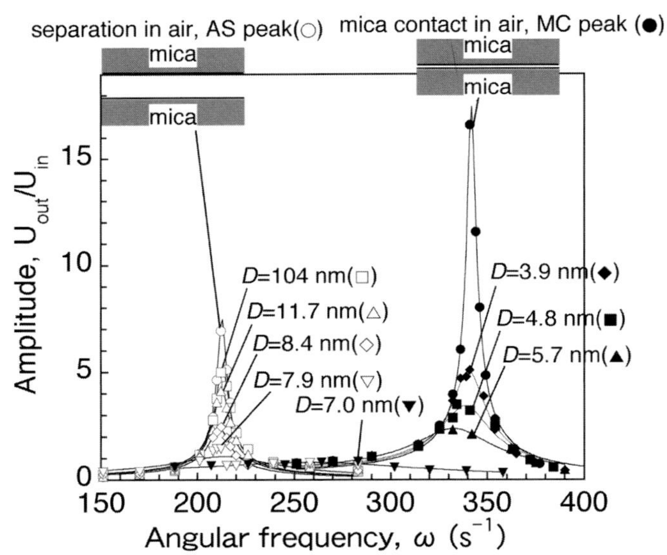

Fig. 2 Resonance curves of 4-cyano-4′-hexylbiphenyl (*6CB*) confined between mica surfaces at various separation distances. The resonance curves for surfaces separated in air, and connected by a mica–mica contact are also shown. The curves appearing at the AS position are denoted by *open symbols*, and others by *filled symbols*

amplitude of the resonance peak started to decrease at $D = 11.7$ nm. It has been reported that 6CB molecules form dimers with a short axis of approximately 1 nm and a long axis of approximately 3 nm; thus, $D = 11.7$ nm corresponds to 16 molecular layers.

In the range of the distance from 7.9 nm (14 molecular layers) to 5.7 nm, which we call the intermediate region, the resonance peak completely disappeared. Here, unlike the resonance in air, the top surface did not move independently owing to the structuring of the liquid crystals. The loose coupling of the two surfaces should result in the disappearance of the resonance peak by the energy dissipation.

Below $D = 5.7$ nm (12 molecular layers), both the resonance amplitude and the frequency started to increase toward the mica contact peak (the frequency of 341 s^{-1}). This indicated that, similar to the mica–mica adhesive contact, the film bound the two surfaces as the structuring of the liquid crystals progressed. However, the resonance peak on the contact side grew by only a limited amplitude and was much lower than the magnitude of the MC peak. A possible mechanism is the lubrication of the mica surfaces, where the internal friction of the liquid crystal sample may play an important role. The surfaces could come as close as only a 4-nm film thickness without damaging the mica surfaces (flat contact). It was difficult to remove eight molecular layers of 6CB owing to the high affinity of 6CB to the mica surfaces and the confinement effect.

At $D < 7.9$ nm, the distance decreased stepwise by about 1 nm, which corresponded to the length of the short axis of the 6CB dimer, indicating that the 6CB dimers were removed from the nanospace one layer at a time. Similar removal behavior was previously observed for thin films of liquids and liquid crystals which accompanied the oscillatory forces [13]. It is interesting that the oscillatory force was reported at $D = 5$ nm (for 8CB), which was the distance of the so-called contact region in our work. The resonance method is able to monitor the properties of a confined liquid for much larger distances than the SFA, indicating the higher sensitivity of this method than the SFA technique.

Properties of OMCTS thin films

The resonance measurement was used to study OMCTS confined between mica surfaces (Fig. 3). OMCTS is a ring molecule with a diameter of approximately 0.9 nm. At $D = 38.5$ nm, the resonance peak appeared at an angular frequency of 220 s^{-1}, which was the same as the frequency for AS. The resonance amplitude was the same as that for the AS peak, which means the viscosity of OMCTS is very low and similar to the value for air within the sensitivity of the measurement. This resonance behavior did not change on decreasing the distance to

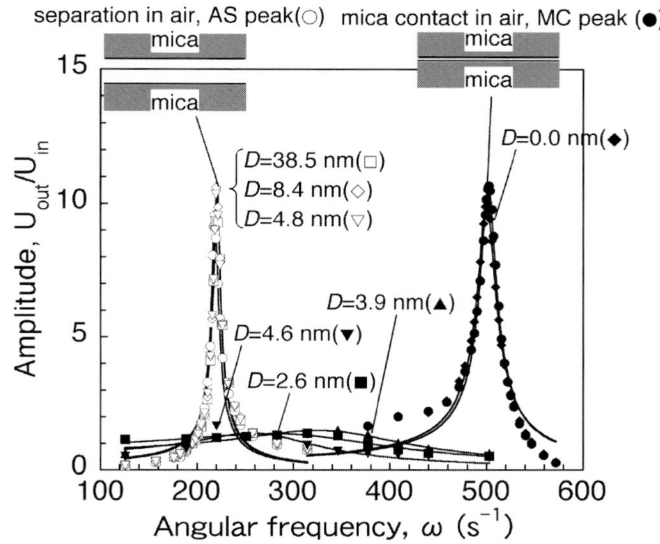

Fig. 3 Resonance curves of octamethylcyclotetrasiloxane (*OMCTS*) confined between mica surfaces at various separation distances. The resonance curves for surfaces separated in air, and connected by a mica contact are also shown. The curves appearing at the AS position are denoted by *open symbols*, and others by *filled symbols*

about 5 nm, where the amplitude of the resonance peak started to decrease. The peak almost disappeared at $D = 4.6$ nm. The distance of 5 nm, corresponding to five molecular layers, was same as the distance where the oscillating force appeared [3].

In the range from $D = 3.9$ nm (four molecular layers) to 2.7 nm (three molecular layers), which we call the intermediate region, the broad resonance peaks appeared at angular frequencies of 250–350 s^{-1}. Here, unlike the resonance in air, the two surfaces were loosely coupled owing to the structuring of OMCTS. The most significant difference from 6CB was that for the resonance curve of OMCTS: the peak frequency never increased close to the value that was measured for the mica adhesive contact as long as the OMCTS molecules remained in the gap. One may note that the higher resonance frequency of the MC peak (500 s^{-1}) shown in Fig. 3 compared with the MC peak shown in Fig. 2 (341 s^{-1}) was caused by the larger spring constant (160 N m^{-1} in the direction normal to the surface) of the cantilever spring, which supported the lower surface, compared with the value, 60 N m^{-1}, used for the 6CB measurement (see the equations in Refs. [7–9]). When we further compressed the surface, the distance suddenly became 0.0 nm, and the resonance peak shifted to an angular frequency of 500 s^{-1}, which was almost identical to the MC peak. This means that OMCTS molecules were completely removed from the gap between the mica surfaces. It was a quite contrast to the fact that a 4-nm-thick 6CB is not possible to remove from the gap under a similar pressure. These differences between OMCTS and

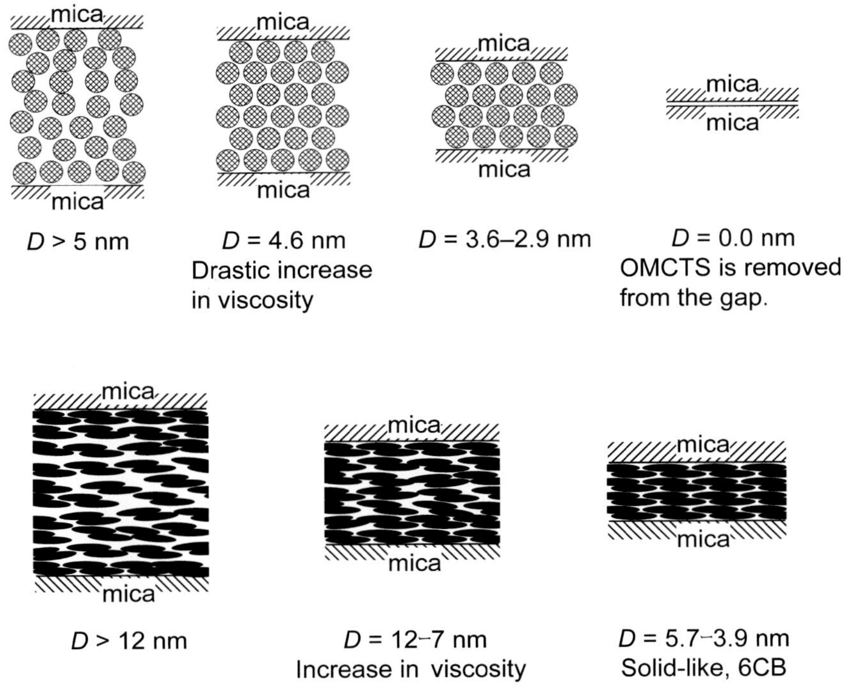

Fig. 4 Schematic drawings of plausible structures of confined OMCTS (*top*) and 6CB (*bottom*)

6CB reflected the weaker interactions between OMCTS molecules and the mica surface than those between 6CB molecules and the mica surface. Plausible models of the confined OMCTS and 6CB between the mica surfaces are shown in Fig. 4.

Conclusion

The structuring of liquids at solid–liquid interfaces is becoming progressively important in both basic science and technology. A novel technique, the resonance shear measurement, was developed for characterizing liquid (or liquid crystal) thin films confined between mica surfaces. This technique is relatively simple and easy to use, and has demonstrated high sensitivity to the rheological and tribological behaviors of thin films of liquids.

It is interesting that two types of molecules, OMCTS and 6CB, exhibited different behavior in spite of the fact that both demonstrated an oscillating force. This is most likely due to differences in their interactions with the mica surfaces as well as with themselves. This difference should have an important bearing for better understanding the properties of confined liquids, such as friction and lubrication. This study definitely demonstrated the high potential of the resonance shear measurement.

Acknowledgements This work was supported by Grant-in-Aids for Scientific Research (14340214) and for the COE project, Giant Molecules and Complex Systems, 2003, from the Ministry of Education, Culture, Sports, Science and Technology; and by the CREST program of the Japan Science and Technology Cooperation.

References

1. Granick S (1991) Science 253:1374
2. Adamson AW, Gast AP (1997) Physical chemistry of surfaces, 6th edn. Wiley, New York
3. Horn RG, Israelachvili JN (1981) J Chem Phys 75:1400
4. Horn RG, Israelachvili JN, Perez E (1981) J Phys 42:39
5. Israelachvili JN, McGuiggan PM, Homola AM (1988) Science 240:189
6. Van Alsten J, Granick S (1990) Langmuir 6:876
7. Dushkin CD, Kurihara K (1997) Colloids Surf A 129/130:131
8. Dushkin CD, Kurihara K (1998) Rev Sci Instrum 69:2095
9. Dushkin CD, Kurihara K (1997) Prog Colloid Polym Sci 106:262
10. Kusakabe K, Kurihara K (1999) Technical report of IEICE OME99–31:51
11. Mori K, Kusakabe K, Haraszti T, Kurihara K (2001) Trans MRS-J 26:909
12. Kurihara K (2002) Prog Colloid Polym Sci 121:49
13. Ruths M, Steinberg S., Israelachivili JN (1996) Langmuir 12:6637

Progr Colloid Polym Sci (2004) 128: 109–112
DOI 10.1007/b97100

Marcelo Henrique Sousa
Francisco Augusto Tourinho
Jérôme Depeyrot
Emmanuelle Dubois
Régine Perzynski

Synthesis and room temperature magnetic and magneto-optical characterization of size-controlled nickel nanoferrite based ferrofluids

M.H. Sousa (✉) · F.A. Tourinho
Complex Fluids Group, Instituto de Química, Universidade de Brasília, Caixa Postal 04478, CEP 70919-970, Brasília, Brazil
e-mail: fralda@unb.br
Tel.: +55-61-3072163
Fax: +55-61-2734149

J. Depeyrot
Instituto de Física, Universidade de Brasília, Caixa Postal 04478, CEP 70919-970, Brasília, Brazil

E. Dubois
Laboratoire Liquides Ioniques et Interfaces Chargées, Université Pierre et Marie Curie (Paris 6), Case 63, 4 Place Jussieu, 75252 Paris Cedex 05, France

R. Perzynski
Laboratoire des Milieux Désordonnés et Hétérogènes, Université Pierre et Marie Curie, Tour 13, Case 78, 4 place Jussieu, 75252 Paris Cedex 05, France

Abstract Nickel ferrite nanoparticles are synthesized bottom up and then peptized in aqueous media resulting in stable magnetic fluids. X-ray powder diffraction and transmission electronic microscopy permit the investigation of the crystalline structure and the morphological aspects of the nanoparticles. Magnetic and magneto-optical properties of the resulting materials are investigated by static measurements performed on individual particle solutions.

Keywords Nickel ferrite · Ferrofluid · Magnetization · Birefringence

Introduction

Ferrite particles have been used extensively with grain sizes of the order of a few nanometers. In fact, at this scale, the size reduction induces the normal macroscopic multidomain structure to convert into a single domain at a critical size, which typically lies below 100 nm. The magnetism of ferrite nanoparticles exhibits properties such as superparamagnetism, enhanced anisotropy and surface effects, which are of great interest for industrial applications, particularly in the field of information storage [1]. Nanoparticles also present some attractive possibilities in biomedicine since they have dimensions that are smaller than or comparable to those of the biological entities. Indeed, they can be coated with biological molecules and dispersed in a liquid medium, forming colloidal solutions called magnetic fluids or ferrofluids. Owing to their liquid properties and sensitivity to an applied magnetic field, such materials can be, for example, made to deliver a package, such as an anticancer drug, or radionuclide atoms, to a targeted region of the body, such as a tumor [2]. In this work, we propose the synthesis and investigation of the magnetic and magneto-optical properties of electrical double-layered magnetic fluids (EDL-MF) based on size-controlled $NiFe_2O_4$ nanoparticles, good candidates and precursors for future biomedical applications [3].

Table 1 Size sample characteristics: d_{XR} is the crystalline size from X-ray diffraction (*XRD*) experiments; d_0 and s are the parameters of the size distribution from transmission electron microscopy (*TEM*) measurements; dV/dt is the rate of addition of the reagents during the particle precipitation; *snt* and *ppt* are, respectively, the supernatant and precipitate fractions after centrifugation

dV/dt (mL/s)	XRD					TEM	
	d_{XR} (nm)					d_0 (nm)	s
125	A 4.2	Centrifugation	A1	3.8	snt	–	–
			A2	4.3	ppt	–	–
10	B 7.1		B1	5.9	snt	5.4	0.18
			B2	7.3	ppt	–	–
3	C 8.2		C1	7.7	snt	6.7	0.23
			C2	8.9	ppt	8.3	0.17

Experimental

Sample preparation and characterization

EDL-MF samples were synthesized by following the procedure described in Ref. [4]. Ferrite nanoparticles were prepared by alkalinizing 1:2 mixtures of Ni^{2+} and Fe^{3+} at 100 °C, under vigorous stirring. Changing the velocity of mixing the reagents led to nanoparticles with different mean sizes. Then particles were peptized electrostatically in aqueous media, using an appropriate particle surface treatment. Then, in order to reduce the polydispersion, the samples synthesized were centrifuged at 4,000 rpm for 15 min. After centrifugation, the resulting precipitate and supernatant were redispersed, forming stable sols of high quality.

Morphological and structural characterization

Micrographs and electronic diffraction patterns were obtained by transmission electronic microscopy (TEM). In this case, the samples were prepared by evaporating very dilute ferrofluid solutions onto carbon-coated grids. The particle size distribution was estimated by measuring the size of about 500 particles, using a lognormal law of size distribution which corresponds to the probability density that a particle has a diameter d, with a standard deviation s. The crystalline structure of the solids, their crystalline size, and the lattice constants were determined by X-ray diffraction (XRD). The average lattice constants were calculated from the five most intense diffraction lines, while the average crystal size (d_{XR}) was deduced by means of the Scherrer formula from the width at half maximum of the diffraction line (311).

Magnetic and magneto-optical characterization

To characterize the ferrofluid samples the magnetization and magneto-optical birefringence were investigated as a function of the applied magnetic field at room temperature. Ferrofluid magnetizations were measured with a Squid device at 300 K. The setup and the method to measure the ferrofluid birefringence are described elsewhere [5].

Results and discussion

Sample preparation and synthesis characterization

The particles sizes as a function of the addition rate (dV/dt) of the metallic ions into the alkaline solution during the precipitation are shown in Table 1. Usually, in colloid synthesis the nucleation starts when the concentration of the species reaches a critical saturation and each nucleus may further grow by a diffusive mechanism [6]. In this way, slow rates of addition yield large particles, while increasing dV/dt generates smaller particles. A series of EDL-MF samples obtained after centrifugation and redispersion are listed in Table 1 and are labeled A1–C2.

The XRD powder patterns presented in Fig. 1 show only one system of lines, typical of nickel ferrite, with a characteristic d spacing of 0.833 ± 0.05 nm, in good agreement with the value of 0.834 nm found in the literature [7]. The same structure was also identified from TEM diffraction measurements. Moreover, the values of the size distribution obtained from TEM analysis (Table 1) compare well with the crystalline mean sizes since the diameter measured by XRD averages the size distribution as $d_{XR} = d_0 \exp(2.5s^2)$ [8]. Here, d_0 is the characteristic diameter, with $\ln d_0$ being the mean value of $\ln d$.

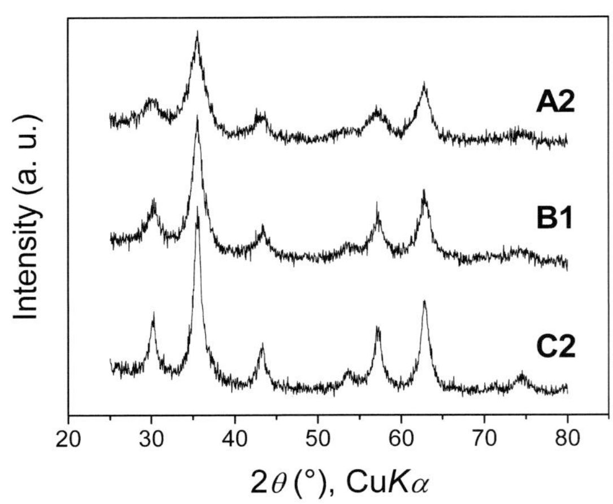

Fig. 1 X-ray diffraction patterns for samples *A2*, *B1* and *C2*

Magnetic and magneto-optical characterization

Let us consider an assembly of independent single-domain grains with a magnetic moment $\mu = \pi m_s d^3/6$, where m_s is the magnetization of the nanoparticle. In the presence of an external field, H, the colloidal solution behaves as an ideal superparamagnet with a magnetization given by the Langevin law: $M = m_s \phi L(\xi)$, $L(\xi) = \coth\xi - \xi^{-1}$, $\xi = \mu_0 \mu H / k_B T$, with ϕ being the ferrite volume fraction, k_B the Boltzmann constant and T the temperature. This equation shows that at zero field, $M = 0$ and as the field is turned on, the magnetic moments tend to align in the field direction so that at high fields, M saturates at $m_s \phi$.

The field-induced birefringence behavior is based on the existence of an optical anisotropy axis along the magnetic anisotropy one [5]. Then, in the presence of a field, one has a birefringence $\Delta n = \phi \delta n_0 [1 - 3L(\xi)/\xi]$, with the optical anisotropy $\delta n_0 = c\phi f(\sigma)$, where C depends on the particle refractive index and shape eccentricity. The function $f(\sigma)$ increases from 0 to 1 as its argument $\sigma = E_a / k_B T$ goes from zero to infinity, E_a being the anisotropy energy.

Nevertheless, since magnetic fluids are always polydisperse, the lognormal law of the size distribution must be taken into account in both measurements by considering a volume-weighted superposition of the contributions of all different particle volumes. Then, if m_S (magnetization measurements) or δn_0 (birefringence measurements) can be determined by the high-field extrapolation, the normalized magnetization or birefringence is reduced to a function of the size distribution parameters, d_0 and s.

Typical normalized magnetization curves are shown in Fig. 2a. The results obtained by fitting the experimental data to the Langevin model are summarized in Table 2. They are in qualitative agreement with the model, except in the case of the samples based on the smallest particles, where the magnetization does not saturate at the maximum value of the field, H_{max}. For these samples, the argument (ξ_{max}) of the Langevin function at the maximum value of the field is not sufficiently large to allow the saturation at H_{max}. Then, in order to compensate the nonsaturation, a correction is made by replacing in the magnetization equation m_S by $m_s / L(\xi_{max})$, where ξ_{max} is calculated by using the mean value of the particle magnetic moment at H_{max}. Thus, the high-field extrapolation gives the m_s values, which are all much lower than 270 kA/m, the bulk magnetization, and vary from 161 kA/m for smaller particles to 183 kA/m for bigger particles. In fact, under size reduction, $NiFe_2O_4$ superparamagnetic particles show a strong reduction of their magnetization when compared with that of the bulk ferrite, probably owing to a

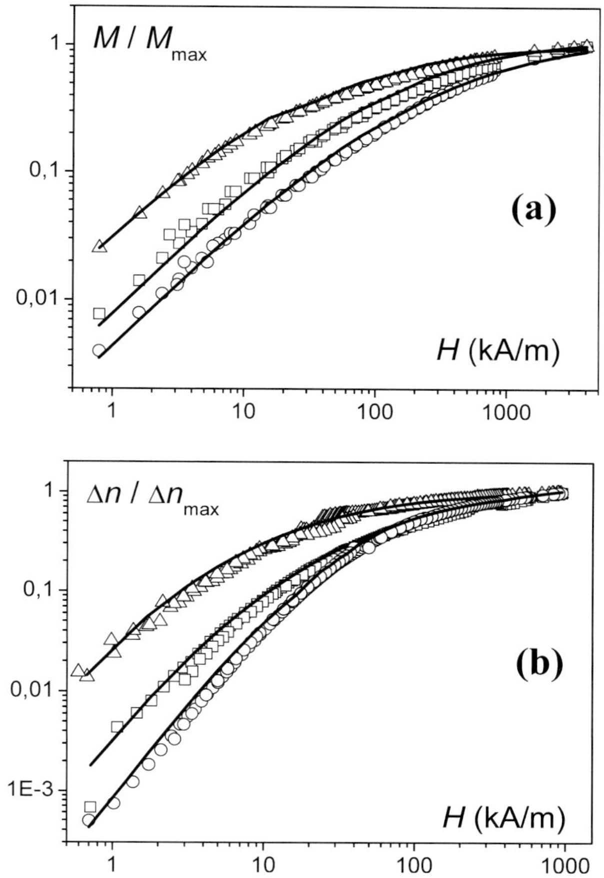

Fig. 2. **a** Reduced magnetization and **b** reduced birefringence of samples A2 (*circles*), B1(*squares*) and C2 (*triangles*) as function of the applied magnetic field. The *full lines* are the best fit obtained by using the Langevin formalism. M_{max} and Δn_{max} are, respectively, the values of the magnetization and the birefringence at H_{max}

Table 2 Birefringence and magnetization parameters: m_s is the saturation magnetization and δn_0 the individual optical anisotropy of the particle, both determined from the high-field behavior

Sample	Magnetization			Birefringence		
	m_s (kA/m)	d_0^{mag} (nm)	s^{mag}	δn_0	d_0^{bir} (nm)	s^{bir}
A2	161	2.9	0.4	0.013	7.7	0.4
B1	180	3.6	0.4	0.018	9.9	0.4
C2	183	6.2	0.4	0.038	14.0	0.4

noncollinearity of the surface spins [9]. Moreover, the characteristic diameters d_0^{mag} and the polydispersion s^{mag} obtained from the fittings compare well with the crystalline mean sizes found by XRD experiments using the relation $d_{XR} = d_0 \exp(2.5s^2)$.

A typical normalized birefringence curve is presented in Fig. 2b. As for the magnetization curves, the observed field-induced birefringence follows well the Langevin model: in particular, the curves readily saturate at high fields, a result which provides the value of the individual particle optical anisotropy δn_0, also listed in Table 2. The δn_0 measurements are close to previous determinations for $NiFe_2O_4$[3]: $\delta n_0 = 0.1$ for nanoparticles with $d_{XR} = 4.4$ nm. This optical anisotropy shows a dependence on the grain size, increasing as the diameter of the particle increases, a result in good agreement with previous birefringence measurements made on γ-Fe_2O_3 [5]. This dependence of δn_0 on d shows that the nanoparticles are soft dipoles with an energy of anisotropy close to k_BT. By adjusting the birefringence curves of our samples, as described in the Experimental section, we can easily determine with standard methods [5] the characteristic diameters $d_0^{bir}$and the polydispersity s^{bir} of the particles (Table 2). The values of d_0^{bir} are larger than the values of d_0^{mag} because the small particles are superparamagnetic and do not contribute to the measured signal; therefore, only the big particles are detected.

Conclusion

Size-controlled nickel ferrite nanoparticles have been synthesized bottom up by chemical precipitation and dispersed in stable colloidal solutions. The structure of the particles, their crystalline size, and the lattice constants were determined by XRD experiments. The magnetization measurements show that m_s determined for the nanoparticles is smaller than the bulk value and the birefringence measurements show that the nanoparticles behave as soft dipoles. These results will be compared in the future with those of measurements of the energy of anisotropy from magnetization measurements as a function of temperature.

References

1. Dormann JL, Fiorani D (1992) Magnetic properties of fine particles. North Holland, Amsterdam
2. Pankhurst QA, Connolly J, Jones SK, Dobson J (2003) J Phys D Appl Phys 36:R167
3. Hasmonay E, Depeyrot J, Sousa MH, Tourinho FA, Bacri JC, Perzynski R (1999) J Magn Magn Mater 201:195
4. Sousa MH, Tourinho FA, Depeyrot J, da Silva GJ, Lara MCFL (2001) J Phys Chem B 105:1168
5. Hasmonay E, Dubois E, Bacri JC, Perzynski R, Raikher YL, Stepanov VI (1998) Eur Phys J B5:859.
6. Hunter RJ (1987) Foundations of colloid science, vol 1. Oxford University Press, Oxford
7. ASTM no 10–0325
8. Tronc E, Bonnin D (1985) J Phys Lett 46:L437
9. Kodama RH, Berkowitz AE, McNiff JE Foner S (1996) Phys Rev Lett 77:394

Progr Colloid Polym Sci (2004) 128: 113–119
DOI 10.1007/b97103

Alexsandro Santos Costa
Toyoko Imae
Katsuhiko Takagi
Koichi Kikuta

Intercalation of dendrimers in the interlayer of hydrotalcite clay sheets

A. Santos Costa · T. Imae
Graduate School of Science,
Nagoya University, Chikusa,
Nagoya 464-8602

T. Imae (✉)
Research Center for Materials
Science, Nagoya University,
Chikusa, Nagoya 464–8602, Japan
e-mail: imae@nano.chem.nagoya-u.ac.jp
Tel.: +81-52-7895911
Fax: +81-52-7895912

K. Takagi, K. Kikuta
Graduate School of Engineering,
Nagoya University, Chikusa,
Nagoya 464-8603, Japan

Abstract Nanocomposites have been prepared through the intercalation of poly(amido amine) dendrimers with carboxylate terminals into hydrotalcite clays. The formation of nanocomposites by anion exchange during incubating at 70 °C was investigated at various compositions with increasing dendrimer content in water and was assessed by Fourier transform IR absorption spectroscopy, X-ray diffraction, small-angle X-ray scattering, and elemental analysis. The dendrimer diffused into the clay galleries: the spacing between the clay layers (sheets) increased from its intrinsic value of 3.2 to 20–23 Å, and the mass fraction of the dendrimer in the composite was around 0.50 for mixing ratios (charge of clay to carboxylate of the dendrimer) higher than 1:2, indicating the saturated intercalation of the dendrimer in the clay. The dendrimers are densely packed with an ellipsoidal structure in the interlayer. At a mixing ratio of 1:8, there was excess adsorption of the dendrimer on the clay surface besides the intercalation and initial adsorption.

Keywords Poly(amido amine) dendrimer · Hydrotalcite · Clay · Intercalation · X-ray diffractometry

Introduction

Dendrimers [1] are highly branched macromolecules consisting of a multifunctional core, to which repeating units are sequentially covalently bonded to create a topologically regular structure. The resultant molecules present a well-defined number of terminal groups and a narrow molecular weight distribution that is truly monodisperse in most cases. Owing to their structural uniformity, dendrimers provide a valuable model for correlating the macromolecular structure to the technological function [2–5]. Thus, dendrimers have a wide range of applications, including drug delivery [6], catalysis, and sensors [7–8] in high-technology fields. The organization of dendrimers in monolayers results in materials that can be applied as devices [7–10] and scaffolds for posterior elaborations [11–17]. Moreover, stratumous-layered composites of a dendrimer and a linear polymer can be prepared by linking terminal groups of a dendrimer to functional groups of a polymer [12, 13, 16, 18–20].

Double-layered hydroxides, also known as ionic clays, are important inorganic materials, and have deserved much attention in recent years. The hydrotalcite is constituted from inorganic sheets that are similar to the basic structure of brucite, $Mg(OH)_2$ [21]. The brucite-like sheets become positively charged owing to the partial substitution of divalent metal cations by trivalent metal cations in the layer structure. In order to balance the extra charges developed by the trivalent cations and to maintain electrical neutrality, organic or inorganic anions and water molecules are intercalated between the interlayer of the clay. Anions like Cl^- and CO_3^{2-} existing in the crude clay can be easily replaced by many other molecules, depending on their charge affinity. The anion-exchange properties of hydrotalcite [22] have been explored for the intercalation of long-chain fatty acids [23], organic dyes [24], phthalocyanine-type molecules [25, 26], aromatic ketocarbonates [27], etc. Intercalation of large anions has also been tried, and the successful intercalation of oxymetalates and phthalocyanines has been reported [28, 29]. The interest in these composites lies in their potential and technological functionality as catalysts or catalyst supports [21], anion adsorbents [30], medicine stabilizers [31], ion exchangers [32], for controlled release of anions [33], and as components of ceramic-based nanocomposites [21].

To date no one has reported the use of a clay framework for the fabrication of dendrimer stratums. Here, we report a new approach to obtain organic/inorganic multilayered nanocomposites making use of a dendrimer and a clay. We investigated the formation of nanocomposites through the intercalation of a dendrimer in the clay interlayer. The structure of the nanocomposites was studied regarding the preservation of the original structural organization of the clay as well as changes of interlayer distance and dendrimer conformation. The poly(amido amine) (PAMAM) dendrimer with carboxylate terminals used in this study is negatively charged in water. Such charges act as the driving force in the anion-exchange process in the hydrotalcite clay. A 4.5th generation (G4.5) dendrimer was used to promote considerable change in the hydrotalcite interlayer spacing. Nanocomposites were prepared at high temperature to assist the hydrotalcite swelling and to facilitate the dendrimer diffusion into the clay interlayer. The preparation of nanocomposites was investigated at different hydrotalcite/dendrimer ratios. In order to assess the intercalation, the structural changes were monitored by X-ray diffraction (XRD) and small-angle X-ray scattering (SAXS). Furthermore, we discuss the packing and the structure of the dendrimer in the nanocomposites at different ratios assessed by X-ray diffraction and elemental analysis. Two-dimensional organization of dendrimers by using clay as a platform should be utilizable in the preparation of dendrimer-containing nanocomposites that are applicable to systems of drug-delivery, catalysis, sensors, and so on.

Experimental

Materials

Hydrotalcite clay, $[Mg_{4.5}Al_2(OH)_{13}Cl_2]\cdot 3.5H_2O$, was donated from Kyowa Chemical Co. and was used as received. Its anion-exchange capacity is 350mEq 100 g^{-1}. The G4.5 PAMAM dendrimer with carboxyl terminal groups (molecular weight 26,258, 128 terminal groups) was purchased from Aldrich and was used as received. MilliQ water was used throughout the experiment.

Nanocomposite preparation

PAMAM dendrimer was dried by purging it with nitrogen for 1 h and then by drying it in vacuo for 30 min. Water and hydrotalcite clay were added to the dendrimer residue in order to prepare hydrotalcite/PAMAM dendrimer mixtures at different ratios. The composition number ratio (anionic charge of clay to carboxylate of the dendrimer) varied from 1:1 to 1:8, where the clay concentration was kept constant. The mixtures were incubated at 70 °C for 24 h under shaking, where the adsorption equilibrium is attained. The resulting materials were filtered using a Millipore filter (0.22 μm pore size), washed a few times with water to remove the possible excess of free dendrimer, and dried.

Measurements

Fourier transform IR (FT-IR) spectra were acquired by using a Bio-Rad FTS 575C FT-IR spectrometer in the range 1,000–4,000 cm^{-1} at 4-cm^{-1} resolution. A few milligrams of hydrotalcite/PAMAM mixtures or hydrotalcite were ground with potassium bromide and afterwards pressed to prepare discs. The spectrum of a dendrimer was measured by dropping a small volume of a dendrimer solution in methanol onto a pure KBr disc.

The XRD patterns were recorded using a Rigaku RINT 2100 V X-ray diffractometer possessing Cu Kα radiation of wavelength 1.54 Å, operating at 60 kV and 50 mA and scanning at a rate of 0.5° min^{-1}.

SAXS measurements were performed with a Rigaku instrument (SAX-LPB), equipped with an ultraX18HB X-ray source and He-protected CMF25-77.5Cu6 confocal multilayer optics. The radiation used was Cu Kα (λ = 1.54 Å) and the supply voltage and current were 45 kV and 60 mA, respectively. Prior to the measurements, specimens were placed between two Kapton films and dried at 60 °C for a few minutes. Two-dimensional SAXS patterns were recorded on a Fuji imaging plate (13×13 cm^2) detector. The sample-to-detector distance was set to 0.5 m and the exposure time to 24 h. Scanned scattering patterns were azimuthally averaged and corrected for a background of two Kapton films in the air gap. The scattering function $I(2\theta)$ was converted to $I(q)$, where 2θ is the scattering angle and $q = 4\pi \sin(2\theta/2)/\lambda$ is the scalar of the scattering vector.

Elemental analysis experiments were done using a LECO CHN-900 C, H, N analyzer. Prior to the analysis, mixtures were dried in vacuo in a desiccator containing P_2O_5 for 16 h.

Results

FT-IR spectra and band assignment of the hydrotalcite, dendrimer, and mixtures of hydrotalcite/dendrimer are presented in Fig. 1 and Table 1, respectively. The hydrotalcite spectrum showed a broad absorption band at 3,474 cm^{-1} which is attributed to an OH stretching vibration mode of intercalated and adsorbed water

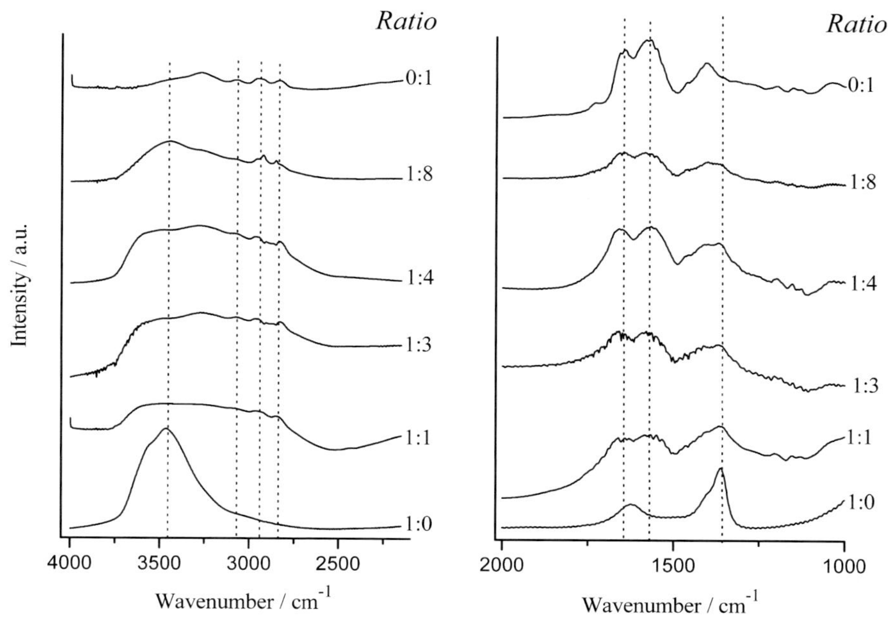

Fig. 1 Fourier transform IR spectra for mixtures at different mixing ratios of hydrotalcite to dendrimer

molecules and/or the hydroxyl groups of brucite-like sheets [34–39]. In addition to this band, the spectrum of hydrotalcite showed an absorption band at 1,624 cm^{-1} assigned to a HOH bending mode. The absorption band at 1,364 cm^{-1} is attributed to carbonate anion impurities [40, 41]. The carbonate anions have high affinity for hydrotalcite cations. Hence, Cl anions are easily replaced with carbonate anions in the hydrotalcite interlayer during the synthesis or the dispersion of hydrotalcite in water.

The FT-IR spectra of all hydrotalcite/dendrimer mixtures displayed very broad absorption bands between 3,700 and 3,300 cm^{-1}, which coincide with the broad IR band of OH groups in hydrotalcite. A HOH bending band of hydrotalcite is not observed in the spectra of the mixtures and this may be caused by the superposition with dendrimer bands or the shift of the HOH band. The band of the carbonate anion impurities shifts to 1,370 cm^{-1} with increasing fraction of the dendrimer in the mixture. The hydrotalcite/dendrimer mixtures also displayed absorption bands that are associated with the dendrimer, as is apparent when comparing the IR spectrum with the IR spectrum of the G4.5 PAMAM dendrimer [42]. The bands at 3,275, 3,078, 2,957, and 2,833 cm^{-1} represent the amide A and B (NH stretching mode) and CH_2 antisymmetric and symmetric stretching

Table 1 Assignments of IR absorption bands of hydrotalcite, dendrimer, and the hydrotalcite/dendrimer mixture. Spectral resolution 4 cm^{-1}

Assignment	Band position/cm^{-1} [a]		
	Hydrotalcite	Dendrimer	Hydrotalcite/dendrimer mixture
OH stretching	3,474(b, s)		3,482(b, s)
Amide A (NH stretching)		3,272(s)	3,275(s)
Amide B (NH stretching)		3,077(m)	3,078(m)
CH_2 antisymmetric stretching		2,957(m)	2,957(m)
CH_2 symmetric stretching		2,834(m)	2,833(m)
C=O stretching		1,730(m)	1,731(w)
Amide I		1,643(vs)	1,645(vs)
HOH bending	1,624(m)		–
COO^- antisymmetric stretching		1,584(vs)	1,585(vs)
Amide II		1,570(vs)	1,572(vs)
COO^- symmetric stretching		1,406(s)	1,407(s)
CO_3^{2-} (impurity)	1,364(vs)		1,370(m)
Amide III		1,358(w)	1,360(w)
C–N stretching		1,040(w)	1,048(w)

a) (vs) very strong, (s) strong, (m) medium, (w) weak, (b) broad modes, respectively, and originated from the PAMAM dendrimer. The absorption bands at 1,731, 1,645, 1,585, 1,572, and 1,407cm^{-1} are attributed to the C=O stretching, amide I, COO^- antisymmetric stretching, amide II, and COO^- symmetric stretching modes, respectively. The bands at 1,360 and 1,048 cm^{-1} were assigned to the amide III and C–N stretching. The presence of IR bands of the PAMAM dendrimer in the specimens after rinsing indicates the complex formation between hydrotalcite and the PAMAM dendrimer. It should be noted that the intensity ratio between the IR bands of carbonate anion impurities and the dendrimer is almost constant at a mixing ratio above 1:2.

Information on the morphology and crystallinity were obtained from analysis of XRD data. The XRD patterns and the indexing of the hydrotalcite and hydrotalcite/dendrimer mixtures are shown in Fig. 2 and Table 2, respectively. The spectra are shown as $I(q) \times q^2$ versus q (Kratky representation). The spacing between the reflection planes was calculated from the indexed Bragg peak position q_{hkl}, where hkl are the Miller indices, and from the equation $dhkl = 2\pi/qhkl$. Typically, the XRD pattern of hydrotalcite presents major diffraction peaks: (1) the large basal spacing (d_{003}); (2) the half-height harmonic spacing (d_{006}) (where $d_{006} = d_{003}/2$); and (3) the d_{009} diffraction (where $d_{009} = d_{003}/3$) [43]. The hydrotalcite spectrum, as seen in Fig. 2, exhibits strong and sharp diffraction peaks at $q = 0.788$, 1.600 and 2.440 Å^{-1}, which correspond to a basal spacing (d_{003}) value of 7.97 Å. The thickness (4.77 Å) of the brucite-like sheet [44] is a good approximation of the thickness of a single Mg···Al···$(OH)_x$ (hydrotalcite) layer. Subsequently, the interlayer distance is simply found by subtracting 4.77 Å from the observed basal spacing. In this case, the interlayer distance was 3.20 Å, which correlates well with the size of the foremost-entrapped anion, Cl^-. All hydrotalcite/dendrimer mixtures present, in addition to the diffraction peaks of the crude hydrotalcite, new diffraction peaks around $q = \sim 0.23$ Å^{-1} and $q = \sim 0.52$ Å^{-1} at mixing ratios larger than 1:1. The new peaks are indexed as the dendrimer intercalated spacing (d_{003}),

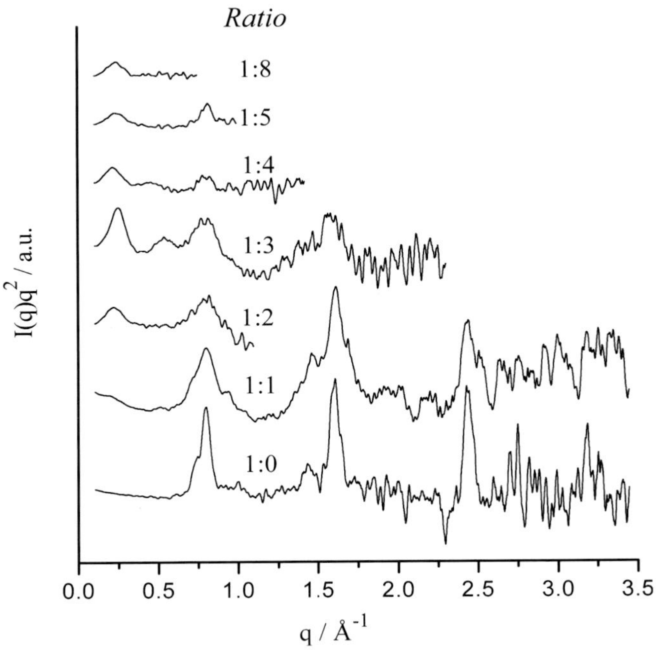

Fig. 2 X-ray diffraction patterns for mixtures at different mixing ratios of hydrotalcite to dendrimer

Table 2 Bragg peak positions (Å^{-1}) for mixtures at different ratios, with the indexes according to rhombohedral R(-3)m symmetry in *parentheses*, and data evaluated from Bragg peaks

Ratio	Basal hydrotalcite	Dendrimer-intercalated hydrotalcite	Dendrimer layer thicknessÅ	Area occupied by dendrimer/Å^2
1:0	0.788 (003)			
	1.600 (006)			
	2.440 (009)			
1:1	0.788 (003)			
	1.607 (006)			
	2.420 (009)			
1:2		0.230 (003)		
	0.790 (003)	–		
1:3		0.250 (003)		
		0.530 (006)		
	0.800 (003)	–	20.4	7.92×10^3
	1.600 (006)	–		
1:4		0.224 (003)		
		0.480 (006)	23.3	6.93×10^3
	0.810 (003)	–		
1:5		0.245 (003)		
		0.520 (006)	20.9	7.73×10^3
	0.820 (003)	–		
1:8		0.240 (003)		
		0.530 (006)	21.4	7.57×10^3

which is around 25–28 Å. The coexistence of diffraction peaks of crude hydrotalcite suggests that there are interlayers in which the Cl^- anions were not replaced by the dendrimer molecules.

To obtain more information on the character of hydrotalcite/dendrimer systems, SAXS measurement was performed and scattering patterns were plotted as $I(q)$ versus q (Fig. 3). The crude hydrotalcite presented no reflection peak, since the basal distances for hydrotalcite intercalated with Cl^- and CO_3^{2-} anions give reflections outside the measured q range in SAXS. Hydrotalcite/dendrimer mixtures prepared at ratios of 1:2–1 T:8 showed broad bands at 0.224–0.250 and 0.480–0.530 $Å^{-1}$, which are the interlayer spacings d_{003} and d_{006} for hydrotalcite intercalated by dendrimer molecules.

Elemental analysis was used to assess the dendrimer content in the hydrotalcite/dendrimer mixtures. Since nitrogen atoms are nonexistent in the hydrotalcite structure, the dendrimer mass fraction, f_D, was calculated directly from the N content in the hydrotalcite/dendrimer mixture, as listed in Table 3, which also includes the results of elemental analysis. Mixtures at ratios of 1:3, 1:4, and 1:5 presented very similar results, where the fraction was constant at 0.495. This numerical value indicates that the dendrimers are almost fully intercalated in the interlayer of the clay and are adsorbed on the surface of the clay. On the other hand, the mixture at a ratio of 1:8 presented a higher fraction (0.558) of dendrimers. This may be due to the additional (excess) adsorption of dendrimers on the adlayer of the dendrimer on the surface of the clay.

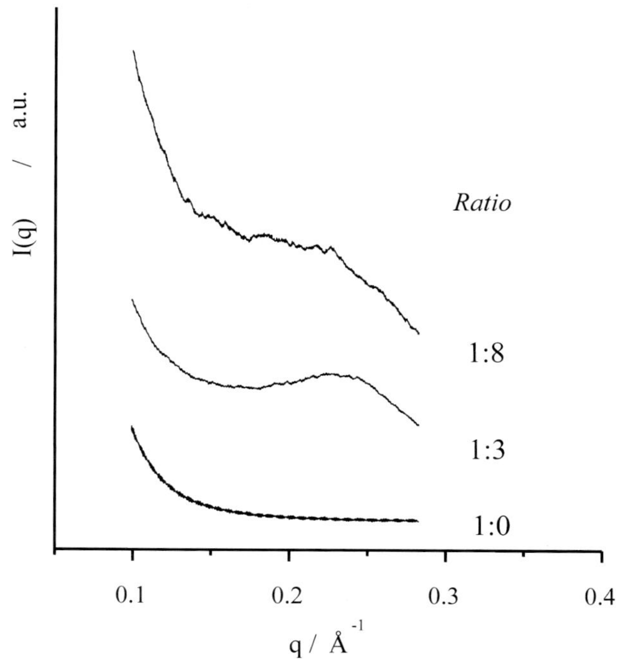

Fig. 3 Small-angle X-ray scattering patterns for mixtures at different mixing ratios of hydrotalcite to dendrimer

Discussion

To ensure the intercalation of the dendrimer into hydrotalcite, the hydrotalcite/dendrimer mixtures were incubated at high temperatures. The IR results evidenced the complex formation after the incubation. There are two possible ways of hydrotalcite/dendrimer composite formation: the intercalation of the dendrimer into hydrotalcite and its adsorption onto hydrotalcite. The X-ray techniques were used to study the morphology of dendrimer/hydrotalcite mixtures. XRD and SAXS indicated that dendrimer molecules were intercalated into hydrotalcite. The lamellar structure is preserved upon the intercalation of the dendrimer but the interlayer distance increased, as illustrated in Fig. 4.

At first glance, the XRD spectra seemed to be similar but some trends were observed with an increase in the

Table 3 Elemental analysis results for mixtures at different ratios and data evaluated from elemental analysis

Ratio		C/wt %	H/wt %	N/wt %	Dendrimer mass fraction f_D	Hydrotalcite area per dendrimer/$Å^2$
1:3	Observed	22.56	5.56	7.22	0.495 ± 008	1.73×10^3
	Calculated using f_D	22.77	3.09	6.60		
1:4	Observed	22.55	5.72	7.17	0.495 ± 0.006	1.73×10^3
	Calculated using f_D	22.78	3.09	6.60		
1:5	Observed	22.66	5.64	6.69	0.494 ± 0.009	1.74×10^3
	Calculated using f_D	22.73	3.08	6.59		
1:8	Observed	25.82	5.57	7.39	0.558 ± 0.004	1.35×10^3
	Calculated using f_D	25.70	3.49	7.45		

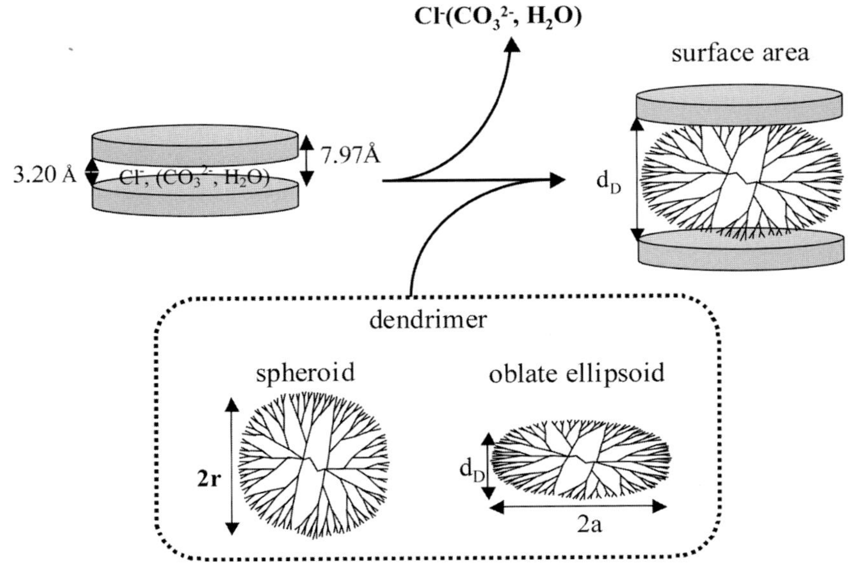

Fig. 4 Schematic representation of a dendrimer intercalated into hydrotalcite via direct ion exchange

amount of dendrimer in the mixtures. The diffraction peak of the basal spacing of dendrimer-intercalated hydrotalcite dispersed within 0.22–0.25 Å^{-1} at mixing ratios of 1:2–1:8, indicating the saturation of intercalation. On the other hand, there was a progressive increase from 0.79 to 0.82 Å^{-1} in the diffraction peak position corresponding to the basal distance of the crude hydrotalcite. These positions coincide with the spacing of hydrotalcite that interlaid Cl^- (7.9 Å) and CO_3^{2-} (7.6 Å), respectively [45]. The carbonate anions are strongly adsorbed on hydrotalcite and/or intercalated in it, as revealed by IR spectroscopy. On the basis of these facts, it was proposed that, even at higher mixing ratios, there are some regions of hydrotalcite intercalated with mainly carbonate anions and possibly chloride anions.

Although the diameter of the spherical G4.5 PAMAM dendrimer is around 59 Å [46], the hydrotalcite interlayer intercalated with dendrimer molecules was not thicker than 24 Å. Dendrimer molecules of lower generation are flexible rather than hard and spherical. This indicates that dendrimer molecules distort their shape from a sphere, when intercalated into hydrotalcite. Similar distortion of the dendrimer structure was observed between lamellar bilayers of didodecyldimethylammonium [47].

The area occupied by a dendrimer molecule was evaluated by making use of the XRD results. The dendrimer in solution presents a spherical shape with a diameter represented by $2r$ (59 Å for the G4.5 PAMAM dendrimer) [46]. On the other hand, from the basal distance obtained from the XRD results, we assume that the dendrimer molecules are intercalated with an oblate ellipsoidal structure. If the volume of the dendrimer does not change during the structural deformation, the volume obtained for the spheroid shape can be matched to that of an oblate ellipsoid in order to obtain the long-axis diameter of the dendrimer denoted as $2a$ in Fig. 4. The dendrimer layer thickness, d_D, that was obtained from the XRD result and that is listed in Table 2 was used as the short-axis diameter in the oblate ellipsoid. Then, the spheroid volume of $(4/3)\pi r^3$ is equal to the oblate ellipsoid volume of $(4/3)\pi a^2(d_D/2)$. Making use of the evaluated long-axis diameter $2a$ of the intercalated dendrimer, we obtained the area (πa^2) occupied by a dendrimer molecule, which is listed in Table 2.

The dendrimer packing within the interlayer of hydrotalcite clay was also studied by making use of the result of elemental analysis and was compared with the intercalation structure obtained from XRD. In this study, a model where all interlayers are intercalated with dendrimer molecules was used, even though the coexistence of chloride and carbonate anions with dendrimers in the hydrotalcite interlayer is more likely to be true from the XRD results. From the combination of the dendrimer mass fraction f_D obtained from elemental analysis, the clay area (50.6 Å^2) [47] per charge site and the anionic exchange capacity (350 mEq 100 g^{-1}), we calculated the hydrotalcite area per dendrimer $[50.6\times350\times10^{-3}\times N_A(1-f_D)/2]/(f_D N_A\times100/M_{den})$, where N_A is the Avogadro number and M_{den} (26,258) is the molecular weight of the PAMAM dendrimer. The hydrotalcite area per molecule, which is listed in Table 3, was smaller than the area occupied by the dendrimer obtained from XRD (Table 2). This discrepancy may occur because (1) the intercalated dendrimers coexist with the intercalated chloride and carbonate anions in the clay interlayer, (2) dendrimer molecules are not only intercalated in the clay interlayer but are also adsorbed on the clay surfaces, or (3) the intercalated dendrimers are overlapped or compressed.

Conclusions

We have prepared nanocomposites composed of a clay and a dendrimer through the intercalation of a PAMAM dendrimer into the interlayer of hydrotalcite clay. The XRD and SAXS results of nanocomposites indicated the increase of the clay interlayer distance after the intercalation. The saturation of dendrimer intercalation was also identified, and the possible dendrimer adsorption on the hydrotalcite clay was assumed. The nanocomposites obtained may be useful as dendrimer storage systems and in processes where dendrimer-controlled release is desirable.

References

1. Tomalia DA, Baker H, Dewald J, Hall M, Kallos G, Martin S, Roeck J, Ryder J, Smith P (1985) Polym J 17:117–132
2. Nagasaki T, Ukon M, Arimori S, Shinkai S (1992) J Chem Soc Chem Commun 608–610
3. Nagasaki T, Kimura O, Ukon M, Arimori S, Hamachi I, Shinkai S (1994) J Chem Soc Perkin Trans 1 75–81
4. Jansen JFGA, de Brabander-van den Berg EMM, Meijer EW (1994) Science 266:1226–1229
5. Jansen JFGA, Janssen RAJ, de Brabander-van den Berg EMM, Meijer EW (1995) Adv Mater 7:561–564
6. Esfand R, Tomalia DA (2001) Drug Discovery Today 6:427–436
7. Wells M, Crooks, RM (1996) J Am Chem Soc 118:3988–3989
8. Tokuhisa H, Crooks RM (1997) Langmuir 13:5608–5612
9. Losada J, Cuadrado I, Moran M, Casado, CM, Alonso B, Barranco M (1997) Anal Chim Acta 338:191–198
10. Koo BW, Song CK, Kim C (2001) Sens Actuators B 77:432–436
11. Watanabe S, Regen, SL (1994) J Am Chem Soc 116:8855–8856
12. Zhou Y, Bruening ML, Bergbreiter DE, Crooks RM, Wells MJ (1996) J Am Chem Soc 118:3773–3774
13. Liu Y, Bruening ML, Bergbreiter DE, Crooks RM (1997) Angew Chem Int Ed Engl 36:2114–2116
14. Tsuruk VV, Rinderspacher F, Bliznyuk VN (1997) Langmuir 13:2171–2176
15. Tsuruk VV (1998) Adv Mater 10:253–257
16. Zhao M, Liu Y, Crooks RM, Bergbreiter DE (1999) J Am Chem Soc 121:923–930
17. Wang J, Chen J, Jia X, Cao W, Li M (2000) Chem Commun 511–512
18. Grohn F, Kim G, Bauer BJ, Amis EJ (2001) Macromolecules 34:2179–2185.
19. Sellner H, Karjalainen JK, Seebach D (2001) Chem Eur J 7:2873–2887
20. Alvarez J, Sun L, Crooks RM (2002) Chem Mater 14:3995–4001
21. Cavani F, Trifiro F, Vaccari A (1991) Catal Today 11:173–301
22. Allmann R (1970) Chimia 24:99–108
23. Raki L, Rancourt DG, Detellier C (1995) Chem Mater 7:221
24. Park IY, Kurado K, Kato C (1990) J Chem Soc Dalton Trans 3071
25. Carrado KA, Forman JE, Botto RE, Winans RE (1993) Chem Mater 5:472
26. Bonnet S, Forano C, de Roy D, Besse JP, Millard P, Momenteau M (1996) Chem Mater 8:1962
27. Takagi K, Harata E, Shicchi T, Kanoh T, Sawaki Y (1997) J Photochem Photobiol 105:47–54
28. Gardner EA, Yun SK, Kwon T, Pinnavaia TJ (1998) Appl Clay Sci 13:479–494
29. Chatti I, Ghorbel A, Grange P, Colin JM (2002) Catal Today 75:113–117
30. Hermosin MC, Pavlovic J, Ulibarri MA, Cornejo J (1996) Water Res 30:171
31. Qian M, Zeng HC (1997) J Mater Chem 7:493–499
32. Xu ZP, Zeng HC (1998) J Mater Chem 7:2499–2506
33. Ambrogi V, Fardella G, Grandolini G, Perioli L (2001) Int J Pharam 220:23–32
34. Thevenot F, Szymanski R, Chaumette P (1989) Clays Clay Minerals 37:396–402
35. Tsuji M, Mao G, Yoshida T, Tamaura Y (1993) J Mater Res 8:1137
36. Chisem IC, Jones W (1994) J Mater Chem 4:1737–1744
37. Hibino T, Yamashita Y, Kosuge K, Tsunashima A (1995) Clays Clay Minerals 43: 427–432
38. Kooli F, Rives V, Ulibarri MA (1995) Inorg Chem 34:5122–5128
39. Kannan S, Swamy CS, (1997) J Mater Sci 32:1323–1630
40. Rey F, Fornes V, Rojo JM (1992) J Chem Soc Faraday Trans 88:2233
41. Hansen HCB, Koch CB, Taylor RM (1994) J Solid State Chem 113:46–53
42. Manna A, Imae T, Aoi K, Okada M, Yogo T (2001) Chem Mater 13:1674–1681
43. Carlino S (1997) Solid State Ionics 98:73–84
44. Ookubo A, Ooi K, Hayashi H (1993) Langmuir 9:1418–1422
45. Yun SK, Pinnavaia T (1995) J Chem Mater 7:348
46. Prosa TJ, Bauer BJ, Amis IJ (2001) Macromolecules 34:4897–4906
47. Li X, Imae T, Leisner D, López-Quintela MA (2002) J Phys Chem 106:12170–12177
48. Itoh, T, Ohta N, Shichi T, Yui T, Takagi K (2003) Langmuir 19:9120–9126

Progr Colloid Polym Sci (2004) 128:120–125
DOI 10.1007/b97106

Zorica Crnjak Orel
Danjela Kuščer
Marija Kosec
Aleksandra Turković

The characterization of nanocrystalline V_2O_5 and mixed V_2O_5/Ce oxide

Z.C. Orel (✉)
National Institute of Chemistry,
Hajdrihova 19, Ljubljana 1000, Slovenia
e-mail: zorica.crnjak.orel@ki.si
Tel.: +386-1-4760236
Fax: +386-1-4760300

D. Kušcer · M. Kosec
Jozef Stefan Institute,
Jamova 39, Ljubljana, Slovenia

A. Turković
Institute "Ruder Boškovic",
P.O. Box 118, 10002 Zagreb

Abstract Thin films of vanadium oxide and new, mixed V/Ce oxide thin films with 55 and 38 atom % of V were prepared on glass substrates covered with SnO_2:F (K-glass) The influence of the added cerium precursors as well as the influence of the substrate material on the formation of the films were studied in order to connect their improved intercalation properties for Li ions. The structural properties and the morphologies of the films were determined with transmission electron microscopy and grazing-incidence small-angle X-ray scattering. Transmission electron microscopy revealed the crystalline structure, with the formation of V_2O_5 (orthorhombic structure) in all the films. The average grain radius, obtained by grazing-incidence small-angle X-ray scattering was correlated with the layer thickness.

Keywords Nano · V_2O_5 · Mixed V/Ce oxide · Films · Structural analysis

Introduction

The preparation of coatings containing nanosized particles belongs to a rapidly growing field of scientific interest. Very often these types of coatings exhibit novel properties, and some nanoscale systems offer interesting device applications as a consequence of their size and properties. In the past decade a lot of interest has been shown in the preparation of nanomaterials, particularly oxides, using the sol–gel process. The most important part of the sol–gel procedure is the preparation of a sol that presents a stable colloidal suspension. From this the films or coatings can be prepared from sols on a substrate by using the dip-coating technique. With this procedure both pure and mixed oxide systems can be easily prepared.

Vanadium oxides (V_2O_5) have received much attention owing to their broad applicability as reversible cathode materials in lithium batteries, as a catalyst, and in electrochromic devices [1]. All these applications are closely connected by the method used for the growth of the films – like chemical vacuum evaporation, pulsed-laser deposition, sputtering, sol–gel, and electron-beam deposition [2]. Their characteristics are strongly influenced by the morphology and the crystallinity of the films obtained. It has also been observed that the nature of the substrate strongly influences the dynamics of nucleation. For films prepared by the plasma-enhanced chemical-vapor-deposition method on a barium borosilicate substrate there is a much greater tendency to form crystalline-oriented materials than is the case with indium tin oxide (ITO) and soda lime glass; much more disordered materials were formed with these substrates. A possible explanation for this phenomenon was given by Losurdo et al. [3], which combines the properties of vanadium oxide and the substrate, and connects the well-known good intercalation properties of V_2O_5 and the less densely packed network of soda lime glasses. As a result of this it is possible that the alkaline ions and the tin migrate into the V_2O_5 network. These researchers showed a 150-nm-wide region where the V_2O_5 films, the ITO films, and the soda lime glass were mixed together. Other authors have made similar observations [4]. Owing to the nanoscale thickness of the films and the possible interaction with the substrate, the oxide layers may show different properties to those of a bulk oxide.

From this it is obvious that we have to know if and how the substrate can influence the final properties of films, and we have to characterize and compare the structural characteristics of films on the same type of substrate.

The aim of this study is to find out if the films of pure vanadium oxide and mixed V/Ce oxide on K-glass incorporate some tin ions, and to characterize the structure of the vanadium oxide and the V/Ce oxides prepared at 55 and 38 atom % of V. The results are compared with the structure which was previously determined using another spectroscopic technique. In this article, with the help of results of transmission electron microscopy (TEM) and grazing-incidence small-angle X-ray scattering (GISAXS) studies, we characterize the substrate, which was in all cases K-glass, and the films prepared on the same substrate at 673 K using the sol–gel dip-coating technique.

Experimental

An aqueous dispersion of hydrated oxides (sols) – V_2O_5 and CeO_2 – was prepared from a solution of the corresponding metal oxide, V_2O_5, (Aldrich 99.6+) and the metal salt, $Ce(NH_4)_2(NO_3)_6$, (Fluka, p.a. ASC > 99%) as previously described [5]. The mixed V/Ce oxides with different atomic percents of V in the V/Ce oxide (55 and 38 atom %) were prepared by adding the appropriate concentration of CeO_2 sol to a V_2O_5 sol. The resulting sol was deposited on K-glass (SnO_2:F covered glass) using the dip-coating technique. After the evaporation of the water, gel/xerogels were formed; and after heating at 673 K stable coatings were obtained.

For TEM, thin films annealed on K-glass were prepared using a standard cross-section procedure. The samples were cut perpendicular to the film, shaped into 3-mm discs, thinned, dimpled, and finally ion-milled using 4 kV Ar^+ ions at a milling angle of 12°. The samples were characterized using a transmission electron microscope (model JEM 2000 FX, JEOL, Tokyo, Japan) operated at 200 kV. The chemical composition was investigated using a Link AN-10000 energy-dispersive X-ray spectroscopy system (EDXS) with an ultra-thin-window Si (Li) detector. During the TEM analysis no beam damage to the material and no specimen contamination were observed.

GISAXS measurements were performed on the new SAXS beamline at the ELETTRA synchrotron, Trieste. Its characteristics are as follows: the photon beam wavelength is $\lambda = 2.3\times10^{-8}$ cm, the energy resolution is $\Delta E/E \leq 2.5\times10^{-3}$, the focal spot size full width at half maximum is 1.2×0.6 mm^2, the spot at the sample is 5.4×1.8 mm^2, and the flux at the sample is 5×10^{12} photons s^{-1} (electron beam energy 2 GeV, beam current 200 mA, X-ray photon energy 16 keV).

Results and discussion

We recently prepared CeO_2/SnO_2 [5–7] and mixed V/Ce oxide [8, 9] films via an inorganic sol–gel route as a counter electrode for a smart window. Our aim was to prepare a stable counter electrode deposited with just a single dipping. Samples with different atomic ratios of V to Ce (78, 55, 38, and 32 atom % of V in V/Ce oxide) were prepared on K-glass. The best capacity (around 30 mC/cm^2 for a single dip layer) for the intercalation/deintercalation of Li^+ in 1 M $LiClO_4$ in propylene carbonate was obtained for samples with 55 and 38 atom % of V. Not only was the capacity of the films improved, so was the electrochemical stability (+2 to −1.5 V versus Ag/AgCl) of pure V_2O_5 films after the addition of CeO_2. The best samples which can serve as a counter electrode were obtained at 55 and 38 atom % of V in V/Ce oxide [5]. Pure V oxide and these two samples as thin films and powder were characterized by IR spectroscopy. The samples in powder form were also analyzed by Raman spectroscopy [10]. The results obtained show the formation of V_2O_5 in samples prepared without the addition of CeO_2. The formation of V_2O_5 and $CeVO_4$ was observed in mixed samples prepared with 55 atom % of V in a V/Ce oxide. Additionally, we established the formation of an amorphous phase with 38 atom % of V in a V/Ce oxide, which contained a tetrahedral species of VO_4^{3-}.

Transmission electron microscopy

Three samples of pure V oxide, and mixed samples with 55 and 38 atom % of V in V/Ce oxide deposited on K-glass were characterized using TEM.

In the transmissive region of the pure V oxide deposited on K-glass the columnar growth of SnO_2:F with a thickness of 350–400 nm could be clearly seen (Fig. 1). The contact between the V oxide film and the SnO_2 was pyramid-shaped. Energy-dispersive spectroscopy analysis proved that the substrate consisted of glass covered with a SnO_2 layer.

The thickness of the V oxide layer deposited on K-glass was between 50 and 100 nm. Using a direct

Fig. 1 Bright-field transmission electron microscope (*TEM*) image of the contact between a V_2O_5 film and K-glass. The pyramid-shaped columnar structure of the SnO_2 layer, and an about 100-nm-thick V_2O_5 layer

method (bright field–dark field imaging), a particle size of about 5 nm was detected (Fig. 2). This layer shows the formation of a crystalline structure without any preferential orientation, which is evident from the selected-area electron diffraction (SAED) ring pattern. Together with EDXS analysis this confirms the formation of V_2O_5 (orthorhombic structure). The crystallites that were randomly oriented within the V_2O_5 film were theoretically simulated from the SAED image with an electron microscopy simulation program [11]. The SAED image was calculated for orthorhombic V_2O_5 (ASTM 89–2483) that gives the position and the intensity of the rings. The intensity of the rings was calculated from structural parameters. The intensity, for example, the width of the rings, depends on the crystallite size. The experimental SAED image was compared with a simulated one taking into account the crystallite size of V_2O_5. The best agreement between the calculated and the experimental SAED images was observed for a V_2O_5 crystallite size of 2 nm (Fig. 3).

Samples with 55 and 38 atom % of V in Ce/V oxide deposited on K-glass substrates are shown in Figs. 4 and 5. In both samples the columnar growth of the SnO_2:F, which is 260–350-nm thick, was again observed. The thickness of the films for the mixed samples was estimated to be between 90 and 150 nm in the case of 55 atom % of V, and between 180 and 280 nm in the case of 38 atom % of V in Ce/V oxide. Both samples crystallized, which is evident from the ring pattern of the SAED image. The theoretical simulation shows that the crystallite size for 55 and 38 atom % of V in Ce/V oxide is 1–2 nm. The EDXS analysis proves that the active layer contains Ce and V, whereas no Sn was detected, at least within the detection limit of the analysis.

The SAED images of all three samples for pure V_2O_5, and mixed V/Ce oxides at 55 and 38 atom % of V are similar. From the results obtained with the SAED images for the samples prepared at 38 and 55 atom % of V in V/Ce oxide it appears that the

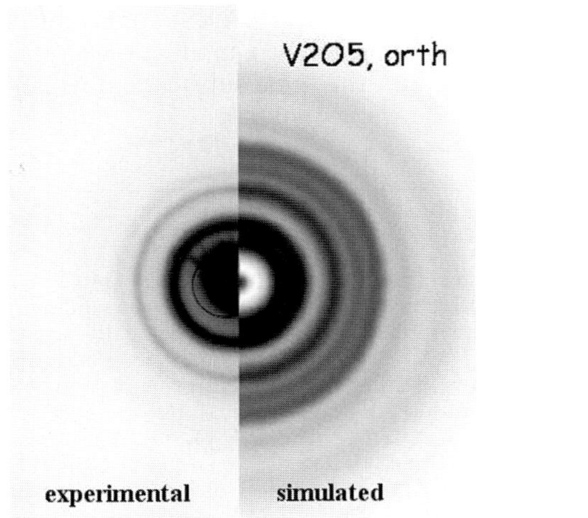

Fig. 3 Selected-area electron diffraction image of a V_2O_5 film deposited on K-glass (*left*). Theoretical simulation for the orthorhombic V_2O_5 structure (*right*). The estimated crystallite size is up to 2 nm

Fig. 2 **a** Bright-field TEM image and **b** dark-field TEM image of V_2O_5 film on K-glass. The 111 reflection was used for dark-field imaging

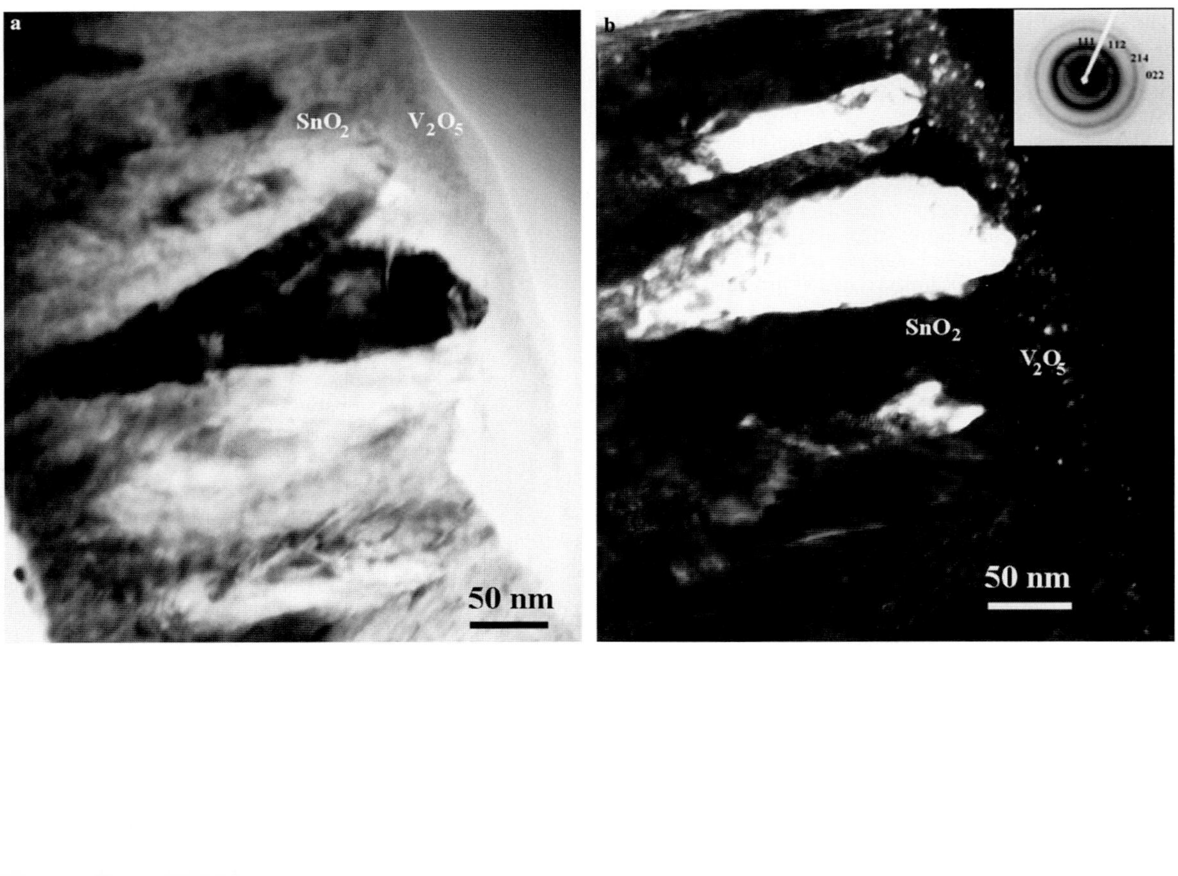

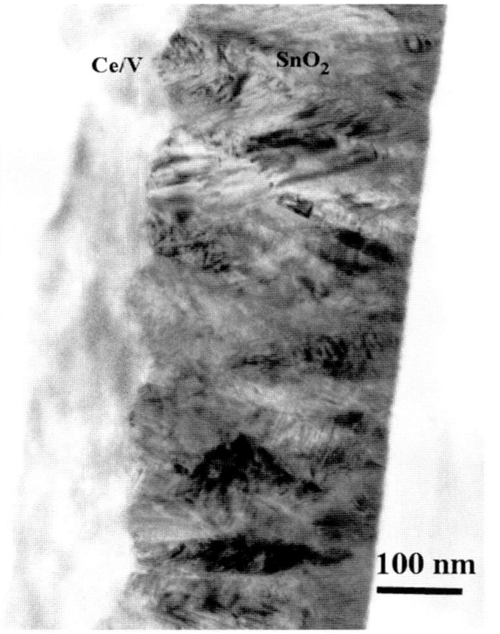

Fig. 4 Thin film with 55 atom % of V in Ce/V oxide deposited on K-glass substrate. The thickness of the film was between 90 and 150 nm. The thickness of the SnO_2:F layer is between 260 and 350 nm

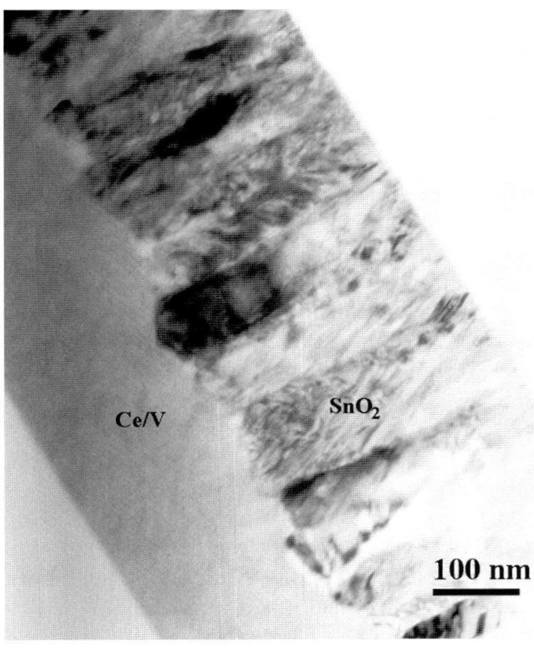

Fig. 5 Thin film with 38 atom % of V in Ce/V oxide deposited on K-glass substrate. The thickness of the film was between 180 and 280 nm. The thickness of the SnO_2:F layer is between 260 and 350 nm

crystallite structure does not correspond to the tetragonal and monoclinic structure of $CeVO_4$, nor to the cubic CeO_2. From the ring-pattern simulation the best agreement was achieved for the orthorhombic V_2O_5.

The formation of crystalline $CeVO_4$ or CeO_2 could not be observed within the detection limit of the transmission electron microscope. With IR and Raman spectroscopy the formation an amorphous phase (VO_4^{3-} tetrahedral species) in mixed powder samples prepared at 55 and 38 atom % of V in V/Ce oxide was confirmed.

Grazing-incidence small-angle X-ray scattering

A vanadium-oxide thin film on a glass substrate can be considered as an aggregate containing V_2O_5 nanosized grains and nanosized pores [9, 12]. In this case, the SAXS is caused by the difference in the electron density within and around the nanosized grains. Using the Guinier approximation [13] the grain size can be readily determined. The limiting angle Θ_m of the small-angle diffuse scattering is approximately $\Theta_m = \lambda/2D$, where D is the largest grain diameter. The Porod approximation [14–16] is suitable for determining the specific surface area of nanosized thin films. With the high intensity of synchrotron radiation sources the scattered intensity is high enough so that we can apply both approximations and obtain both, i.e., the grain size and the specific surface of the nanosized thin films.

The measurements were performed using a small-angle geometry, where the scattering from the substrate is small. The background was 2 orders of magnitude smaller than the signal of the sample and was neglected during the analysis. Our analysis by grazing-incidence reflectivity [17] showed that this material had a single layer. The thicknesses of the layers in the V_2O_5 and mixed oxides were determined as 108 nm for V_2O_5, 219 nm for V/Ce at 55 atom % of V and 217 nm for V/Ce at 38 atom % of V. Photographs obtained using atomic force microscopy [9] show the surface of this material with a surface roughness changing from 2 to 100 nm.

Plotting the data in a log$I$ versus q^2 plot, where $q = 4\pi\Theta/\lambda$, we tested [12] whether one can apply the Guinier law: $I(q) = (\Delta\rho)^2 \exp(-R_g^2 q^2/3)$ for small q, where $\Delta\rho$ is the difference in the electron density between the grain and the surrounding medium. This allowed us to estimate the average grain radius from the radius of gyration R_g in the Guinier formula.

This was calculated from the slope of the linear fit of log$I$ versus q^2 (per square nanometer). From these fitting lines we obtained R_g and the average grain radius using $= (5/3)^{1/2} R_g$ (for a spherical shape). The Gaussian form is valid up to $q = 1.5\ nm^{-1}$. The approximation is very good (5%) up to $qR_g = 2\pi \times 0.4$ for grains with a spherical shape. The grain size was determined for the V/Ce oxides for sets of fixed grazing angles. They were varied in the range from 0.3° to 1.1° with a step size of 0.01°.

Table 1 V_2O_5/CeO_2 samples, annealing temperature, grazing angle, "average grain radius" $<R>$ and "specific surface area" S_s, as determined from grazing-incidence small-angle X-ray scattering (*GISAXS*)

Sample	Annealing temperature/K	Grazing angle/°	GISAXS $<R>$/nm	GISAXS S_s/nm^{-1}
V_2O_5	673	0.4	6.44 ± 1.17	0.187
V_2O_5	673	0.5	5.39 ± 1.27	0.304
V_2O_5	673	0.6	4.16 ± 1.30	0.449
V_2O_5	673	0.7	3.51 ± 1.37	0.674
V_2O_5	673	0.8	3.13 ± 1.07	0.946
V_2O_5	673	0.9	2.94 ± 0.74	1.574
55%	673	0.4	4.95 ± 1.42	0.095
55%	673	0.5	4.24 ± 1.19	0.168
55%	673	0.6	3.63 ± 1.37	0.193
55%	673	0.7	2.43 ± 0.79	0.244
55%	673	0.8	2.29 ± 0.92	0.344
55%	673	0.9	2.09 ± 0.76	0.472
38%	673	0.4	5.02 ± 1.37	0.150
38%	673	0.5	3.43 ± 1.28	0.190
38%	673	0.6	2.63 ± 1.31	0.240
38%	673	0.7	1.93 ± 0.94	0.500
38%	673	0.8	1.90 ± 1.49	0.820
38%	673	0.9	1.45 ± 0.59	1.150

Supported by atomic force microscopy photographs [9] the "specific surface area" S_s was calculated using the Mittelbach–Porod formula [14–16]. The average grain sizes $<R>$ and specific surface S_s for all the samples at chosen grazing angles are shown in Table 1. Our GISAXS values for $<R>$ are in agreement, within the error limits, with the results from the TEM analysis.

Our analysis of the measurements at the synchrotron light source was done by a graphical method of calculation for the grain size and the specific surface, which determines porosity. The average radius of the mixed oxides shows a difference in the values of pure V_2O_5. For the sample of V/Ce oxide at 38 atom % of V, as presented in Table 1, S_s is 1.150 nm^{-1}, which is smaller than the value of 1.574 nm^{-1} for pure V_2O_5 [17]. Pure V_2O_5 for a grazing angle at 0.9° has bigger grains $<D> = (5.88 \pm 1.48)$ nm and a higher porosity than the mixed V/Ce oxide at 38 atom % of V with $<D> = (2.90 \pm 1.18)$ nm. Pure V_2O_5 for all grazing angles has a higher porosity than the mixed V/Ce oxides at 38 and 55 atom % of V. The grain size determined by XRD [8] is in accordance with our results. All the values for $<R>$ and S_s are in agreement, within the error limits, with results from previous work [12]. The addition of cerium dioxide to vanadium pentoxide increased the stability of vanadium pentoxide during 1,400 cycles of charge–discharge with a constant current. The V/Ce oxide is the intercalation electrode in the electrochemical cell with the lithium electrolyte ($LiClO_4$ in propylene carbonate). The V/Ce oxide with 38 atom % of V shows the biggest capacity for the intercalation of Li ions in comparison with the other V/Ce oxides and is 300 C m^{-2} [9].

Conclusions

The present study shows that only the combination of structural analyses with morphological characterization will provide a complete understanding of the properties of the materials prepared. We showed that GISAXS combined with grazing-incidence reflectivity at the ELETTRA light source could be applied for the layer-thickness determination in nanosized films of V_2O_5 and V/Ce oxides on a K-glass substrate. Comparing the GISAXS grain sizes and the porosity of the samples with results from TEM measurements, we could reveal the complex layered and nanostructured nature of these materials. No solid-state reaction occurred between the films and the substrate. It is clear that we always have to compare and characterize materials prepared in the same way, and on the same type of substrate.

The particular morphology obtained for the samples prepared at 38 atom % of V in V/Ce oxide is suitable for application in electrochromic devices, in an advanced electrochemical cell concept, and in efficient new solar cells.

Acknowledgements The Ministry of Education, Science and Sport of the Republic of Slovenia, under contract number P-024512, is thanked for its support of this work, as is the Ministry of Sciences and Technology of the Republic of Croatia, under contract number 0098026.

References

1. Murphy DW, ChristianPA, DiSalvo FJ, Cardies JN (1979) J Electrochem Soc 126:497
2. Julien C, Haro-Pontiatovski E, Camacho-Lopez MA, Escorbar-Alarcon L, Jimenez-Jarquin (1999) J Mater Sci Eng 65:170
3. Losurdo M, Barreca D, Bruno G, Tondello E (2001) Thin Solid Films 384:58
4. Julien C, Guesdon JP, Gorenstein A, Khelfa A, Ivanov I (1995) Appl Surf Sci 90:389
5. Orel ZC, Orel B (1995) J Mater Sci 30:2284
6. Turkovic A, Orel ZC (1996) Solid State Ionics 89:255
7. Orel ZC (1999) Appl Spectrosc 53:241
8. Orel ZC (1999) Solid State Ionics 116:105
9. Orel ZC, Muševic I (1999) Nano struct Mater 12:399
10. Orel ZC, Turkovic A, Gaberscek M Sol Energy Mater Sol Cells (accepted 2004)
11. Stadelmann PA (1987) Ultramicroscopy 21:131
12. Turkovic A, Orel ZC, Dubcek P (2001) Mater Sci Eng B 79:11
13. Guinier A (1939) Ann Phys Paris 12:161
14. Porod G (1951) Kolloid Z 124:83
15. Porod G (1952) Kolloid Z 125:108
16. Porod G (1953) Kolloid Z 133:51
17. Posedel D, Turkovic A, Dubcek P, Orel ZC (2002) Mater Sci Eng B 90:154

Progr Colloid Polym Sci (2004) 128: 126–130
DOI 10.1007/b97108

Marcela Mohallem Oliveira
Daniela Zanchet
Daniel Ugarte
Aldo José Gorgatti Zarbin

Synthesis and characterization of silver nanoparticle/polyaniline nanocomposites

M.M. Oliveira · A.J.G. Zarbin (✉)
Departamento de Química,
Universidade Federal do Paraná,
CP 19081, CEP 81531-990,
Curitiba, PR, Brazil
e-mail: aldo@quimica.ufpr.br
Tel.: +55-41-3613297
Fax: +55-41-3613186

D. Zanchet · D. Ugarte
Laboratório Nacional de Luz Síncrotron,
CP 6192, CEP 13984-971, Campinas, SP,
Brazil

Abstract Novel dodecanethiol-capped silver nanoparticle (NP)/polyaniline (PANI) nanocomposites were prepared via the in situ polymerization of aniline in a metal NP containing medium. Sample PANI/Ag-1 was obtained by the polymerization of aniline in an aqueous solution in the presence of powdered NPs, and sample PANI/Ag-2 was obtained by a two-phase reaction, in which a toluene solution of silver NPs and aniline were mixed with an acidic aqueous solution of ammonium persulphate (that acts as an oxidizing agent in the aniline polymerization). Fourier transform IR, Raman and UV–vis spectroscopy confirmed PANI formation in both samples. Transmission electron microscopy and high-resolution transmission electron microscopy images indicate that silver NPs were homogeneously dispersed on a polymer mass, and new Ag/PANI core–shell material was observed on PANI/Ag-2.

Keywords Silver nanoparticles · Nanocomposites · Polyaniline

Introduction

Nanocomposites of conducting polymers (such as polyaniline and polypyrrole) and inorganic materials have received much attention during the last few years [1–6]. They have shown various interesting properties that can be expected to be different from those of the polymers or the inorganic particles themselves [1]. One particular research objective has been the dispersion of metal nanoparticles (NPs) within the polymer matrix [1, 7–12], since the incorporation of metal NPs is known to enhance the conductivity of the polymer [10]. A great number of different applications (catalysts, sensors) can be expected for these materials [13–15].

Current methods used to incorporate inorganic NPs within polymer matrices are based on blending or mixing the individual components, using a polymer solution or a melt. However, these approaches are not easily applicable to conducting polymers [1] owing the fact that they are infusible and generally insoluble in common solvents. Metal NP/conducting polymer nanocomposites have thus been obtained by using different and more complex methods: reduction of metal salts by the polymer [8, 12], metal NP and polymer formation within a porous structure [9], electrochemical methods leading to the incorporation of metal NPs during the electrosynthesis of polymers [16] or the electrodeposition of metal NPs on conducting polymer electrodes [17]. In this work, we report a new in situ method to prepare nanocomposites in which silver NPs are homogeneously distributed within a bulk polyaniline matrix.

Experimental

All analytical grade reagents were used as received except aniline, which was vacuum-distilled before use. Dodecanethiol-capped silver NPs were prepared as described by Brust et al. [18], based on a two-phase reduction method. Typically 30 mL of an aqueous $AgNO_3$ solution (0.03 mol L^{-1}) was added to 80 mL of a toluene solution containing octylammonium bromide (0.05 mol L^{-1}) to form a two-phase system. The system was maintained under vigorous stirring for 10 min to transfer the metal ion into the organic phase, which was then separated. A 403 μL aliquot of dodecanethiol was subsequently added to the solution while stirring, followed by rapid injection of 25 mL of a 0.4 mol L^{-1} aqueous solution of $NaBH_4$ (reducing agent). The system was maintained under stirring for 3 h. The silver NPs were precipitated by the addition of excess ethanol, subsequently isolated by centrifugation, washed with ethanol and redissolved in toluene. The nanocrystal obtained consisted of a metal core capped by thiol ligands that passivate the surface and prevent flocculation. The silver NPs had an average diameter of 4.7 nm.

Ag NP/polyaniline nanocomposites were prepared by two different routes:

1. Ammonium persulphate (36.8 mg) was added to 30 mL of a 1 mol L^{-1} H_2SO_4 aqueous solution. Ag NPs prepared as described earlier were precipitated and the resulting solid (about 0.1 g) was added to the flask containing the ammonium persulphate/sulphuric acid solution. The dispersion was maintained under magnetic stirring and subsequently 58 μL of aniline was added. After 5 h the resulting green solid was isolated by centrifugation, washed several times with water and ethanol and dried in air at 40 °C. The solid is referred to here as PANI/Ag-1;
2. Aniline (58 μL) was mixed with 20 mL toluene solution of Ag NPs obtained as described earlier (about 0.1 g passivated silver NPs). This solution was mixed with 30 mL of a 1 mol L^{-1} H_2SO_4 aqueous solution containing 36.8 mg ammonium persulphate. The two-phase system was maintained under magnetic stirring for 5 h. After this period the brown colour of the organic phase disappeared and a highly disperse green powder was formed on the aqueous phase. The two phases were separated and the solid on the aqueous phase was isolated by centrifugation, washed several times with water and ethanol and dried at 40 °C. This product is referred to here as PANI/Ag-2.

Pure polyaniline was also prepared following the same procedures as described earlier. The experimental steps were realized exactly as described, but without the presence of silver NPs. The samples obtained are referred to as PANI-1 (similar synthetic procedure to PANI/Ag-1) and PANI-2 (similar synthetic procedure to PANI/Ag-2).

X-ray diffraction (XRD) patterns were obtained with a Shimadzu difractometer, using Cu Kα radiation with 40 kV and 40 mA, at 0.2° scan rate (in 2θ). Powder silicon reflections were used for 2θ calibration.

The Fourier transform IR spectra of the samples were obtained with a Bomem MB-100 spectrophotometer in the 4,000–400-cm^{-1} range with 32 scans. The samples were prepared in KBr pellets.

The Raman spectra were obtained using a Renishaw Raman image spectrophotometer, coupled to an optical microscope that focuses the incident radiation down to an approximately 1-μm spot. A He–Ne laser (emitting at 632.8 nm) was used, with an incidence intensity of 2 mW over the 2,000–100-cm^{-1} region.

The UV–vis spectra were collected with a Scinco S-1150 spectrophotometer, using air as the reference, in the 190–1,000-nm range.

Transmission electron microscopy (TEM) measurements were done with a JEOL JEM 120 KV instrument. The samples for observation were suspended in water and allowed to settle for 15 min. Then a drop of the supernatant dispersion was placed onto a carbon film supported by a copper grid. High-resolution TEM (HRTEM) was performed with a JEOL JEM 3010 microscope operating at 300 kV (1.7-Å resolution).

Results and discussion

The synthetic approach used to form both nanocomposites described here is based on the in situ polymerization of aniline in the presence of dodecanethiol-capped silver NPs. In the first method, a powder of silver NPs was added to an aqueous medium in which the aniline polymerization was carried out. In this case the powder of NPs was dispersed in the reaction medium and the polymer was supposed to be formed around the NPs, in a heterogeneous process. In the second method, the NPs were dissolved in an organic solution containing aniline and mixed with an aqueous solution containing the oxidizing (ammonium persulphate). In this case, the polymerization took place at the interface of both phases. The disappearance of the brown colour of the toluene solution after the reaction was strong evidence that the silver NPs were carried out of the organic phase concomitantly with the aniline polymerization.

The green colour of the two products obtained was evidence that the conducting emeraldine salt of polyaniline was formed. The presence of emeraldine salt was confirmed by IR (not shown), Raman and UV–vis spectroscopy. Beyond the bands attributed to the emeraldine salt, the IR spectra of the samples also showed sulphate bands (indicating that sulphate is acting as a counteranion in the emeraldine salt structure) and characteristic dodecanethiol bands, similar to those observed for pure dodecanethiol-capped silver NPs, indicating that there was no degradation or desorption of the ligand during the nanocomposite synthesis. The resonance Raman spectra of the two nanocomposites and the pure polyaniline are shown in Fig. 1. The spectra of the pure polyaniline and the nanocomposites are very similar and are in agreement with the previously described spectra of pure emeraldine salt [19, 20], with the following main bands: about 1,620 (ν C=C of the benzenoid rings), 1,586 cm^{-1} (ν C=C of the quinoid rings), about 1,485 cm^{-1} (ν C=N of the quinoid diimina units), about 1,336 cm^{-1} (ν C–N radical cation), about 1,318 cm^{-1} (ν C–N radical cation) about 1,255 cm^{-1} (ν C–N benzene diamine units) and about 1,166 cm^{-1} (C–H bending of the quinoid rings).

UV–vis spectra of the samples are shown in Fig. 2. Clearly the nanocomposite samples present bands characteristic of polyaniline–emeraldine salt at about 350 nm, about 430 nm and about 760 nm, which are attributed to π–π^*, polaron–π^* and π–polaron transitions, respectively [21]. The spectrum of pure silver NPs is also shown in Fig. 2. We can see a noticeable plasmon band at 452 nm, which is characteristic of metallic NPs.

The X-ray diffractogram of the nanocomposite samples (not shown) confirms the presence of metallic silver. Also, the XRD patterns of both samples contain many strong and sharp equally spaced reflections at lower angles, as observed for the pure silver NP sample, which is attributed to the formation of a superlattice owing to the self-organization of dodecanethiol-capped

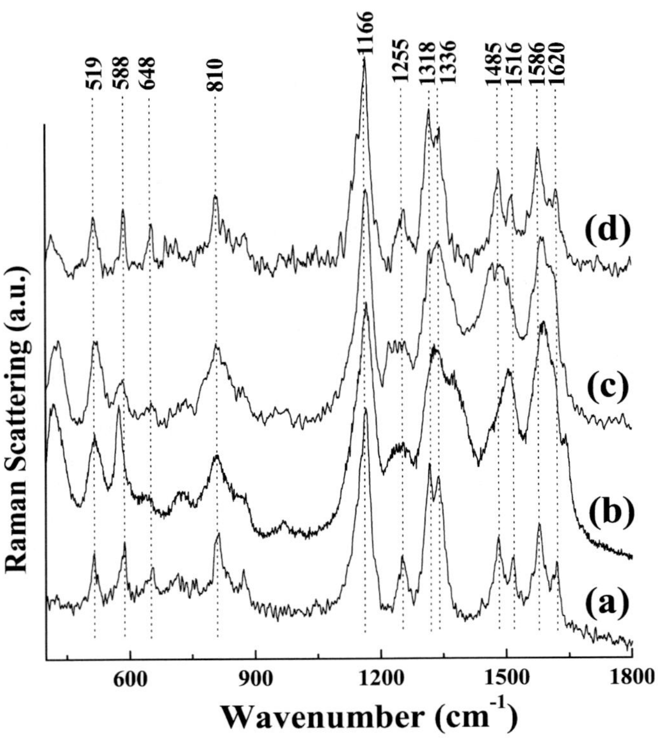

Fig. 1 Raman spectra of the following samples: PANI-1 (*a*); PANI-2 (*b*); PANI/Ag-1 (*c*); PANI/Ag-2 (*d*)

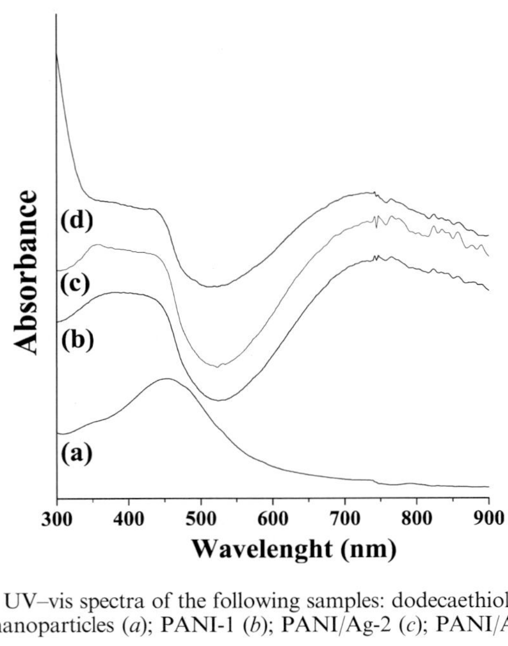

Fig. 2 UV–vis spectra of the following samples: dodecaethiol-capped silver nanoparticles (*a*); PANI-1 (*b*); PANI/Ag-2 (*c*); PANI/Ag-1 (*d*)

silver NPs [22]. The occurrence of the superlattice peaks in the XRD patterns of the nanocomposite samples is an indication that the silver NPs are highly organized in the polymer mass. To confirm this hypothesis and to verify how the silver NPs are distributed in the polymer, TEM measurements were done for both nanocomposite samples. The results are shown in Fig. 3.

TEM images of the PANI/Ag-1 sample are shown in Fig. 3a and b. Clearly we can see spherical metal NPs homogeneously dispersed on a polymer mass, indicating that the method was successful in obtaining a true nanocomposite. The size of the silver NPs presented a bimodal distribution, as observed in Fig. 3a and b, and is representative of the entire sample. Figure 3a shows silver NPs with a mean size of 4.7 nm, similar to the size of the original dodecanethiol-capped NPs. We can clearly see that the particles are distributed on the polymer mass with a high degree of organization, forming a kind of superlattice that was embedded in the polymer, which explains the lower angle peaks observed in the XRD patterns, as discussed earlier. Figure 3b also shows a homogeneous distribution of silver NPs on a polymer mass, but the mean size of the particles is approximately 13.5 nm. The growth of some of the NPs could result from aggregation of some of them after the polymerization step caused by the drying of the sample, or could be due to some effect caused by the polymerization reaction. Work is in progress to clarify this point.

TEM images of the PANI/Ag-2 indicate that this sample was formed with two well-differentiated regions, which are representatively shown in Fig. 3c and d. Figure 3c shows homogeneously distributed silver NPs on a polymer mass, with a high degree of organization and an average size of around 5 nm, similar to the PANI/Ag-1 sample. However, more than 50% of the PANI/Ag-2 sample was formed as illustrated in Fig. 3d, where we can note individual silver NPs surrounded by a polymer shell, forming well-defined silver/polyaniline core–shell nancomposites (see, e.g., particles marked with A in Fig. 3d). In a great number of particles the polymer was formed as a half-shell (B in Fig. 3d). These very interesting features can be directly related to the fact that the polymerization reaction was done in a two-phase system. Since the silver NPs were homogeneously mixed with aniline in the toluene solution, they could be acting as a nucleus to promote the polymerization. Once the polymerization occurs at the interface of the organic (containing the aniline and the silver NPs) and aqueous (containing the oxidizing) phase, the polymer can grow around the silver NPs, forming the structures shown in Fig. 3d. As we conducted the reaction for a long time, the excess of aniline previously added to the system possibly produced a large amount of polyaniline that grew around the several core–shell NPs forming a polymeric mass, resulting in regions like that presented in Fig. 3c. These results show that this is a very promising route to produce core–shell

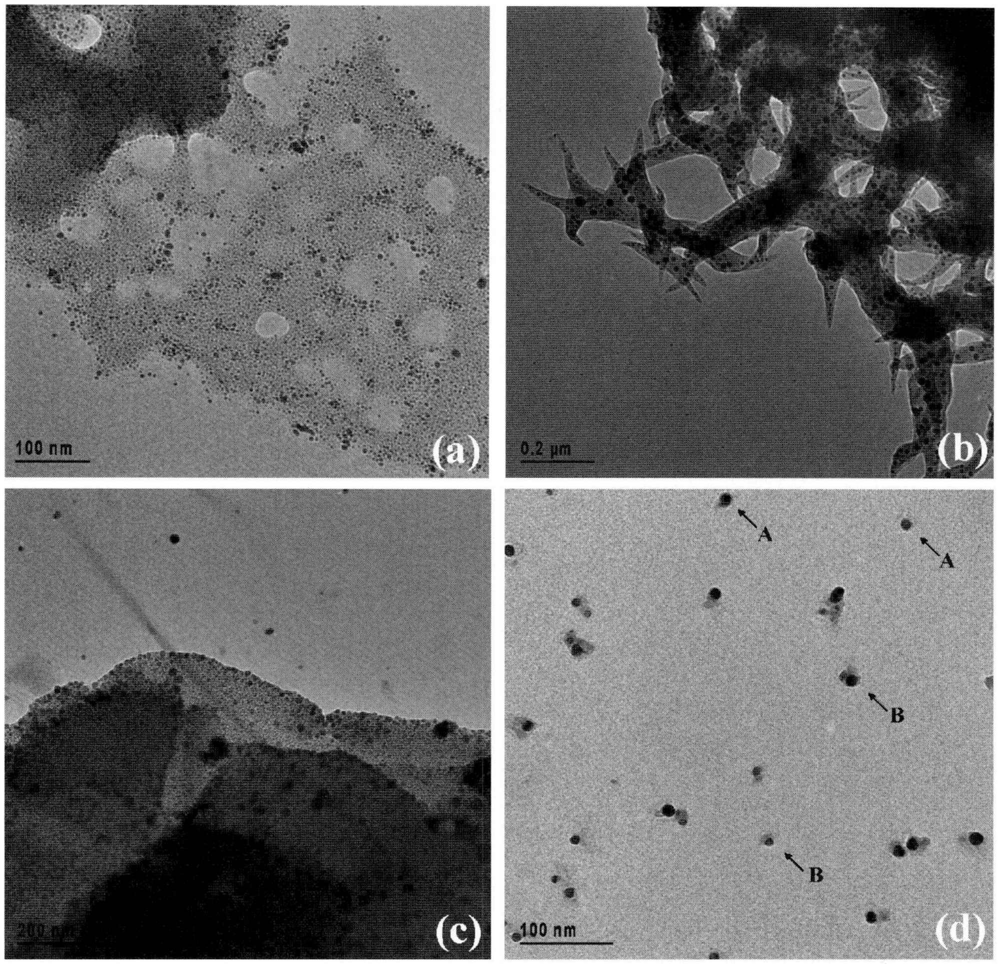

Fig. 3 Transmission electron microscopy images of **a**, **b** PANI/Ag-1 and **c**, **d** PANI/Ag-2

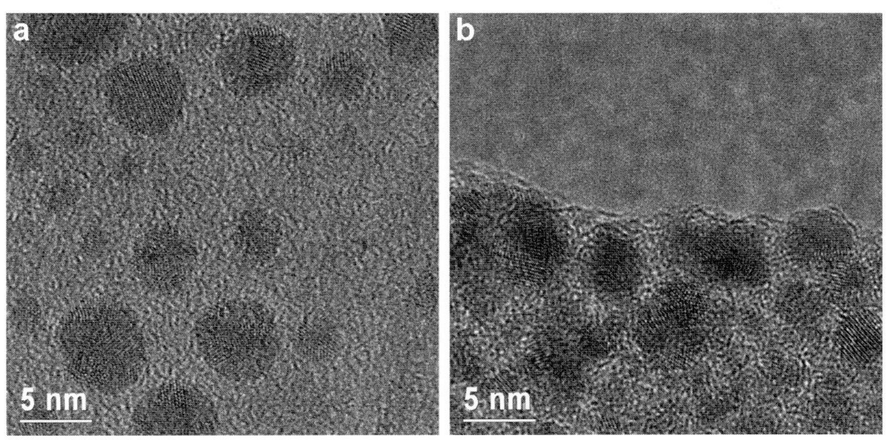

Fig. 4 High-resolution transmission electron microscopy image of PANI/Ag-1

structures, by a rigorous control of the synthetic parameters, such as the concentration of aniline and silver NPs and the polymerization time. Experiments are under way to prove this point.

The HRTEM image of the silver NPs on the PANI/Ag-1 nanocomposite is shown in Fig. 4. Lattices fringes show the crystalline structure of the NPs and can be attributed to the (111) planes of face-centred-cubic metallic silver.

Conclusion

Two new approaches are effective in producing novel dodecanethiol-capped silver NP/polyaniline nanocomposites, based on the in situ polymerization of aniline in a silver NP containing medium. In both approaches polyaniline was formed in its conducting form, the emeraldine salt. Silver NPs were homogeneously dispersed in the polymer mass. The synthetic route based on a two-phase reaction is a promising way to produce silver NP/polyaniline core–shell materials.

Acknowledgments The authors thank CNPq and Rede de Materiais Nanoestruturados (MCT/FINEP) for financial support, Centro de Microscopia Eletrônica–UFPR and Laboratório de Microscopia Eletrônica–LNLS for the TEM and HRTEM images, LEM-IQ-USP for the Raman spectra. M.M.O. thanks CAPES for a fellowship.

References

1. Gangopadhyay R, De A (2000) Chem Mater 12:608
2. Greenham NC, Peng XG, Alivisatos AP (1996) Phys Rev B 54:17628
3. Colvin VL, Schlamp, MC, Alivisatos, AP (1994) Nature 370:354
4. Cassagneau, T, Mallouk, TE, Fendler, JH (1998) J Am Chem Soc 120:7848
5. Zarbin AJG, De Paoli, MA, Alves OL (1999) Synth Met 99:227
6. Gonçalves AB, Mangrich AS, Zarbin AJG (2000) Synth Met 114:119
7. Hatchett DW, Josowicz M, Janata J (1999) Chem Mater 11:2989
8. Zhou Y, Itoh H, Uemura T, Naka K, Chujo Y (2002) Langmuir 18:277
9. Marinakos SM, Shultz DA, Feldheim DL (1999) Adv Mater 11:34
10. Neoh KG, Young TT, Looi, NT, Kang ET, Tan KL (1997) Chem Mater 9:2906
11. Sarathy KV, Narayan KS, Kim J, White JO (2000) Chem Phys Lett 318:543
12. Wang J, Neoh KG, Kang ET (2001) J Colloid Interface Sci 239:78
13. Ficicoglu F, Kadirgan F (1997) J Electroanal Chem 430:179
14. Lui EKW, Beattie PD, Holderaft S (1997) Synth Met 84:87
15. Langmaier J, Janata J (1992) Anal Chem 64:523
16. Vork FTA, Janssen LJJ, Barendrecht E (1986) Electrochim Acta 31:1569
17. Holdcroft S, Funt BL (1988) J Electroanal Chem 240:89
18. Brust M, Walker M, Bethell D, Schiffrin DJ, Whyman R (1984) J Chem Soc Chem Commun 801
19. Quillard S, Berrada K, Lovarn G, Lefrant S, Lapkowiski M, Pron A (1995) New J Chem 19:365
20. Louarn G, Lapkowski M, Quillard S, Pron A, Buisson JP, Lefrant S (1996) J Phys Chem 100:6998
21. Xia Y, Wiesinger M, MacDiarmid AG (1995) Chem Mater 7:443
22. Sandhyarani N, Resmi MR, Unnikrishnan R, Vidyasagar K, Shuguang M, Antony MP, Sevan GP, Visalakshi V, Chandrakumar N, Pandian K, Tao YT, Pradeep T (2000) Chem Mater 12:104

Progr Colloid Polym Sci (2004) 128: 131–134
DOI 10.1007/b97110

Marina S. Leite
Varlei Rodrigues
Daniela Zanchet

Structural effects on Au and Ag colloidal nanoparticles

M.S. Leite · V. Rodrigues
D. Zanchet (✉)
Brazilian Synchrotron Light Laboratory,
Cx.P. 6192, 13084-971 Campinas,
SP, Brazil
e-mail: zanchet@lnls.br
Tel: +55-19-32874632
Fax: +55-19-3287-4520

M.S. Leite
IFGW-UNICAMP, Cx. P. 6163,
13084-971 Campinas, SP, Brazil

Abstract On the nanometer scale, size reduction effects affect materials properties. Au and Ag nanoparticles are interesting cases owing to a structural transition that occurs at nanometric sizes; this has been explored, for example, to change material properties. Here, we present structural studies on dodecanethiol-capped Au and Ag colloidal nanoparticles synthesized at room temperature and 80 °C. For the same synthesis conditions, Au nanoparticle sizes are smaller than Ag ones. From the structure point of view, the synthesis temperature has a minor effect on Au nanoparticles but it induces stacking faults and twin planes in the Ag case. These results suggest that hexagonal close-packed symmetry may be favored for Ag for certain synthesis conditions, but this has not been identified for Au.

Keywords Colloidal gold · Nanoparticle · High-resolution transmission electron microscopy · X-ray diffraction

Introduction

On the nanometer scale, size reduction effects play an important role and greatly affect material properties. In this aspect, unique properties, such as the catalytic activity and optical properties of Au nanoparticles, have been reported [1–3]. In fact, Au colloids have been known for more than a century but they have attracted special attention in the last few years owing to the improvement of synthetic methods and the development of novel applications based on their characteristic surface plasmon resonance, such as surface-enhanced Raman spectroscopy and biological labeling [2–4]. In particular, strong modifications of plasmon peaks have been found for anisotropic particles, such as nanorods and nanoprisms, increasing the potential applications of these nanosystems [3, 5, 6]. In a similar way, Ag nanoparticles have also been the subject of many studies and the development of new technologies [6].

From the structure point of view, Au and Ag are also interesting materials since a structure transition has been found for sizes below 4 nm [7]. Since the atomic arrangement affects most particle properties and it is directly related to morphology, structural modifications could be better explored to tune material properties and shape-control.

Au and Ag have the same bulk crystallographic structure (face-centered cubic, fcc, $a=4.07$ Å) and are expected to follow similar structural modifications as a function of particle size, i.e., decahedra in the 1–3-nm range and icosahedra for sizes below 1 nm [7]. Icosahedra and decahedra are noncrystallographic structures known as multiply twinned particles and they are characterized by a fivefold axis symmetry. The origin of this structural transition is based on surface energy that becomes important at nanometric size, and that is minimized by structural modifications. The difference in energy among these structures, however, is small and a real sample usually presents more than one type of structure. Kinetic effects may also play an important role during particle formation and they have been used to produce metastable structures [8, 9]. Although several works can be found on Au nanoparticles [7, 8, 10–12], much less is known about the structure of Ag nanoparticles [8, 13].

Here, we address structural effects on Au and Ag nanoparticles produced by colloidal methods. Despite their structural similarity in bulk, Au and Ag nanoparticles have shown different structural aspects when synthesized under the same conditions. An important result is the tendency of Ag nanoparticles to present defects (stacking faults and twins), which are not observed in Au nanoparticles.

Experimental

Au and Ag nanoparticles were produced by the method developed by Brust et al. [4]. Briefly, hydrogen tetrachloroaurate or silver nitrate is reduced by sodium borohydrate in a two-phase system (water and toluene) in the presence of dodecanethiol (capping molecules). Tetra-*n*-octylammonium bromide is used as phase-transfer agent. Samples were synthesized under Ar at room temperature (Au-RT and Ag-RT, 25 °C) and at 80 °C (Au-80 °C and Ag-80 °C), keeping the other parameters constant. Size and structural characterization were performed by a combination of a local probe technique, such as high-resolution transmission electron microscopy (HRTEM), and average-type techniques, such as small-angle X-ray scattering (SAXS) and X-ray diffraction (XRD). HRTEM was performed with a JEM-3010, operating at 300 kV (1.7-Å point resolution) at the Electron Microscopy Laboratory (LME) at the Brazilian Synchrotron Light Laboratory (LNLS). SAXS experiments were performed in dilute solutions at the SAXS beamline-LNLS, with $\lambda = 1.608$ Å and $(0.01) < q < 0.7$ Å^{-1} ($q = 4\pi \sin\theta/\lambda$; q is the scattering vector, 2θ is the scattering angle and λ is the X-ray wavelength). XRD measurements were performed in dried samples either at the XRD1 beamline-LNLS ($\lambda = 1.06$ Å) or using a conventional source ($\lambda = 1.54$ Å), operating in θ–2θ geometry. XRD data were compared with simulated XRD patterns, using the Debye formula, of nanoparticles of different sizes and structures [7, 8, 10].

Results and discussion

The low-magnification TEM images of Au-RT and Ag-RT nanoparticles and the corresponding size distributions are shown in Fig. 1a. Both samples present spherical particles with average diameters (d_m) of 2.0 and 4.6 nm, respectively, and a narrow size distribution ($\sigma = 0.4$ nm). SAXS data of the same samples are presented in Fig. 2; the dashed lines correspond to theoretical scattering curves of spherical particles with a Gaussian size distribution, showing good agreement with the TEM results [Au-RT $d_m = 1.7$ nm ($\sigma = 0.34$ nm); Ag-RT $d_m = 4.0$ nm ($\sigma = 0.6$ nm)]. Increasing the synthesis temperature to 80 °C has a different effect on the Au and Ag nanoparticle size; it increases the particle size for Au ($d_m = 3.0$ nm for Au-80 °C), while it does not have a significant effect for Ag (data not shown).

Concerning the particle structure, HRTEM images have shown that all the samples present structure distributions, as mentioned before. Representative HRTEM images for the four samples (more than 100 images were analyzed for each one) are shown in Fig. 3.

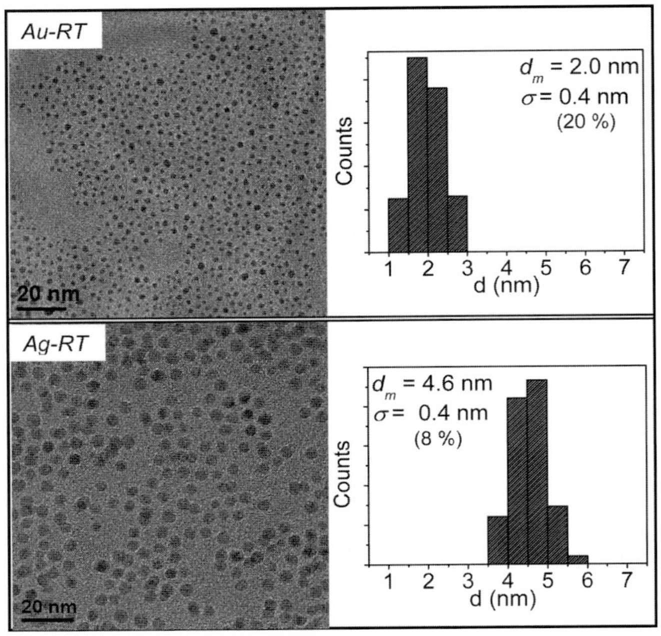

Fig. 1 *TEM* images of Au-RT and Ag-RT nanoparticles and corresponding size distributions. More than 300 particles were evaluated for each sample

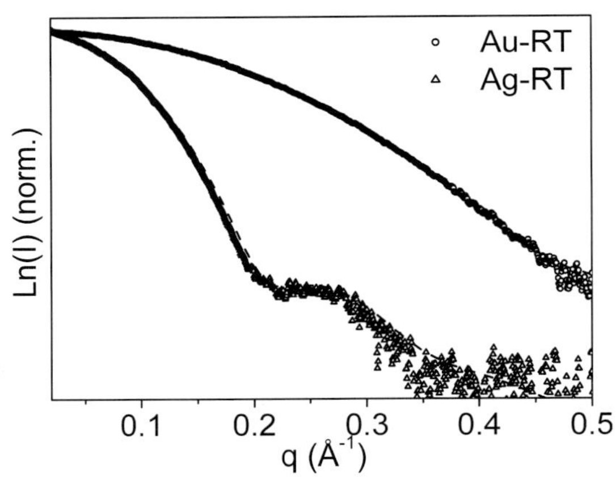

Fig. 2 Small-angle X-ray scattering data for the same samples as in Fig. 1. The *dashed lines* correspond to theoretical scattering curves of spherical particles with a Gaussian size distribution

In both the Au-RT (Fig. 3a, b) and the Au-80 °C (Fig. 3c, d) samples decahedral and fcc particles could be frequently identified. In the case of Ag-RT (Fig. 3e, f) and Ag-80 °C (Fig. 3g, h), the HRTEM images show a large number of perfect fcc particles and fcc particles with defects. In this case, however, the increase in the synthesis temperature seems to result in an increase in the number of defects per particle. This can be seen in Fig. 3 h, which shows a very defective particle with several stacking faults and twin planes. As a

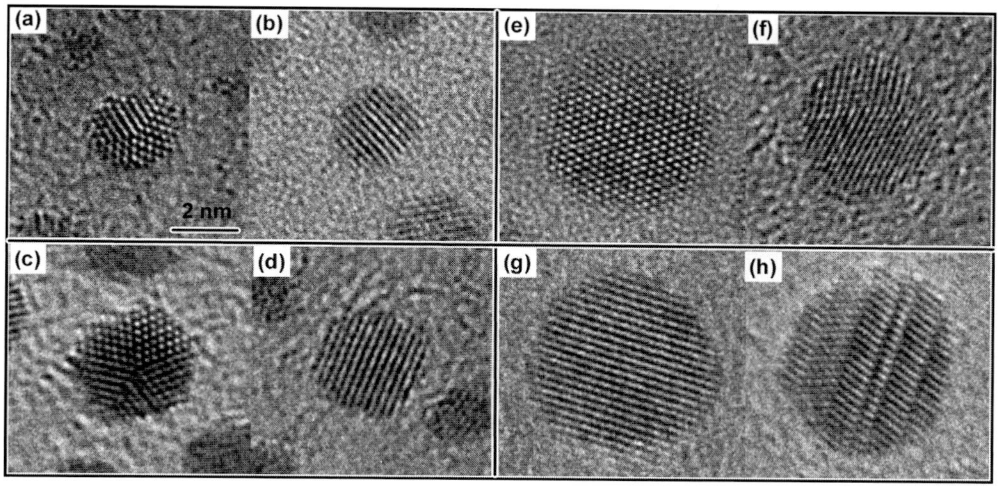

Fig. 3 Representative *HRTEM* images of **a** decahedral Au-RT, **b** face-centered cubic (*fcc*) Au-RT, **c** decahedral Au-80 °C, **d** fcc Au-80 °C, **e** fcc Ag-RT, **f** fcc Ag-RT with a twin plane, **g** fcc Ag-80 °C and **h** defective fcc Ag-80 °C. Note the several stacking faults and twin planes in **h**

consequence, from the structural point of view the synthesis temperature seems to have more of an effect for Ag than for Au. It is important to remark that in the 2–3-nm size range a decahedron is the energetically favored structure, whereas for larger particles, fcc is the expected structure.

The XRD patterns were analyzed to confirm the HRTEM results. In the case of Au, a careful analysis of the XRD patterns (data not shown) confirms that the majority population is indeed decahedral particles for Au-RT and Au-80 °C. The decahedral pattern can be easily identified by the higher relative intensity of the (111) peak compared with the fcc pattern and the convolution of the (220) and (311) peaks [7, 10].

The XRD patterns of Ag-RT and Ag-80 °C are shown in Fig. 4. It can be seen that most of the peaks match the fcc pattern (dotted lines) and that the peak broadening is similar for both samples, confirming the close nanoparticle size. In Ag-80 °C, however, additional features at 2.8, 3.6 and 4.7 $Å^{-1}$ can be clearly seen (marked by arrows). These features cannot be explained by fcc, decahedral or icosahedral patterns, indicating an additional type of structure. HRTEM had already pointed out the existence of defective fcc particles, mainly for the Ag-80 °C sample. A HRTEM image of a defective fcc particle (inset) and the corresponding simulated XRD pattern are shown in Fig. 4b. The dashed curve corresponds to a perfect fcc particle with the same size. It can be clearly seen that the additional features that appear in the experimental XRD pattern are a consequence of stacking faults and twin planes. Looking carefully at the XRD pattern of Ag-RT, we could identify the same features, but with much lower intensity, showing that in

▶

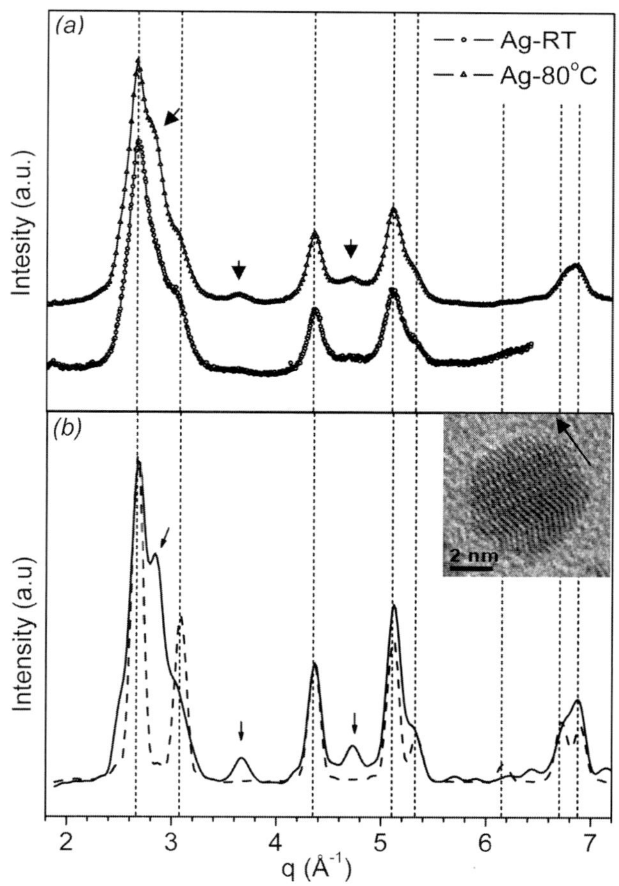

Fig. 4a X-ray diffraction pattern of Ag-RT and Ag-80 °C samples. The *dashed lines* correspond to the fcc peak positions. The *arrows* show additional features that are enhanced in the Ag-80 °C pattern. **b** *Inset*: HRTEM image of a defective particle found in the Ag-80 °C sample (the *arrow* indicates the hexagonal close-packed region of this particle). The graphic shows the corresponding theoretical X-ray diffraction pattern of this particle (*solid line*). Note the same additional features present in the experimental pattern, marked by arrows. The *dashed curve* corresponds to the diffraction pattern of a perfect fcc particle of the same size

the Ag case there is indeed a change in the structure distribution when the synthesis temperature is increased.

One interesting point to note is that the presence of stacking faults indicates a transition from fcc to hexagonal close-packed (hcp). In particular, the HRTEM image in the inset of Fig. 4 corresponds to a particle that is half hcp (indicated by the arrow) and half fcc with a twin plane. These results suggest that a higher synthesis temperature may enhance the hcp symmetry in Ag nanoparticles, and additional studies are under way to corroborate this point. Nevertheless, this is a very interesting result since several experiments have been reported on the Au system, and the hcp symmetry has not been identified [10–12]. A hypothesis to explain the structural differences found in Au and Ag could be based on the faster growth rate of Ag nanoparticles. Additional data, using higher synthesis temperatures, will help to address this issue. In addition, the energy stability of hcp Ag and Au particles as a function of particle size has still to be evaluated by theoretical calculations to determine whether different behavior should be expected. The hcp system is an interesting one that might be explored, for example, to induce anisotropic-type growth and produce nanorods or nanodisks with small diameters (below 10 nm).

Conclusions

Au and Ag colloidal nanoparticles have been synthesized at room temperature and 80 °C. The increase in synthesis temperature leads to larger particle sizes in the case of Au, but it does not have a significant effect on Ag. By using HRTEM in association with XRD measurements and simulations, we have been able to point out structural differences on Ag and Au colloidal nanoparticles. Whereas Au nanoparticles preferentially form decahedra at both synthesis temperatures, Ag nanoparticles have the tendency to form several stacking faults, and this is enhanced at 80 °C. The sequence of stacking faults in the fcc structure leads to a hcp structure and particles with half hcp character have been identified in the Ag-80 °C sample. These results suggest that hcp symmetry may be favored in Ag under certain synthesis conditions and additional studies are under way to address this issue.

Acknowledgements LNLS is acknowledged for the use of LME (HRTEM); SAXS and XRD1 beamlines. This work was funded by FAPESP (01/07715-8), CNPq (470904/2001-1) and CAPES (M.S.L. scholarship).

References

1. Sanchez A et al (1999) J Phys Chem A 103:9573
2. Mulvaney P (1996) Langmuir 12:788
3. El-Sayed MA (2001) Acc Chem Res 34:257
4. Brust M et al (1994) J Chem Soc Chem Commun 801
5. Link S, El-Sayed, MA (1999) J Phys Chem B 103:8410
6. Jin R et al (2001) Science 294:1901
7. Hall BD (2000) J Appl Phys 87:1666, and references therein
8. Murray CB, Kagan CR, Bawendi MG (2000) Annu Rev Mater Sci 30:545
9. Dinega DP, Bawendi MG (1999) Angew Chem 38:1788
10. Zanchet D, Hall BD, Ugarte D (2000) J Phys Chem B 104:11013
11. Whetten RL et al (1996) Adv Mater 8:428
12. Vogel W et al (1999) J Phys Chem B 102:10853
13. Baletto F, Mottet C, Ferrando R (2000) Phys Rev Lett 84:5544

Progr Colloid Polym Sci (2004) 128: 135–138
DOI 10.1007/b97113

Dotchi Exerowa

Thin liquid films from phospholipids: formation, stability and phase transitions

D. Exerowa
Institute of Physical Chemistry,
Bulgarian Academy of Sciences,
Acad. G. Bonchev Str. Bl. 25-A,
1113 Sofia, Bulgaria
e-mail:exerowa@ipchp.ipc.bas.bg
Tel.: +359-2-971-2662
Fax: +359-2-9712662

Abstract Phospholipid bilayers (Newton black films and bilayer lipid membranes) are obtained from thin liquid films by means of increasing either the electrolyte concentration or the capillary pressure. Both formation and stability of phospholipid bilayers are given uniform consideration on the basis of the Kashchiev–Exerowa theory of nanoscopical hole nucleation. The phospholipids used are dimyristoylphosphatidylcholine (DMPC), dipalmitoylphosphatidylcholine (DPPC) and their natural mixtures: egg lecithin, amniotic fluid, alveolar surfactant. The microinterferometric method of Scheludko and Exerowa in its updated version is used to investigate the equilibrium behaviour. This method allows for direct or indirect measurement of parameters characterizing the formation and stability of bilayers, for example, critical surfactant concentration for foam bilayer formation, equilibrium concentration and binding energy, Q, characterizing the short-range interactions in the phospholipid bilayers. The two-dimensional chain-melting phase transition in phospholipid bilayers has been observed for the first time. The Q values for the liquid-crystalline state are about 17–19 kT for DMPC, DPPC and natural mixtures egg phosphatidylcholine and amniotic fluid, and about 50–60 kT for the gel phase of DMPC and DPPC bilayers. The parameter studied – the threshold concentration – is crucial for a very successful assessment of the risk of respiratory distress syndrome in newborns and could be applied in medicine for assessment of other respiratory disturbances.

Keywords Amphiphile bilayer · Phospholipid bilayer · Phase transition in a bilayer · Binding energy · Thin liquid films

Introduction

Investigation of phospholipids at interfaces is a largely used and undoubtedly justifiable approach with respect to the crucial role they play in many processes in live organisms. The most common are the investigations of monolayers, of Langmuir–Blodget layers, and of liposomes. Till now thin liquid films have not been used systematically as models of phospholipid investigation at interfaces. Thin liquid films are an effective study tool not only for surface phenomena but also for the interaction between two phases through a liquid layer, including bilayer contact. In that way the transition from long-range to short-range interactions (surface forces) can be observed.

These interactions in microscopic foam films were considered in a series of systematic works [1–7]. In this paper investigations with phospholipid bilayers (Newton black films and bilayer lipid membranes), obtained from microscopic thin liquid films by means of increasing either the electrolyte concentration or the capillary pressure are presented.

Theoretical

The formation and stability of phospholipid bilayers are both considered from a uniform point of view on the basis of the Kashchiev–Exerowa theory of nanoscopical hole nucleation [1, 9, 10]. The amphiphile bilayer is regarded as a two-dimensional ordered system (Fig. 1a). The bilayers are in contact with the same ambient phase (air/air, oil/oil or water/water). The bilayer is in contact with a liquid phase containing surfactant molecules. The interactions between first-neighbour molecules are short-range interactions, like in a crystal, ε and ε_0 in the lateral and the normal direction, respectively. We have already described thermodynamically and statistically the nucleation process of hole formation [1, 8, 9]. The driving force for bilayer rupture is the supersaturation $\Delta\mu$ of the nucleation process:

$$\Delta\mu(c) = kT\ln(c_e/c), \quad (1)$$

where c is the actual total concentration of amphiphiles in the solution and c_e is the equilibrium concentration.

In terms of the lattice model [9]

$$c_e = c_0\exp(-Q/2kT), \quad (2)$$

where c_0 is a constant and Q is the binding energy of the amphiphile molecules in the bilayers.

The foam film stability can be presented by the bilayer film lifetime as a function of bulk surfactant concentration c [1, 9, 10]:

$$\tau(c) = A\exp[B/\ln(c_e/c)], \quad (3)$$

where A is a constant accounting for the nucleation kinetics and is measured in seconds and

$$B = \pi A_0\gamma^2/2(kT)^2$$

where A_0 is the area per molecule and γ (in units of joules per metre) is the specific hole linear energy. This parameter demonstrates the interaction between first-neighbour molecules in the hole edge.

The probability W for amphiphile bilayer formation is given by [1, 9, 10]

$$W = \exp[-t_r/\tau(c)], \quad (4)$$

where t_r is the resolution time of the apparatus.

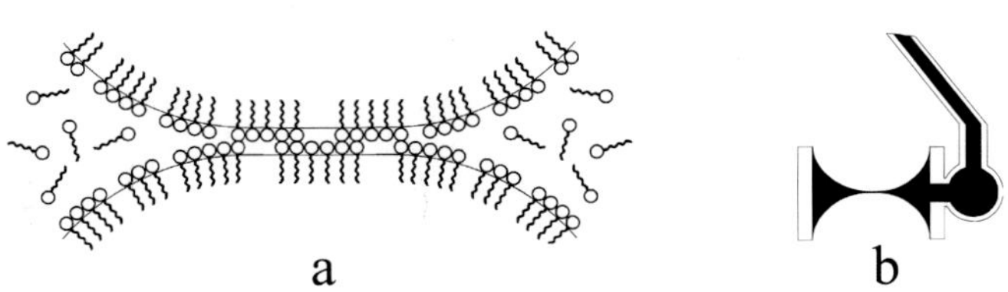

Fig. 1 Molecular model of foam bilayer. **b** Element of the measuring cell for the microinterferometric investigation of foam bilayers

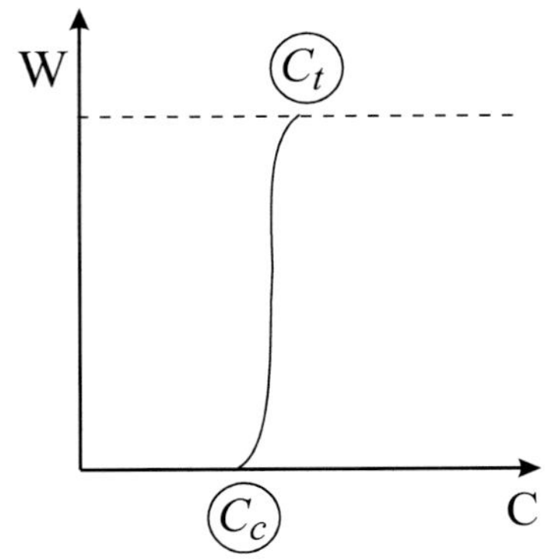

Fig. 2 Probability W of observing a foam bilayer on the total amphiphile concentration c. c_{cr} is the critical amphiphile concentration for bilayer rupture and c_t is the threshold amphiphile concentration at which there is 100% probability of obtaining a foam bilayer

Equations (3) and (4) suggest a very sharp rise in τ or W with an increase in concentration. Such a steep course is typical for nucleation processes. A schematic presentation of the dependence of the bilayer probability of observing a foam bilayer on the total amphiphile concentration is given in Fig. 2.

This allows us to determine a very important parameter c_c, i.e. the critical concentration at which bilayer formation starts. The following equation can be used for c_c:

$$c_c = c_e\exp[-B/\ln(\tau_c/A)], \quad (5)$$

where

$$\tau_c = \tau(c_c).$$

It also is possible to define c_t – the threshold concentration at which there is a 100% probability to obtain a bilayer film (Fig. 2).

Experimental

Materials

Nonionic phospholipids dimyristoylphosphatidylcholine (DMPC) and dipalmitoylphosphatidylcholine (DPPC), as well as natural products such as lung surfactant (LS), pig lung lavage [11], exogenic therapeutical surfactant Infrasurf (In) and amniotic fluid (AF) were used.

Methods

The experimental investigation of foam bilayers was carried out with an updated version of the microinterferometric method of Scheludko and Exerowa [1, 10]. The microscopic horizontal foam films were formed in a special measuring cell. A schematic presentation of an element of this cell is given in Fig. 1b. The foam film was formed in the middle of a biconcave drop in a glass tube by drawing out the solution from the drop through the capillary tube. The biconcave drop from which the foam film was formed was connected to an auxiliary volume of a liquid glass reservoir. This auxiliary volume was measured so as to ensure reaching a sufficiently high bulk phospholipid concentration, required for the equilibrium film/meniscus.

With the microinterferometric method it is possible to measure directly and indirectly the parameters characterizing the formation and stability of bilayers and to determine precisely the bilayer thickness.

Formation and stability of phospholipid foam bilayers

The $\tau(c)$ and $W(c)$ dependences (see "Theoretical") were determined experimentally . For DMPC and DPPC foam films these dependences were very steep and a quantitative fit with theory was not possible . However, the critical concentrations c_c for formation of DMPC and DPPC bilayers were very precisely determined: $c_c = 320\ \mu g/cm^{-3}$ for DMPC and $c_c = 70\ mg/cm^{-3}$ for DPPC at 25 °C [12, 13]. Since the $W(c)$ curves were very steep, we could not state that in these cases the c_c valueswere very close to the equilibrium concentrations c_e [12].

Phase transition in phospholipid foam bilayers

Foam bilayers from individual and mixed phosphatidylcholines were studied experimentally at different temperatures. The occurrence of a chain-melting phase transition in the foam bilayers was detected with two independent parameters: the critical concentration for the formation of foam bilayers and the foam bilayer thickness. Can foam bilayers be of different thickness? Yes, they can. This happens when they are at different temperatures. Combining the results obtained for the phospholipid microscopic foam films and the hole-nucleation theory, we registered different phase states and different thickness at these temperatures. A thickness leap of about 0.7 nm seen in the thickness/temperature dependence that occurred at 23 °C for films from DMPC, and at 41 °C for films from DPPC. These temperatures correspond to the phase transition in fully hydrated water dispersions of phosphatidylcholines [13, 14].

The Arrhenius plot of the critical concentration of a DPPC foam bilayer versus temperature is shown in Fig. 3 for experimental results and the theoretical dependence according to Kashchiev–Exerowa theory. A bend in the curve was observed at 41 °C. From the slope it is possible to determine the binding energy Q at temperatures below 41 °C and at higher temperatures. The change in the binding energy was due to the chain-melting phase transition of DPPC foam bilayers. Obviously the higher values of Q refer to the more ordered gel state, while the lower ones refer to the liquid-crystalline state. Thus, for the first time a phase transition in foam bilayers was observed. Figure 3 also presents the c_c $(1/t)$ dependence for natural mixtures: LS-B, In and AF. As can be seen in these cases there was no phase transition, even though they contain 80% phosphatidylcholines. It is obvious that the phase transition has to be expected at a temperature of about 41 °C.

Binding energy of molecules in the bilayer

The binding energy was determined from the experimental data from the c_c $(1/t)$ dependence and involving Kashchiev–Exerowa theory that assumed $c_c = c_e$ (Eq. 5) [12]. The values of Q obtained are presented in Table 1.

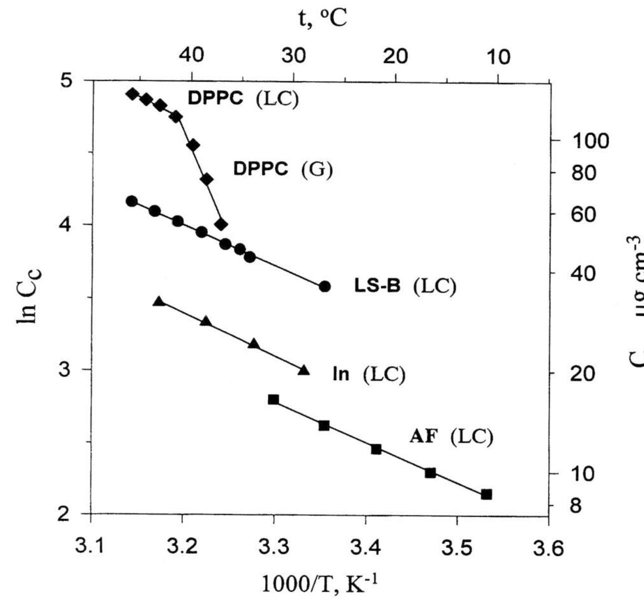

Fig. 3 Dependence of the critical concentration for formation of a dipalmitoylphosphatidylcholine (*DPPC*) foam bilayer on temperature. Experimental data (*circles*), theoretical dependence according to Eqs. (2) and (5) (*straight line*)

For the liquid-crystalline state the values of Q were about 17–19 kT for DMPC, DPPC and natural mixtures AF, In and LS-B and about 50–60 kT for the gel phase of DMPC and DPPC bilayers.

The short-range molecular interactions were also demonstrated by the hole-specific edge energy, calculated according to the theory. Data on γ can be found in Refs. [1, 9, 10]. It is worth noting that the γ values for liquid-crystalline phospholipid foam bilayers were similar to the ones obtained from experiments on the rupture of bilayer lipid membranes [15–17].

Applications

Foam bilayers from phospholipids could be used as a model for investigating short-range forces in biological structures, interaction between membranes, etc. Particularly interesting is a new in vitro model introduced to study the alveolar surface and stability. This is the study of a microscopic foam film under the conditions of the lung alveolus: capillary pressure, film radius, electrolyte concentration and temperature. Under these conditions

Table 1 Binding energy Q of dimyristoylphosphatidylcholine (*DMPC*) and dipalmitoylphosphatidylcholine (*DPPC*) foam bilayers

	Surfactant						
	DMPC	DMPC	DPPC	DPPC	Amniotic fluid	LS-B	Infrasurf
Q (kT)	18	48	17	60	19	19	18
Phase state	Liquid crystal	**Gel**	Liquid crystal	**Gel**	Liquid crystal	Liquid crystal	Liquid crystal

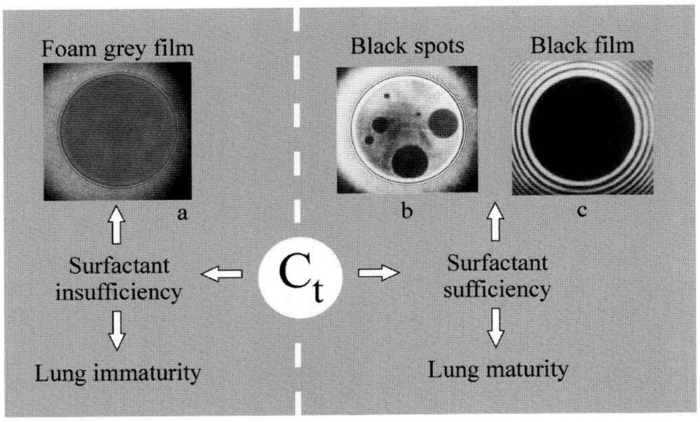

Fig. 4 Parameter c_t used for the assessment of surfactant insufficiency/sufficiency and lung immaturity/maturity in newborns. Photographs of microscopic films in the cases of lung immaturity and maturity

the foam film from AF is a bilayer. This gave us reason to introduce a new parameter – the threshold concentration c_t (Fig. 4) – that characterizes surfactant insufficiency in newborns [1, 18]. Surfactant insufficiency leads to lung immaturity and causes respiratory distress syndrome in newborns. Based on this we developed a new physicochemical clinical method for assessment of lung maturity [18]. The parameter c_t (in that case $c_t = c_c$, owing to the very steep W/c curve) is the concentration at which the formation of the black foam film from AF (by black spots!) starts and this can be observed very clearly [1, 10, 18]. A picture of a microscopic foam film below and above c_t is presented in Fig. 4. The prediction rate of the method and the clinical results for respiratory distress syndrome fit very well thus leading us to the development of a new hypothesis for the alveolar structure [19, 20]. This opens new perspectives for the study of the main physiological process – breathing.

References

1. Exerowa DR, Kruglyakov PM (1998) Foam and foam films. Elsevier, Amsterdam
2. Cohen R, Koynova R, Tenchov B, Exerowa D (1991) Eur Biophys J 20:203
3. Cohen R, Exerowa D, Kolarov T, Yamanaka T, Muller VN (1992) Colloids Surf 65:201
4. Cohen R, Exerowa D (1994) Colloids Surf 85:271
5. Cohen R, Exerowa D, Yamanaka T (1997) Langmuir 12:5419
6. Cohen R, Exerowa D, Kolarov T, Yamanaka T (1997) Langmuir 13:3172
7. Exerowa D (2002) Adv Colloid Interface Sci 96:75
8. Kashchiev D (2000) Nucleation: basic theory with applications. Butterworth-Heinemann, Oxford
9. Kashchiev D, Exerowa D (1980) J Colloid Interface Sci 77:501
10. Exerowa D, Kashchiev D, Platikanov D (1992) Adv Colloid Interface Sci 40:201
11. Lalchev Z, Todorov R, Christova Y, Wilde P, Mackie A, Clark D (1996) Biophys J 71:2591
12. Exerowa D, Nikolova A (1992) Langmuir 8:3102
13. Nikolova A, Koynova R, Tenchov B, Exerowa D (1997) Chem Phys Lipids 83:111
14. Nikolova A, Exerowa D, Lalchev Z, Tsonev L (1994) Eur Biophys J 23:145
15. Exerowa D, Scheludko A (1964) In: Overbeek JTG (ed) Chemistry, physics and application of surface active substances, vol 2. Gordon & Breach, London, pp 1097–1108
16. Exerowa D (1977) In: Materialy z seminarium – Fizikochemia Zjawisk Powierzchniowych, Baranow Sandomierski. Polska Akademia Nauk, pp 1–33
17. Nikolov A, Exerowa D, Zhelyazkova L (1976/77) Ann Univ Sofia Fac Chim 71:17
18. Exerowa D, Lalchev Z, Marinov B, Ognianov K (1986) Langmuir 2:664
19. Exerowa D, Lalchev Z (1986) Langmuir 2:668
20. Kashchiev D, Exerowa D (2001) Eur Biophys J 30:34
21. Larsson M, Larsson K, Andersson S, Kakhar J, Nylander T, Ninham B, Wollmer P (1999) J Disperson Sci Technol 20:1

Progr Colloid Polym Sci (2004) 128: 139–142
DOI 10.1007/b97115

C. Raffournier
C. Dubernet
P. Couvreur
M. Ollivon

Phospholipid/triglyceride mixtures: analysis and comparison of the solubilisation profiles

Abstract In excess water, mixtures of phospholipids (PL) and triglycerides (TG) form different types of aggregates, principally vesicles and emulsion droplets, the respective amounts of which depend on *R*, the PL-toTG concentration ratio. When *R* is high, liposomes are the major aggregate species and when this ratio is low, emulsion droplets coexist with liposomes. The objective of the present study was to determine the proportion of vesicles in mixtures of low PL/-to-TG ratio. It has been previously described that time-resolved selective solubilisation of high PL-to-TG ratio mixtures with Triton X-100 allowed the detection of the presence of emulsion droplets in a liposomal suspension. The method has been adapted here by using another detergent, octylglucoside, allowing us to obtain sharper solubilisation profiles thanks to the fact that it is a pure compound. A large range of PL-to-TG ratio mixtures were then solubilised by octylglucoside and two ratios were more closely compared: 10:4, at which liposomes are the major components, and 10:40, at which emulsion droplets are the major species. For a PL-to-TG ratio of 10:4, liposomes could be selectively solubilised (which was already known), but we demonstrated that emulsion droplets could be selectively solubilised as well. Hence, we evidenced two solubilisation steps: first the liposomes and then the emulsion droplets. For a PL-to-TG ratio of 10:40, the solubilisation was more complex and depended on the initial PL concentration. An event disrupting the solubilisation profile could be identified corresponding to a temporary aggregation/fusion of the droplets owing to the destabilisation of the water interface. Defined conditions where this event will not interfere with the solubilisation profile have to be found to precisely determine the solubilisation steps. In this way, it will be possible to achieve the quantification of the proportion of bilayers in an emulsion (low PL-to-TG ratio).

Keywords Liposomes · Emulsion droplets · Solubilisation · Phospholipids/triglycerides mixtures

C. Raffournier (✉) · C. Dubernet
P. Couvreur · M. Ollivon
Faculté de Pharmacie,
Université de Paris-XI,
UMR CNRS 8612,
5 rue JB Clément,
92296 Châtenay-Malabry, France
e-mail: michel.ollivon@cep.u-psud.fr
Tel.: +33-1-46835629
Fax: +33-1-46835312

Introduction

Recently, some bicompartmental structures (BS) have unexpectedly been evidenced by cryogenic transmission electron microscopy in a cationic submicronic oil-in-water emulsion [1]. The cryomicroscopy pictures of these BS were analysed as follows. The BS are believed to be stabilised by phospholipids and are composed of both an oily compartment (triglycerides, TGs) covered by a phospholipid (PL) monolayer and a water phase delimited by a PL bilayer. The present study is part of a larger work whose aim is to determine how these objects could form. Cryomicroscopy examinations have helped to highlight the influence of the nature of the lipids used on the number of BS observed. However, this technique could not allow the determination of the exact proportion of BS in the emulsion samples. Therefore, there was a need for a specific method to be developed and one way might be to quantify the proportion of bilayers in the emulsion (they are supposed to delimit the aqueous compartment of BS). Data from the literature [2, 3] have shown that at high PL-to-TG ratios, emulsion droplets could form and be detected by optical density (OD) monitoring. More precisely, when such a mixture of emulsion droplets and liposomes was submitted to the action of a detergent (Triton X-100), liposomes were completely solubilised through mixed micelles, whereas emulsion droplets were not significantly modified. We have applied this approach to emulsions (low PL-to-TG ratio) in order to selectively quantify the bilayer membranes present and thus to propose a possible way to quantify BS in an emulsion.

Methodology

Solubilisation of PL/TG mixtures was followed at 25.0 ± 0.1 °C by turbidity monitoring at 400 nm with a double-beam UV–vis spectrophotometer (Lambda 2, PerkinElmer). For each PL-to-TG ratio, initial preparations were diluted (3/4, 1/2, 1/4, 1/10) to a total volume of 1.2 ml and placed in a glass cuvette (10 × 10 mm). To each of these suspensions, octylglucoside (OG) solution (200 mM) was added continuously at a constant rate under stirring, using a syringe pump as previously described [4]. The influence of the detergent addition rate was determined (data not shown). Different rates (0.1157, 0.2135, 0.4630, 0.9260, 1.852 μl/s) were examined. The solubilisation profiles were not significantly modified. This means that the combination of the OG concentration and the addition rate used is low enough to maintain a dynamic equilibrium in detergent/PL molecule rearrangements occurring during OG addition. The intermediate rate of 0.463 μl/s was chosen for all experiments.

The total volume of OG solution added, which depends on the respective concentrations of PL and TG in the cuvette, was between 0.5 and 1.5 ml. The solubilisation curves were not corrected for dilution since turbidity does not necessarily follow Beer–Lambert linearity especially at high lipid concentrations (at OD > 1) and detergent concentrations at breakpoints are not affected by eventual turbidity changes. Moreover, it has been demonstrated that (1) turbidity is a complex function of droplet size, (2) some size changes are expected during solubilisation as mentioned in the text and (3) the OG concentration at breakpoints is not affected by changes in specific turbidity [5]. Solubilisation curves, representing the evolution of OD during the surfactant addition, allow us to visualise the different steps of the solubilisation of the aggregates. All experiments were recorded at least in triplicate.

Results and discussion

The solubilisation curves were obtained by reporting OD versus OG concentration. Selective solubilisation of a mixture with a PL-to-TG ratio of 10:4 in which liposomes represented the largest part in the suspension, but with some emulsion droplets however being present too, is shown in Fig. 1.

For the two PL concentrations shown, the solubilisation profile was the same and four breakpoints could be identified (I, II, III, IV). Data from the literature [2, 3] have shown that upon addition of a detergent to a liposome suspension, the vesicles are progressively transformed through mixed micelles. The scattering properties of both types of aggregates are very different, and the solubilisation profile observed during this liposome to mixed micelle transition corresponds to a progressive decrease of the OD during detergent addition. This information allowed us to identify the first jump of the OD, ending in breakpoint III (Fig. 1), as the vesicle to mixed micelle transition occurred. It was shown that between breakpoints II and III open vesicles coexist with and are progressively converted into mixed micelles. That means that breakpoint III corresponded to the end of liposome solubilisation (called S), the emulsion droplets not yet having been modified. At this stage, the aggregates present in the suspension were then mixed micelles and emulsion droplets. The second jump of the OD upon OG addition finishing at breakpoint IV corresponded to the completion of emulsion droplet solubilisation (called S′). At breakpoint S′, emulsion droplets have been completely solubilised by OG and mixed micelles are the only objects remaining. Hence, these experiments brought a new result: all the aggregates formed in PL and TG mixtures can be solubilised by OG,

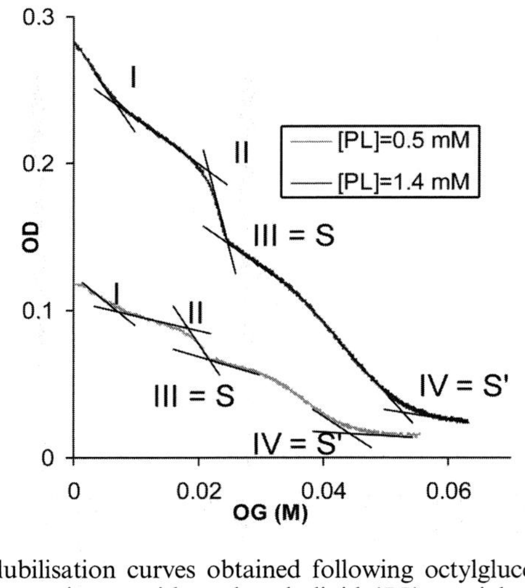

Fig. 1 Solubilisation curves obtained following octylglucoside (*OG*) addition to a mixture with a phospholipid (*PL*)-to-triglyceride (*TG*) ratio of 10:4. These curves describe the evolution of the optical density as a function of the detergent concentration in the suspension for two different initial PL concentrations, 1.4 and 0.5 mM. The four breakpoints (*I*, *II*, *III*, *IV*) correspond to the different steps of the aggregate evolution under detergent addition. The main events are the end of liposome solubilisation (III), called *S*, and the completion of emulsion droplet solubilisation (IV), called *S′*

first liposomes then emulsion droplets in the same solubilisation experiment. If emulsion droplet solubilisation has not been described before this must be due to the surfactant used, here OG, which is clearly more effective than Triton X-100. Now, as for high PL-to-TG ratios OG allowed us to clearly distinguish between liposome and emulsion droplet solubilisation, and the same type of result was expected for higher TG proportions.

However, when emulsion droplets were the major species in the suspension (PL-to-TG ratio 10:40), the solubilisation profile was more complex (Fig. 2). Additional breakpoints could indeed be identified and, moreover, some of them depended on the PL concentration.

A careful comparison between the solubilisation profiles obtained for PL-to-TG ratios of 10:4 and 10:40 (Figs. 1, 2, respectively) clearly shows that for the 1.4 mM PL concentration, the two curves are very close: four breakpoints identified with liposomes solubilisation ending in breakpoint III (S = III) and emulsion droplet solubilisation ending in breakpoint IV (S′ = IV). The only difference was the importance of the OD jump between breakpoints III and IV because of the huge proportion of TG in the emulsion (PL-to-TG ratio 10:40, Fig. 2). Conversely, for a lower initial PL concentration (0.5 mM), the solubilisation curves differed in the region of breakpoint III. Indeed, a new additional event could be detected for a PL-to-TG ratio of 10:40, leading to a temporary increase of the OD. This increase which is absent for a PL-to-TG ratio of 10:4 is also observed for dilutions corresponding to PL concentrations close to 0.5 mM (data not shown). A set of experiments was performed to understand what could happen at these new breakpoints, which we called IIIa and IIIb. For example, the influence of the emulsion dilution was evaluated in more detail (data not shown). Events IIIa and IIIb were found to be more pronounced when the emulsion was less dilute, meaning that they strongly depended on the aggregate concentration. Then, the aggregate size during the detergent addition was determined by quasielastic light scattering and allowed us to correlate the OD increase beginning at IIIa and ending at IIIb with a size increase (data not shown). Finally, different PL-to-TG ratios for which emulsion droplets were the major species were compared. The results clearly indicated that when the TG proportion was increased events IIIa and IIIb were more pronounced. These data finally led to the conclusion that the jump from breakpoint IIIa to breakpoint IIIb corresponds to the destabilisation of the interface of the emulsion

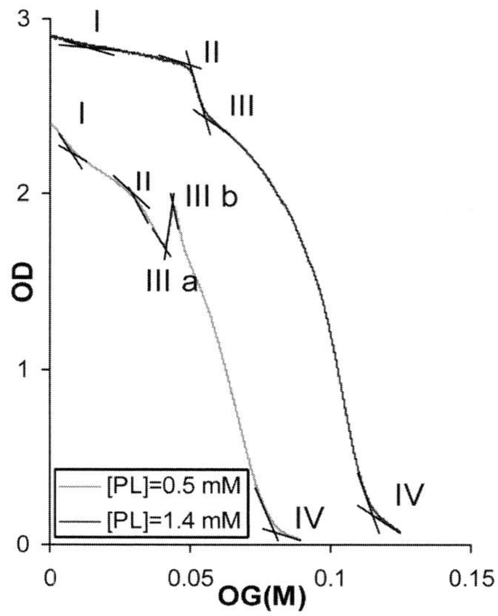

Fig. 2 Solubilisation curves obtained following OG addition to a mixture with a PL-to-TG ratio of 10:40. These curves describe the evolution of the optical density as a function of the detergent concentration in the suspension for two different initial PL concentrations, 1.4 and 0.5 mM. The breakpoints depend on the initial PL concentration: four identified at 1.4 mM (*I*, *II*, *III*, *IV*), and five at 0.5 mM (*I*, *II*, *IIIa*, *IIIb*, *IV*). The liposome solubilisation S cannot be easily identified here

droplets. When emulsion droplets were indeed depleted from PL stabilising their interface by the addition of OG, they tended to aggregate and fuse. Therefore the solubilisation of TG by OG could begin.

Conclusion

The results show that the solubilisation profile of PL and TG mixtures strongly depends on the TG proportion. For high PL-to-TG ratios, the distinction between liposome (S) and emulsion droplet (S′) solubilisation was easy. For low PL-to-TG ratios, the solubilisation profile was more complex. Indeed, transitory aggregation fusion disrupts the solubilisation profile in the region corresponding to both the end the of liposome solubilisation and the beginning of the emulsion solubilisation. Further experiments are under way to optimise experimental conditions and determine from the solubilisation profiles the compositions of all types of aggregates formed and solubilised, as well as to apply this technique to the BS cationic emulsion and to quantify the occurrence of bilayers.

References

1. Teixeira H, Dubernet C, Rosilio V, Benita S, Lepault J, Erk I, Couvreur P (2000) Pharm Res 17:1329–32
2. Dahim M., Gustafsson J, Puisieux F, Ollivon M (1998) Chem Phys lipids 97:1–14
3. Hamilton JA, Small DM (1981) Biophysics 78:6878–6882
4. Ollivon M, Walter A, Blumenthal R (1988) Biochemistry 27:1695–1703
5. Lesieur S, Grabielle-Madelmont C, Paternostre M, Moreau J-M, Handjani-Vila R-M, Ollivon M (1990) Chem Phys Lipids 56:109–121

Progr Colloid Polym Sci (2004) 128: 143–145
DOI 10.1007/b97116

Peter Schiller
Mirco Wahab
Frank Bagusat
Uwe Reimer
Hans-Jörg Mögel

Adsorption of colloidal particles on mixed membranes

Abstract The adsorption of polypeptides such as peripheral proteins disturbs the homogeneous distribution of the lipid components which form membranes. The lipid redistribution leads to an enhanced surface density of adsorbed polypeptides. For lipid mixtures with a tendency towards demixing this effect is more pronounced than for ideal mixtures. Then the adsorbate density strongly increases with increasing correlation length of thermal concentration fluctuations.

Keywords Biological membranes · Fluctuations · Peptide adsorption

P. Schiller (✉) · M. Wahab, F. Bagusat, U. Reimer · H.-J. Mögel
Institute of Physical Chemistry, TU Freiberg, Leipziger Strasse 29, 09599 Freiberg, Germany
e-mail: peter.schiller@chemie.tu-freiberg.de

Introduction

The lipid matrix of biological membranes consists of a mixture containing various lipid compounds. These amphiphilic molecules, which have a hydrophilic head group and one or two hydrocarbon chains, move freely in the membrane plane forming a two-dimensional fluid. Natural membranes are a mixture of electrically neutral, for example, zwitterionic, and charged lipid species. In addition to the lipids, biological membranes contain a large number of embedded and adsorbed polymers, especially proteins consisting of polypeptide chains folded in compact rigid particles [1].

In several cases, the investigation of structural and thermodynamic properties of multicomponent membranes revealed the nonideal behavior of a lipid mixture [1, 2]. The formation and decay of temporary domains with different lipid concentrations is an indication of a tendency towards demixing of lipid components. For simplicity, let us consider a membrane consisting of a binary lipid mixture with components A and B. Concentration fluctuations produce regions with enhanced and reduced concentrations of one component, say component B. If these concentration fluctuations remain small, they can be characterized by a correlation length ξ, which roughly characterizes the mean diameter of a region with enhanced concentration of compound B (Fig. 1).

The lipid distribution is not only disturbed by spontaneous fluctuations but also by the polypeptides adsorbed on the membrane surface or embedded in the membrane. In this paper we consider the influence of peripheral polypeptides, for example, small peptides or large globular proteins adsorbed on the membrane surface. Usually, there is at least a small difference in the interaction energies between the different components of the lipid mixture and an adsorbed polypeptide. The interaction energy ε_A of lipid A molecules should be different from the interaction energy ε_B of lipid B molecules in contact with the polypeptide. If $\varepsilon_B < \varepsilon_A$, then the adsorbed polypeptide prefers component B, which has an enhanced concentration in the region around the lipid–protein contact area (Fig. 1) [3]. The region with inhomogeneous lipid distribution increases with increasing correlation length ξ. The redistribution of the lipid components due to the presence of the

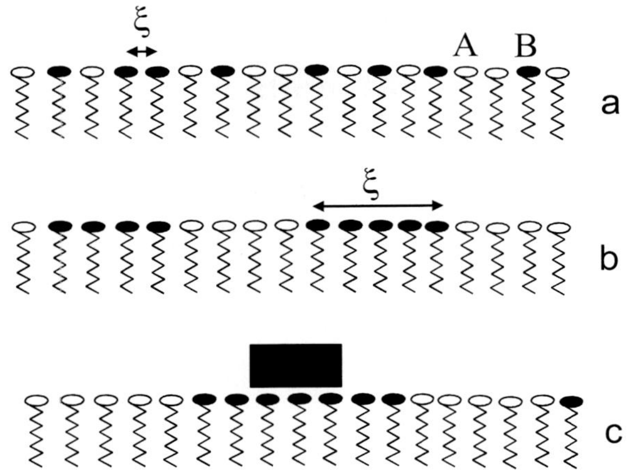

Fig. 1a–c Distributions of lipid molecules in a membrane. **a)** For an ideal mixture the correlation length ξ is small. **b)** A large correlation length in nonideal mixtures leads to clusters. **c)** An adsorbed polypeptide attracts lipid component B

adsorbed polypeptide leads to an increase of the effective adsorption energy and thus to an enhanced surface concentration of adsorbed molecules. In this paper we investigate how the adsorption of polypeptides from the aqueous solution depends on the correlation length of concentration fluctuations.

Theory

Let us consider a mixed membrane consisting of two lipid components A and B. If M_A and M_B denote the particle numbers of the binary mixture, its composition can be characterized by the ratios $X_A = M_A/M$ and $X_B = M_B/M$, where $M = M_A + M_B$. Concentration fluctuations are taken into account by introducing local particle number densities $\rho_A(x)$ and $\rho_B(x)\rho$, which may depend on the coordinates x defined for the membrane plane. For a homogeneous membrane these number densities can be expressed as $\rho_A^0 = M_A/S$ and $\rho_B^0 = M_B/S$, where S denotes the membrane area. If fluctuations of the total lipid density $\rho = \rho_A(x) + \rho_B(x)$ are neglected, the concentration field $\psi(x) = \rho_B^0 - \rho_B(x) = \rho_A(x) - \rho_A^0$ characterizes local deviations from the equilibrium surface density. Fluctuations $\psi(x) \neq 0$ increase the free energy of the membrane. According to the linearized version of the Cahn–Hilliard theory the increase of the free energy due to concentration fluctuations is obtained from the integral [4]

$$f_1 = \frac{1}{2}\int_S d^2x\left\{A[\nabla\psi(x)]^2 + B[\psi(x)]^2\right\}, \quad (1)$$

where the coefficients $A > 0$ and $B > 0$ do not depend on the coordinate x. The expression $\xi = \sqrt{A/B}$ defines the correlation length of spontaneous thermal concentration fluctuations. In the vicinity of the spinodal or the critical point of demixing coefficient B tends to zero and thus ξ diverges, while coefficient A remains regular. Then the product $B\xi^2 = A$ can be considered as a constant. Sufficiently far away from these singularities the lipid mixture should be nearly ideal. Then coefficient A can be estimated as $A = \xi_0{}^2kT/(\rho X_A X_B)$ where $\xi_0 \cong 0.3$ nm defines the correlation length of the concentration fluctuations of an almost ideal lipid membrane mixture. As A behaves regularly at the spinodal or the critical point, it is reasonable to assume that A remains approximately constant. Apart from spontaneous fluctuations, the homogeneity of the membrane mixture is disturbed by the adsorbed polypeptides. Neglecting lateral interactions between the adsorbing molecules, we consider the Henry regime at low surface densities. Let ε_A and ε_B be adsorption energies which refer to a polypeptide bound to membranes consisting of the pure lipid components A and B, respectively. Assuming short-range interactions between the molecules, the polypeptide–lipid coupling is restricted to the lipids which are in direct contact with the adsorbed polymer. The polypeptides are assumed to cover a circular membrane region C which contains $n = \rho\pi a^2$ lipid molecules, where a denotes the protein radius and ρ is the density of the lipid molecules in the membrane plane. Then the contribution to the free energy due to the coupling of an adsorbed molecule to the membrane can be written as [5]

$$f_2 = (X_A\varepsilon_A + X_B\varepsilon_B) + \frac{(\varepsilon_B - \varepsilon_A)}{n}\int_C d^2x\psi(x). \quad (2)$$

The term $e = (X_A\varepsilon_A + X_B\varepsilon_B)$ corresponds to the adsorption energy of a polypeptide to a randomly distributed lipid mixture. In the second contribution to f_2 the perturbation of the homogeneous mixture is taken into account, because in the case $\varepsilon_A \neq \varepsilon_B$ the preferred lipid component migrates into the contact region repelling the unfavourable component. The explicit evaluation of the free energy $f = f_1 + f_2$ for a polypeptide associated with the membrane is a straightforward problem, which can be solved analytically [5] for peptides which cover a circular membrane region with radius a. Using Eqs. (1) and (2) as a starting point, the result of this evaluation can be expressed as $f = e + u/2$, where the second term (denoted as self-energy in Ref. [5]) is defined as

$$\frac{u}{2} = -\frac{(\varepsilon_A - \varepsilon_A)^2}{2\rho^2 A} \times \frac{1}{(\pi r^2)^2}\int_{C^*} d^2y \int_{C^*} d^2\hat{y}\frac{K_0(|\,y - \hat{y}\,|)}{2\pi}, \quad (3)$$

where $r = a/\xi$, K_0 denotes the modified Bessel function [6] and the integration area C^*, expressed in reduced coordinates $(y = x/\xi)$, comprises the circular membrane area covered by a polypeptide.

Finally, let us consider the adsorption equilibrium at low surface densities of the polypeptides $\Gamma = N/S$, where N is the number of adsorbed molecules. In the Henry limit of small surface concentrations, the chemical potential $\mu'' = f + kT\ln[\Gamma/\Gamma_0(T)]$ results for the adsorbed polypeptides. As the concentration c of the polypeptides in the aqueous solution is also small, their chemical potential can be written as $\mu' = kT\ln[c/c_0(T)]$, where $c_0(T)$ denotes a function of the temperature T (k, Boltzmann constant). Phase equilibrium between adsorbed and dissolved polypeptides requires that the condition $\mu' = \mu''$ is satisfied. This condition is rewritten as

$$\frac{\Gamma}{\Gamma_0(T)} = \frac{c}{c_0(T)}\gamma\exp\left(-\frac{X_A\varepsilon_A + X_B\varepsilon_B}{kT}\right), \quad (4)$$

where $\gamma = \exp(-u/kT)$.

Results and discussion

Equations (3) and (4) provide a relation between the surface and the bulk concentrations of polypeptides. If the adsorption energies ε_A and ε_B are equal for both lipid components A and B, we obtain $u = 0$ and $\gamma = 1$. In this case a plot of ln Γ versus X_B would show a straight line

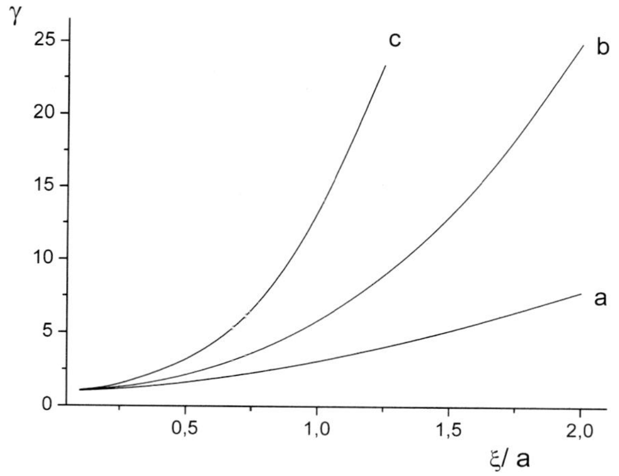

Fig. 2 Amplification factor γ for the adsorption versus reduced correlation length ξ/a of spontaneous concentration fluctuations for a) $|\varepsilon_A - \varepsilon_B|/n = 0.2kT$ b) $|\varepsilon_A - \varepsilon_B|/n = 0.25kT$ and c) $|\varepsilon_A - \varepsilon_B|/n = 0.3kT$ $n = 20$, $\xi_0 = 0.3$ nm, $\rho \cong 2$ nm^{-2} and $X_A = X_B = 0.5$

with slope $(\varepsilon_A - \varepsilon_B)/kT$ The same behaviour should be observed when the polymers adsorb on a frozen membrane with randomly distributed components A and B in the membrane plane. However, if the membrane is in the fluid state and if $\varepsilon_A \neq \varepsilon_B$, the redistribution of the lipid components leads to an increase of the surface concentration Γ of adsorbed molecules. This effect is characterized by the multiplier $\gamma = \exp(-u/2kT)$ in Eq. (4). Because of $-u/2kT > 0$, the condition $\gamma > 1$ is always satisfied. The value of γ depends on the ratio ξ/a of the correlation length ξ and the polypeptide radius a. The factor γ describes the amplification of adsorption due to the redistribution of the lipid membrane mixture. The dependence of this amplification factor on the reduced correlation length ξ/a is illustrated in Fig. 2. Obviously, the density Γ of the adsorbed molecules increases strongly with increasing correlation length.

In conclusion, if the lipid mixture which forms the membrane matrix is nonideal (with a tendency towards demixing), the adsorption of polypeptides on the membrane surface becomes more effective. The adsorbate density increases strongly with increasing correlation length of spontaneous thermal concentration fluctuations.

Acknowledgement Financial support by the DFG (Sonderforschungsbereich 285 and project SCHI 368/3–1) is gratefully acknowledged.

References

1. Sackmann E (1995) In: Lipowsky R, Sackmann E (eds) Structure and dynamics of membranes. Handbook of biological physics, vol 1A. Elsevier, Amsterdam, pp 213–304
2. Garidel P, Blume A (2000) Langmuir 16: 1662
3. Heimburg T, Angerstein B, Marsh D (1999) Biophys J 76:2575
4. Chaikin PM, Lubensky TC (1997) Principles of condensed matter physics. Cambridge University Press, Cambridge
5. Schiller P, Mögel HJ, Wahab M, Reimer U (2002) J Phys Chem B 106:12323 Schiller P, Mögel HJ, Wahab M, Reimer U (2003) J Phys Chem B 107:4478
6. Arfken GB, Weber HJ (1995) Mathematical methods for physicists. Academic, San Diego

Progr Colloid Polym Sci (2004) 128: 146–150
DOI 10.1007/b97118

Cesar Perez Zapata
Herley Casanova
Johans Restrepo

Monte Carlo simulation of the self-assembly process of nonionic surfactants (type H_XT_Y) considering second-neighbor interaction between surfactant molecules

C.P. Zapata · H. Casanova (✉)
Grupo de Coloides, Universidad de Antioquia, Medellin, Antioquia, Colombia
e-mail: Casanova@mathematics.udea.edu.co
Tel: +57-4-2110647; Fax: +57-4-2330120

J. Restrepo
Grupo de Estado Sólido, Universidad de Antioquia, Medellin, Antioquia, Colombia

Abstract Monte Carlo simulation for the self-assembly process of nonionic surfactant type $H_X\mathrm{T}_Y$ in water on a two-dimensional lattice, considering second-neighbor interactions, was carried out. The model surfactant molecules selected for the simulation were H_1T_3, H_1T_4, H_2T_4, H_2T_5, H_3T_4, H_3T_6 and H_4T_7, and their aggregation behavior was evaluated between 0.4 and 3 vol% at a reduced temperature between 1 and 3. By setting second-neighbor interactions between surfactant tails at values lower than 20% of the interaction energy magnitude (ε) and a reduced temperature of $t = 1$, no surfactant aggregation was observed. At a value of 20% of ε, micelles were formed with aggregation numbers between 2 and 27. Higher values for the second-neighbor interactions induced a micellation process where dimers were the main species. The values of the critical micelle concentration obtained in the simulation were dependent on the head and tail length of the surfactant molecule following the tendencies reported in the literature. Increasing the reduced temperature to over 1.25 induced the destruction of the surfactant aggregates, suggesting that the present model is highly sensitive to such changes. The simulation also permitted calculation of Gibbs free energies for the micellation process from the micelle size distribution function.

Keywords Self-assembly · Nonionic surfactant · Critical micelle concentration

Introduction

The self-assembly characteristics of surfactant molecules have important implications for processes like detergency, compound separation in ion-exchange chromatography and emulsion preparation and stability [1]. Among surfactant solution properties the micelle size and the critical micelle concentration (cmc) are probably the most important and, therefore, the most studied [2]. Experimental and theoretical work has shown that the surfactant molecule length and the surfactant composition in terms of the number of hydrophylic and hydrophobic groups determine the values of the micelle size and the cmc. Long hydrophobic chains promote bigger micelles and lower cmcs, while linear hydrophilic groups promote smaller micelles than branched groups [3].

Monte Carlo simulation has been the most widely used method to study surfactant self-assembly processes since it allows simulations from low surfactant concentrations (0.01 vol%) up to high concentrations (75 vol%), and evolution of the system normally reaches the equilibrium state [4]. Monte Carlo simulation also allows the change of surfactant characteristics, such as the length of the hydrophilic group or the hydrophobic group, enabling their effect on micelle size, cmc and thermodynamical properties to be determined [5]. A three-dimensional simulation by Rodriguez-Guadarrama et al. [6] presented cmc data for amphiphilic molecules with different compositions of hydrophilic groups (H_X) and hydrophobic (T_Y) groups: from H_1T_4 to H_4T_7. Girardi and Figueiredo [7] have studied the effect of reduced temperature in the micelle process for H_1T_3 amphiphilic molecules showing the inhibition of micelle formation at reduced temperatures of over 1.53. Those studies were carried out taking into account only nearest-neighbor interactions: Since the micellation process is a self-assembly phenomenon, the second or third sphere of coordination could have an important effect on micelle characteristics. A simulation model that considers second-neighbor interactions for surfactant molecules at different reduced temperatures and with different hydrophobic and hydrophilic group lengths is yet to be described.

This study deals with the self-assembly process of nonionic surfactants, type H_XT_Y, and the effect of temperature on their cmc considering second-neighbor interactions. A two-dimensional Monte Carlo simulation was carried out for X values between 1 and 4, and Y values between 3 and 7, and reduced temperatures between 1 and 3.

Simulation model

The model used in this work is a square lattice of linear dimension $L = 300$ with a coordination number of $z = 4$, taking into consideration nearest-neighbor interactions. Periodic boundary conditions were implemented. Surfactant molecules are represented by monomers occupying adjacent lattice sites, with water molecules occupying one lattice site. Seven different surfactant molecule compositions were studied here: H_1T_3, H_1T_4, H_2T_4, H_2T_5, H_3T_4, H_3T_6 and H_4T_7. Each molecule was built by self-avoiding walks in order to fulfill the principle of excluded volume [8]. Interaction energies between molecules present in the system (i.e., surfactant and water molecules) were as follows: three nearest-neighbor attractive interactions [tail-to-tail (ε_{TTNN}), head-to-head (ε_{HH}) and head-to-water (ε_{HW})], one second-neighbor attractive interaction [tail-to-tail (ε_{TTSN})] and two nearest-neighbor repulsive interactions [tail-to-head (ε_{TH}) and tail-to-water (ε_{TW})]. In this study all nearest-neighbor interaction energies have the same magnitude, i.e., $\varepsilon_{TTNN} = \varepsilon_{HH} = \varepsilon_{HW} = \varepsilon$ and $\varepsilon_{TH} = \varepsilon_{TW} = -\varepsilon$; the second-neighbor attractive interaction energy has a magnitude of $\varepsilon_{TTSN} = -\alpha\times\varepsilon$, with α taking values between 0 and 1. The magnitude of the interaction energy was set to a value of $\varepsilon = 1$ and the reduced temperature $t = k_BT/\varepsilon$ was used, where k_B is the Boltzmann constant and T is the temperature. The reduced temperatures evaluated in the present work were in the range 1–3. The total internal interaction energy for the model is given by

$$\frac{E}{k_BT} = \frac{\varepsilon}{k_BT}(n_{TH} + n_{TW} - n_{TTNN} - n_{HH} - n_{HW}) - \frac{\alpha \times \varepsilon}{k_BT}(n_{TTSN}),$$

where n_{TH}, n_{TW}, n_{TTNN}, n_{HH}, n_{HW} and n_{TTSN} are the numbers of tail-to-head contacts, tail-to-water contacts, tail-to-tail nearest neighbors, head-to-head contacts, head-to-water contacts and tail-to-tail second neighbors, respectively.

The simulation was carried out according to the following steps. First, surfactant molecules were randomly generated on the square lattice avoiding double occupancy; the remaining sites were considered to be occupied by water molecules. Movement of surfactant molecules was performed by reptation; we allowed the movement to be guided by both the terminal head surfactant group and the terminal tail surfactant group. At each Monte Carlo step (MCS) and for each molecule, random numbers were generated with a 0.5 probability in order to determine which group was the guide, following the methodology of Talsania et al. [9]. Reptation movements guaranteed the ergodicity of the algorithm. Acceptance of reptation movements was determined using the Metropolis algorithm, whose acceptance probability is given by

$$P = \min\left\{1, \exp\left(-\frac{\Delta E}{k_BT}\right)\right\},$$

where ΔE is the difference in the total internal energy between the trial and the old configuration. If a move is not accepted, the old configuration is maintained and a new move is tried in the next MCS. One MCS is taken as a complete visit over all surfactant molecules. After typically 1×10^5 MCS, the system reaches equilibrium, producing over 1×10^8 Boltzmann-weighted configurations. The optimal number of MCS depends on the concentration of the surfactant molecules and the temperature. The surfactant concentration was varied between 0.4 and 3 vol % at reduced temperatures between 1 and 3 for all surfactant molecules studied here. Estimation of the micellation Gibbs free energy was carried out according to a phenomenological approximation proposed by Goldstein [10]. The micellation Gibbs free energy is defined as

$$\frac{\Delta G_n^0}{k_BT} = \ln X_1 - \frac{1}{n}\ln X_n,$$

where X_1 and X_n are the mole fraction of free surfactant and aggregates of size n respectively, and ΔG_n^0 is the difference in the standard Gibbs free energy between a surfactant molecule in an aggregate of size n and a free surfactant in the solvent.

Results

Preliminary simulations were carried out considering both nearest-neighbor and second-neighbor interactions at different levels of energy interaction (i.e., $\alpha = 0.0$–1.0) and a concentration of 2 vol% in surfactant molecules, a concentration that is considered high enough to promote some surfactant aggregation. The preliminary results presented in Fig. 1 show that no surfactant aggregation was observed at a value of $\alpha = 0$, i.e., the energy involved in nearest-neighbor interactions was not sufficient to guarantee micelle stability. At a value of $\alpha = 0.2$ effective aggregation took place and micelles with aggregation numbers between 2 and 27 were observed. Higher values of α (i.e., above 0.5) showed that dimers are the main species in the system at the equilibrium state, which is due to the high interaction energy between molecules upon collision; therefore, each collision has a high probability of generating an aggregate. Such a high interaction energy also shows that the equilibrium state between aggregated and nonaggregated molecules is reached very early and consequently the probability of obtaining bigger micelles is considerably reduced. According to the results a value of $\alpha = 0.2$ was chosen to carry out the following

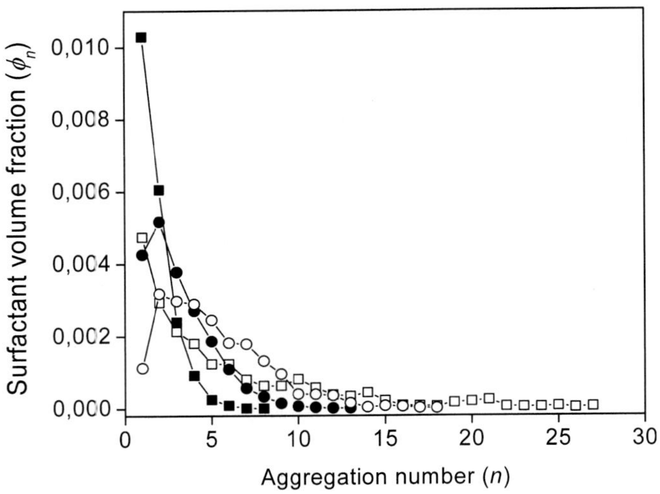

Fig. 1 Micelle size distribution for a H_1T_3 surfactant molecule ($t = 1$) having different energy magnitudes for the second-neighbor interaction between surfactant tails (ε_{TTSN}): $\varepsilon_{TTSN} = 0.0$ (*black squares*); $\varepsilon_{TTSN} = 0.2\varepsilon$ (*white squares*); $\varepsilon_{TTSN} = 0.5\varepsilon$ (*black circles*); $\varepsilon_{TTSN} = 1.0\varepsilon$ (*white circles*)

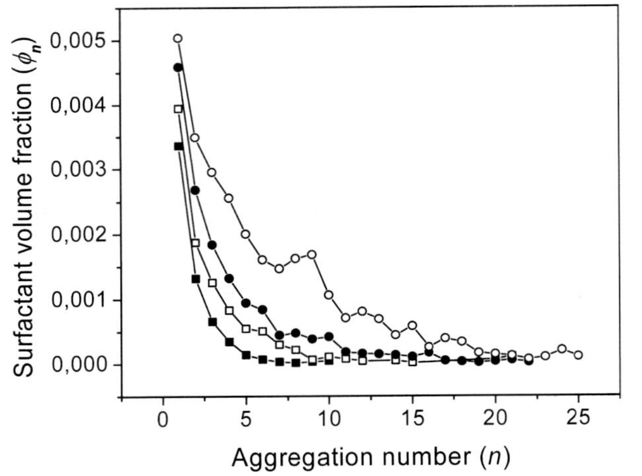

Fig. 3 Micelle size distribution for a H_1T_3 surfactant molecule ($t = 1$) at different surfactant concentration: 0.6 vol % (*black squares*); 1.0 vol % (*white squares*); 1.5 vol % (*black circles*); 3.0 vol % (*white circles*)

simulations in the present work. A typical snapshot for the final state of the simulation for a H_1T_3 surfactant molecule (2 vol%, $t = 1$) is shown in Fig. 2.

Micelle formation is confirmed, as suggested by Girardi and Figueiredo [7], by two system characteristics: a local minimum and maximum on the micelle size distribution plot and a well-defined micelle concentration. Evidence for the first micellation process characteristic for surfactant type H_1T_3 is presented in Fig. 3. At low surfactant concentrations (below 1.5 vol%) a local minimum and maximum is present in the micelle size distribution curve. On increasing the concentration up to 3 vol%, such local minima and maxima are observed at an aggregation number of around 9. Similar behavior was obtained for the other surfactant molecules. Evidence for the second micellation process characteristic of the different surfactant molecules simulated in this work is presented in Fig. 4. For all surfactant molecules the free surfactant concentration reaches a plateau, indicating that any new surfactant molecule added to the system preferably undergoes micellation instead of remaining as a monomer in the continuous phase. The free surfactant molar fraction plateau value decreases with an increase in the surfactant molecule tail length, indicating that micellation is taking place at a

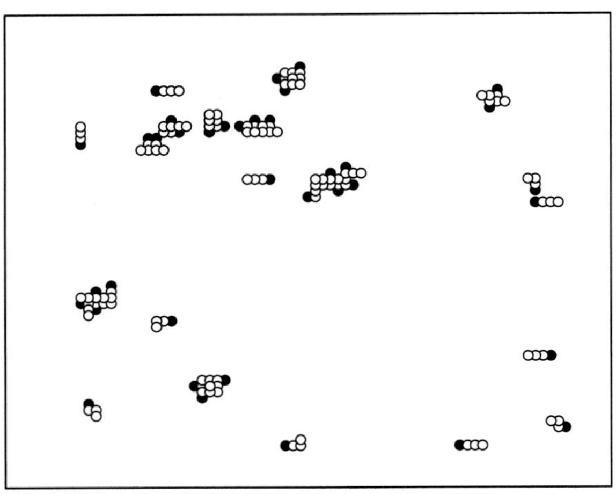

Fig. 2 Snapshot of the surfactant solution (H_1T_3) at 2 vol % and $t = 1$. Surfactant head beads are represented by *black circles* and the tail beads are represented by *white circles*

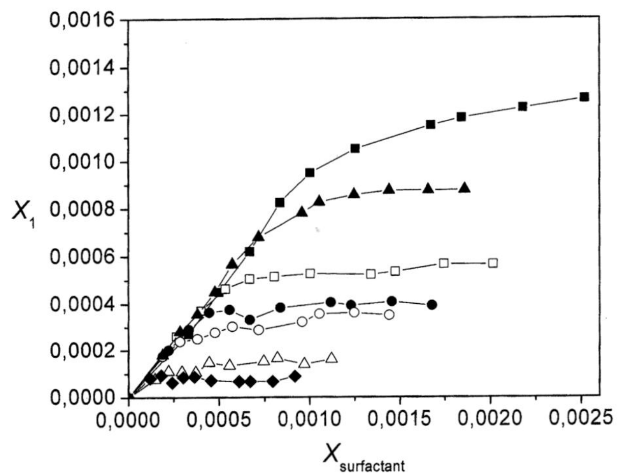

Fig. 4 Free surfactant mole fraction (X_1) as a function of the total surfactant concentration ($X_{surfactant}$) for the surfactant molecules simulated: H_1T_3 (*black squares*); H_1T_4 (*white squares*); H_2T_4 (*black circles*); H_2T_5 (*white circles*); H_3T_4 (*black triangles*); H_3T_6 (*white triangles*); H_4T_7 (*diamonds*)

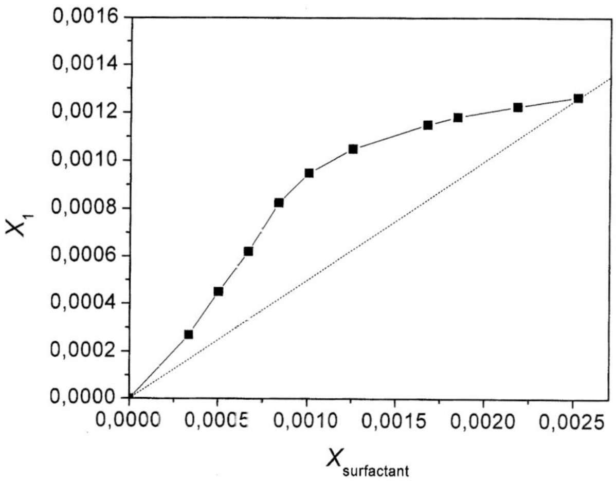

Fig. 5 Free surfactant mole fraction (X_1) as a function of the total surfactant concentration ($X_{surfactant}$) to determine the critical micelle concentration for the surfactant molecule H_1T_3 at $t = 1$

lower surfactant concentration, which is in agreement with experimental evidence [11]. For the case in which the length of the tail remains constant and the length of the head increases, i.e., for surfactant molecules H_1T_4, H_2T_4 and H_3T_4 it was expected that the free surfactant mole fraction would increase with increasing head length. However, when the head length changed from 1 to 2 this behavior was not observed; it only occurred when the head length was increased to a value of 3. That could be explained by the effect of second-neighbor interactions which may have a predominant effect on the H_1T_4 and H_2T_4 molecules, therefore increasing their probability of rearrangement in the micelle. In the case of H_3T_4, geometrical restrictions of the surfactant head dominate over the second-neighbor interactions, giving the expected tendency. The observations are corroborated by the cmc calculated using the method of Stauffer et al. [12]. That methodology defines the cmc as being the total amphiphile concentration for which half of the surfactant is in the form of clusters of two or more amphiphiles, i.e., $X_1 = \sum_{n=2}^{n_{max}} nX_n$. The results of a cmc calculation for a H_1T_3 surfactant molecule at a reduced temperature of $t=1$ are shown in Fig. 5, where the surfactant volume fraction at the cmc (X_{cmc}) corresponds to the intersection of the straight line defined by $X_1 = X_{surfactant}/2$ and the simulation data obtained for X_1. A summary of X_{cmc} for the different simulated molecules at $t=1$ is presented in Table 1.

The effect of reduced temperature on the micellation process for a H_1T_3 surfactant molecule is shown in Fig. 6. At a reduced temperature of $t=1$, the free surfactant mole fraction reaches a plateau at a value of around 0.00125, which corresponds to 16% of free monomers in the system at the equilibrium state. On increasing the reduced temperature to values of 1.25 and above the plateau disappears, and a continuous increase

Table 1 Critical micelle concentration of H_XT_Y surfactant molecules at reduced temperature $t=1$

Surfactant molecule	X_{cmc}
H_1T_3	0.00250
H_1T_4	0.00104
H_2T_4	0.00066
H_2T_5	0.00060
H_3T_4	0.00193
H_3T_6	0.00022
H_4T_7	0.00018

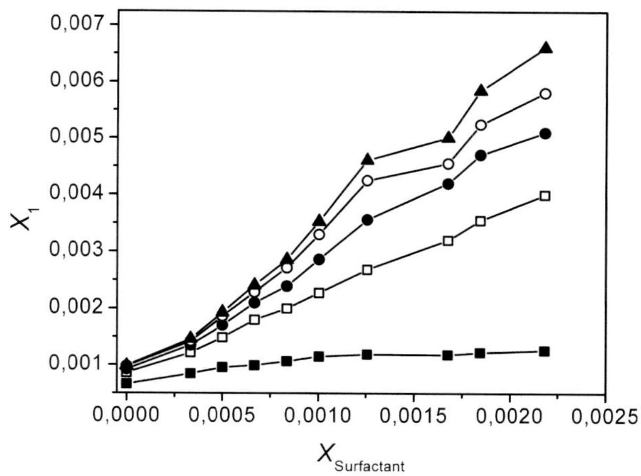

Fig. 6 Free surfactant mole fraction (X_1) as a function of the total surfactant concentration ($X_{surfactant}$) for the surfactant molecule H_1T_3 at different reduced temperatures: $t = 1$ (*black squares*); $t = 1.25$ (*white squares*); $t = 1.5$ (*black circles*); $t = 2.0$ (*white circles*); $t = 3$ (*diamonds*)

in the free surfactant mole fraction is observed, suggesting that no micellation is taking place in the system. With reduced temperature values of $t=2$ and $t=3$, the trend observed for the free surfactant mole fraction against total surfactant is very similar, indicating that most of the surfactant molecules are in the free state; this is confirmed by the micelle size distribution plots (results not shown) where the percentage of monomers is 79% at $t = 2$, and 88% at $t = 3$. The low reduced temperature (i.e., 1.25) at which the aggregation between amphiphilic molecules is not observed suggests that the present simulation model is more sensitive to temperature changes when compared with other models, such as that of de Moraes and Figueiredo [13], where the critical reduced temperature has a value of about 1.53. Therefore, a better approach to real systems can be achieved by using the second-neighbor interaction model proposed here.

Equilibrium Gibbs free energy values for the micellation process of the different molecules simulated in the present work are presented in Table 2. The highest values were observed for the H_1T_3 and the H_3T_4 surfactant molecules, which is in agreement with the high X_1 values

Table 2 Gibbs free energy of micellation for $H_X T_Y$ surfactant molecules at reduced temperature $t=1$

Surfactant molecule	ΔG ($k_B T$)
H_1T_3	−6.303
H_1T_4	−6.847
H_2T_4	−7.181
H_2T_5	−6.701
H_3T_4	−5.723
H_3T_6	−7.128
H_4T_7	−7.755

found in Fig. 4 and their high cmc values (Table 1), indicating that molecular aggregation is less probable. The highest value of ΔG for the micellation process corresponded to the molecule H_4T_7, which also has the lowest cmc value (Table 1). The other surfactant molecules showed similar trends in agreement with cmc values.

Conclusions

Monte Carlo simulation of amphiphilic surfactant molecules in solution, that considers second-neighbor interaction, has been shown to be an alternative way to study their process of aggregation. An energy value of 0.2ε for the second-neighbor interaction was shown to be an appropriate figure to induce controlled aggregation between surfactant molecules; higher interaction energy values produced surfactant solution systems where the predominant species were dimers. The two main characteristics of the micellation process were observed for the model proposed here, i.e., the local minimum and maximum on the micelle size distribution plot and a well-defined micelle concentration. The cmc tendencies of simulated molecules with increasing surfactant tail length were in agreement with the experimental evidence. The cmc tendencies of simulated molecules with increasing surfactant head length showed some differences compared with the experimental evidence. The present aggregation surfactant model showed itself to be more sensitive to changes in the reduced temperature, giving no aggregation between amphiphilic molecules at temperatures lower than those reported by other authors. This could give a better approach to modeling real systems.

Acknowledgements We thank the Comité para el Desarrollo de la Investigación of the Universidad de Antioquia, and Instituto Colombiano para el Desarrollo de las Ciencias.

References

1. Casero I, Sicilia D, Rubio S, Perez-Bendito D (1999) Anal Chem 71:4519
2. Shelley JC, Shelley MY (2000) Curr Opin Colloid Interface Sci 5:101
3. Hunter RJ (1986) Foundations of colloid science. Clarendon Press, Oxford
4. Larson RG (1997) Curr Opin Colloid Interface Sci 2:361
5. Mackie AD, Panagiotopoulos AZ, Szleirfer I (1997) Langmuir 13:5022
6. Rodriguez-Guadarrama LA, Talsania SK, Mohanty KK, Rajagopalan R (1999) Langmuir 15:437
7. Girardi M, Figueiredo W (2000) J Chem Phys 112:4833
8. Gould H, Tobochnik J (1988) An introduction to computer simulation methods, part 2. Addison-Wesley, Reading, MA
9. Talsania SK, Wang Y, Rajagopalan R, Mohanty KK (1997) J Colloid Interface Sci 190:92
10. Goldstein, RE (1986) J Chem Phys 84:3367
11. Mukerjee P, Mysels KJ (1971) Critical micelle concentration. National standards data series. National Bureau of Standards, Washington, DC
12. Stauffer D, Jan N, He Y, Pandley RB, Marangoni DG, Palmer S (1994) J Chem Phys 100:6934
13. de Moraes JNB, Figueiredo W (2001) Phys Status Solidi A 187:57

Progr Colloid Polym Sci (2004) 128: 151–155
DOI 10.1007/b97120

Renaud Toussaint
Jørgen Akselvoll
Geir Helgesen
Eirik G. Flekkøy
Arne T. Skjeltorp

Interactions of magnetic holes in ferrofluid layers

R. Toussaint (✉) · J. Akselvoll
E.G. Flekkøy · A.T. Skjeltorp
Department of Physics,
University of Oslo,
P.O. Box 1048 Blindern,
0316 Oslo, Norway
e-mail: renaud.toussaint@fys.uio.no

J. Akselvoll · G. Helgesen · A.T. Skjeltorp
Physics Department, Institute for Energy
Technology, 2027 Kjeller, Norway

Abstract Nonmagnetic microspheres in a ferrofluid layer interact as dipoles in an external magnetic field. With a constant field normal to the layer and an oscillating in-plane field, the spheres stabilize at fixed separations, and can thus be trapped at tunable finite distances. We show analytically how the susceptibility contrast at the system boundary is responsible for this secondary minimum of a pair interaction potential, obtain the effective interaction potential and equilibrium separation as a function of the applied field, and experimentally validate this theory.

Keywords Ferrofluids · Magnetic holes · Rotating magnetic fields · Effective interactions · Confined system

Introduction

Magnetic holes in a carrier ferrofluid are micrometric spherical nonmagnetic particles, whose size is orders of magnitudes above that of magnetites (0.01 μm), so the ferrofluids appears homogeneous on their scale. In the presence of an external magnetic field, they generate dipolar magnetic perturbations, whose moment is the opposite of that of the ferrofluid displaced [1, 2]. This induces dipolar interactions between them, which can be tuned through the type of imposed external field. This system, first invented by Skjeltorp [1], is confined between two nonmagnetic plates, spaced by a fixed distance ranging from one to ten particle diameters. The ability to design and modify the effective interactions in this system enables us to grow crystals of such holes and induce order/disorder transitions in them [1], to study aggregation dynamics of these particles [3], or to study the nonlinear dynamics of such magnetic holes in low-frequency rotating magnetic fields [4]. Understanding these systems is important for ferrofluid industrial applications [5], or their possible use in biomedicine [6]. Nonetheless, there was no satisfactory theoretical description of the effective interactions between holes in magnetic fields composed of a high-frequency rotating in-plane component and a constant normal one, and the existence of stable configurations of particles with a finite distance between them [2] remained unexplained.

Focusing on the boundary conditions of the magnetic fields along the confining plates, we will derive analytically the pair interaction potential in such oscillating fields, and demonstrate for a wide range of them the existence of a secondary minimum whose position depends continuously on the ratio between out-of-plane and in-plane field magnitudes. We will then compare this theory with experiments where the motion of a particle pair is followed.

System presentation and derivation of an effective pair interaction potential

In the presence of a far-range field $\vec{H}$ in a ferrofluid of susceptibility χ, each hole generates a dipolar perturbation of the dipolar moment equal and opposite to that of the displaced ferrofluid, $\vec{\sigma} = -V\chi_e \vec{H}$, where V is the volume of the particle, and $\chi_e = 3\chi/(3+2\chi)$ is the effective susceptibility including a demagnetization factor rendering for the boundary conditions of the magnetic field along the spherical particle boundary [1, 7]. The susceptibility contrast between the ferrofluid and the two planar nonmagnetic confining plates leads to a different dipolar field perturbation from that in the infinite medium expression. According to the image method [8], the boundary conditions for the magnetic field along the plates are fulfilled if in addition to the infinite space expression, the dipolar field emitted in an unbounded medium by an infinite series of dipole images is taken into account. This series consists of mirror images in the plane boundaries of the initial dipoles or of some previous image, by multiplying the magnitude of the dipole at each mirror symmetry operation by an attenuation factor $\kappa = \chi/(\chi+2)$ – see Fig. 1.

The instantaneous interaction potential between a pair of confined particles can then be expressed as [7]

$$U = \frac{\mu}{8\pi}\sum_{i\neq j}\left[\frac{\vec{\sigma}_i\cdot\vec{\sigma}_j}{r_{ij}^3} - 3\frac{(\vec{\sigma}_i\cdot\vec{r}_{ij})(\vec{\sigma}_j\cdot\vec{r}_{ij})}{r_{ij}^5}\right] \quad (1)$$

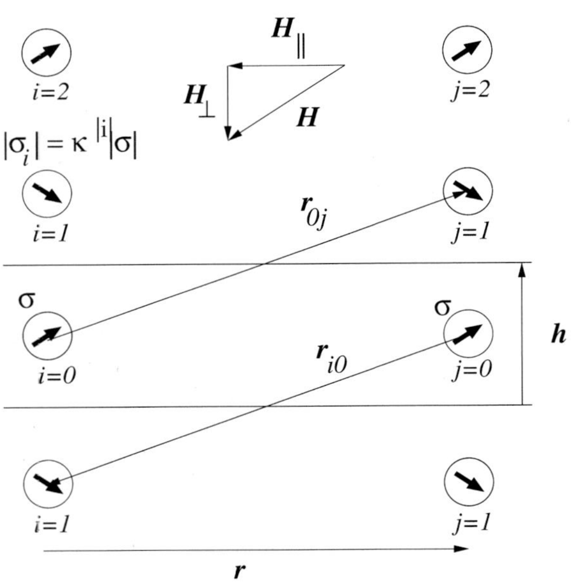

Fig. 1 Pair of nonmagnetic particles in a ferrofluid layer, viewed along the confinement plates, and a series of dipole images accounting for the boundary conditions of the magnetic field along the plates

where μ is the ferrofluid's permeability, the i-index runs over both the source and image dipoles, and the j-index runs only over the two source particles. A detailed analysis shows that the dominant effect for the components of the forces normal to the plates is the interaction between a particle and its own mirror images, which stabilizes the particles midway between the plates. This constraint is therefore adopted throughout this paper and in Fig. 1. Buoyancy forces have also been checked to be negligible for the ferrofluid and particles used in this work.

Decomposing the instantaneous far-range field into its in-plane and normal components $\vec{H}_\perp$ and $\vec{H}_\parallel$, we define the ratio of their magnitudes as $\beta = H_\perp/H_\parallel$, the angle between in-plane component and separation vector as φ, the particle diameter and interplate separation, respectively, as a and h, and the scaled separation as $x = r/h$.

A straightforward analysis of Eq. (1) in the configuration of Fig. 1 leads up to an additive constant to

$$\frac{144h^3U}{\mu\pi a^6\chi_e^2H_\parallel^2} = \sum_{l=-\infty}^{+\infty}\kappa^{|l|}\left[\frac{1+(-1)^{|l|}\beta^2}{(x^2+l^2)^{3/2}} - 3\frac{(y\cos\varphi+l\beta)\left(y\cos\varphi+(-1)^{|l|}l\beta\right)}{(x^2+l^2)^{5/2}}\right]. \quad (2)$$

The term $l=0$ corresponds to the source–source interaction term, already used in previous studies [1, 2], the other terms correspond to interactions between a particle and the images of the other one. For all existing ferrofluids, κ is sufficiently smaller than unity, so the prefactor ensures that the three first images are enough to get a relative precision better than 1% for the potential and its derivatives.

The imposed fields considered consist of a constant component $\vec{H}_\perp$, while $\vec{H}_\parallel$ rotates uniformly at sufficiently high frequency (10–100 Hz in this work). For the micrometric particles considered here (a = 50 μm), inertial terms can be neglected, and a characteristic viscous relaxation time can be obtained as the time to separate two particles initially in contact by their size, balancing the Stokes drag with the magnetic interaction forces derived from the potential. Retaining the main term $l=0$ in Eq. (1) leads to an estimate of $T_c = 144\eta/(\mu\chi_e^2 H^2) \approx 5$s, for the ferrofluid [$\eta = 9\times10^{-3}$ Pa s, $\mu = \mu_0(1+\chi)$, $\chi = 1.9$, $\mu_0 = 4\pi\times10^{-7}$ H m^{-1}] and the typical field $H = 10$ Oe considered here. At field rotation frequencies significantly exceeding the inverse relaxation time, the particle's motion can be neglected during a field oscillation, and an effective interaction potential averaged over time can be obtained by replacing simply the φ-dependent terms in Eq. (2) by their average over an oscillation, while the slowly varying separation vector is maintained constant, which leads to the dimensionless effective interaction potential:

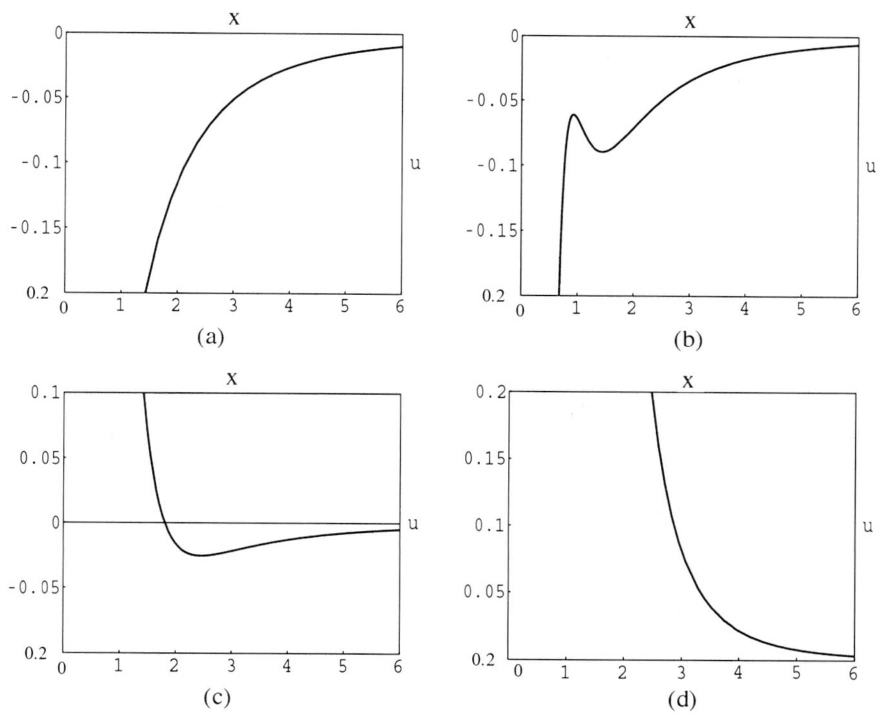

Fig. 2 Four possible types of interaction potentials: **a** $\beta < \beta_m$ purely attractive interactions, **b** $\beta_m < \beta < \beta_c$ interactions with a secondary minimum, **c** $\beta_c < \beta < \beta_u$ interactions with a single equilibrium position at a finite distance, **d** $\beta_u < \beta$ purely repulsive interactions

$$u(x) \equiv \frac{144h^3\overline{U}}{\mu\pi a^6 \chi_e^2 H_\parallel^2} = \sum_{l=-\infty}^{+\infty} \kappa^{|l|} \left[\frac{1+(-1)^{|l|}\beta^2}{(x^2+l^2)^{3/2}} - 3\frac{(-1)^{|l|}l^2\beta^2 + y^2/2}{(x^2+l^2)^{5/2}} \right]. \tag{3}$$

For a given ferrofluid and field, this central potential can be of four possible types as illustrated in Fig. 2.

At low normal field $\beta < \beta_m$, the interactions are purely attractive up to contact; at higher fields $\beta_m < \beta < \beta_c$, a secondary minimum at finite distance appears, later on in the regime $\beta_c < \beta < \beta_u$ this minimum becomes the only one, and ultimately interactions are purely repulsive for $\beta_u < \beta$. For the ferrofluid used, the equilibrium separation for these effective interactions as a function of β are displayed as a continuous line in Fig. 3.

From Eq. (3), the separating values can be shown to be $\beta_c = 1/\sqrt{2}$, $\beta_u = \beta_c(1+\kappa)/(1-\kappa)$, which is a growing function of the susceptibility diverging to infinity when the susceptibility does, and β_m a function of the susceptibility decreasing regularly from β_c at zero susceptibility to 0 at infinite susceptibility – for the ferrofluid used here, $\beta_m \approx 0.55$ and $\beta_u \approx 2.05$. Neglecting the susceptibility contrast along the plates – terms $l \neq 0$ in Eq. (3) – would correspond to the limiting case $\kappa = 0$, where these three separating values merge, and the interactions are either purely attractive or repulsive. The presence of this susceptibility contrast is thus essential to trap the particles at a given equilibrium distance in this type of field.

In order to confront this theory with experiments, the motion of particle pairs initially in contact in a given field was recorded through time. The dynamical equation ruling the particles in this overdamped regime is obtained by balancing the Stokes drag from the embedding fluid with the magnetic interactions, which leads to $\dot{x} = -u'(x)$, where the dot refers to derivation with respect to $t' = t/T$, the dimensionless time, with $T = 3T_c h^5/a^5$, the unit time. The function $t'(x)$ was then

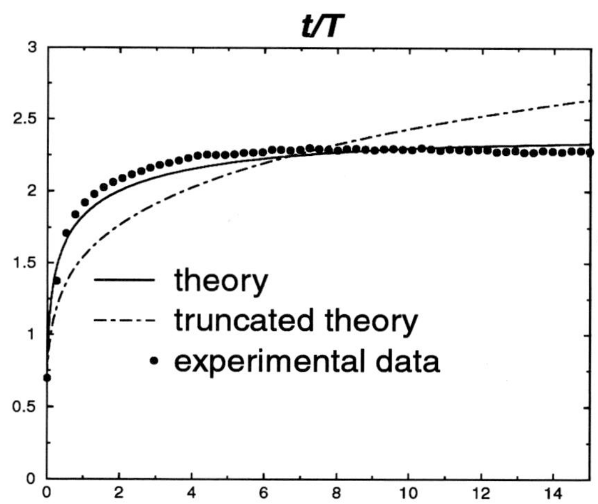

Fig. 3 Scaled separation between a pair of particles as a function of the scaled time, comparison between theory and experiment obtained with an oscillating in-plane field of magnitude $H_\parallel = 14.2$ Oe and a constant normal field $H_\perp = 0.8H_\parallel$

numerically evaluated as the integral of $-1/u'$ from the initial a/h to the actual x value of the scaled separation, to obtain the dashed lines of Fig. 3 and the full line of Fig. 4.

Experiments and results

The experiments were carried using $a = 50$-μm diameter neutral nonmagnetic polystyrene particles designed by Ugelstad's technique [9] and produced under the trademark Dynospheres by Dyno Particles, Lillestrøm, Norway, in a kerozene-based ferrofluid (type EMG905, Ferrofluidics, Nashua, NH, USA). The confining cell consisted of two glass plates 70 μm from each other, and was obtained by slightly pressing the cells together with a few $h = 70$-μm-diameter spacers in-between (of the same composition as the holes). The cell was placed in three pairs of Helmholtz coils, and the particle motion was recorded and digitized from optical microscopy data. The particles were initially brought into contact by applying a fast oscillating purely in-plane field, after which the constant normal field component was added at time zero. The particle pair observed was separated from any other particle or spacer by more than 20 diameters, so as to avoid perturbations.

A typical record of the scaled distance as a function of the scaled time, obtained for a field $H_{\|} = 14.2$ Oe, at $\beta = 0.8$, is shown in Fig. 3.

In this case, the unit time is $T = 32$ s, and particles come to the predicted equilibrium distance $r = 2.35h$ after a few minutes. Comparing this data with the present theory (full line) and with the preexisting expression [1, 2] ignoring the effect of the boundary conditions along the nonmagnetic plates (dashed line) we conclusively show the primary importance of this susceptibility contrast in explaining the existence of this secondary minimum.

A range of geometric β parameters of the applied field was explored in a series of experiments with the same setup, and the resulting scaled separation as a function of β, for various characteristic scaled times, is displayed in Fig. 4.

The error bars correspond to an experimental error in β reflecting a possible angle up to 2.5° of the constant field over the normal direction, which is the maximum variation of the field orientation over the cell, calculated from its precise geometry with respect to the external coils. This error effect was calculated to be the major one owing to the slight inhomogeneity of the field along the experimental cell. Once again, in the range $\beta > \beta_m \approx 0.55$ experiments and theory agree fairly well.

The main discrepancy lies in the small but nonzero separation ($x > 50/70$) for $0.3 < \beta < \beta_m \approx 0.55$. This is believed to be due to the effect of the non-point-like character of the magnetic holes for the magnetic field, which should generate higher-order terms in a multipolar expansion at moderate separations r/a – qualitatively, the dipolar perturbation field of a hole does not fulfill proper boundary conditions along the surface of another hole that is close enough, and a repulsive term corresponding qualitatively to taking into account images of one sphere in the other one, similar to the repulsive effect of images in the plane boundaries on its source particle, becomes sensitive at these short distances.

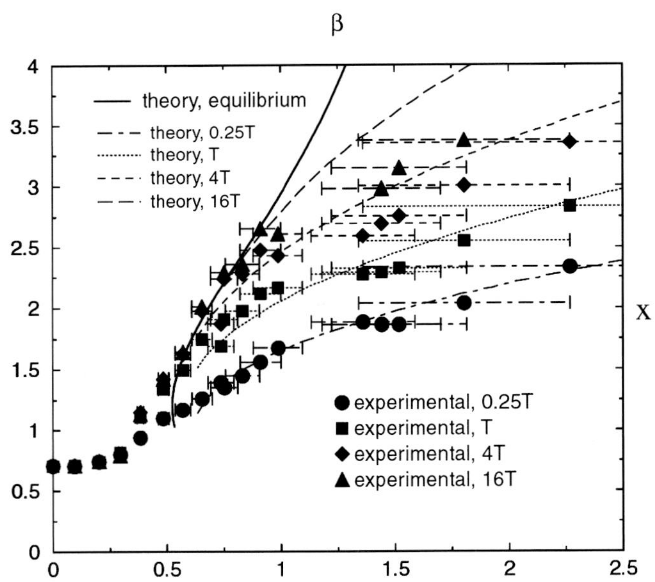

Fig. 4 Scaled distance as a function of the field geometric parameter β, at various values of the scaled time

Conclusions

We have established the effective pair interactions of nonmagnetic particles embedded in a ferrofluid layer confined between two nonmagnetic plates, submitted to magnetic fields including constant normal components and high-frequency oscillating in-plane components. Owing to the susceptibility contrast along the glass plates, a family of potentials displaying a secondary minimum at a finite separation distance has been proven to exist, which should allow us to trap nonmagnetic bodies at tunable distances via the external field.

A system with interactions such as those described here should be relevant for any colloidal suspension of electrically or magnetically polarizable particles constrained in layers. The realization of the detailed effective interaction potentials of this system also makes it a good candidate as an analog model system to study phase transitions [1], aggregation phenomena in complex fluids [4], or fracture phenomena in coupled granular/fluid systems.

References

1. Skjeltorp AT (1983) Phys Rev Lett 51:2306
2. Helgesen G, Skjeltorp AT (1991) J Magn Magn Mater 97:25
3. Helgesen G, Skjeltorp AT, Mors PM, Botet R, Jullien R (1988) Phys Rev Lett 61:1736
4. Helgesen G, Pieranski P, Skjeltorp AT (1990) Phys Rev Lett 64:1425
5. Chantrell R, Wohlfart E (1983) J Magn Magn Mater 40:1
6. Hayter T, Pynn R, Charles S, Skjeltorp AT, Trewhella J, Stubbs G, Timmins P (1989) Phys Rev Lett 62:1667
7. Bleaney B, Bleaney B (1978) Electricity and magnetism. Oxford University Press, Oxford
8. Weber E (1950) Electromagnetic fields: theory and applications, vol 1. Wiley, New York
9. Ugelstad J, Mork P (1980) Adv Colloid Interface Sci 13:101

Progr Colloid Polym Sci (2004) 128: 156–158
DOI 10.1007/b97122

Luiz Carlos Salay
Shirley Schreier

Effect of a kosmotropic ion on doxorubicin self-assembly and interaction with biomimetic systems

L.C. Salay · S. Schreier (✉)
Laboratory of Structural Biology,
Department of Biochemistry,
Institute of Chemistry,
University of São Paulo,
C.P. 26077, 05513-970 São Paulo,
Brazil
e-mail: schreier@iq.usp.br
Tel.: +55-11-30912179
Fax: +55-11-30912179

Abstract Ions of the Hofmeister series alter water structure. Kosmotropes (e.g. sulfate) increase, while chaotropes decrease water structure. Hofmeister ions modulate various properties, such as the solubility and the partitioning of solutes between water and organic solvents and into model membranes. The effect of the kosmotropic ion sulfate on the self-assembly and interaction of the anticancer drug doxorubicin (DOXO) with lipid bilayers was studied by employing fluorescence spectroscopy. As expected, sulfate decreased DOXO's water solubility. In the absence of sulfate, DOXO binds more strongly to large unilamellar vesicles of the anionic phospholipid dioleoyl phosphatidylglycerol (DOPG) than to large unilamellar vesicles of the zwitterionic phospholipid dioleoyl phosphatidylcholine (DOPC). While increasing sulfate concentrations promoted increased DOXO binding to DOPC, the addition of salt caused an initial decrease in drug binding to DOPG, followed by an increase at higher sulfate concentrations. The results indicate that water structure effects prevail in the effect of sulfate upon DOXO solubility and partitioning into DOPC bilayers, whereas a more complex series of events contributes to the observed effect of the ion on the partitioning of the drug into DOPG bilayers.

Keywords Doxorubicin · Large unilamellar vesicles · Kosmotropic ion · Fluorescence spectroscopy · Molecular self-assembly

Introduction

Many pharmacologically active compounds are amphiphilic or hydrophobic molecules. As a result, they tend to self-associate and to interact with biological membranes. The modulation of the physical properties of drugs by self-assembly and membrane binding is of fundamental importance with regard to both academic research and bionanotechnological applications [1].

The anthracycline doxorubicin (DOXO) (Fig. 1) is a potent anticancer drug employed in chemotherapy [2]. However, multidrug resistance mediated by a membrane P-glycoprotein causes decreased drug intracellular accumulation, jeopardising its effectiveness [3]. The ability of anthracyclines to overcome multidrug resistance depends largely on the kinetics of their passive influx. The creation of new strategies to improve a drug's passive influx is a crucial task. The incorporation of DOXO into phospholipid vesicles has been extensively studied, both from the basic and from the applied points of view [4, 5].

Ions of the Hofmeister series alter water structure. Kosmotropes (sulfate, phosphate, citrate) increase, while chaotropes (thiocyanate, perchlorate, trichloroacetate) decrease water structure. For this reason, Hofmeister ions modulate various phenomena in aqueous solutions, playing a role in fundamental properties, such as the solubility of relatively nonpolar compounds, molecular and mesoscopic organization of aggregates, and partitioning of solutes between water and organic solvents [6] and into model membranes (detergent micelles and

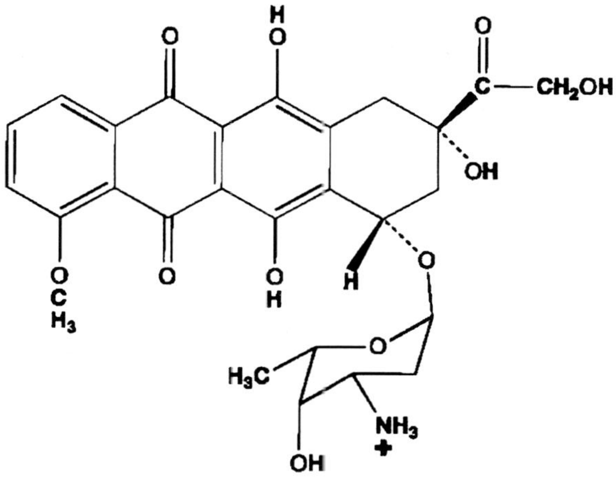

Fig. 1 Chemical structure of doxorubicin (*DOXO*). The p*K* of the charged group is 8.3

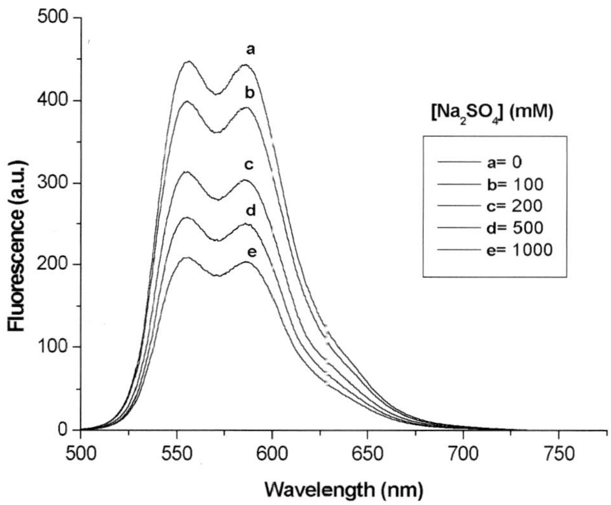

Fig. 2 Fluorescence spectra of 25 μM DOXO in phosphate–borate–citrate buffer, pH 7.2, at different Na_2SO_4 concentrations (M): 0 (*a*), 0.1 (*b*), 0.2 (*c*), 0.5 (*d*), 1.0 (*e*)

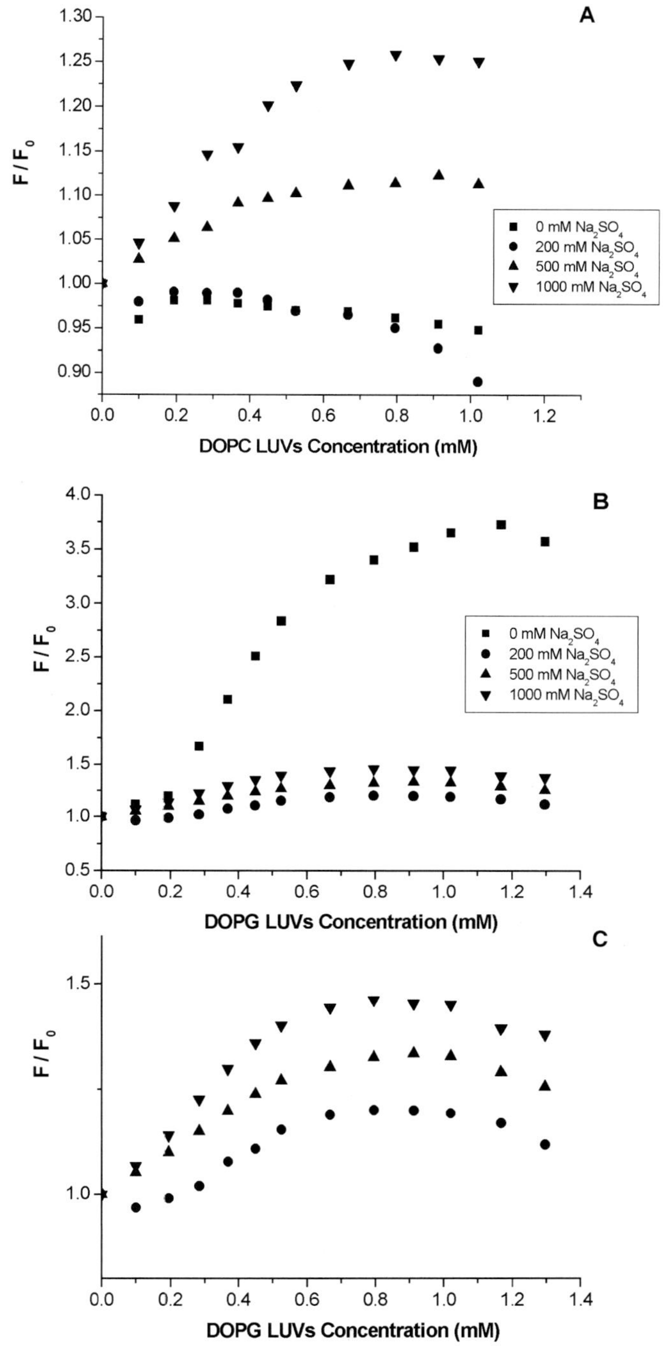

Fig. 3 F/F_0 as a function of a dioleoyl phosphatidylcholine (*DOPC*) and b, c dioleoyl phosphatidylglycerol (*DOPG*) concentration, for different Na_2SO_4 concentrations. In **c** the ordinate was expanded to clarify the differences for the different sulfate concentrations. Na_2SO_4 (M): 0.2 (*circles*); 0.5 (*up triangles*); 1 (*down triangles*). DOXO concentration 25 μM

phospholipid bilayers) [7]. Kosmotropes decrease solute solubility and favor partitioning of solutes into hydrophobic phases, while chaotropes exert opposite effects. In this work we investigated the effect of the kosmotropic ion sulfate on the self-association of DOXO and on its interaction with model membranes by means of fluorescence spectroscopy.

Materials and methods

Pure DOXO was a gift from Phamacia-Upjohn, Italy. DOXO stock solutions (25 μM) were freshly prepared in 5 mM phosphate–borate–citrate buffer, pH 7.2. Sodium sulfate was from Merck (Rio de Janeiro, RJ, Brazil). Dioleoyl phosphatidylcholine (DOPC) and dioleoyl phosphatidylglycerol (DOPG) (Sigma

Chemical Co., St. Louis, MO, USA) large unilamellar vesicles (LUVs) were obtained by evaporating stock chloroform solutions under a stream of nitrogen. The samples were left under vacuum for no less than 2 h, phosphate–borate–citrate buffer was added, and the samples were vortexed for 5 min. LUVs were obtained by extruding the dispersions 30 times through 100-nm-pore polycarbonate filters. The LUVs were kept in ice and used the same day. DOXO binding was assessed by measuring its intrinsic fluorescence [8] with a Hitachi F-4500 spectrofluorimeter. Excitation was at 480 nm and the spectra were scanned from 500 to 750 nm at 22 ± 2 °C.

Results and discussion

Sulfate caused pronounced DOXO self-association, as indicated by a decrease in fluorescence intensity (Fig. 2)

The results are in agreement with reports describing ionic-strength-dependent anthracycline self-association [9]. Extensive DOXO aggregation has been found in the presence of high (of the order of a hundred to a thousand millimolar) sulfate and citrate concentrations. Large supramolecular fibrous bundles were formed at DOXO concentrations 100 times lower than its aqueous solubility [4, 5].

The fluorescence of anthracyclines is enhanced upon binding to lipid bilayers [8]. Making use of this property, we assessed the interaction of DOXO with LUVs by calculating the F to-F_0 ratio, where F and F_0 correspond to the fluorescence intensity in the presence and absence of LUVs, respectively (Fig. 3). It should be recalled that F_0 decreased with increasing sulfate concentration (Fig. 2).

In the absence of sulfate, a much greater increase in F/F_0 was observed in the presence of DOPG (Fig. 3b) than in the presence of DOPC (Fig. 3a). This indicated a much more pronounced interaction of the drug with negatively charged than with zwitterionic membranes, in agreement with previous observations [10, 11]. Sulfate affected the binding of DOXO to the bilayers in different ways. As found for the partitioning of the local anesthetic tetracaine into zwitterionic micelles [7], increasing sulfate concentrations promoted an enhancement of DOXO binding to DOPC (Fig. 3a).

In contrast, the binding to DOPG was reduced upon addition of 200 mM sulfate (Fig. 3b). In this case, a competition occurs between DOXO and sodium ions for binding to the negatively charged phospholipid via electrostatic interactions, promoting a decrease in drug binding. Indeed, this behavior has been found for other anthracyclines [12]. Ionic strength effects should cause a decrease in drug binding with increasing salt concentrations. However, the opposite effect is seen in Fig. 3b and c; in Fig. 3c, the scale was expanded for the data for 200, 500, and 1,000 mM sulfate to show more clearly that DOXO binding to DOPG increases with increasing sulfate concentration. Thus, the ionic strength effect is counteracted by the Hofmeister effect, that promotes enhanced binding.

By exerting its effect on water structure the kosmotropic ion sulfate was expected to decrease DOXO's solubility and to increase its partition into the less polar environment of the bilayers. This hypothesis was confirmed with regard to the drug's solubility. However, the interplay between several phenomena determines the final result with respect to partitioning. Thus, while DOXO's partitioning into zwitterionic DOPC LUVs increased with increasing sulfate concentration, in the case of DOPG the initial increase in ionic strength caused a decrease in DOXO binding. Nevertheless, at higher sulfate concentrations, the Hofmeister effect prevailed, leading to a larger extent of drug binding.

In conclusion, the results show that, by affecting water structure, the kosmotrope sulfate can modulate both drug solubility and partitioning into lipid membranes. These phenomena are of potential use in the design of drug delivery systems.

Acknowledgements This work was supported by a grant from FAPESP to S.S., by a FAPESP postdoctoral fellowship to L.C.S., and by a CNPq research fellowship to S.S. We also thank Phamacia-Upjohn, Italy, for kindly providing pure DOXO.

References

1. Schreier S, Malheiros SV, De Paula, E (2000) Biochim Biophys Acta 1508:210
2. Capranico G, Butelli E, Zunico F (1995) Cancer Res 55:312
3. Mankhetkorn S, Dubru F, Hesschenbrouck J, Fiallo M, Garnier-Suillerot A (1996) Mol Pharmacol 46:532
4. Li X, Hirsh DJ, Cabral-Lilly D, Zirkel A, Gruner SM, Janoff AS, Perkins WR (1998) Biochim Biophys Acta 1415:23
5. Lasic DD, Ceh B, Stuart MCA, Guo L, Frederik PM, Barenholz Y (1995) Biochim Biophys Acta 1239:145
6. Collins KD, Washabaugh MW (1985) Q Rev Biophys 18:323
7. Ferreira GSS, Périgo DM, Politi MJ, Schreier S (1996) Photochem Photobiol 63:755
8. Gallois L, Fiallo M, Laigle A, Priebe W, Garnier-Suillerot A (1996) Eur J Biochem 241:879
9. Menozzi M, Valentini L, Vannini E, Arcamone F (1984) J Pharm Sci 73:766
10. Speelmans G, Staffhorst RWH, de Kruijff B (1997) Biochemistry 36:8657
11. Henry N, Fantine E, Bolard J, Garnier-Suillerot A (1985) Biochemistry 24:7085
12. Burke, TG, Sartorelli AC, Tritton TR (1988) Cancer Chemother Pharmacol 21:274

Progr Colloid Polym Sci (2004) 128: 159–162
DOI 10.1007/b97124

Farid Belaroui
Yves Grohens
Pascal Marie
Yves Holl

Recent results and perspectives on some aspects of latex film formation: drying, coalescence and distribution of additives

F. Belaroui · Y. Grohens · P. Marie
Y. Holl (✉)
Institute Charles Sadron, CNRS and
University of Strasbourg,
6, Rue Boussingault,
67083 Strasbourg, France
e-mail: holl@ics.u-strasbg.fr
Tel.: +33-3-88414116;
Fax: +33-3-88414099

Abstract Four monodisperse core–shell latices were synthesized for small-angle neutron scattering studies, differing in the acrylic acid content in the particle shell (1 or 4 wt %) and the glass-transition temperature of the acrylic core (around –40 or 10 °C). The coalescence kinetics of the surfactant free latices was studied. It was shown that coalescence was hindered by an increase in the acrylic acid content of the shell, the pH of the latex and the glass-transition temperature of the core. These results could be interpreted in terms of chain mobility in the shell and in the core. Labeled SDS was used to follow desorption of the surfactant during film formation. It was shown that desorption occurred early in the film formation process, when the latex still contained around 20% of water. A small fraction of the surfactant remained irreversibly adsorbed at the particle surface.

Keywords Latex · Film formation · Coalescence · Small-angle neutron scattering · Surfactant desorption

Introduction

The complexity of the thin film formation mechanism from a polymer colloid (latex) is unanimously recognized. This subject has been extensively studied in the academic and industrial scientific communities. Excellent review articles can be found in the literature [1]. Several aspects of the process remain insufficiently understood as attested by examples of works published recently. Some of the main ones are (1) drying [2] and particle deformation [3], (2) particle coalescence, and (3) distribution of low molecular weight species [4]. All these aspects are interdependent and can hardly be discussed independently. This explains the complexity referred to earlier.

We intended to readdress two important issues related to the topics mentioned. The first one is coalescence of core–shell particles. In nascent films, these particle shells form a continuous phase. They have to rupture to allow proper particle coalescence. Upon rupturing, they are expelled in separated domains, forming a discontinuous phase. The final morphology of the film will depend on this phase-inversion mechanism. The process is kinetically controlled and it is interesting to investigate the parameters influencing the kinetics. Interestingly enough, some steps in drying of systems consisting of soft particles show analogies with coalescence of core–shell particles. Our second topic of interest pertains to the problem of the distribution of small molecules in latex films. In order to gain better insight into the complex phenomena leading to the distribution of surface-active small molecules, it is important to determine at what time and to what extent these species are desorbed from the particle–water interface when particles get very close in the film formation process.

Only a few experimental methods are available for such kinds of problems. Small-angle neutron scattering (SANS) is a powerful technique well suited for this purpose.

Experimental

The latex syntheses were performed by a semicontinuous process, in a double-wall 1.5 l glass reactor, under a nitrogen blanket. The

Table 1 Main characteristics of the latices used in this study

Latex	Composition of the solids content			Mean particle diameter (nm ±1 nm)	Size polydispersity index	T_g (°C)
	Butyl acrylate (wt %)	Methyl methacrylate (wt %)	Acrylic acid (wt %)			
BuA 1	99	0	1	97	1.03	−341
BuA 4	96	0	4	101	1.09	−37
BuA/MMA 1	59.6	39.4	1	96	1.02	8
BuA/MMA 4	57.9	38.1	4	95	1.06	15

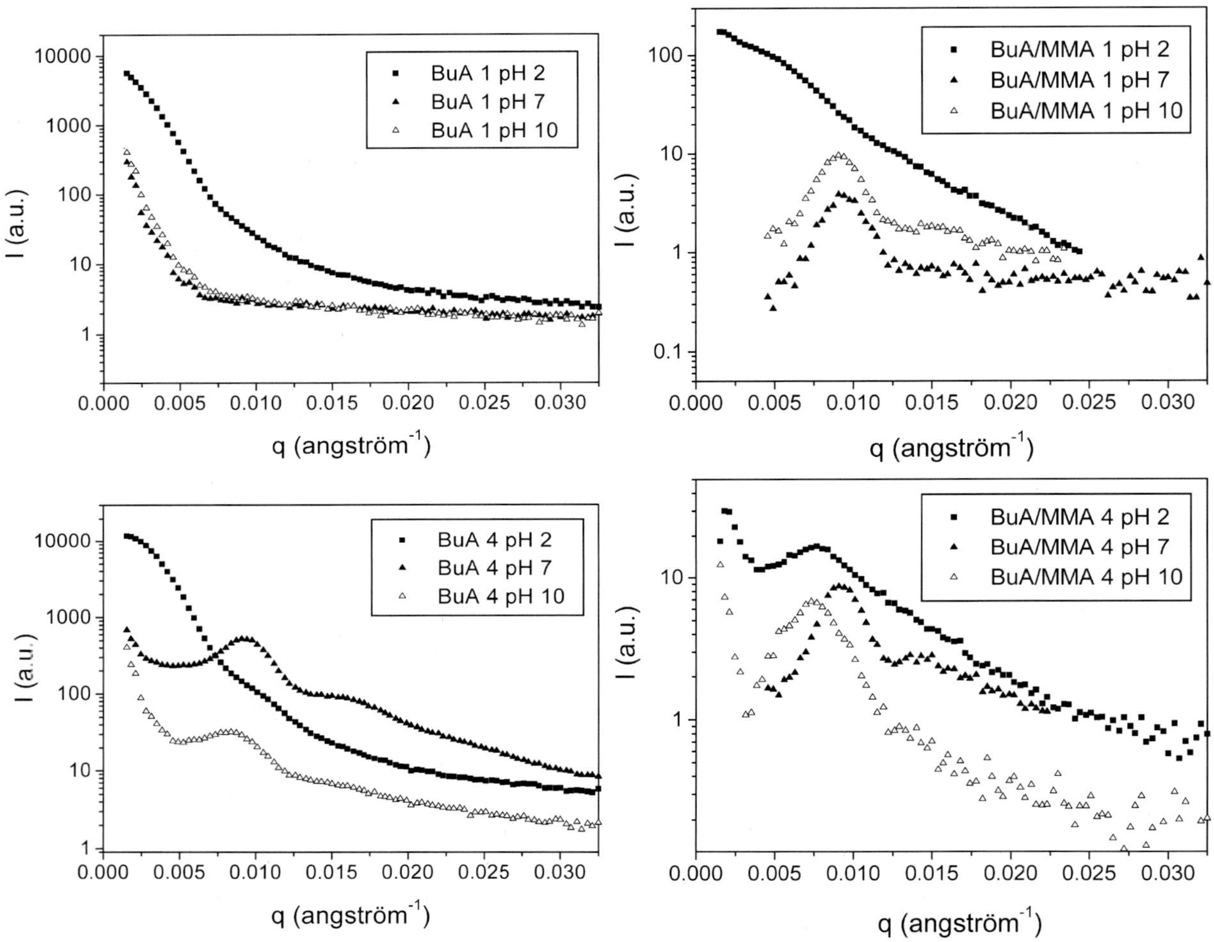

Fig. 1 Scattered intensity versus q for surfactant-free latex films, at various pH. Drying conditions 23 °C, 50% relative humidity, 10 days

reaction temperature was maintained at 75 °C for 3 h and then at 82 °C for 2 h. The four latexes are called BuA 1, BuA 4, BuA/MMA 1, and BuA/MMA 4, indicating the composition of the core (pure butyl acrylate or butyl acrylate/methyl methacrylate copolymer) and the acrylic acid content (Table 1).

Neutron scattering experiments were carried out with the D11 and D22 diffractometers at the Institute Laue Langevin at Grenoble, France. The wavelengths were 6 or 10 Å, and two sample-to-detector distances of 35.7 or 18 m were used, ensuring a large q range of 1.54×10^{-3} Å^{-1} $< q < 3.4\times10^{-2}$ Å^{-1}. More details about the latex synthesis and the SANS experiments (especially on contrast conditions) can be found in Ref. [5].

Results and discussion

Neutron scattering spectra of the films formed from surfactant free latices are shown in Fig. 1. When structure peaks are present at q values around 0.01 Å^{-1}, the expected value for particle diameters close to 100 nm (see, e.g., BuA 4 pH 7), the system has not coalesced, and the hydrophilic acrylic acid rich phase is continuous. When coalescence has occurred, no peak is visible (see, e.g., BuA 1 or BuA4 at pH 2). The pH, acrylic acid content and core mobility which allow or prevent coalescence are summarized in Table 2. It appears that coalescence is hindered by an increase in the acrylic acid content, the pH and the glass-transition temperature of the core.

These results can be rationalized by saying that the higher the chain mobility in both core and shell, the faster the coalescence process. A more extensive discussion of these findings, in particular a comparison with earlier work [6], can be found elsewhere [5].

The results of sodium dodecyl sulfate (SDS) desorption studies devoted to BuA 1 are shown in Fig. 2. The film containing 1 wt % of deuterated SDS exhibits a strong scattering peak at a q value corresponding to the particle diameter after 45 min of drying at 23 °C (solids content 80%). This means that the surfactant is still at the particle surface at that time. At 60 min (solids content 85%), the peak has almost disappeared; it is only faintly detectable. In this system, most of the surfactant leaves the particle surface at solids contents between 80 and 85%; however, small amounts of SDS remain at the surface, also after 90 min. Desorption of the major part of the surfactant occurs in a narrow time (or solids content) slot. No peak is visible after annealing at 75 °C, indicating that this treatment leads to the total desorption of the surfactant.

It might seem surprising that (1) SDS adsorbs onto a polyacrylic acid layer and (2) it desorbs at that particular time in the drying process. In fact, the particle surface is not pure polyacrylic acid. It is an acrylic acid rich copolymer that also contains some hydrophobic moieties (butyl acrylate and methyl methacrylate) on which SDS can adsorb. If SDS was not adsorbed, i.e., if it was only present in water, the results would have been different. We have performed a control experiment [5] with hydrogenated (ordinary) SDS and pure D_2O as the serum which showed that the water phase was still continuous and produced the well-known honeycomb structure after SDS desorption, proving that SDS was not in the water phase.

Desorption takes place when head-to-head repulsion overcomes the adsorption energy. It is not surprising that it occurs at a given layer-to-layer distance, corresponding to a relatively precise state of drying. We have further experiments in progress to get better insight into this "forced desorption" process.

Table 2 Particle coalescence in films formed from latices of different composition and pH. + indicates coalescence has occurred; – indicates coalescence has not occurred

	pH 2	pH 7	pH 10
BuA 1	+	+	+
BuA 4	+	–	–
BuA/MMA 1	+	–	–
BuA/MMA 4	–	–	–

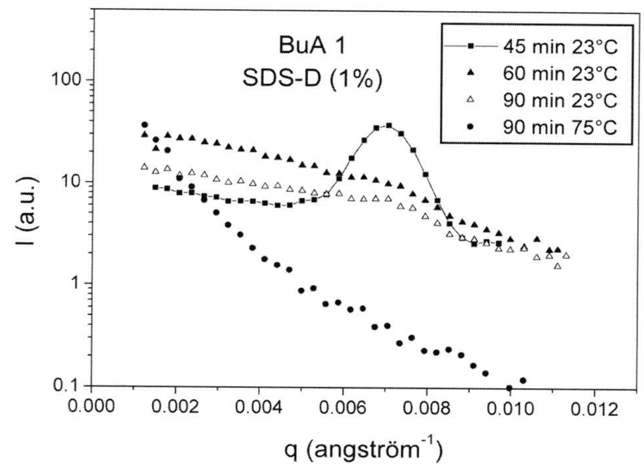

Fig. 2 Scattered intensity versus q for a *BuA 1* film containing 1% of deuterated sodium dodecyl sulfate. The drying conditions are as indicated. The latex serum, at pH 10, is a H_2O/D_2O (80/20) mixture ensuring "index-matching" between the particles and the liquid phase (thus the neutron spectra revealed the repartition of the surfactant in the drying film)

Conclusion

The desorption results presented in this paper will be very useful in the interpretation of distribution data of low molecular weight species (surface-active or not) in these kinds of latex films [7].

On the other hand, the precise knowledge of the morphology of the films, whether the rigid hydrophilic phase remains continuous despite its low volume fraction or becomes discontinuous, and the thermal conditions required to switch from one to the other will allow us to establish nice structure–property relationships. The properties we shall study first are dynamic mechanical properties and adhesion. The effect of low molecular weight species such as surfactants on these properties will be usefully readdressed as well.

More data would be required to get more precise insight into the kinetics and to build models accounting for phase inversion. SANS is a versatile and powerful technique; however, the outcome of neutron scattering results is necessarily slow. Alternative methods should be aimed at. Our project is to investigate the possibilities offered by modern dielectric measurements.

References

1. Keddie JL (1997) Mater Sci Eng R21:101
2. Holl Y, Keddie JL, McDonald PJ, Winnik MA (2001) In: Provder T, Urban MW (eds) Film formation in coatings; mechanisms, properties, and morphology. ACS symposium series 790. American Chemical Society, Washington, DC, p 2
3. Pashley RM, Karaman ME (2001) Colloids Surf A 191:253
4. Zhao Y, Urban MW (2001) Langmuir 17:6961
5. Belaroui F, Cabane B, Dorget M, Grohens Y, Marie P, Holl Y (2003) J Colloid Interface Sci 262:409
6. Joanicot M, Wong K, Richard J, Maquet J, Cabane B (1993) Macromolecules 26:3168
7. Belaroui F, Hirn MP, Grohens Y, Marie P, Holl Y (2003) J Colloid Interface Sci 261:336

Progr Colloid Polym Sci (2004) 128: 163–168
DOI 10.1007/b97125

Eloi Feitosa
Paula Cristina Alves Barreleiro

The effect of ionic strength on the structural organization of dioctadecyldimethylammonium bromide in aqueous solution

Abstract High-curvature and stabilized vesicles of dioctadecyldimethylammonium bromide (DODABr) can be formed spontaneously in aqueous electrolytic solution. It is shown by cryo-transmission electron microscopy that 5.0 mM DODABr molecules associate in water at a temperature above its gel-to-liquid-crystalline phase transition temperature (T_m≈45 °C) in a variety of complex bilayer structures. However, in the presence of NaCl the preferred structures formed are unilamellar and bilamellar vesicles with high curvature and the dispersion is polydisperse in size and geometry, but the main vesicle population contains spherical, flattened and smoothed structures. It is, however, less polydisperse than the corresponding salt-free dispersion, and the size polydispersity and the vesicle curvature radius tend to decrease with NaCl concentration. Long cylindrical bilamellar vesicles, with a very thin water layer separating the bilayers are also formed in the presence of 10 mM NaCl. The effect of the ionic strength on T_m, obtained by differential scanning calorimetry, is shown to depend on the nature of the counterion: Br^- decreases, whereas Cl^- increases T_m of DODABr, indicating different affinity of these counterions for the vesicle surfaces.

Keywords Cryo-transmission electron microscopy · Micro differential scanning calorimetry · Cationic vesicles · Vesicle curvature · Gel-to-liquid-crystalline phase transition temperature

E. Feitosa (✉)
Physics Department, IBILCE/UNESP,
Rua Cristóvão Colombo, 2265,
15054-000, São Jos do Rio Preto,
SP, Brazil
e-mail: eloi@df.ibilce.unesp.br,
Tel.: +55-17-2212240
Fax: +55-17-2212247

P.C.A. Barreleiro
Physical Chemistry 1, Center of Chemistry and Chemical Engineering,
University of Lund, P.O. Box 124,
221 00 Lund, Sweden

Introduction

Dioctadecyldimethylammonium salts (DODA*X*, *X* being usually Br^- or Cl^-) are long (C_{18}) double chain cationic surfactants which associate in water, above the gel-to-liquid-crystalline phase transition temperature, T_m, as bilayer structures (T_m=45 °C, for DODABr) [1–14]. The molecular organization of DODA*X* depends on the surfactant concentration, the temperature and the solvent condition. Since DODA*X* molecules are poorly soluble in water at room temperature, energy-input-based methods are required to form high-curvature vesicles. The most common vesicle preparation protocols consist of sonicating [8] or extruding [14] the homogeneous surfactant aqueous dispersion, or dissolving the surfactant in an appropriate organic solvent (e.g., chloroform, dichloromethane or ethanol) and then injecting it in the desired solvent (usually water or electrolytic aqueous solution) above the evaporation temperature of the organic solvent [8]. These protocols usually introduce contaminants in the surfactant dispersion. Exceptions are bath sonication and extrusion since the flask or syringe containing the dispersion may be sealed or isolated, but the loss of surfactant, retained in the extrusion membrane, is a problem to be overcome. Thus, the formation of spontaneous vesicles seems to be the best way to form vesicles with good control of the solution contents. The curvature of such spontaneous vesicles can be controlled by the presence of additives, such as inert salts for ionic vesicle formation.

When dissolved in water, above its T_m and at low concentration (up to about 1 mM), DODA*X* molecules associate spontaneously as large unilamellar vesicles, whose physical and thermal properties have already been reported [7, 11–13]. At higher concentrations, however, they tend to associate as more complex structures. Extrusion or sonication of the surfactant dispersion yields bilayer fragments or vesicles with high curvature, which depends on factors like the sonication power and time and the nominal filter pore dimension used in the extrusion.

Alternatively, high-curvature and stabilized vesicles, in the semidilute regime of DODABr concentration (e.g., 5 mM) can be prepared spontaneously in the presence of monovalent salt, as shown in this communication. The process consists simply in dissolving DODABr crystals in an appropriate electrolytic aqueous solution (we have used aqueous solutions of NaCl or NaBr up to 10 mM) above its T_m. In the presence of iso-osmolar NaCl, and depending on the salt concentration, the preferred structures formed are closed unilamellar and bilamellar vesicles with higher curvature relative to those in pure water. The morphology of the 5.0 mM DODABr aggregates in the presence of up to 10 mM iso-osmolar NaCl was investigated by cryo-transmission electron microscopy (cryo-TEM), and its thermotropic phase behavior was monitored by differential scanning calorimetry (DSC).

The effect of salts on the aggregation of DODACl vesicles prepared by sonication and chloroform injection methods (i.e., not spontaneously) have been reported before [9, 10]. In this report, however, many more details are given about the structural organization, as well as the thermotropic phase behavior, of DODABr spontaneous vesicles upon a systematic change in the ionic strength.

Experimental

Materials

DODABr with purity higher than 99% was used as received from Avanti Polar Lipids. NaCl and NaBr (Merck) were analytical grade. The vesicle aqueous dispersions were prepared by simply diluting DODABr in water or water plus salt (NaBr or NaCl), at 55 °C (i.e., above the surfactant T_m), to a final surfactant concentration of 5.0 mM and at selected salt concentrations up to 10 mM. After obtaining a homogeneous solution, the dispersions were allowed to stand at room temperature for at least 2 h before the measurements. The vesicles so formed are referred to as spontaneous. After 24 h of storage some of the salt-containing samples showed some precipitate.

Apparatus and methods

Cryo-transmission electron microscopy

Cryo-TEM allows the direct visualization of vesicle structures in dispersions [15]. Thin (10–500-nm) films of DODABr dispersions were prepared under controlled temperature and humidity conditions in a custom-built environmental chamber. The films were vitrified from 25 °C by quickly freezing them in liquid ethane and they were then transferred to a Zeiss EM 902 transmission electron microscope for examination. To prevent sample perturbation and the formation of ice crystals, the samples were kept below 108 K during the transfer and viewing procedures. All observations were made in a zero-loss bright-field mode and at an accelerating voltage of 80 kV.

Differential scanning calorimetry

The high-sensitivity differential scanning calorimeter MicroCal MC-2 (Microcal, Northampton, MA, USA) was used to obtain the DSC thermograms. It is equipped with twin total-fill cells of 1.2 ml for the reference and sample solutions. The solutions were degassed using a Nueva II stirrer (Termolyne) before being transferred to the cells to avoid bubble formation. The instrument measures the power required to keep the temperature of the sample and reference cells equal while the temperature is raised (upscan) or lowered (downscan) at a constant rate, 1 °C/min in this work. Lower scanning rates were also used to verify transition kinetics. The reference thermogram for both cells filled with the same solvent was recorded under the same conditions. The difference between the sample and reference thermograms is proportional to the excess heat capacity, ΔC_p. The position of the main peak corresponds to T_m, the area under the peak equals the enthalpy of the transition, ΔH, and the peak width is inversely proportional to the transition cooperativity [16].

Results and discussion

Salt-free DODABr dispersions

Spontaneously formed dispersions of 5.0 mM DODABr in water exhibit a rich variety of bilayer structures. Accordingly, the cryo-TEM micrographs shown in Fig. 1 reveal small (dimension up to around 90 nm), large (dimension around 500 nm) and giant (dimension up to around 12,500 nm) unilamellar vesicle structures, as well as open extended bilayers, multilamellar vesicles with the inner bilayers being not well-defined vesicle structures, flattened vesicles and the interesting coalesced structures consisting of large densely packed bilayers enveloped by a giant unilamellar structure. Interestingly, such coalesced structures were observed only inside giant unilamellar vesicles, and the reason for that is not clear. Furthermore, such structures were observed only for DODABr but not for DODACl (results not shown) in water, i.e., in the absence of NaCl (see later), indicating that specific electrostatic and screening effects play important role in the formation of DODABr aggregate structures. Similar (but not enveloped) aggregates referred to as "snake-balls" have been reported by

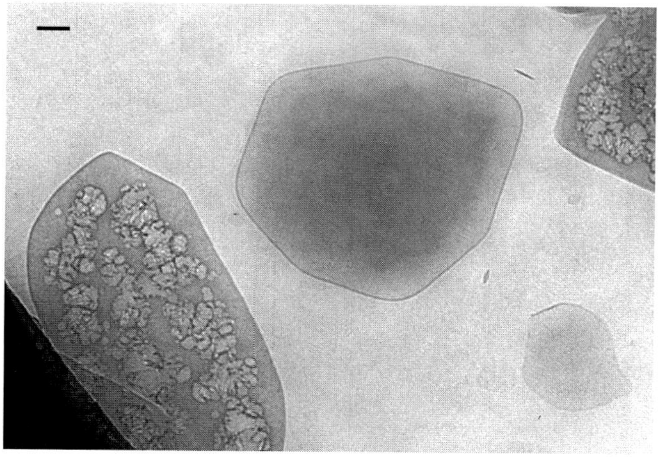

Fig. 1 Cryo-transmission electron microscopy (*cryo-TEM*) micrographs of 5.0 mM dioctadecyldimethylammonium bromide (*DODABr*) in water showing two areas of the same specimen. The dispersions were prepared spontaneously by simply mixing the surfactant and water at 55 °C. The sample was vitrified from 25 °C. The *bar* represents 200 nm

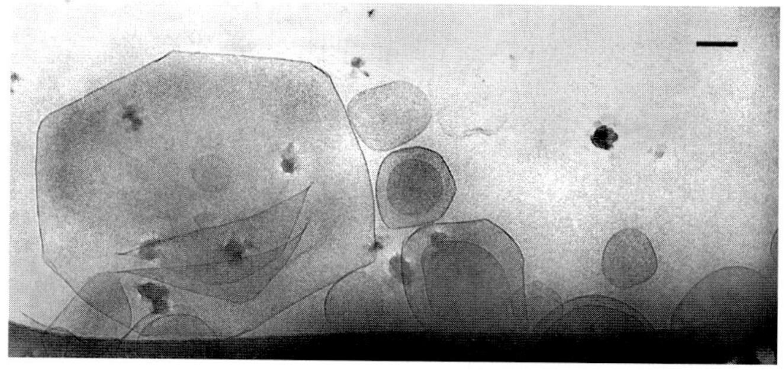

Fig. 2 Cryo-TEM micrographs of the same specimen of 5.0 mM DODABr in 1.2 mM NaCl iso-osmolar aqueous solution. The sample was vitrified from 25 °C. The *bar* represents 200 nm

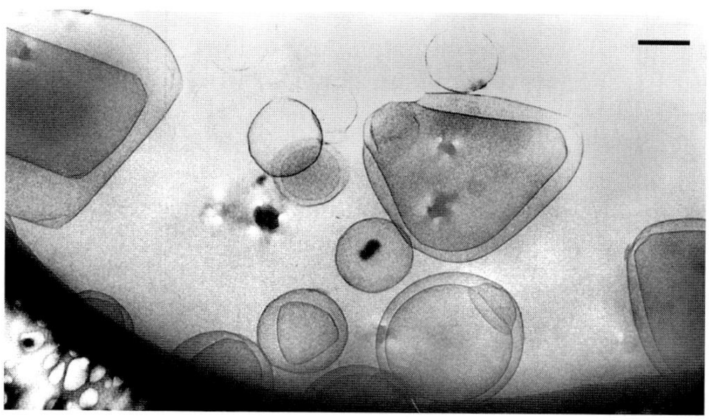

Fig. 3 Cryo-TEM micrographs of the same specimen of 5.0 mM DODABr in 5.0 mM NaCl iso-osmolar aqueous solution. The sample was vitrified from 25 °C. The *bar* represents 200 nm

Zheng et al. [17] for a mixture of cationic cetyltrimethylammonium chloride with anionic sodium 3-methylsalicylate in aqueous solution.

Among the unilamellar vesicles we can distinguish spherical, globular and lenslike vesicles; this later structure being characteristic of sonicated DODABr vesicles [6]. The open lamella structures are small and large, flat and undulated, and sometimes they appear branched and coiled (Fig. 1). Most probably these structures are very dynamic and able to be transformed from one form to another.

Salt-containing DODABr dispersions

In the presence of NaCl, DODABr molecules tend to associate as vesicles with a salt-dependence curvature and are more homogeneous in size and structure relative to those in pure water. The micrograph for an aqueous dispersion of 5.0 mM DODABr in 1.2 mM iso-osmolar NaCl is shown in Fig. 2, where it can be seen that the dominant structures in the dispersion are large unilamellar and bilamellar vesicles. The unilamellar and bilamellar vesicles are mainly distributed within two populations of sizes in the range 100–500 and 140–380 nm. Surprisingly the bilamellar vesicles have higher curvatures than the unilamellar ones. Most of the vesicles are spherical or quasi-spherical, and the largest ones are faceted. The thickness of the water layer separating the surfactant bilayers of the bilamellar structures ranges from 15 to 40 nm. A faceted vesicle with a mean diameter of about 540 nm seems to enclose two large "semi-lens"-shaped vesicles; alternatively, the enclosure of these vesicles may be a matter of overlap of the projections of the images of these vesicles. It is worth stressing that bilamellar vesicles can be smaller in size than unilamellar vesicles. These pictures also clearly show that in the presence of salt the coagulated structures present in the salt-free dispersions are not formed, stressing the importance of electrostatic interactions in DODABr vesicle formation.

In the presence of 5.0 mM iso-osmolar NaCl, bilamellar vesicles are the dominant structure, with diameters in the range 140–420 nm, but some large and giant unilamellar vesicles still remain, with diameters in the range 140–750 nm (Fig. 3). The smaller unilamellar and bilamellar vesicles are spherical, while the larger ones are faceted. In most of the bilamellar vesicles the inner vesicle has the same geometry as the outer vesicle. Exceptions are, however, for a spherical vesicle (diameter of 170 nm) that encloses a faceted vesicle, and another spherical vesicle with a diameter of 290 nm that encloses two vesicles that are in a fusion stage, or, alternatively, a large vesicle in the process of separation to form two smaller vesicles (Fig. 3).

The micrograph of 5.0 mM DODABr in a 10 mM iso-osmolar NaCl aqueous dispersion are shown in Fig. 4. Clearly most of the vesicles have a curvature higher than those at lower ionic strengths or in pure water, and the diameter of most of the vesicles does not exceed 300 nm. There are, however, some characteristic long cylindrical bilamellar structures, as large as about 850 nm, and with diameters of 115–125 nm, with a very thin water layer of 8–12 nm separating the surfactant bilayers. At this ionic strength some trilamellar vesicles are formed with a 17-nm-thick water layer between the surfactant bilayers. Most of the trilamellar vesicles consist of ordinary concentric bilamellar vesicles enclosing a very small lenslike vesicle in the center. The larger vesicles may be formed by fusion of two vesicles.

Effects of ionic strength and counterion on the thermotropic phase behavior of DODABr

Another interesting point to be mentioned is that despite the great variety of structures in the salt-free DODABr dispersion its T_m is about the same as that obtained for very dilute dispersions in which the vesicle structure is more homogeneous [12]. The surfactant tails form different microdomains, which are sensitive to spectroscopic probes like fluorescence [11] or electron spin resonance probes [13]. The effects of NaCl and NaBr on the uspscan DSC thermograms of 5.0 mM DODABr in aqueous dispersions are shown in Figs. 5 and 6, respectively, and the corresponding parameters related to the main transition obtained from these diagrams are summarized in Tables 1 and 2. Accordingly, T_m of DODABr increases with the addition of NaCl, but decreases at a lower rate with addition of NaBr (Fig. 7). Two additional peaks in the thermograms are observed, as reported before [12]. T_s and T_p denote the pretransition (or secondary) and posttransition temperatures ($T_s \approx 36$ °C and $T_p \approx 53$ °C), which are probably related to undulation of the vesicle bilayer and changes in vesicle structures, respectively, as suggested by the cryo-TEM experiments. T_p is related to the dispersion flow birefringence, indicating the formation

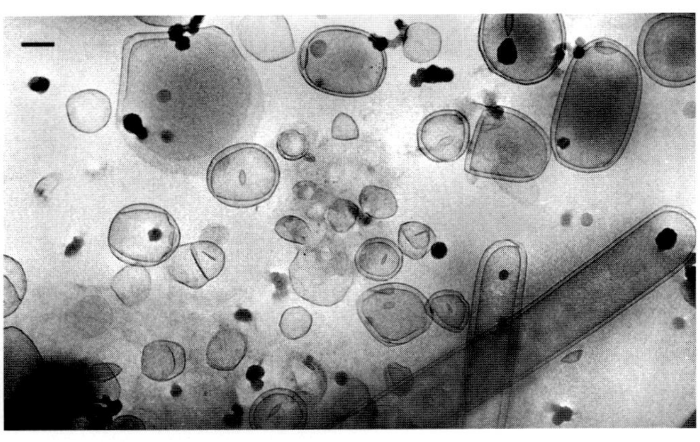

Fig. 4 Cryo-TEM micrographs of the same specimen of 5.0 mM DODABr in 10.0 mM NaCl iso-osmolar aqueous solution. The sample was vitrified from 25 °C. The *bar* represents 200 nm

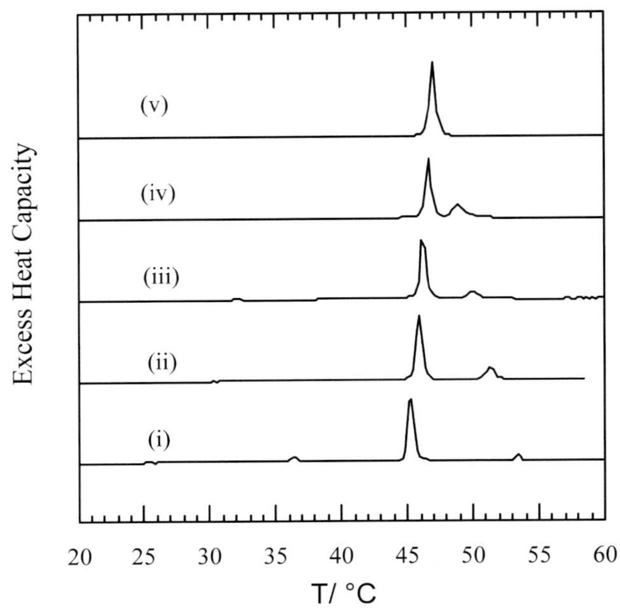

Fig. 5 Differential scanning calorimetry thermograms obtained at the upscan rate of 60 °C/h for DODABr aqueous vesicle dispersions prepared in pure water and in NaCl: 0 mM (*i*), 1 mM (*ii*), 5 mM (*iii*), 10 mM (*iv*), and 20 mM (*v*)

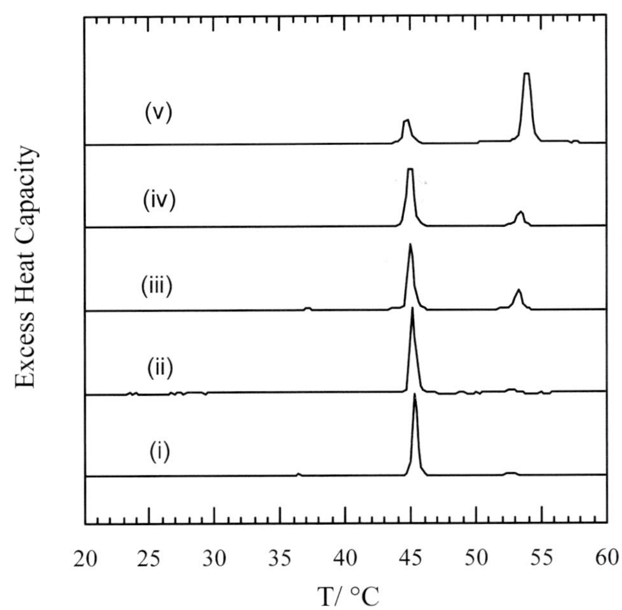

Fig. 6 Same as Fig. 5, except for NaBr instead of NaCl

Table 1 Differential scanning calorimetry (*DSC*) parameters related to the main transition, obtained from the thermograms in Fig. 5, for 5.0 mM dioctadecyldimethylammonium bromide (*DODABr*) in the presence of increasing concentration of NaCl. The estimated errors in T_m and ΔH are ± 0.1 °C and ± 0.5 kcal/mol, respectively

NaCl concentration (mM)	T_m (°C)	$\Delta T_{1/2}$ (°C)	ΔH (kcal/mol)
0	45.3	0.63	10.4
1	45.4	0.67	9.5
5	45.9	0.77	8.7
10	46.2	0.84	8.4
20	46.8	0.82	10.8

Table 2 DSC parameters related to the main transition, obtained from the thermograms in Fig. 6, for 5.0 mM DODABr in the presence of increasing concentration of NaBr. The estimated errors in T_m and ΔH are ± 0.1 °C and ± 0.5 kcal/mol, respectively

NaBr concentration (mM)	T_m (°C)	$\Delta T_{1/2}$ (°C)	ΔH (kcal/mol)
0	45.3	0.63	10.4
1	45.1	0.66	10.0
5	44.9	0.67	9.7
10	45.0	0.69	9.8
20	44.7	0.72	3.8

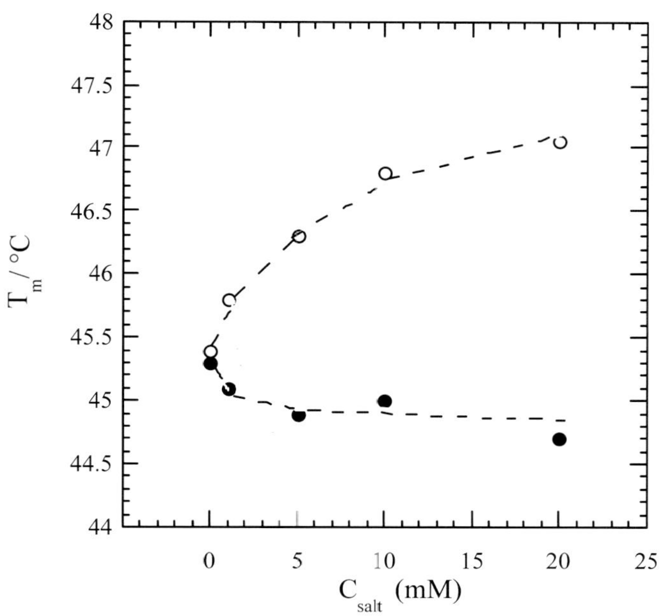

Fig. 7 Effect of NaCl (*open circles*) and NaBr (*closed circles*) concentration on the main transition temperature of DODABr vesicles

of local lamellar phases. This is in agreement with the fact that DODACl dispersions exhibit no posttransition peak, meaning that the addition of NaCl to DODABr dispersions yields DODACl-like dispersions, i.e., without flocculation.

The pretransition temperature peak is very small and tends to disappear upon addition of a small amount of salt. The posttransition temperature decreases with salt concentration, indicating a stabilization effect of DODABr vesicles by NaCl. T_p is about 53 °C without any salt, and about 49 °C in the presence of 10 mM NaCl. Like the pretransition peak it can be very weak in salt-free DODABr dispersions, but its intensity tends to increase with the ionic strength. The addition of NaCl has a similar effect on the thermograms as the extrusion of the spontaneously formed DODABr vesicle dispersions [12]: it broadens the transition peak, increases $\Delta T_{1/2}$, and decreases the transition enthalpy, ΔH, meaning that the transition becomes less cooperative as the NaCl concentration is increased. Unlike extrusion (which decreases the T_m of DODABr), NaCl (but not NaBr) increases the T_m of DODABr. The reason for the opposite effect of NaCl and NaBr (see later) on T_m might be related to the stronger affinity of the Br^- counterion relative to Cl^- for the vesicle interfaces, and reflects the lower T_m of DODABr relative to DODACl [13].

NaBr has a stronger effect on the intensity of the posttransition peak than NaCl, indicating that NaBr yields vesicle aggregation more than NaCl. For 20 mM NaBr the intensity of the posttransition peak is higher than the main transition peak. It is worth observing that NaBr does not change considerably T_s and T_p, whereas NaCl decreases T_p and inhibits T_s. For 5 mM DODABr in the presence of an increasing concentration of NaBr, T_m and ΔH of the main transition decrease whereas $\Delta T_{1/2}$ increases slightly, indicating that the transition becomes less cooperative as the ionic strength is increased (Table 2).

Concluding remarks

We have investigated by cryo-TEM and DSC the effect of NaCl and NaBr (up to 20 mM) on the structural organization and on the main, pretransition and posttransition temperatures, T_m, T_s and T_p, of 5.0 mM DODABr spontaneously formed vesicles in aqueous dispersions. Under this condition, NaCl increases T_m (but not T_s and T_p) and the vesicle curvature, and decreases the size polydispersity. Both in the absence and in the presence of salt, there is a rich variety of bilayer structures in the dispersions. Open bilayers and densely packed bilayers enveloped by a giant unilamellar vesicle were observed only in pure DODABr water dispersions. Unilamellar and bilamellar vesicles are formed in the absence and in the presence of salt. Trilamellar vesicles, however, are formed only in the presence of salt. The vesicle fusion phenomenon can be visualized in the presence of 10 mM NaCl. The effect of ionic strength on the DODABr T_m depends on the counterion: Cl^- increases, while Br^- decreases T_m, probably owing to the stronger binding of Br^- to the vesicle interfaces resulting from their different hydration radii and polarizabilities [12]. Both counterions inhibit the T_s peak; Cl^-, on the other hand, inhibits the T_p peak, whereas Br^- increases its intensity, which becomes, at the highest salt content, larger than the main transition peak, indicating that NaCl has a more pronounced stabilizing effect on vesicles than NaBr. That is in agreement with the fact that 5.0 mM DODACl in water exhibits no posttransition temperature (results not shown).

As a general conclusion, owing to specific electrostatic and screening effects, NaCl modifies the physical properties of DODABr (5 mM) spontaneously formed dispersions, such as the structure and size dispersity, and tends to increase the vesicle curvature. It also inhibits formation of coalesced aggregates characteristic of 5.0 mM DODABr in water.

Acknowledgements E.F. wishes to thank CNPq (grant 303563/85–0). P.C.A.B. acknowledges the PRAXIS XXI, JNICT, for financial support, scholarship BD/13788/97. We express our gratitude to K. Edwards (Uppsala University) for kindly supplying the cryo-TEM equipment, and G. Karlsson (Uppsala University) for operating the cryo-TEM equipment.

References

1. Kunitake T, Okahata Y (1977) J Am Chem Soc 99:3860
2. Kunitake T, Okahata Y, Tamaki K, Kumamaru F, Takayanagi M (1977) Chem Lett 387
3. Yogev D, Guillaume BCR, Fendler JH (1991) Langmuir 7:623
4. Fendler JH (1980) Acc Chem Res 13:7
5. Pansu RB, Arrio B, Roncin J, Faure J (1990) J Phys Chem 94:796
6. Andersson M, Hammarström L, Edwards K (1995) J Phys Chem 99:1431
7. Feitosa E, Brown W (1997) Langmuir 13:4810
8. Carmona-Ribeiro AM (1992) Chem Soc Rev 21:209
9. Carmona-Ribeiro AM, Yoshida LS, Chaimovich (1986) J Phys Chem 89:621
10. Carmona-Ribeiro AM, Chaimovich (1985) Biophys J 50:2928
11. Benatti CR, Tiera MJ, Feitosa E, Olofsson g (1999) Thermochim Acta 328:137
12. Feitosa E, Barreleiro PCA, Olofsson G (2000) Chem Phys Lipids 105:201
13. Benatti CR, Feitosa E, Fernandez RM, Lamy-Freund MT (2001) Chem Phys Lipids 111:93
14. Cuccovia IM, Sesso A, Abuin EB, Okino PF, Tavares PG, Campos JFS, Florenzano FH, Chaimovich H (1997) J Mol Liq 72:323
15. Bellare JR, Davis HT, Scriven LE, Y. Talmon Y (1988) J Electron Microsc 10:87
16. Bach D (1984) Chem Phys Lipids 35:385
17. Zheng Y, Zakin JL, Talmon Y, Davis HT, Scriven LE (2001) J Phys Chem 104:5263

Progr Colloid Polym Sci (2004) 128: 169–174
DOI 10.1007/b97126

Stephen A. Holt
Karen J. Edler
Cristina Fernandez-Martin

Formation of surfactant–silica mesophase films at a silica interface

Work originally presented at the 11th International Conference on Colloid and Interface Science

S.A. Holt (✉)
ISIS, Rutherford Appleton Laboratory, Chilton, Didcot, OX11 0QX, UK
e-mail: s.a.holt@rl.ac.uk
Tel.: +44-1235-446081
Fax: +44-1235-445720

K.J. Edler · C. Fernandez-Martin
Department of Chemistry, University of Bath, Claverton Down, Bath, BA2 7AY, UK

Abstract Neutron reflectometry has been used to study the development of mesoporous silicated films at a hydrophilic silicon–solution interface. We find the time for film formation was altered in a manner similar to that seen at the air–liquid interface as the silica source-to-surfactant ratio was altered. Interestingly for films grown at the silica–liquid interface in a cell with no air space only one growth mechanism is observed, that of the solution assembly of small particles which then pack at the interface to form the film. We see no evidence of any growth mechanism mediated by interfacial processes.

Keywords Mesoporous films · Neutron reflectometry

Introduction

Surfactant–silicate mesophase composites self-assemble from dilute solutions containing surfactant micelles and a silica precursor to form a variety of mesoscale and macroscale morphologies, dependent on solution conditions. These materials contain ordered arrays of surfactant micelles within the silica matrix and are of interest for the production of high-surface-area mesoporous materials with pores that are very uniform, with controllable sizes and shapes for use as separations media, catalyst supports, in sensor arrays, and to produce nanocomposites. In acidic syntheses a range of well-defined secondary structures are produced ranging from thin films at interfaces, to monodisperse spheres to "single crystals" depending on synthesis conditions [1]. We are interested in understanding and controlling the growth of mesostructured films on solid substrates since this geometry raises possible applications of these materials for sensors and functional surface coatings. Growing the film onto a substrate from solution allows surfaces of arbitrary shape to be coated and allows alignment of the mesoscale structure [2] and patterning of film growth on larger scales [3].

Previously one study was carried out on the in situ growth of mesostructured silica–surfactant films at solid interfaces using a single concentration of silica precursor/surfactant and the growth of the film at hydrophilic and hydrophobic interfaces was observed. In that work, despite the difference in interface structure causing observable differences in the initial surfactant–silica interactions with the substrate, the structure of the final film was unchanged [4]. In the current paper we report on results from experiments where films were grown from solution onto a hydrophilic interface using a range of silica precursor concentrations, at two different values of solution pH.

Experimental

The experiments were carried out using neutron reflectivity on the beamline SURF at ISIS (UK). A solid-solution cell was used consisting of a silicon block clamped against an O-ring sealed facing disk machined out of polyetheretherketone with a fluid reservoir 5-mm deep. The cell was temperature-controlled by circulating water at 28 °C through the top and bottom clamp sections. Silicon is relatively transparent to neutrons so the neutron beam can be transmitted through a silicon block to impinge upon the horizontal silicon–solution interface from above. The silicon block was pretreated to form a hydrophilic silica interface (by immersion in H_2SO_4–H_2O_2, followed by rinsing in water) and the reacting solution was injected into the cell underneath the block. The solution was a dilute acidic solution (0.2 M HCl in D_2O) of a silica source, tetramethoxysilane (TMOS) and the surfactant template cetyltrimethylammonium bromide (CTAB) in D_2O. Molar ratios of TMOS to CTAB to HCl to D_2O of X:1.53×10^{-3}:3.66×10^{-3}:1 were used and the TMOS molar ratio was varied between 1.66×10^{-2} and 2.76×10^{-3} (i.e. 0.80 and 0.133 M solution concentration). In a separate set of experiments the TMOS molar ratio was held constant at 1.10×10^{-2}, while the ratio of HCl was varied from 3.66×10^{-3} to 1.10×10^{-2} (0.2 M–0.6 M HCl). Time-resolved specular neutron reflectivity data were collected in 15-min shots to follow the formation of the first-order peak from the ordered mesostructure. Film growth occurs over 30 min to 1 h after a lengthy induction period of several hours, so 15-min intervals were sufficient to provide time-resolved information on the processes occurring at the silica-solution interface. The data were scaled against the pattern of D_2O collected in the same cell prior to injection of the reacting solution.

It is possible to compare the peak positions of different runs and the time required for the peak to grow as a function of solution conditions. This though only accesses part of the available information; to understand the interfacial processes more fully it is necessary to fit a model to the measured reflectivity data. This was performed via the optical transfer method [5], as described in Ref. [6]. This method depends on dividing the interface into a series of slabs parallel to the interface typically with three adjustable parameters: layer thickness (d/angstroms), scattering length density (SLD/per square angstrom) and a Gaussian roughness between layers. From an initial starting point the calculated reflectivity is compared with the dataset and a least-squares minimisation method is the used to find the best fit to the data. In all cases the simplest model which adequately described the data was chosen. This method is best suited to systems with a smaller number of layers; it has therefore been used primarily to study the interfacial structure prior to film growth.

Results and discussion

The time evolution of a typical set of reflectivity profiles is shown in Fig. 1. From this it can be seen that from the earliest patterns collected a broad fringe appears in the data which changes little over the first 2–3 h of the reaction during the induction period. At the end of the induction period, a diffraction peak, positioned on top of the fringe, grows in rapidly, reaching a constant intensity after roughly 30 min. The fringe is due to the formation of surfactant and silica layers at the solid interface during the induction period, while the peak is similar to that observed in silica–surfactant films grown at the air–water interface. In the air–water interface case the composite film which develops has been shown to have a structure composed of elongated cylindrical micelles arranged in a hexagonally close-packed array with the long axis aligned parallel to the air–solution interface [7, 8]. It is likely that in the present case a similar structure has formed, although from the specular reflectivity patterns,

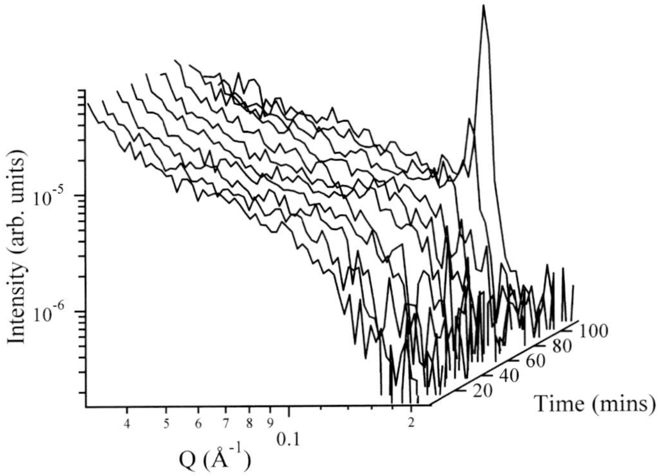

Fig. 1 A series of time-resolved reflectivity patterns taken with an incident angle of 1.0° for a solution at 0.3 M HCl with a tetramethoxysilane (*TMOS*)-to-cetyltrimethylammonium bromide (*CTAB*) ratio of 3.60

which give information only on the film structure in the direction perpendicular to the substrate, we can only say that a layered structure forms at the interface with a repeat unit of 46.5 Å. From the peak width, which is limited by instrumental factors, using a Scherrer analysis we have at least 16 layers of alternating silica and surfactant in the film which deposits rapidly at the end of the induction period. From our current solid–liquid cell data we can obtain both direct information on the length of the induction period (i.e. the time until the peak appears) under a given set of experimental conditions

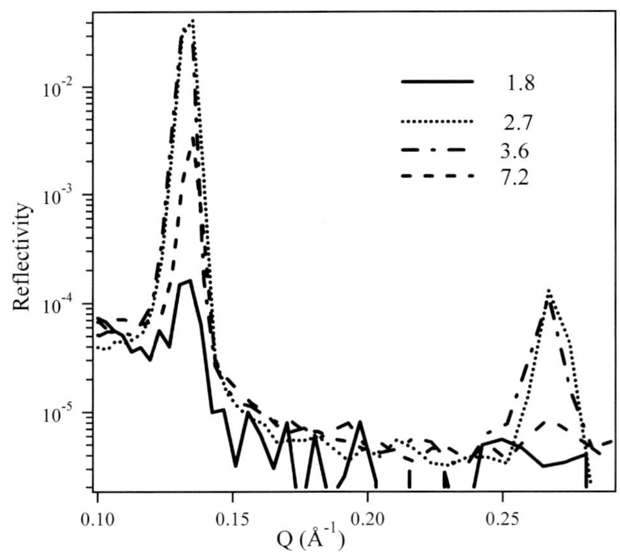

Fig. 2 Final reflectivity patterns from films grown over a range of TMOS-to-CTAB ratios at 0.2 M HCl (error bars not shown to aid in clarity)

and also, through fitting of the underlying fringe, information on the layering that develops during the induction period.

Prior to fitting any of the datasets we note from Fig. 2 that changing the TMOS-to-CTAB ratio did not change the position of the diffraction peaks in the final film. We therefore assume that at the TMOS-to-CTAB ratios where a film did form that the final structure was essentially the same in all cases. It can be seen from this figure that while the peak positions are the same the absolute intensities of the peaks are not the same for all ratios. At ratios of 2.84 and 3.80 the intensities are the same, there is then a decrease in intensity when the ratio is increased to 7.2 and finally the lowest intensity is observed when the ratio is 1.9. A mesoporous film did not grow when the ratio was 10.8, and an example dataset from this solution is not included on the graph.

Some examples of the fitting of the reflectivity data from a sample with a TMOS-to-CTAB ratio of 7.2 (0.2 M HCl) are shown in Fig. 3. The data and the fit for this particular system when the solution was first placed in the cell and just as the diffraction peak is appearing are displayed in Fig. 3a. It is important to note that there is very little change in the underlying shape of the reflectivity dataset, a fact that is highlighted when the real space SLD profile in Fig. 3b is examined. Here we see the first layer (surfactant) has not changed in thickness during the induction time, although some subphase has been excluded, as evidenced by the decreased SLD. The major difference between the two SLD profiles is that a second layer with a SLD not too different from that of the subphase has formed at the surfactant subphase interface. If one continues fitting as the peak begins to grow a two-layer model (virtually identical to that seen here) continues to fit the underlying structure. This suggests that processes taking place during the induction time are predominantly solution based.

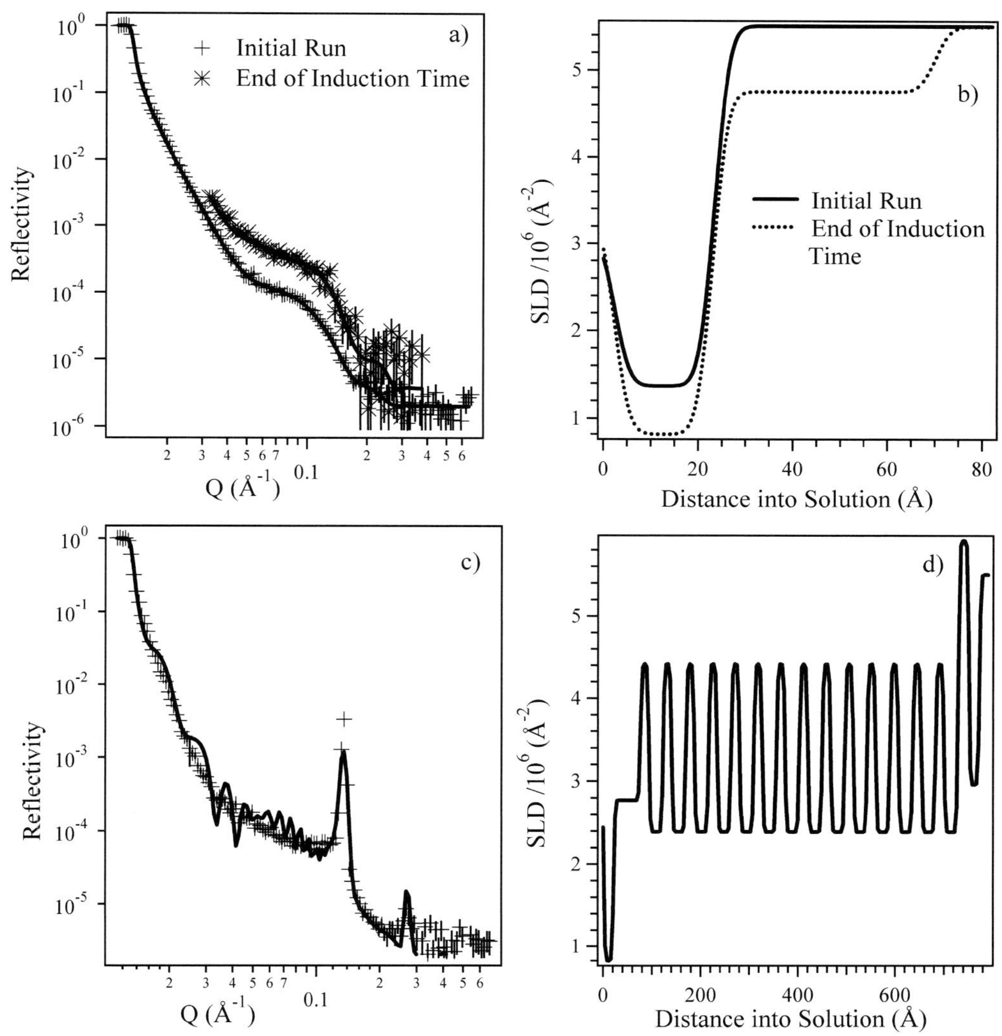

Fig. 3 Fitting of the reflectivity profiles for a solution at 0.2 M HCl and with a TMOS-to-CTAB ratio of 7.20. **a** The data (*symbols*) and the fit (*solid line*) for the initial dataset and immediately prior to the appearance of the diffraction peak, data offset for clarity. **b** The real space scattering length density profile corresponding to the fits in **a**. **c** The data (*symbols*) and the fit (described in the text) for the final dataset collected. **d** The real space scattering length density profile according to the fit in **c**

That the changes taking place during the induction time are solution-based is emphasised when the plots in Fig. 3c and d are examined. This shows the final dataset collected on this sample: there are two obvious diffraction peaks to be seen in the reflectivity. Using a 33-slab model we reproduced the peak positions and most of the intensity; the wiggles and undulations at the lower-Q region of the fit are Keissig fringes and are not seen in the experimental data owing to a combination of interfacial roughness and instrumental resolution. This fit was made by fitting the thickness and SLD of the first and last two layers and then constraining alternating layers in the stack between these two extremes. The first two layers in Fig. 3d are virtually the same as that seen in Fig. 3b: the thicknesses are almost the same but there are some changes in density. The alternating layers then correspond to a surfactant-rich layer 30-Å thick with a SLD of 2.8×10^{-6} Å^{-2} and a silica–aqueous layer 17-Å thick with a SLD of 4.4×10^{-6} Å^{-2}. These values are very similar to that seen at the air–liquid interface [6, 8].

Table 1 Induction times and fitting parameters for film growth at a solid–liquid interface at a constant tetramethoxysilane (*TMOS*)-to-cetyltrimethylammonium bromide (*CTAB*) ratio of 7.2

Sample	Induction time (min)	Layer 1		Layer 2	
		Thickness (Å)	Scattering length density	Thickness (Å)	Scattering length density
0.2 M HCl	438	21	0.8	47	4.8
0.3 M HCl	110	34	1.4	46	4.6
0.4 M HCl	6	26	0.5	39	4.6
0.6 M HCl	<4	18	0.7	–	–

Increasing the acid concentration in the reacting solution causes the diffraction peak corresponding to the ordered mesostructured silica–surfactant film at the solid interface to appear faster (Table 1). This decrease in the induction period for lower pH values is in line with our earlier work on the formation of similar films at the air–solution interface [9]. At lower pH, further from pH 2, the isoelectric point for silica, the condensation rate of TMOS increases, increasing the rate of the self-assembly processes. The thickness and scattering length density of the adsorbed layer of surfactant is roughly constant at all acid concentrations except for 0.2 M HCl, where the layer is significantly thicker. However, as the acid concentration increases, the second layer of silicate deposited on top of the adsorbed surfactant becomes thinner, and at 0.6 M HCl this layer does not form, with the diffraction peak of the multilayered film forming directly on the fringe belonging to the surfactant layer. This implies that the formation of the silica layer at the headgroups of the adsorbed surfactant layer is a slow process, in competition with the processes forming the mesostructured film. Where the film grows quickly, the interface is covered in the organised mesophase before the slower process has a chance to complete the formation of a second, well-defined layer.

Table 2 Induction times and fitting parameters for film growth at a solid–liquid interface at constant acid concentration, where the *top line* of *numbers* for each TMOS-to-CTAB ratio represents the initial adsorbed state and the *bottom row* either the last run if there was no film growth or the structure at the point of film growth

TMOS-to-CTAB molar ratio	Induction time (min)	Layer 1		Layer 2	
		D (Å)	Scattering length density/10^{-6} (Å^{-2})	D (Å)	Scattering length density/10^{-6} (Å^{-2})
0.2 M HCl					
10.8	–[a]	15	0.5	47	4.7
		15	1.0	50	4.5
7.2	438	21	1.4	–	–
		21	0.8	47	4.8
3.6	425	18	0.9	–	–
		18	0.1	52	4.6
2.7[b]	<180	–[b]	–[b]	–[b]	–[b]
		19	0.5	47	4.2
1.8	<530	21	0.8	–	–
		17	0.4	–	–
0.3 M HCl					
7.2	110	32	1.1	–	–
		34	1.4	46	4.6
3.6	91	32	0.9	–	–
1.8	152	28	0.9	–	–
		22	1.0	–	–

[a] No peak grew in this sample
[b] For this sample no initial run was performed

In experiments where the acid concentration is constant but the TMOS concentration is changed, fitting of the reflectivity fringe underlying the diffraction peak shows a decreasing thickness of the silica (second) layer with decreasing TMOS concentration, but an increasing thickness of the first layer of adsorbed surfactant, see Table 2. The decreasing silica layer thickness is easily understood as a lack of sufficient free silica in solution to allow formation of the secondary layer. The increasing thickness of the surfactant layer may be due to the fact that as the TMOS concentration is decreased there will also be less of the TMOS hydrolysis product, methanol, present in the reaction mix. Methanol is known to hinder formation of CTAB micelles owing to the greater solubility of the surfactant in this solvent than in water [10], and so it may also hinder adsorption of the surfactant to the silica interface.

The induction period, however, varies in an unusual manner with increasing TMOS concentration. As the TMOS concentration in the reacting solution is increased up to 0.199 M the peak due to the formation of the ordered film occurs more quickly, as expected. However, further increases in TMOS concentration after this point causes the appearance of the peak to be increasingly delayed, and, at the highest concentration studied, only the fringe due to the adsorbed surfactant layer was observed at the interface for the entire period studied (16.25 h). This behaviour was observed at both of the acid concentrations studied – 0.2 M HCl and 0.3 M HCl. The data are plotted in Fig. 4 for the two acid concentrations, shown as induction time versus silica - to-surfactant ratio. The most well-ordered films, showing two or more diffraction peaks in the final reflectivity patterns, also form with the shortest induction periods (Fig. 2, Table 2), consistent with observations at the air–liquid interface [9].

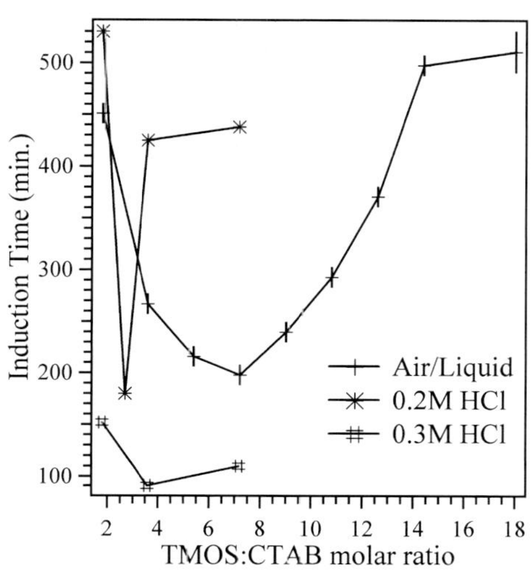

Fig. 4 Induction time at the solid–solution interface for 0.2 and 0.3 M HCl and corresponding data for the appearance of the first-order diffraction peak at the air–solution interface [9] for films grown with a HCl concentration of 0.2 M

In our recent experiments on films grown at the air–solution interface we have also observed that the formation mechanism is strongly concentration-dependent with a horse-shoe or U-shaped film formation time dependence on the silica-to-surfactant ratio also shown in Fig. 4. At intermediate silica-to-surfactant ratios ($\sim 6 < \text{TMOS:CTAB} < 10$) the film grows via the nucleation of mesostructured particles in the bulk solution which rise and pack at the surface to form the film. At high and low ratios, film growth occurs at the surface with the mesostructure growing down from the interface [11]. This changing mechanism results in a time for film formation which shows a U-shaped dependency with the film forming quickly at the intermediate ratio, in the particle-driven film formation region and more slowly at high and low ratios where the film forms by accretion of layers at the interface [9]. The experiments have encompassed the range of TMOS-to-CTAB ratios where a change in the mechanism from particle-driven to surface-driven would be expected at the air–liquid interface. While the induction time changes and the film ordering varies in line with that seen at the air–liquid interface there is no evidence of a changes of mechanism. We do find that for a TMOS-to-CTAB ratio of 10.8 that no film formed on the timescale of the experiment. We propose that this is because the cell is sealed and that the surface-mediated mechanism has been "turned off" in a system where there is no access to air and therefore no surface evaporation is possible.

From the current set of data on film formation at the solid–solution interface it seems that the particle-driven film formation mechanism dominates across all TMOS-to-CTAB ratios investigated. The diffraction peaks corresponding to ordered mesostructured films appear rapidly at the interface at the end of the induction period without changing the structure of the underlying broad fringe which is due to adsorbed surfactant at that interface. This implies that there is no slow accretion of layers forming the film, but that mesostructure formation is occurring elsewhere in the solution and it is the arrival of these preformed particles at the interface, packing to form a surface layer, that causes the appearance of the diffraction peak. Interestingly the minimum in the U-shape curve is offset to lower silica-to-surfactant ratios in the case of the solid–solution interface than is observed at the air–solution interface but the induction periods are longer at equivalent pH. This may be due to the complete absence of evaporation in the case of the closed solid interface cell, meaning that methanol cannot escape from the system by evaporation, as occurs in open air–solution interfaces, and so hinders micelle formation and slows the self-assembly of the mesostructured particles in solution which are the film precursors. However, methanol in solution also decreases the solubility of the silica [12], leading to phase separation and mesostructured particle formation at lower silica concentrations. The films themselves, when removed from the cell, are translucent when thin, but increasingly opaque if allowed to grow thicker. They are easily removed from the hydrophilic surface by washing in water and chloroform; however, since these films were grown for only short periods at room temperature the degree of silica polymerisation would be expected to be low, and the fragility of the films is therefore not unexpected.

Conclusions

Film growth at the air–solution interface proceeds via a solution or interfacially driven mechanism [11]. Films grown at a hydrophilic solid–solution interface in an enclosed cell, where the interfacial mechanism is disabled, appear to grow solely by a mechanism involving the formation of mesostructured particles in solution which pack at the interface to form the film. The formation of a layer of adsorbed surfactant on the hydrophilic interface occurs immediately upon introduction of the reactant solution into the cell, and formation of a subsequent layer of silica on this adsorbed layer occurs only for high silica concentrations, and low acid concentrations where the induction period is long. The ordered mesostructure appears rapidly at the end of the induction period, and does not perturb the underlying layer structure formed during the induction period. The time-dependence for diffraction-peak appearance shows a U-shaped dependence on the silica-to-surfactant ratio, similar to that observed for films grown at the air–solution interface.

The differences between the air–liquid and solid–liquid film growth systems can be understood in the following manner. We suggest that at the air–liquid interface the interfacially mediated film growth occurs over the range of TMOS-to-CTAB ratios at which a film has been observed to grow, but there is a restricted range over which the solution-mediated mechanism is much faster and solution processes dominate during the induction time, to the extent that the surface growth is not seen at all. With the solid–liquid interface system the film growth mechanism is restricted to solution processes only. At a high TMOS-to-CTAB ratios film growth was not observed as the solution process is not favoured and the interfacial mechanism is not operating in this experimental configuration.

References

1. (a) Yang H, Coombs N, Sokolov I, Ozin GA (1996) Nature 381:5035; (b) Yang H, Coombs N, Ozin G (1997) Nature 386:692; (c) Aksay IA, Trau M, Manne S, Honma I, Yao N, Zhou L, Fenter P, Eisenberger PM, Gruner SM (1996) Science 273:892
2. Miyata H, Kuroda K (2000) Chem Mater 12:49
3. Yang H, Coombs N, Ozin GA (1997) Adv Mater 9:811
4. Holt SA, Reynolds PA, White JW (2000) Phys Chem Chem Phys 2:5667
5. Parratt LG (1954) Phys Rev 9:359
6. Brown AS, Holt SA, Reynolds PA, Penfold J White JW (1998) Langmuir 14:5532
7. Brown AS, Holt SA, Dam T, Trau M, White JW (1997) Langmuir 13:6363
8. Brennan T, Roser SJ, Mann S, Edler KJ (2003) Langmuir 19:2639
9. Edler KJ, Brennan T, Roser SJ, Mann S, Richardson RM (2003) Microporous Mesoporous Mater 62:165
10. Anderson MT, Martin JE, Odinek JG, Newcomer PP (1998) Chem Mater 10:1490
11. Brennan T, Roser SJ, Mann S, Edler KJ (2002) Chem Mater 14:4292; Brennan T, Hughes AV, Roser SJ Mann S Edler KJ (2002) Langmuir 18:9838
12. Iler RK (1979) The chemistry of silica. Solubility, polymerization, colloid and surface properties, and biochemistry. Wiley, New York

Progr Colloid Polym Sci (2004) 128: 175–177
DOI 10.1007/b97059

Luis Fernando Pacheco
Débora Braga Vieira
Felipe Manoel Correia
Ana Maria Carmona-Ribeiro

Interaction between cationic bilayers and *Candida albicans* cells

L.F. Pacheco · D.B. Vieira
F.M. Correia
A.M. Carmona-Ribeiro (✉)
Departamento de Bioquímica, Instituto de Química, Universidade de São Paulo, Caixa Postal 26077, 05513-970 São Paulo, SP, Brazil
e-mail: mcribeir@iq.usp.br
Tel.: +55-1130913810 ext. 237
Fax: +55-11-38155579

Abstract The high-affinity adsorption of a cationic dioctadecyldimethylammonium bromide (DODAB) bilayer to the yeast form of the fungal cell *Candida albicans* was studied as a function of DODAB dispersion method, DODAB concentration and cell number density (n) in pure water. As previously observed for bacteria, the number of liposomes adsorbed per cell decreased as a function of n, possibly owing to increased cell aggregation and reduction of the total cell area with n. The DODAB limiting adsorption was 1.9×10^9 and 9.5×10^8 molecules per cell for bilayer fragments and large vesicles, respectively, in water solution; the extraattractive hydrophobic interaction between hydrophobic regions of biomolecules and hydrophobic borders of bilayer fragments possibly accounting for this result. Furthermore, at 1 mM cationic lipid or surfactant, the percentage cell phosphate leakage induced by bilayer fragments, large vesicles or cetyltrimethylammonium bromide micelles was small, about 4–10%, and was consistent with previously published data on the absence of bacterial cell lysis. Electrophoretic mobilities for cells in the presence of both DODAB dispersions were consistent with desorption of large vesicles from cells; this desorption was absent for the bilayer fragments.

Keywords Dioctadecyldimethylammonium bromide bilayers · *Candida albicans* · Cell leakage · Adsorption · Electrophoretic mobility

Introduction

Synthetic cationic bilayers have received increased attention as potential drug, gene and vaccine delivery systems owing to their potential capability of interacting with oppositely charged biomolecules and biological structures as cell membranes [1]. Dioctadecyldimethylammonium bromide (DODAB) has been shown to be particularly interesting regarding drug delivery to microorganisms owing to its antimicrobial activity [2, 3]. Furthermore, DODAB bilayer fragments solubilized antifungal drugs such as amphotericin B [4] and miconazole [5] and improved colloid stability of hydrophobic drug particles [5] allowing also antifungal therapy in vivo [6]. In this work, physicochemical aspects of the DODAB/*Candida albicans* interaction in pure water were evaluated and found to be consistent with previous data for the interaction between DODAB and bacteria.

Experimental

C. albicans exponential growth was stopped at about 10^8 cells/mL. DODAB was dispersed above the main phase-transition temperature for the bilayer (43–45 °C) either ultrasonically to yield open bilayer fragments (BF) or by vortexing to yield large and closed vesicles (LV); 74- and 500-nm mean diameter, respectively, from dynamic light scattering. Electrophoretic mobility (EM) was measured using a Rank Brothers microelectrophoresis apparatus with a flat cell (25 °C/ ± 60 V). The percentage phosphate leakage from the cells (2×10^8 cells/mL) was determined at 1 mM

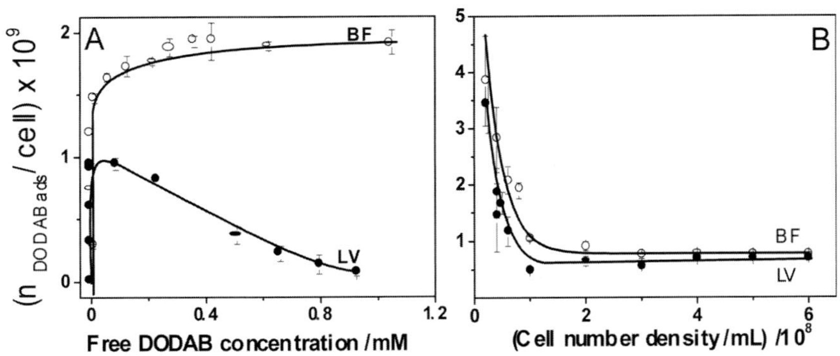

Fig. 1 a Dioctadecyldimethylammonium bromide (*DODAB*) adsorption isotherm from open bilayer fragments (*BF*) or large and closed vesicles (*LV*) onto *Candida albicans*, 1×10^8 cells/mL in water at 25 °C. **b** DODAB adsorption as a function of cell number density at 1 mM DODAB

DODAB, cetyltrimethylammonium bromide (CTAB) or sodium dodecyl sulfate (SDS) after 1-h interaction by centrifuging and analysing inorganic phosphate in the supernatant.

Results and discussion

DODAB adsorption isotherm on cells from DODAB BF showed high affinity and nonreversible character in contrast to those obtained from LV (Fig. 1a). Apparently, BF presented a deeper penetration in the dense forest of biomolecules on the cell surface than LV possibly owing to their smaller sizes and to less reversible interaction with the cell surface than in the case of LV. Desorption of LV from the cells was favoured by the increase of free LV in the supernatant, suggesting that there was a significant intervesicle interaction of hydrophobic nature between the DODAB adsorbed and free LV (Fig. 1a).

At 1 mM DODAB, the effect of increasing the cell number density, *n*, was decreasing DODAB adsorption both for BF and LV; possibly, increased cell/cell aggregation with *n* accounted for this result. At any *n*, DODAB adsorption from BF was above that from LV (Fig. 1b).

The effect of DODAB concentration on electrophoretic mobility of *C. albicans* was similar for BF and LV over the low concentration regime but some significant differences occurred above 0.1 mM DODAB: mobilities were higher for the mixtures with LV than for those with BF (Fig. 2a).

The EM increased as a function of increasing DODAB concentration for DODAB LV in the absence of cells, as shown in Fig. 2b. The same curve over a range of low DODAB concentrations is shown in Fig. 2c. Over this range, LV mobilities linearly increased as a function of DODAB concentration (Fig. 2c). This effect was previously explained as due to LV/LV electrostatic repulsion accelerating mobilities in a crowded population of free large vesicles [7]. In order to eliminate this concentration effect on LV mobilities, the EM curve was extrapolated to a DODAB concentration of zero, a situation where infinite dilution would yield the absence of intervesicle interaction (Fig. 2c). However, when LV interacted with cells and cell mobilities were measured as a function of added DODAB it was not possible to dilute the LV/cells mixture without changing the adsorption equilibrium, so it was not feasible to eliminate the DODAB-induced increase in EM. Therefore, in the DODAB LV/cells mixtures (Fig. 2a), the higher EM values for LV as compared with BF over the highest DODAB concentration range could well be due to an increase in concentration of free LV. One should also

Fig. 2 a Electrophoretic mobility for *C. albicans* (10^6 cells/mL) as a function of DODAB concentration for BF and LV. **b** The effect of DODAB concentration on electrophoretic mobility of DODAB LV in pure water and in the absence of cells. **c** The same as in **b** over a range of very low DODAB concentrations to extrapolate electrophoretic mobility at infinite dilution, i.e., at a DODAB concentration of zero

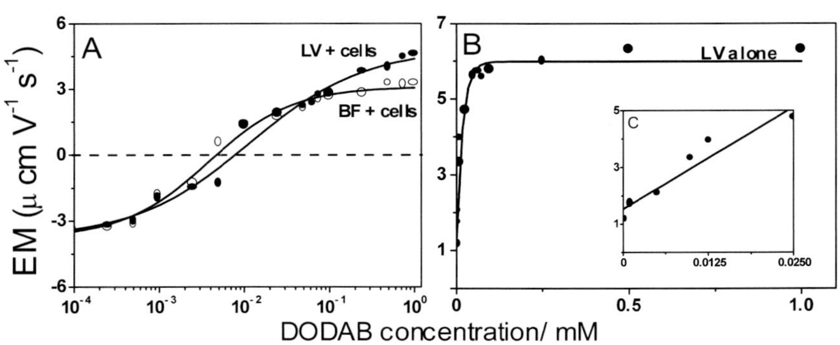

Table 1 Effect of dioctadecyldimethylammonium bromide (*DODAB*) large and closed vesicles (*LV*), DODAB open bilayer fragments (*BF*), cetyltrimethylammonium bromide (*CTAB*) and sodium dodecyl sulfate (*SDS*) at 1 mM amphiphile on the percentage leakage from internal contents of *Candida albicans* cells (2×10^8 cells/mL) after 1-h interaction. As a control, the cell suspension without surfactant was pelleted and the supernatant analysed. The percentages are mean values with a mean standard deviation within 5–10% of the mean value

Sample	Percentage leakage
C. albicans (control)	4
DODAB LV + *C. albicans*	12
DODAB BF + *C. albicans*	14
CTAB + *C. albicans*	14
SDS + *C. albicans*	8

recall the competitive adsorption obtained for LV (Fig. 1a); free LV possibly scavenges adsorbed LV from cells and also contributes to an increased free DODAB concentration and thereby to increased mobilities.

The percentage leakage of the internal contents of the cells at 1 mM amphiphile is shown in Table 1. The percentages for single chained, micelle-forming amphiphiles or double-chained cationic lipids were similar, amounting to 4–10% after discounting the control. This could be understood from the low amphiphile concentration employed, so 1 mM CTAB or SDS induced about the same *C. albicans* cell leakage as 1 mM DODAB BF or LV. At least these results were consistent with the leakage percentages of 5–8% previously determined for two different bacteria species, *E. coli* and *Staphylococcus aureus*, by using the same method at the same final amphiphile concentration for the three amphiphiles tested [8]. In fact, large leakage percentages such as 50–60% had been obtained previously for the micelle-forming amphiphiles only over the 4–10 mM SDS or CTAB concentration range, i.e., at concentrations well above the critical micellar concentration for each surfactant [8]. For DODAB, the preparation of bilayer dispersions above 4 mM DODAB by vortexing yielded a more turbid and viscous phase exhibiting a second phase-transition temperature of about 52 °C; possibly this additional phase was related to interconnected bilayers forming a network (Carmona-Ribeiro and coworkers, unpublished results). For this reason, in this work, the DODAB final concentrations in the mixtures were kept lower than 2 mM. Basically, the cells (10^6 cells/mL) attained zero charge over the micromolar range of DODAB concentrations but in the case of LV, LV desorption interfered with EM measurements causing overestimated mobilities of cells.

At 1 mM amphiphile, no significant differences between leakage from cells induced by micelle-forming and bilayer-forming amphiphiles could be detected.

Conclusions

1. There was a large affinity between DODAB BF or LV and *C. albicans* cells; however, upon increasing the DODAB concentration in the supernatant, LV desorption took place, whereas BF adsorption attained a maximum. This difference in behaviour was understandable from the smaller BF sizes and the larger hydrophobic interaction with cell surface components.
2. The effect of increasing cell concentration was decreasing DODAB adsorption possibly owing to a reduction of the available cell surface area for adsorption as a consequence of cell aggregation.
3. High affinity between DODAB dispersions and cells was reconfirmed from the effect of the DODAB concentration on EM; cells (10^6 cells/mL) attained zero charge over the micromolar range of DODAB concentrations but in the case of LV, LV desorption interfered with EM measurements, causing overestimated mobilities of cells.
4. DODAB adsorption on *C. albicans* cells was similar to that previously described for bacteria regarding high affinity and absence of cell lysis.

Acknowledgments Financial support from FAPESP and CNPq is gratefully acknowledged. L.F.P. thanks CAPES for a PhD fellowship. D.B.V. and F.M.C. were recipients of undergraduate FAPESP fellowships.

References

1. Carmona-Ribeiro AM (2003) Curr Med Chem 10:2425
2. Campanhã MTN, Mamizuka EM, Carmona-Ribeiro AM (1999) J Lipid Res 40:1495
3. Campanhã MTN, Mamizuka EM, Carmona-Ribeiro AM (2001) J Phys Chem B 105:8230
4. Vieira DB, Carmona-Ribeiro AM (2001) J Colloid Interface Sci 244:427
5. Pacheco LF, Carmona-Ribeiro AM (2003) J Colloid Interface Sci 258:146
6. Lincopan N, Mamizuka EM, Carmona-Ribeiro AM (2003) J Antimicrob Chemoth 52:412
7. Carmona-Ribeiro AM, Midmore BR (1992) J Phys Chem 96:3542
8. Martins LMS, Mamizuka EM, Carmona-Ribeiro AM (1997) Langmuir 13:5583

Progr Colloid Polym Sci (2004) 128: 178–183
DOI 10.1007/b97066

Sabrina Montero Pancera
Rosangela Itri
Denise Freitas Siqueira Petri

The effect of poly(ethylene glycol) on the activity, structural conformation and stability of yeast hexokinase

S.M. Pancera · D.F.S. Petri (✉)
Instituto de Química,
Universidade de São Paulo,
P.O. Box 26077,
05513-970, São Paulo, SP, Brazil
e-mail: dfsp@iq.usp.br
Tel: +55-11-30913831
Fax: +55-11-38155579

R. Itri
Instituto de Física,
Universidade de São Paulo,
P.O. Box 66318,
05315-970, São Paulo, SP, Brazil

Abstract In the dilute range (3 mg/L), the effect of poly(ethylene glycol) (PEG) on the enzymatic activity of hexokinase (HK) was investigated by means of spectrophotometry in the absence and presence of PEG400, PEG1500 and PEG4000 in tris(hydroxymethyl)aminomethane–HCl buffer. Comparing the values obtained in pure buffer with those in the polymer solutions, we found that the enzymatic activity increased 6% in the presence of PEG400 and decreased in the range 14–7% in the presence of PEG1500 and PEG4000, respectively. Circular dichroism measurements revealed that in the presence of PEG the α-helix content in HK decreased up to 15% in relation to that measured in pure water. Small-angle X-ray scattering measurements were performed in a higher concentration regime (10 g/L). Data analysis performed from small-angle X-ray scattering curves by means of the intraparticle distance distribution function yielded for HK in water solutions and in PEG400 solutions a radius of gyration, R_g, of (32 ± 1) Å, which is compatible with that obtained from the crystalline dimeric structure of HK. In contrast, in the presence of PEG1500 or PEG4000 large aggregates (without precipitation) were observed. In the presence of glucose and polymer, the aggregation of HK was always favored.

Keywords Poly(ethylene glycol) · Hexokinase · Activity · Small-angle · X-ray scattering

Introduction

Poly(ethylene glycol) (PEG) is a nontoxic water-soluble synthetic polymer, which is widely used in biotechnological processes, in chemical and biomedical industries [1]. Although it has been considered as an inert material [1–3], PEG might strongly influence the enzymatic activity of some enyzmes. The activity of phosphofructokinase and glyceraldehyde-3-phosphate dehydrogenase in the presence of PEG was, respectively, 1.3 and 1.55 times higher than in water [4]. Recently [5], the effect of PEG on the activity of glucose-6-phosphate dehydrogenase (G-6-PDH) in the oxidation of glucose-6-phosphate (G-6-P), using β-nicotinamide adenine dinucleotide phosphate ($NADP^+$) as a coenzyme was investigated. The presence of 400 and 4,000 g/mol PEG led to an increase of 20% in the enzymatic activity in relation to that measured in pure buffer. It was shown by calorimetric measurements that PEG increases the enzymatic activity owing to favorable interactions between the coenzyme $NADP^+$ and PEG and not owing to the interactions between G-6-PDH and PEG.

The effect of PEG on the protein behavior has been attributed to protein hydrophobicity [6–9], water activity [10, 11], steric exclusion [12–14] and osmotic stress [15–18]. Timasheff and coworkers [12–14] proposed that the presence of PEG favors the stability of native proteins owing to steric exclusion and preferential hydration, but PEG can also contribute to the aggregation of denaturated proteins due to hydrophobic interactions. The osmotic stress method measures changes in the number of water molecules associated with single macromolecules undergoing conformational changes [15–18]. This method was successfully applied to explain how PEG of different molecular weights and water affect glucose binding to hexokinase (HK). This has a site in a cleft whose domains close on glucose [21], and the enzyme discriminates against water as a substrate [22]. Water activity affects both glucose equilibrium and HK turnover in such a way that the affinity for glucose increases with decreasing water activity [18]. In the presence of PEG, this effect is even stronger. The number of water molecules released when HK closes the cleft increases from 50 to 326 per HK molecule with increasing PEG molecular weight [16]. These studies performed in the HK concentration range of 0.0145 g L^{-1} [18] to 0.0384 g L^{-1} [16] provided remarkable understanding of the effect of PEG on the enzymatic activity mechanism. The present work deals with the influence of PEG on the enzyme activity, conformation and stability of HK in very dilute (0.0029 g L^{-1}) and concentrated (10.0 g L^{-1}) ranges. In the very dilute regime, the enzymatic activity was determined by means of UV–vis spectrophotometry and the structural conformation was measured by circular dichroism (CD). In the concentrated regime, the structural conformation and stability were investigated by small-angle X-ray scattering (SAXS). HK was chosen owing to its key role in glucose metabolism (glycolysis), which allows this enzyme to be used in the construction of biosensors for glucose [23].

Materials and methods

Materials

PEG samples with nominal molecular weight of 400, 1,500 and 4000 Da were purchased from Merck (Darmstadt, Germany). $NADP^+$ (N0505), adenosine 5′-triphosphate (ATP, A3377), β-D-(+)-glucose (G5250), G-6-PDH type IX from bakers yeast (G4134), tris(hydroxymethyl)aminomethane (Tris)–HCl buffer of pH 7.5 with $MgCl_2$ (5 mM) and HK type III from bakers yeast (H5000) were obtained from Sigma (St. Louis, USA) and used without further purification.

HK, in the presence of Mg^{2+}, catalyses the transfer of the γ-phosphoryl group from ATP to the hydroxyl group on carbon 6 of some hexoses, mainly glucose [24, 25]. HKs from different species differ in molecular weight and tissue distribution, and the enzyme often exists as a mixture of isoenzymes that also differ in kinetic characteristics and molecular weight [25]. In the present work HK from yeast, which has three isoenzymes, P_I (or A), P_{II} (or B) and glucokinase, was used. In solution and at pH 5.5–6.5, yeast HK B is a dimer of identical subunits that amount to 104 kDa [25–27]. These dimers tend to dissociate into monomers as the ionic strength, temperature or pH is raised but both forms are enzymatically active [27].

Enzyme activity measurement

The HK activity assay is based upon the reduction of $NADP^+$ to NADPH, which absorbs at a wavelength of 340 nm, through a coupled reaction with G-6-PDH:

$$\mathrm{D-Glucose + ATP \rightarrow G-6-P + ADP}, \quad (1)$$

$$\mathrm{G-6-P + NADP^+ \rightarrow gluconate-6-P + NADPH + H^+}. \quad (2)$$

HK activity was determined by spectrophotometry using a Beckmann Coulter DU 640 UV–vis spectrophotometer at 25 °C by measuring the increase in absorbance at a wavelength of 340 nm as a function of time, according to the method described by Bergmeyer [28]. The volume and concentration of each reagent mixed in the spectrophotometric cell were the following: 15 μL G-6-PDH solution (0.12 g L^{-1}), 5 μL 0.5 mM $NADP^+$ solution, 15 μL 0.1 mol L^{-1} ATP solution, 50 μL 1 mol L^{-1} glucose solution and 10 μL HK solution (2.5×10^{-6} mol L^{-1}). The total volume of 850 μL was achieved with buffer and PEG, so that the PEG concentration was kept constant as 2.94×10^{-3} mol L^{-1}. All solutions used in this experiment were prepared in Tris–HCl buffer (50×10^{-3} mol L^{-1}, pH 7.5) with 5×10^{-3} mol L^{-1} $MgCl_2$. The final HK concentration in all the samples amounted to 5.8×10^{-8} mol L^{-1} or 0.0029 g L^{-1}. Higher concentrations of HK resulted in too much NADPH being produced, thus overloading the detector. Spectrophotometric measurements were performed immediately after the sample preparation. The activity was calculated considering one HK unit (U) as the amount of enzyme catalyzing the phosphorylation of 1 μmol of glucose per minute under the assay conditions. Using the Beer–Lambert equation, the amount of NADPH formed per minute can be obtained from the linear range in the absorbance curve (340 nm) as a function of time, considering the NADPH molar absorptivity [27] as $\varepsilon = 6.22$ mol^{-1} L cm^{-1} and the optical pathway length as 1 cm. The absorbance was measured at time intervals of 5 min. The activity value reported for each system represents the mean value of triplicates.

Circular dichroism

CD spectra were recorded with a Jobin-Yvon CD6 spectropolarimeter at (25 ± 1) °C, using a 2-cm quartz cell in the region of the characteristic absorption bands of the amide groups, between 190 and 260 nm. Data were collected using a scan speed of 100 nm min^{-1} and a sensitivity of 0.01°. HK solutions in the absence and presence of PEG400, PEG1500 or PEG4000 were prepared in deionized water at HK and PEG final concentrations of 5.8×10^{-8} and 2.94×10^{-3} mol L^{-1}, respectively. These concentration values correspond to those used in the enzymatic activity measurements. The α-helix content was estimated by using Eq. (3) [29, 30]:

$$[\Phi]_{220\mathrm{nm}} = \frac{-40000}{\left(1 - \frac{k}{n}\right)}, \quad (3)$$

where $[\Phi]_{220\ \mathrm{nm}}$ is the molar ellipticity at 220 nm, k is a wavelength-dependent constant and n is the number of residues. For the assessment of the HK α-helix content values of $k = 2.6$ [30] and $n = 457$ [26] were considered.

Small-angle X-ray scattering

SAXS experiments were carried out at National Laboratory of Synchrotron Light (LNLS, Campinas, Brazil) at room temperature of (25 ± 1) °C, with radiation of wavelength $\lambda = 1.608$ Å. The samples consisted of HK/water and HK/water/PEG (20 wt %), with HK at a concentration of 10 g L^{-1} (pH 6.5), in the absence and presence of glucose (10 g L^{-1}). PEGs with different molecular weights (400, 1,500 and 4,000 Da) were investigated. The measurements of the glucose-containing samples were done after 5 h of interaction in order to observe possible kinetic effects due to glucose binding. It was not possible to measure solutions with protein concentrations lower than 10 g L^{-1} (10^{-4} mol L^{-1}) because the particle volume fraction was too low to give measurable X-ray scattering (reasonable signal-to-noise ratio). The data acquisition of freshly prepared samples took typically 15–30 min. In the presence of glucose the measurements were performed after 5 h of incubation. The experimental data were corrected for detector homogeneity (one-dimensional position-sensitive detector) and normalized by taking into account the decrease of the X-ray beam intensity during the experiment. The parasitic background (water) was subtracted considering the sample's attenuation.

It is well known that the scattering intensity, $I(q)$, in the absence of interference effects, is described by a Fourier transform of the pair distance distribution function [31–33], $p(r)$, that provides information about the shape of the scattering particle, and is directly related to its radius of gyration. Moreover, the particle maximum dimension, D_{max}, can be evaluated at the r value for which $p(r)$ goes to zero. In the current work, we made use of the Gift package, developed by Glatter [33] to calculate $p(r)$ from the SAXS curves.

Results and discussion

Influence of PEG on the enzymatic activity

In order to estimate the influence of PEG on the enzyme activity four different media were considered: (1) pure Tris–HCl buffer and solutions containing (2) PEG400, (3) PEG1500 and (4) PEG4000. The catalytic action of HK followed the Michaelis–Menten model. The enzymatic activity of HK was calculated from the slopes in the very early stages and was expressed in terms of the number of moles of NADPH formed per minute, as shown in Table 1. Comparing the enzymatic activity measured in pure buffer with that in the presence of PEG400, we observed an increase of about 6%. In the presence of PEG1500 or PEG4000 the enzymatic activity decreased 14% and 7%, respectively. Although these opposite changes in the enzymatic activity are close to the reproducibility margin of error (5–10%), they indicate a general trend. Moreover, they are in agreement with the osmotic stress results, which indicated that the number of water molecules released upon glucose binding increased from 50 to 326, when the PEG molecular weight increased from 300 to 1,000 g/mol [16]. When the PEG molecular weight amounts to 1,000 g/mol, the chains are excluded from larger aqueous compartments around the protein. Larger chains ($M_w > 1{,}000$) are supposed to dehydrate the protein before the glucose binding, reducing the water activity and leading to a reduced number of released water molecules upon glucose binding.

Comparing the results in Table 1 with those reported for G-6-PDH [5], also an enzyme from the glycolytic pathway, we observed different behavior. The activity of G-6-PDH increased similarly in the presence of PEG400 or PEG4000, i.e., the chain length of PEG was an inert parameter. Recently [3] spectroscopic studies on PEG–lysozyme interactions showed that regardless of the molecular weight, PEG exerted no influence on the enzymatic activity. It is well known that the enzymatic activity is related to the active site and the conformation of each enzyme, which are complex parameters; therefore, predicting the effect of PEG on the behavior of enzymes in a general way is not a trivial task.

Table 1 Mean rate of reduced nicotinamide adenine dinucleotide phosphate (*NADPH*) formation due to hexokinase activity and hexokinase α-helix content in the absence and in the presence of poly(ethylene glycol) (*PEG*)

	Mean rate (μmol NADPH L^{-1} min^{-1})	α-Helix content (%)
Tris(hydroxymethyl) aminomethane–HCl buffer	15,732 ± 797	–
Water	–	47
PEG400	16,676 ± 1,610	32
PEG1500	13,596 ± 1,088	37
PEG4000	14,674 ± 1,100	35

Circular dichroism

The comparison between CD spectra obtained for HK/water and for HK/water/PEG (Fig. 1) shows that the secondary structure of HK in PEG solution differs slightly from that of HK in water. Glucose was not added to the systems because its own optical activity would interfere in the measurement of the spectra of the enzymes.

Crystallized native yeast HK A has 35% α-helix, 28% β-sheet, 19% β-turns and 18% random structure [34]. HK can be classified as an α/β protein; in this case the

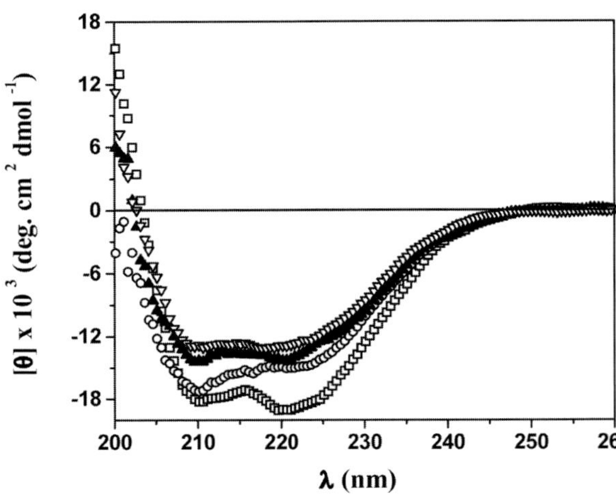

Fig. 1 Circular dichroism spectra for the systems hexokinase (HK)/water (*squares*), HK/water/PEG400 (*down triangles*), HK/water/PEG 1500 (*circles*) and HK/water/PEG4000 (*up triangles*), where *PEG* is poly(ethylene glycol)

intensities of the CD spectra related to the α-helix usually predominate compared with those of the β-sheet [34, 35]. The HK α-helix contents in water and in the presence of PEG estimated by using Eq. (3) are presented in Table 1. In the presence of PEG1500 and PEG4000 the decrease in the α-helix content in relation to that in pure water of 10% and 12%, respectively, might explain the reduction in the enzymatic activity in both cases. In the case of PEG400, the HK α-helix content decreased 15% in comparison with that in pure water, but a small increase in the enzymatic activity was observed. These findings indicate that there is no direct correlation between the variation of 10–15% in the HK α-helix content and the enzymatic activity mechanism.

Small-angle X-ray scattering

HK/water/PEG system

The $p(r)$ functions obtained from the SAXS curves of HK in aqueous solutions in the absence and presence of PEG1500 and PEG4000 are shown in Fig. 2; the corresponding theoretical fittings to the experimental data are shown in Fig. 3. The scattering curves, and hence $p(r)$, of HK/water/PEG 400 are identical to those observed for polymer-free solution and, therefore, are not shown. Some measurements were done with PEG of higher molecular weight (20,000 and 35,000 g mol^{-1}) but the solutions became turbid, probably owing to the presence of very large aggregates or denaturated HK (data not shown).

From Fig. 3a, it is clear that the scattering is due to the contribution of two distinct types of scatters in solution. In the inner-q region (marked by a dashed line in the figure), there is an accentuated increase in the intensity related to large protein aggregates; therefore, the $p(r)$ evaluation was performed without these data. The SAXS profile in the other q region is consistent with the scattering of the HK dimer, since the $p(r)$ analysis yielded values for R_g and D_{max} of 32 and 110 Å (Fig. 2), respectively. The R_g value of 32 Å obtained for the HK/water system, in the absence and presence of PEG400, is consistent with that found in the literature for the HK dimer [26, 36].

The scattering pattern of HK in PEG1500 (Fig. 3b) is rather different from the others: it presents a higher intensity and a much larger slope in the low-q region. This feature is the first evidence that HK in PEG1500 solutions presents larger structures, which must be attributed to the HK aggregation process. This finding is confirmed by a $p(r)$ analysis that resulted in values of $D_{max} = 350$ Å and $R_g = 120$ Å (Fig. 2). It should be remarked that these values could be underestimated,

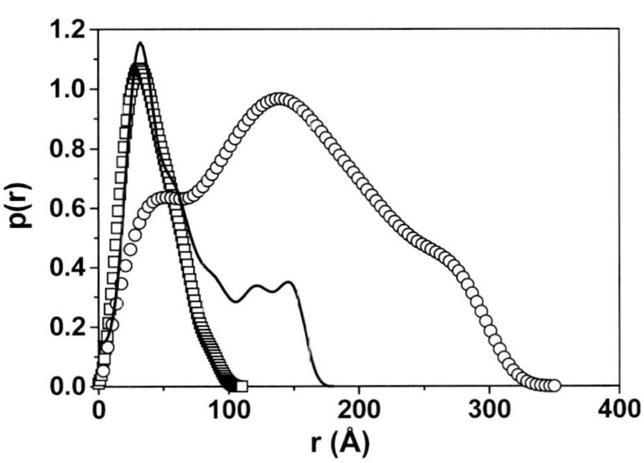

Fig. 2 The distance distribution functions $p(r)$ for the systems HK/water (*squares*), HK/water/PEG 1500 (*circles*) and HK/water/PEG4000 (*line*)

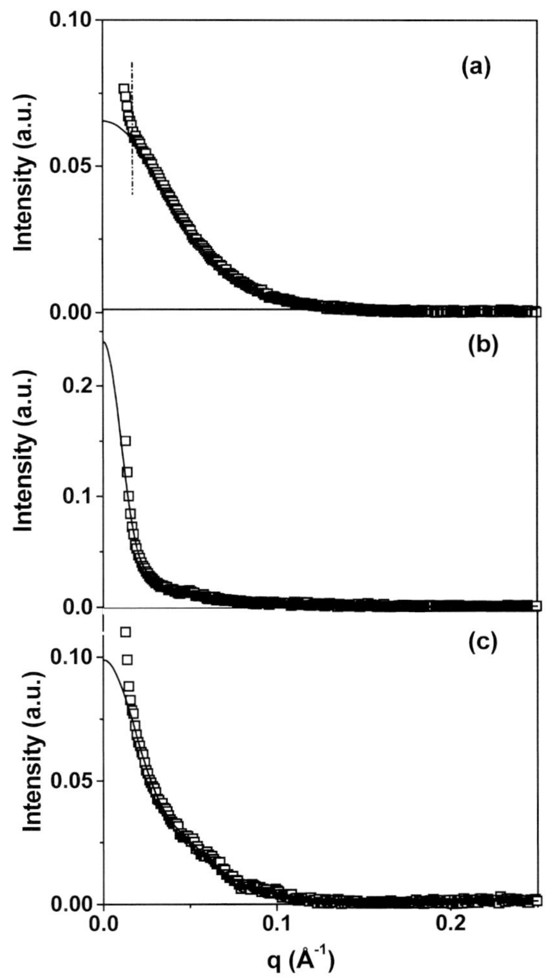

Fig. 3 The theoretical scattering intensities, obtained with the Glatter program (*solid lines*), along with the experimental data from **a** HK/water, **b** HK/water/PEG1500 and **c** HK/water/PEG4000

because larger dimensions are not accessed in our measurements owing to the resolution of the experimental setup. Moreover, the dimeric HK feature (high frequency of distances at $r \approx 30$ Å) does not prevail on the corresponding $p(r)$ function whose greater frequency of distances occurs at larger r values compared with the other systems studied (Fig. 2). Therefore, it seems that some enzyme unfolding also takes place under the influence of PEG1500, inducing, concurrently, the formation of larger HK aggregates in solution.

As far as the influence of PEG4000 on the enzyme solution is concerned, the corresponding SAXS curve (Fig. 3c) showed dimeric HK, as in distilled water and in PEG400, and some aggregates evidenced again by the larger slopes in the very low q region. This is confirmed by the $p(r)$ analysis that reflects the same dimeric HK behavior with low frequencies at large distances with $D_{max} = 180$ Å and $R_g = 57$ Å.

Considering that at the HK concentration of 10.0 g L^{-1} the dominant species are dimers, a figure of 6×10^{19} dimers 1,000 cm^{-3} can be estimated. Since the volume of one dimer can be calculated (using $R_g = 32$ Å) as 137.3×10^{-21} cm^3, 7.3×10^{21} dimers can occupy a volume of 1,000 cm^3 without overlapping. This approximation shows that there is free space among the dimers. PEG400 is probably small enough ($R_g = 4.5$ Å) [14] to occupy this available room without being excluded, while PEG1500 ($R_g = 7.6$ Å) and PEG4000 ($R_g = 24$ Å), being much larger, might be excluded from the interparticle region, leading to reduced attraction between the dimers [12–14].

The addition of glucose to the HK/water/PEG system

The scattering intensities obtained from HK/water in the absence (squares) and presence of glucose after 5 h (circles) are shown in Fig. 4a, whereas the corresponding $p(r)$ functions are displayed in Fig. 4b. Similarly to the system without glucose (Figs. 2, 3), the $p(r)$ evaluation was performed without the scattering data from the inner-q region (marked by a dashed line in the figure) owing to the presence of aggregates. The $p(r)$ function yielded evidence that when glucose is added to the HK/water solution, the HK R_g and D_{max} values decreased from 32 to 30 Å and from 110 to 85 Å, respectively. This effect is in agreement with the HK's cleft closure during glucose binding [15]. The SAXS curves obtained for HK/water/PEG400 in the presence of glucose (Fig. 4a, triangles) showed that PEG400 induced the aggregation of HK, as evidenced by the steep slope in the very low q range. This is interesting because this effect was not observed in the absence of glucose.

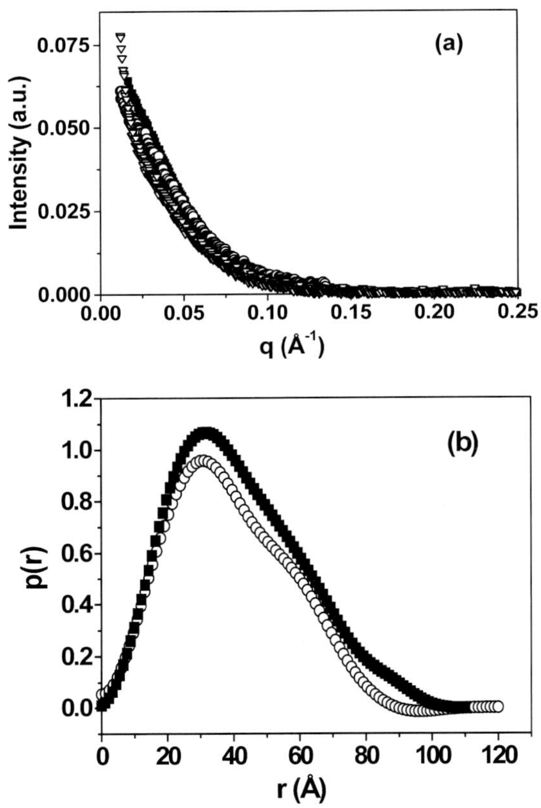

Fig. 4 **a** The scattering intensities and **b** the distance distribution functions for the systems HK/water (*squares*), HK/water/glucose (*circles*) and HK/water/PEG400/glucose (triangles)

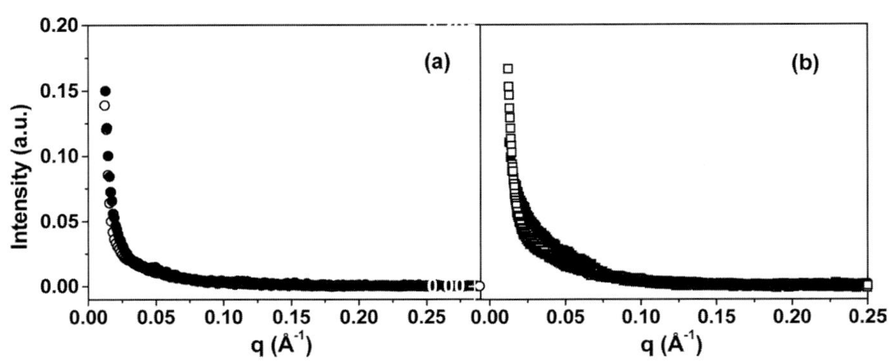

Fig. 5 The scattering intensities for the systems **a** HK/water/PEG 1500 (*closed circles*) and HK/water/PEG 1500/glucose (*open circles*) and **b** HK/water/PEG4000 (*closed squares*) and HK/water/PEG4000/glucose (*open squares*)

The SAXS curves for the HK/water/PEG1500 and HK/water/PEG4000 systems, respectively, in the presence and absence of glucose after 5 h of interaction are presented in Fig. 5a and b. In the absence of glucose, the polymer induced the aggregation of HK (Fig. 3). This aggregation is approximately the same in the presence of glucose and PEG1500, but is strongly intensified in the presence of PEG4000 and glucose, as evidenced by the steeper slope. Although the HK aggregation is evident, precipitation was not achieved. As a control, the PEG/glucose system (without HK) was prepared at the same concentrations and no phase separation was observed either.

Conclusions

In the very dilute range, the effect of PEG on the HK activity is marginal. Small changes in the HK secondary structure could not be directly correlated with the enzymatic activity. In the concentrated regime, SAXS results showed that in the absence of glucose PEG400 has no influence on the HK structure, while PEG1500 induced unfolding and aggregation and PEG4000 favored aggregation. In the presence of glucose, the typical R_g value for HK decreased by 2 Å, corroborating the cleft closure mechanism. The presence of polymer and glucose favored the aggregation and unfolding of HK.

Although PEG has been largely used in biotechnological process, the present results warn that each protein behaves in a different way in the presence of PEG. Predicting the effect of PEG on the behavior of proteins and enzymes over a broad concentration range is not trivial, because of their complex nature with variable contents of hydrophilic and hydrophobic residues on the surface and on the binding site.

Acknowledgements The authors thank the LNLS for the use of their facilities and S. Schreier for the use of CD equipment. This work was supported by research grants from FAPESP and CNPq.

References

1. Johansson G (1992) In: Harris M (ed) Poly(ethylene glycol) chemistry: biotechnical and biomedical applications. Plenum, New York, chap 5
2. Albertsson PA, Johansson G, Tjerneld F (1990) In: Asenjo JA (ed) Separation process in biotechnology. Dekker, New York
3. Malzert A, Boury F, Renard D, Robert P, Lavenant L, Benoit JP, Proust JE (2003) Int J Pharm 260:175–186
4. Grimonprez B, Johansson G (1995) J Chromatogr B 680:55–63
5. Pancera SM, Silva LHM, Loh W, Itri R, Pessoa A Jr, Petri DFS (2002) Colloids Surf B 26:291–300
6. Forciniti D, Hall CK, Kula MR (1991) Biotechnol Bioeng 38:986–994
7. Lee JC, Lee LLY (1981) J Biol Chem 256:625–631
8. Furness EL, Ross A, Davis TP, King GC (1998) Biomaterials 19:1361–1369
9. Jonsson M, Skepo M, Tjerneld F, Linse P (2003) J Phys Chem B 107:5511–5518
10. Kulkarni AM, Chatterjee AP, Schweizer KS, Zukoski CF (2000) J Chem Phys 113:9863–9873
11. Ninni L, Camargo MS, Meirelles AJA (1999) Thermochim Acta 328:169–176
12. Timasheff SN (2002) Biochemistry 41:13473–13482
13. Bhat R, Timasheff SN (1992) Protein Sci 1:1133–1143
14. Arakawa T, Timasheff SN (1985) Biochemistry 24:6756–6762
15. Parsegian VA, Rand RP, Rau DC (2000) Proc Natl Acad Sci USA 97:3987–3992
16. Reid C, Rand RP (1997) Biophys J 72:1022–1030
17. Parsegian VA, Rand RP, Rau DC (1995) Methods Enzymol 259:43–94
18. Rand RP, Fuller NL, Butko P, Francis G, Nicholls P (1993) Biochemistry 32:5925–5929
19. Patel PD, Russel WB (1989) J Colloid Interface Sci 131:192
20. Atha DH, Ingham KC (1981) J Biol Chem 256:12108–12117
21. Bennett WSJ, Steitz TA (1978) Proc Natl Acad Sci USA 75:4848–4852
22. Bennett WSJ, Steitz TA (1980) J Mol Biol 140:211–230
23. Eggins B (1997) Biosensors: an introduction. Wiley, New York
24. Albig W, Entian KD (1988) Gene 73:141–152
25. Cárdenas ML, Cornish-Bowden A, Ureta T (1998) Biochim Biophys Acta 1401:242–264
26. Schulze I, Gazith J, Gooding R (1966) Methods Enzymol 9:376
27. McDonald RC, Engelman DM, Steitz TA (1979) J Biol Chem 254:2942–2943
28. Bergmeyer HU (1984) In: Bergmeyer HU, Bergmeyer J, Grals M (eds) Methods of enzymatic analysis, vol 2, 3rd edn. VCH, Weinheim, pp 222–223
29. Chen Y-H, Chen JT, Chau KH (1974) Biochemistry 13:3350–3359
30. http://dicroprot-pbil.ibcp.fr
31. Guinier A, Fournet G (1955) Small-angle scattering of X-rays. Wiley, New York
32. Glatter O (1982) In: Glatter O, Kratky O (eds) Small-angle X-ray scattering. Academic, London
33. Orthaber D, Bergmann A, Glatter O (2000) J Appl Crystrallogr 33:218–225
34. Olmo R, Blanco MD, Teijón C, Socorro JM, Teijón JM (2002) J Inorg Biochem 89:107–114
35. Fasman GD (1996) In: Fasman GD (ed) Circular dichroism and the conformational analysis of biomolecules. Plenum, New York, chap 3
36. www.rcsb.org

Progr Colloid Polym Sci (2004) 128: 184–186
DOI 10.1007/b97069

Angela M.F. Monteiro
Nathalie Fa
Elizabeth P.G. Arêas
André Schröder
Carlos Marques

Gliadin effect on giant vesicle elastic modulus

Abstract Wheat gluten consists of a complex supramolecular protein assembly of very high molecular mass, with over a million daltons, that displays important viscoelastic and interfacial properties. Its involvement in an immunologically based nutritional disease, the celiac disease, is thought to be somehow related to nontrivial structural features of the protein complex, apparently involving gliadin, one of its fractions. Despite its medical relevance, the involvement of gliadin in the etiology of the celiac disease is nevertheless not understood in terms of the mechanisms involved. In this context, gliadin was investigated relative to its effect on fluid bilayer elasticity. Dioleoylphosphatidylcholine giant unilamellar vesicles (about 50-μm diameter) were produced by electric field pulsing over metal-covered plates. The technique of micropipette suction allowed vesicle invagination into a micropipette by applying a suction pressure. A video camera coupled to a microscope permitted the suction process to be monitored and the different relevant geometric quantities to be measured: vesicle projection length into the pipette, for different applied pressures; vesicle and pipette diameters. The vesicle projection length into the pipette as a function of pressure, for a given vesicle and pipette diameter, is related to the bilayer bending modulus k_c. Assayed samples comprised vesicles with and without ethanol-dissolved gliadin incorporated in the membrane, at two distinct gliadin/phospholipid mass fractions. We observed a significant increase in k_c of the phospholipid bilayer with the largest gliadin content (0.02 gliadin/dioleoylphosphaditylcholine w/w), from about $k_c = 21\ k_BT$, to $k_c = 52\ k_BT$, where k_BT is the thermal energy of the solution. The results reveal the affinity between the protein and the bilayer, suggesting possible correlations between membrane mechanical properties and the celiac disease etiology.

Keywords Gliadin · Bending modulus · Celiac disease · Gluten · Giant vesicles

A.M.F. Monteiro · E.P.G. Arêas (✉)
Instituto de Química, Universidade de São Paulo, Caixa Postal 26077,
CEP 05513-970 São Paulo, SP, Brazil
e-mail: epgareas@usp.br
Tel.: +55-11-30912165
Fax: +55-11-38155579

N. Fa · A. Schröder · C. Marques
Laboratoire de Dynamique des Fluides Complexes, UMR 7506 CNRS/ULP,
4, rue Blaise Pascal, 67070 Strasbourg Cedex, France

Introduction

Gliadin is a component of wheat gluten, a complex supramolecular protein assembly of very high molecular mass [1]. Besides the importance of gluten in biotechnology, owing to its viscoelastic properties and interfacial activity, it is also known to be involved in a disease of autoimmune character. The condition, known as the celiac disease, comprises an intolerance to gluten ingestion and is characterised by remarkable histological changes of the intestinal mucosa involving the complete loss of its microvillosities and their conversion into a smooth, flattened surface. Such changes lead to the failure of the mucosa absorption capacity, with devastating consequences for the nutritional status of the patient [2]. Neurological implications of the disease have also been more recently reported in the literature [3].

Considering the peculiar changes displayed by the celiac intestinal mucosa, we investigated possible alterations induced by gliadin on mechanical properties of a model cell membrane. In particular, we determined the effect of gliadin on the bending elasticity of dioleoylphosphaditylcholine (DOPC) vesicles. The study revealed a sizeable dependence of the membrane rigidity on protein concentration. Our results point to the importance of accounting for membrane elasticity modifications when the etiology of the celiac disease is studied.

Methodology

Vesicles in a solution exhibit thermal fluctuations due to Brownian motion [4, 5]. A micropipette apparatus, first developed by Evans [6, 7], measures the relation between the excess area stored in the fluctuations and the corresponding value of the membrane tension [6]. This relation, known as the Helfrich law, provides for the determination of the membrane bending modulus k_c. In practice one determines α, the relative amount of the membrane surface stored in the fluctuating modes and the membrane tension σ by measuring both the geometrical parameters related to the invaginated vesicles and the aspiration pressure. From the Helfrich relation [8],

$$\alpha = \frac{A^a - A_0^a}{A_0^a} = \frac{k_B T}{8\pi k_C} \ln\left(\frac{\sigma}{\sigma_1}\right), \tag{1}$$

the relation of the excess area, α, and the bending modulus, k_c is established.

In Eq. (1), A_0^a and A^a are the initial and the final apparent projected areas, σ_1 is the value of the first measured tension, T is the absolute temperature and k_B is the Boltzmann constant. The apparent areas in Eq. (1) can be experimentally estimated by measuring parameters R_0, R_L and r, as indicated in the schematic drawing in Fig. 1.

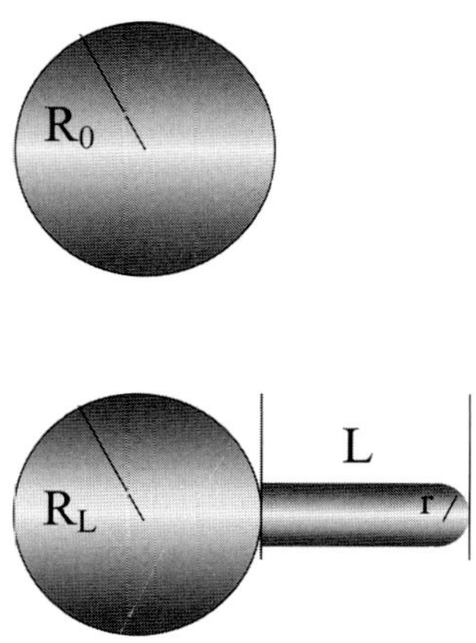

Fig. 1 Schematic drawing indicating parameters analyzed for the vesicles in the micropipette suction experiments

The experimentally measured parameters, indicated in the schematic drawing in Fig. 1, are related to the apparent areas by simple geometrical relations (Eqs. 2, 3):

$$A^a = 4\pi R_L^2 - \pi r^2 + 2\pi r(L - r) + 2\pi r^2, \tag{2}$$

$$A_0^a = 4\pi R_0^2. \tag{3}$$

The excess area is therefore given by

$$\alpha = \left[\frac{R_L^2 + \left(\frac{rL}{2}\right) - \left(\frac{r^2}{4}\right)}{R_0^2}\right] - 1 \tag{4}$$

which allows the determination of k_c [8, 9].

In this work, we were particularly concerned in obtaining mechanical information relative to k_c of a model membrane system, as affected by the presence of gliadin, by means of the Evans micropipette technique. With that purpose, DOPC giant vesicles with an average diameter of 50 μm were produced by electroformation [4]. Phospholipid films were prepared from DOPC dispersed in chloroform, followed by vacuum drying overnight. Samples containing gliadin (whole gliadin fraction from wheat gluten, a Sigma-Aldrich product) were prepared by addition of ethanol-dispersed gliadin to the DOPC/chloroform mixture. A sucrose solution was used to hydrate the phospholipid films during vesicle formation. Transference of vesicles to a glucose solution of same osmolarity provided for the vesicle integrity and for the necessary visual contrast for microscopic observation.

Results and discussion

The micropipette technique devised by Evans and collaborators has proved very relevant in investigations of surface and micromechanical aspects of model surfaces in several works undertaken during the last decade. Examples include studies of cell adhesion and microtubular formation [10], elaboration of membrane rupture models through the investigation of spontaneous frequencies of defect formation and its correlation with membrane thickness and bending modulus [11], investigation of membrane water permeability [12] and studies on the dependence of bilayer elasticity on lipid chain length and degree of unsaturation [13], among others.

The results in the present work indicating the variation of α as a function of the logarithm of the membrane tension σ, for DOPC vesicles containing gliadin at two different concentrations and for a blank sample of pure DOPC, are displayed in Fig. 2. The values of k_c determined and their respective standard deviations are shown in Table 1.

Table 1 Effect of wheat gliadin on the bending modulus, k_c, of dioleoylphosphatidylcholine (*DOPC*) giant vesicles

Sample	$k_c/k_B T$	Standard deviation
(1) gliadin/DOPC 0.01 (w/w)	32.80	14.23
(2) gliadin/DOPC 0.02 (w/w)	52.16	18.86
Blank sample (1)	21.59	4.50
Blank sample (2)	20.22	3.68

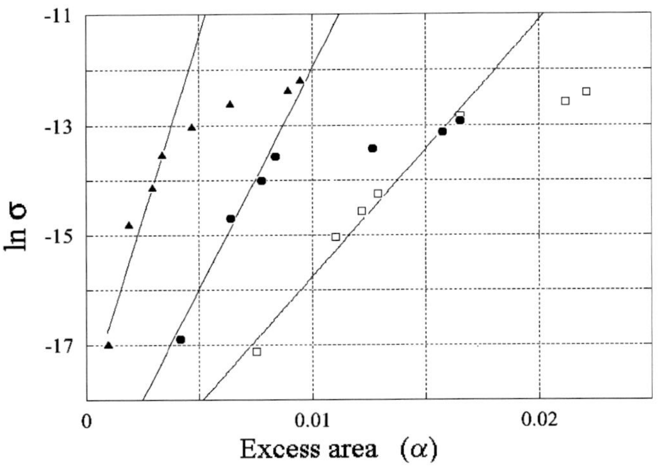

Fig. 2 Variation of the logarithm of the tension σ as a function of the excess area α: blank sample (vesicle of pure dioleoylphosphatidylcholine, *DOPC*) (*squares*); vesicle made of 99% DOPC + 1% w/w gliadin (*circles*); vesicle made of 98% DOPC + 2% w/w gliadine (*triangles*)

According to Eq. (1), the ascending linear region on the graph, attained for the smallest tension values, is the region of interest for assessing the value of k_c. The average values for ten experiments show an increase in k_c from around 21 k_BT for the vesicles of pure DOPC to 32.80 and 52.16 k_BT for protein-containing vesicles at 1% and 2% protein concentration, respectively (Table 1). These results point to the affinity of gliadin for the vesicle bilayer, as expected for a protein with highly hydrophobic character. It is clear that changes in the bending modulus of phospholipid membranes must be a reflection of structural changes of the phospholipid bilayer induced by its interaction with the protein. As an hydrophobic entity, gliadin is likely to penetrate the lipophilic membrane core, where it may change patterns of lipid ordering and flexibility. Changes of around 10 k_BT have been reported by Evans and Rawicz [14] for the bending rigidity of membranes in the presence of grafted water-soluble polymers in the semidilute regime, with the membrane stretching modulus remaining unaltered. A considerably more pronounced effect on the bending modulus is verified for membrane k_c in the presence of minor amounts of gliadin (1–2% relative to the DOPC content), as reported in this work. Considering the pronounced hydrophobic character of that protein and peculiarities of its three-dimensional structure, as revealed in modelling work [15, 16], we argue that its interaction with DOPC should involve protein penetration into the apolar region of the bilayer, rather than a simple anchoring on its external surface. Membrane mechanical properties may be affected by the peculiar viscoelastic behaviour of the interacting protein. Permeation across the bilayer might present alterations as well, leading ultimately to significant morphological changes of the membrane. In an in vivo context, where a specific immunological response takes place, these modifications may contribute to the observed morphological and functional changes in surface properties, such as those displayed by the celiac mucosa.

Conclusions

Gliadin showed an effect of increasing k_c of the phospholipid bilayer in giant vesicles of DOPC. Such an effect was noticeable for minor amounts of the protein present in the preparations (0.01–0.02 w/w protein-to-DOPC ratio). Results discussed on the basis of gliadin structural and mechanical properties suggest possible correlations between changes in elasticity and aspects of the etiology of the celiac disease.

Acknowledgements The authors thank CAPES/COFECUB, a joint Brazilian/French project, project no. 312/00-II, for support. A.M.F.M. is grateful to CAPES (Brazil) and to CAPES/COFECUB, for a doctorate and a sandwich-doctorate scholarships, respectively. E.P.G.A. acknowledges support from FAPESP (Brazil) and C.M. is grateful for support from the CNRS chemistry department under AIP "Soutien aux Jeunes Equipes".

References

1. Örnebro J, Nylander T, Eliasson A-C (2001) J Cereal Sci 31:195
2. Sollid LM (2000) Annu Rev Immunol 18:53
3. Chin RL, Sander HW, Brannagan TH, Green PHR, Hays AP, Alaedini A, Latov N (2003) Neurology 60:1581
4. Manneville J-B, Bassereau P, Ramaswamy S, Prost J (2001) Phys Rev E 64:021908
5. Rawicz W, Olbrich KC, McIntosh T, Needham D, Evans E (2000) Biophys J 79:328
6. Evans E, Needham D (1987) J Phys Chem 91:4219
7. Kwok R, Evans E (1981) Biophys J 35:63711
8. Helfrich, W (1973) Naturforsch 28c:693.7
9. Hulin J, Petit L, Mitescu, C (2001) Physical hydrodynamics. Oxford University Press, Oxford
10. Evans, E Ritchie, K, Merkel, R (1995) Biophys J 68:2580
11. Evans E, Heinrich V, Ludwig F, Rawicz W (2003) Biophys J 85:2342
12. Olbrich K, Rawicz W, Needham D, Evans E (2000) Biophys J 79:321
13. Rawicz W, Olbrich KC, McIntosh T, Needham D, Evans E (2000) Biophys J79:328
14. Evans E, Rawicz W (1997) Phys Rev Lett 79:2379
15. Arêas EPG, Cassiano, MM (2001) Biophys Chem 90:135
16. Arêas EPG, Cassiano, MM (2001) J Mol Struct (THEOCHEM) 572:107

Progr Colloid Polym Sci (2004) 128: 187–192
DOI 10.1007/b97070

Herley Casanova
Carlos Ortíz
Alejandra Vallejo
Pedronel Araque

Nicotine oleate dispersions as botanical insecticides

H. Casanova (✉) · C. Ortíz · A. Vallejo
P. Araque
Grupo de Coloides, Universidad de Antioquia, Medellin, Antioquia, Colombia
e-mail: casanova@matematics.udea.edu.co
Tel.: +57-4-2110647
Fax: +57-4-2330120

Abstract Two alternative extraction media for botanical insecticide nicotine were evaluated in order to increase its recovery efficiency from tobacco leaves: oleic acid oil-in-water (O/W) emulsions and oleic acid. The use of oleic acid O/W emulsions stabilized by calcium caseinate as extraction media in a percolation process showed a recovery efficiency between 62% (25 °C) and 68% (60 °C). Oleic acid O/W emulsions stabilized by sodium dodecyl sulfate (SDS), Tween 20, Tween 80 or Triton X100 were inappropriate as extraction media since they presented flocculation or phase separation after the extraction procedure. Bioassays of nicotine oleate O/W emulsions stabilized by sodium caseinate showed a lethal time 50 (LT_{50}) of 10 min. Nicotine extraction by a modified effleurage method showed a recovery efficiency between 85% (25 °C) and 95% (60 °C). Nicotine oleate O/W emulsions (45 vol %) prepared with the effleurage extract and using SDS showed the highest bioactivity ($LT_{50}=20$ min). Nicotine oleate O/W emulsions prepared with the effleurage extract and stabilized by sodium caseinate, Tween 20, Tween 80 or Triton X100 showed LT_{50} values of over 50 min. Interfacial tension measurements of nicotine effleurage extract-aqueous surfactant solution interfaces, droplet size analysis of nicotine oleate emulsions and nicotine concentrations at emulsion aqueous phases suggest that nicotine oleate in O/W emulsions stabilized by SDS was encapsulated in oil droplet emulsions, while in sodium caseinate, Tween 20, Tween 80 or Triton X100 emulsions the nicotine oleate acts as a cosurfactant, decreasing its insecticide activity.

Keywords Nicotine · Dispersions · Formulations · Extraction · Surfactants

Introduction

In the early 1990s, consumers' concern about the quality of food products and the integration of environmental conservation into agriculture gave organic farming the opportunity for development and consolidation, especially in Europe, the USA, Canada, Australia and Japan [1]. The production system for organic crops requires an effective control of pests with the minimal environmental impact. Different pest control strategies are considered safe for organic farming: crop rotation, biological diversity, resistant varieties, and pesticides [2]. According to organic farming regulations, the use of pesticides is restricted to cases of immediate threat to the crop, where botanical pesticides are preferred over synthetic pesticides [1, 3]. Among botanical pesticides, pyrethrum (from *Chrysanthemum cinerariaefolium*), rotenone (from *Derris* spp. and *Lonchocarpus* spp.), neem (from *Azadirachta indica* A. Juss.) and nicotine (from *Nicotiana tabacum*) have shown to be very effective as pest control alternatives [4]. However, nicotine lost its regulatory status in the European Community in March 2002 [5] and has restricted use in the USA, where it is only allowed for cucumber, lettuce and tomato crops [6]. Such severe restriction is mainly explained by high nicotine mammalian toxicity ($LD_{50} = 50$ mg/kg), which increases health risks during handling and application in the field [7]. Still, nicotine mammalian toxicity could be reduced by formulating it as emulsions where emulsifiers and stabilizers adsorbed at the oil–water interface could act as a protecting barrier. Emulsions prepared with nicotine oleate as the oil phase and stabilized by sodium caseinate (NaCn) have shown to be a possible alternative to formulate nicotine without losing its biological activity [8]. In addition to the issue of mammalian toxicity, nicotine has another problem related to its low process recovery (about 40%) when extracted by organic solvents in percolation or Soxhlet extraction procedures. Therefore, nicotine, as a botanical insecticide, has to solve these two main issues, process efficiency and mammalian toxicity, in order to be competitive in the pesticide market.

This work addresses the problem of process efficiency on nicotine extraction, presenting two alternative extracting media: oleic acid and oil-in-water (O/W) emulsions where the oil phase is oleic acid. Nicotine oleate extracts were formulated as O/W emulsions stabilized by different surfactants and their colloidal stability and bioactivity were determined.

Experimental

Materials

Dry blond tobacco leaves were provided by the Colombian Tobacco Company, Coltabaco, Medellin, Colombia. The material was ground and sieved through 60 mesh. Commercial petroleum ether (50–70 °C), sodium hydroxide and oleic acid were purchased from Protokimica (Medellin, Colombia). Perchloric acid, acetic acid, chloroform, crystal violet indicator, silicotungstic acid, Tween 20, Tween 80, sodium dodecyl sulfate (SDS) and Triton X100 were provided by Sigma Chemical Co. (St. Louis, MO). Celite 545 was purchased from J.T. Baker (Phillipsburg, NJ). NaCn and calcium caseinate (CaCn) (more than 82 wt % protein, less than 6 wt % moisture) were from DMV International (Veghel, Netherlands).

Extraction procedures

For nicotine Soxhlet extraction, a Soxhlet apparatus was filled with ground tobacco leaves following a previous digestion process with a 4 wt % sodium hydroxide solution in a 1:1.25 weight ratio for 2 h. Five washing procedures were carried out with petroleum ether; the nicotine extract was evaporated until removal of the extraction solvent.

For nicotine oleic acid extraction a 200-mL beaker containing 60 g ground tobacco leaves and 50 g oleic acid was placed in a IKAMAG RET magnetic stirrer provided with an automatic temperature control system (IKATRON ETS). Extraction procedures were carried out at temperatures between 25 and 60 °C for up to 3 h. High-concentration nicotine extracts were obtained by repeating the extraction procedure.

Emulsion extraction was carried out using oleic acid O/W emulsions (20 vol %, 3 wt % surfactant) as extraction media. A 200-mL beaker containing 90 g emulsion and 10 g ground tobacco was placed in a IKAMAG RET magnetic stirrer. Extraction procedures were carried out at temperatures between 25 and 60 °C for 3 h. High-concentration nicotine extracts were obtained by repeating the extraction procedure

Nicotine quantification

The alkaloids content (as nicotine) of oleic acid extracts was determined using the AOAC 960.08 method [9]. Total alkaloids are referred to as nicotine content in the present document. The nicotine concentration of O/W emulsions was determined using the AOAC 920.35 method [10].

Emulsion preparation and characterization

Oleic acid O/W emulsions (3 wt % emulsifier, 20 vol % oleic acid) were prepared in a type W2 colloidal mill from the Dumore Company, working at 10,000 rpm and 25 °C. O/W emulsions prepared from nicotine oleate extracts were obtained by emulsification on a jet homogenizer working at 300 bar. NaCn, CaCn, Tween 20, Tween 80, SDS and Triton X100 were used as emulsifiers. The average volume droplet size d_{43} was determined with a Malvern Mastersizer 2000; a refractive index of 1.500 was set for the emulsion oil phase. The creaming behavior was determined by visual observation of the height of a discrete layer formed in the dispersed system thermostated at 25 °C at regular time intervals during a 1-month time period.

Interfacial studies

Interfacial tension measurements were carried out for oleic acid and effleurage nicotine extracts in contact with surfactant aqueous solutions (0.5 wt %) using a Krüss K12 processor tensiometer at a controlled temperature of 25 °C.

Bioassays

The bioassay for insecticidal activity against adults of *Drosophila melanogaster* was carried out following the procedure of Casanova et al. [8].

Results and discussion

Extraction by Soxhlet

Extraction of nicotine from tobacco leaves by the Soxhlet method showed a recovery efficiency of around 40% despite the fact that five washing procedures were carried out. Probably the polar nature of nicotine induced a partition between the organic phase and the matrix of ground tobacco leaves, which reduces the efficiency of the process. In order to increase that recovery efficiency two alternative extraction media were evaluated: oleic acid O/W emulsions and oleic acid.

Extraction by oleic acid O/W emulsions

Droplet size analysis of oleic acid O/W emulsions (Fig. 1) shows that the use of small surfactant molecules (SDS, Tween 20, Tween 80 and Triton X100) for the emulsification process produced emulsions with lower d_{43} values (below 3 μm) than that obtained by using CaCn as an emulsifier ($d_{43} = 4.96$ μm). This is to be expected on the basis of their differences in interfacial configuration and capacity to reduce interfacial tension [11]. Extraction procedures with the different emulsion systems showed that only the oleic acid O/W emulsions stabilized by CaCn and by Tween 80 were stable after extraction; the other emulsions presented flocculation and oil-phase separation. It is important to notice that the CaCn-stabilized emulsion presented a decrease in the average droplet size (d_{43}) from 4.96 to 3.94 μm, while the Tween 80-stabilized emulsion had the same d_{43} value before and after extraction. Owing to their colloidal stability during the extraction procedure, CaCn- and Tween 80-stabilized emulsions were used to carry out nicotine extractions at different temperatures in order to determine their effect on nicotine recovery efficiency; the results are presented in Fig 2. At 25 °C, CaCn-stabilized emulsions showed a 62% nicotine recovery efficiency, which increased to 68% at 60 °C; the nicotine concentration in the emulsion was 0.34 wt % (25 and 40 °C) and 0.37 wt % (60 °C). The recovery efficiency for the Tween 80-stabilized emulsion was 53% at 25 °C and 58.5% at 40 °C, showing a lower nicotine concentration than the CaCn emulsion. At 60 °C, Tween 80-stabilized emulsions presented flocculation; therefore, nicotine quantification was not carried out. Such instability could be explained by the increased hydrophobic attraction between nonionic surfactant molecules when the temperature is increased; this reduces the steric stability of emulsion droplets. Other destabilization mechanisms are also possible considering the complexity of the tobacco matrix.

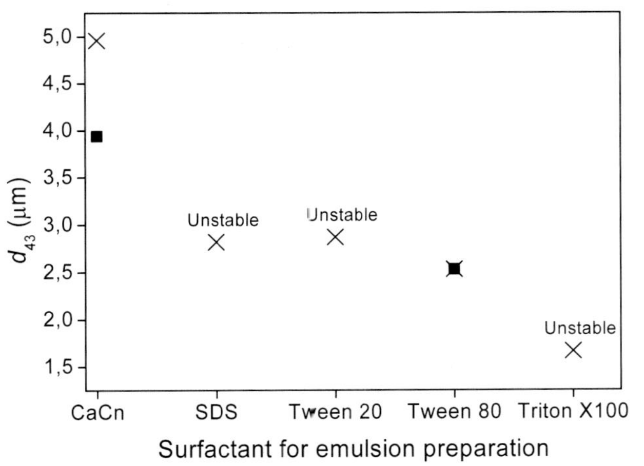

Fig. 1 Average droplet size in volume (d_{43}) of oleic acid oil-in-water (*O/W*) emulsions (20 vol %, 3 wt% emulsifier) using different emulsifying agents for preparation before nicotine extraction (*crosses*) and after nicotine extraction (*squares*)

Owing to the low nicotine concentration in oleic acid O/W emulsion systems (0.37 wt % or less), a multiple extraction procedure was implemented in order to increase the nicotine concentration. Two multiple extractions were carried out with only CaCn-stabilized emulsions owing to their higher recovery efficiency. Nicotine quantification on that emulsion system showed that the nicotine concentration increased to 0.70 wt % with the second extraction using new tobacco leaves. Bioassays with one and two extraction procedures showed a lethal time 50 (LT_{50}) of 50 and 10 min, respectively. These results suggest that at least two extraction procedures are required in order to obtain nicotine insecticide emulsions with adequate bioactivity against soft body insects. Nicotine quantification at the aqueous phase of the emulsion showed that with one extraction procedure 82% of nicotine was in that phase, and after the second extraction that figure decreased to 68%. It is probable that the bioactivity of these emulsions can be correlated with the percentage of nicotine encapsulated in the oil droplet emulsions. This CaCn-stabilized emulsion was stable only for 1 week, and showed flocculation but not phase separation.

Extraction by oleic acid

Oleic acid O/W emulsions used as extraction media showed recovery efficiencies of up to 70% at 60°C, which,

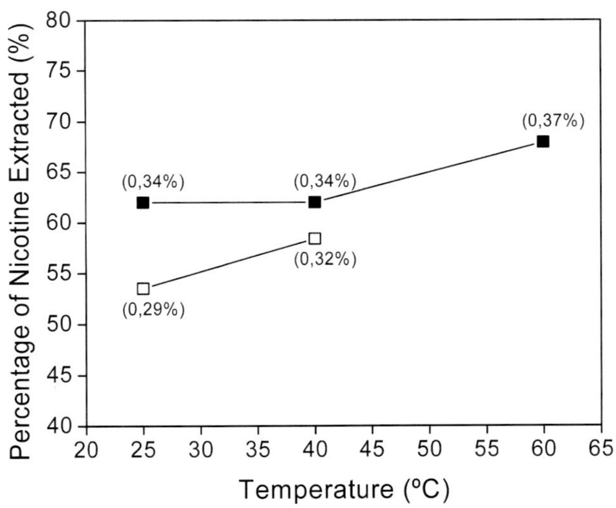

Fig. 2 Effect of temperature on the percentage of nicotine extracted from tobacco leaves using oleic acid O/W emulsions stabilized by calcium caseinate (*closed squares*) or by Tween 80 (*open squares*) as extracting media. The *numbers* in *parentheses* indicate the nicotine concentration in the emulsion. The *solid lines* serve as a guide to the eye

although high, could be a drawback if they are intended to be used an industrial scale. Therefore, oleic acid was evaluated as at alternative extraction medium to seek a higher nicotine recovery from tobacco leaves. This procedure resembles an effleurage method commonly used to extract essential oils and other compounds present in plants. Considering the basic character of nicotine a high recovery efficiency was expected owing to the possible acid–base reaction during the extraction procedure. However, nicotine recovery efficiencies, under different extraction times (1–10 h) and different temperatures (25–60 °C), were not higher than 40%. Other process variables such as the agitation speed and the oleic acid-to-tobacco leaf ratio showed no effect on the recovery efficiency. After realizing that the acid–base reaction between oleic acid and nicotine requires a polar medium for the reagents to show their acid (base) character, water was incorporated into the extraction medium by wetting the tobacco leaves with it prior to extraction. A tobacco leaf-to-water ratio by weight of 5:2 was found to be appropriate for nicotine extraction, and nicotine recovery efficiencies of 84.5% at 25 °C and 95% at 60 °C were obtained (Fig. 3). Although the modified effleurage method showed itself to be very efficient for nicotine extraction, the total nicotine concentration in the oil phase (i.e., oleic acid) is low (e.g., 0.78 wt% at 60 °C), and would be even lower after preparing an emulsion with that nicotine extract. Therefore, it was necessary to carry out multiple extraction procedures in order to enrich the oil phase in the nicotine present as nicotine oleate after neutralization with the oleic acid.

The increase in nicotine concentration after carrying out up to seven multiple extractions is shown in Fig. 4. It is apparent that after the first extraction the percentage of nicotine recovered is lower than the expected value, which can be determined from the line for 95% recovery efficiency. After four consecutive extractions the nicotine concentration was 80% of the expected value and after seven extractions that figure dropped to 70%. This loss in extraction efficiency is not due to a lack of oleic acid required for nicotine neutralization since oleic acid is present in excess. It is more feasible to consider that high concentrations of nicotine oleate during consecutive extractions could push the neutralization reaction backward considering the low acid (base) character of the reacting species.

The nicotine oleate extract obtained after seven extractions (3.36 wt % nicotine) was used to prepare O/W emulsions (45 vol %, 3 wt % emulsifier) employing different emulsifiers: NaCn, SDS, Tween 20, Tween 80 and Triton X100. The average droplet sizes (d_{43}) for emulsions obtained after high-pressure homogenization are shown in Fig. 5.

The emulsifier SDS produced the emulsion with the lowest d_{43}, followed by Tween 20, NaCn, Tween 80 and Triton X100. By comparing the d_{43} values of emulsions prepared from the nicotine oleate extract (less than 0.72 μm) with those obtained for oleic acid emulsions (1.5–5.0 μm), it is possible to suggest an emulsifying role of nicotine oleate in the former emulsions. Such a role is especially notable in the NaCn-stabilized emulsion whose d_{43} decreased from 4.96 to 0.65 μm. Interfacial tension measurement of nicotine oleate extract (3.36 wt % nicotine) and oleic acid against surfactant solutions (0.5 wt %) give some additional support to the nicotine oleate emulsifier properties (Fig. 6).

The nicotine oleate extract has an interfacial tension against water of 1.54 mN m^{-1}, while the oleic acid–water

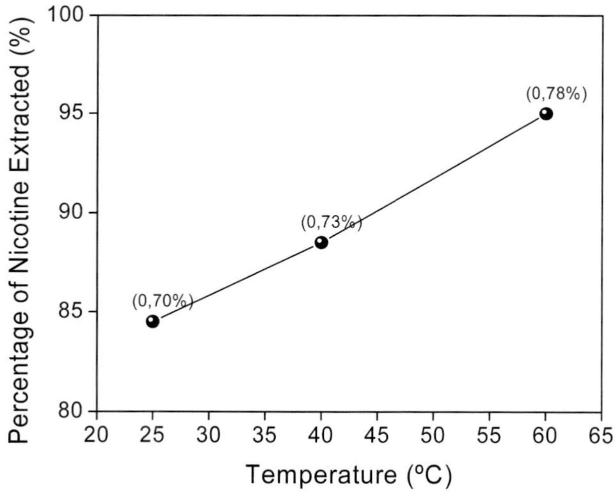

Fig. 3 Effect of temperature on the percentage of nicotine extracted from tobacco leaves using oleic acid as the extracting medium. The *numbers* in *parentheses* indicate the nicotine concentration in the nicotine extract. The *solid line* serves as a guide to the eye

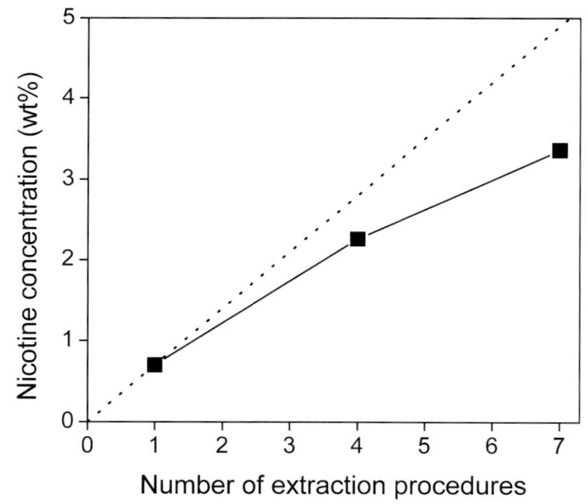

Fig. 4 Effect of the number of extraction procedures on the nicotine concentration for oleic acid extraction (effleurage method). The *dotted line* shows the expected values considering 95% nicotine recovery efficiency. The *solid line* serves as a guide to the eye

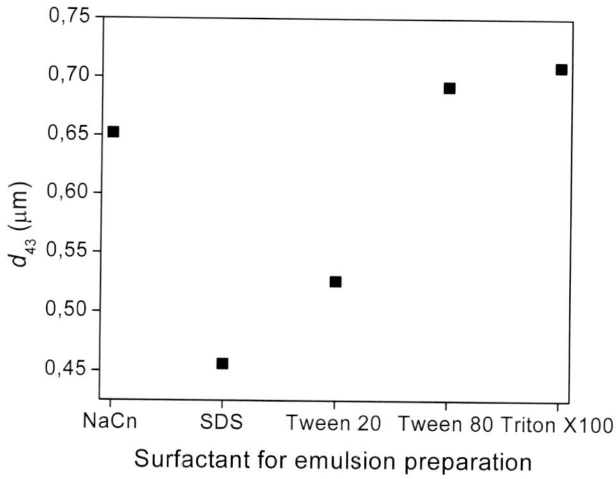

Fig. 5 Average droplet size (d_{43}) of emulsions (45 vol %, 3 wt% emulsifier) prepared from nicotine oleate extracts using different emulsifying agents

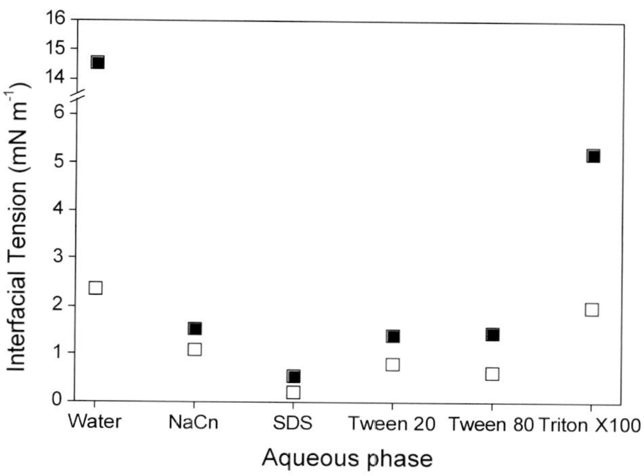

Fig. 6 Interfacial tension measurements for different aqueous phases for oleic acid (*closed squares*) and against nicotine oleate extract (*open squares*)

system has an interfacial tension of 14.54 mN m^{-1}. The lower interfacial tension for the nicotine oleate extract compared with oleic acid is also observed when the different surfactant aqueous phases (0.5 wt %) were put in contact with them. Such a low interfacial tension for nicotine oleate–surfactant aqueous phases also suggests that nicotine oleate is not displaced from the oil–water interface, probably because it associates with the different surfactants, therefore working as a cosurfactant [12]. Comparing the tendency of the average droplet size and the interfacial surface tension against surfactant aqueous phases, we could also observe that the interfacial tension measurements reasonably predict (except for Tween 80 emulsions) the emulsifying power of surfactant solutions.

Bioassays carried out with the O/W emulsions prepared with the nicotine oleate extract showed an LT_{50} of 20 min for the SDS-stabilized emulsion, while the NaCn-, Tween 20-, Tween 80- and Triton X100-stabilized emulsions showed LT_{50} values higher than 50 min, which make them useless for insecticide purposes. That was an unexpected result considering that the nicotine concentration in those emulsions was relatively high (1.4 wt %) compared with the nicotine concentration that oleic acid O/W emulsions, used previously, had (i.e., 0.7 wt %). More surprising was the fact that the only emulsion that presented some bioactivity was that prepared with SDS. After comparing the particle size distribution of nicotine oleate emulsions with those obtained for oleic acid it was possible to observe that all emulsions, except the SDS-stabilized emulsion, were undoubtedly affected by the presence of nicotine oleate in the emulsifying system helping them to reduce their average droplet size. Quantification of nicotine in the emulsion serum layer showed that in the SDS-stabilized emulsion 66% of the total nicotine was present on that layer, while in the other emulsions that figure increased to 100%. That confirms the primary emulsifying role of nicotine oleate in NaCn, Tween 20, Tween 80 and Triton X100. The results suggest that the encapsulation of some of the nicotine (i.e., 34%) into the emulsion oil droplets when SDS is used to stabilize the emulsion could be correlated to its higher bioactivity.

Conclusions

The oleic acid O/W emulsions showed to be the more appropriate extraction media for nicotine extraction from tobacco leaves and produced a highly bioactive insecticide formulation through a simple, economical and environmentally friendly procedure. Still, additional research has to be done in order to increase its nicotine recovery efficiency. The modified effleurage method using oleic acid as an extracting medium was shown to be very efficient to extract nicotine from tobacco leaves; however, emulsions prepared with those nicotine oleate extracts presented low bioactivity, which could be explained by its role as cosurfactant during the emulsification process. Only the SDS-stabilized emulsion showed some bioactivity, probably owing to the partial encapsulation of nicotine in oil emulsion droplets. If the nicotine encapsulation ratio can be increased, it is probable that emulsion bioactivity would also increase; therefore, different surfactants or emulsification conditions have to be tried in order to make such a formulation feasible for crop protection.

Acknowledgements We thank the Comité para el Desarrollo de la Investigación and the Universidad de Antioquia, Grupo Interdisciplinario de Estudios Moleculares, Universidad de Antioquia.

References

1. Guillou LG, Scharpé A (2000) Organic farming, guide to community rules. European Commission, Luxembourg
2. Brown T, Haward R (2001) Pesticide Outlook 2:64
3. US Environmental Protection Agency (1999) Implementing the food quality protection act: progress report EPA 735-R-99001. US Environmental Protection Agency, Washington, DC
4. Sugavanam B, Copping LG (1998) In: Valkenburg VW, Sugavanam B, Khetan KS (eds) Pesticide formulation: recent developments and their applications in developing countries. New Age, New Delhi, pp 3–29
5. CONSLEG (2002) Council regulation (ECC) 1991R2092–23/03/2002. Consleg system of the office for official publications of the European Communities, Luxembourg
6. US Environmental Protection Agency (2002) Fed Regist 67:35912
7. Isman MB (1997) Phytoparasitica 25:339
8. Casanova H, Ortiz C, Peláez C, Vallejo A, Moreno ME, Acevedo M (2002) Agric Food Chem 50:6389
9. AOAC official method 960.08 (1995) In: Cunniff P (ed) Official methods of analysis, 16th edn. AOAC, Arlington, p 31
10. AOAC official method 920.35 (1995) In: Cunniff P (ed) Official methods of analysis, 16th edn. AOAC, Arlington, p 50
11. McClements DJ (1999) Food emulsions: principles, practice and techniques. CRC, Boca Raton
12. Dickinson E (2001) Colloids Surf B 20:197

Progr Colloid Polym Sci (2004) 128: 193–198
DOI 10.1007/b97071

Tatiana Prieto
Otaciro R. Nascimento
Iseli L. Nantes

Microperoxidase-9 cycle in the presence of cetyltrimethylammonium bromide micelles: *tert*-butyl hydroperoxide as both an oxidizing and a reducing agent

T. Prieto · I.L. Nantes (✉)
Centro Interdisciplinar de Investigação Bioquímica, CIIB Universidade de Mogi das Cruzes–UMC, Mogi das Cruzes, Prédio I, Sala 1S-15, CEP 13560-250, SP, Brazil
e-mail: ilnantes@umc.br
Tel.: +55-11-47987102
Fax: +55-11-47987068

O.R. Nascimento
Grupo de Biofisica, Instituto de Física de São Carlos, Universidade de São Paulo, USP–São Carlos, São Carlos, SP, Brazil

Abstract Microperoxidases are attractive water-soluble models to study the reaction mechanism of peroxidases because they are heme peptides able to form the same enzyme intermediates along the reaction with peroxides. In this work, we have demonstrated that cetyltrimethylammonium bromide micelles provide a microenvironment with an alkaline interface and a hydrophobic core that favors the peroxidase activity of microperoxidase-9. The addition of *tert*-butyl hydroperoxide to a system containing microperoxidase-9/cetyltrimethylammonium bromide micelles led to the occurrence of a peroxidase cycle, indicating that this aggregate behaves as a true enzyme. This lipoenzyme is an interesting model for studies of the peroxidase reaction mechanism and for applications in nanotechnology.

Keywords Microperoxidase-9 · Horseradish peroxidase cycle · *t*-Butyl hydroperoxide · Free radicals

Introduction

Microperoxidases are heme peptides obtained by sequential pepsin- and trypsin-catalyzed hydrolysis of horse-heart cytochrome c, which retains the heme group covalently attached to cysteine residues 14 and 17 [1–3]. According to the sequence of native cytochrome c, from which these peptide fragments are derived, four microperoxidase types were identified: microperoxidase-6 (CAQCHT), microperoxidase-8 (MP-8) (CAQCHTVE), microperoxidase-9 (MP-9) (KCAQCHTVE) and micr000operoxidase-11 (VQKCAQCHTVE). Microperoxidases exist in the ferric resting state, retain histidine as a fifth heme iron ligand but not the methionine residue at the sixth coordination position that, in this case, at neutral pH, is occupied by a water molecule [1–3]. Because of these structural characteristics, microperoxidases have been useful to study structural and catalytic properties of peroxidases and cytochromes [3–5] as well as to develop electrode sensors [6]. Although it had been demonstrated that the association of microperoxidases in aqueous micellar solution provides to the heme group a protein-like hydrophobic environment, detailed information about the catalytic properties of these complexes is not still available [7].

In this work, for the first time, we reproduced a typical peroxidase cycle with MP-9 associated to cetyltrimethylammonium bromide (CTAB) micelles.

Experimental

MP-9, *t*-BuOOH and CTAB were obtained from Sigma Chemical Co. (St Louis, MO, USA).

CTAB micellar solutions were prepared by dissolving CTAB in 5 mM phosphate buffer and stirring at 37 °C. The critical micelle concentration was determined as usual from plots of the surface tension versus the logarithm of the CTAB concentration. Surface tensions were measured with a DeNoüy tensiometer equipped with a Pt ring. Solutions of Fe(III)MP-9 and *t*-BuOOH were both buffered with 5 mM sodium phosphate (pH 7.4). The Fe(III)MP-9 concentration was determined by using $\varepsilon_{397} = 1.57 \times 10^5\ M^{-1}\ cm^{-1}$ for Fe(III)MP-9 at pH 7.0 [8]. The Fe(III)MP-9 concentration in CTAB micelles was calculated by using $\varepsilon_{400} = 1.13 \times 10^5\ M^{-1}\ cm^{-1}$and was determined from a standard curve obtained from the spectra of Fe(III)MP-9/CTAB at different heme peptide concentrations. All measurements were performed at 25 ± 0.3 °C. Spectra were recorded using a Shimadzu model 1501 MultiSpec (Tokio, Japan), employing the photodiode array scan mode.

Direct continuous-wave electron paramagnetic resonance (EPR) measurements of Fe(III)MP-9 (100 μM) were obtained using a Bruker ELEXSYS EPR E-580 system under the following conditions: gain 5×10^3, modulation amplitude 1.0 mT, microwave power 4 mW, temperature 4 K, time constant 10.24 ms and conversion time 81.92 ms.

Results

The interaction of MP-9 with CTAB micelles

The spectra of 100 μM MP-9 in a homogeneous medium (5 mM phosphate buffer, pH 7.4, dotted line) and in 60 mM CTAB, equivalent to a micelle-to-MP-9 molar ratio of approximately 6.0, are shown in Fig. 1a. The comparison of the MP-9 spectra shows that the association of MP-9 with CTAB micelles led to Soret band redshift. In order to achieve the alterations in the spin state of MP-9 determined by its association with CTAB micelles, EPR measurements were made for 100 μM MP-9 in the presence and absence of CTAB micelles (Fig. 1b).

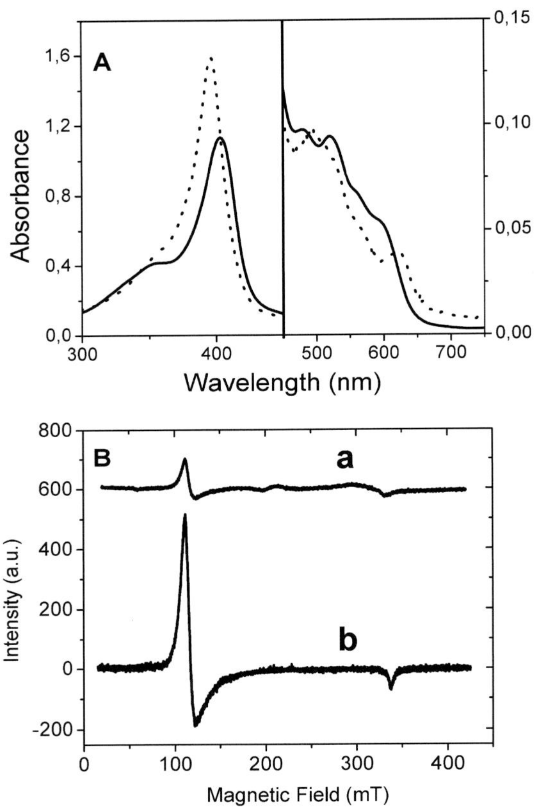

Fig. 1a Microperoxidase-9 (*MP-9*) spectral changes after the association to cetyltrimethylammonium bromide (*CTAB*) micelles. Fe(III)MP-9 in 5 mM phosphate buffer, pH 7.4 (*dotted line*) and Fe(III)MP-9 associated to CTAB micelles in 5 mM phosphate buffer, pH 7.4 (*solid line*). The experiments were carried out with 60 mM CTAB and 100 μM MP-9, at 25 °C using a 0.1-cm optical path length cuvette. **b** Continuous-wave electron paramagnetic resonance (*EPR*) of MP-9 and the complex MP-9/CTAB. Fe(III)MP-9/CTAB EPR spectrum in the absence of CTAB (*line a*); EPR spectrum in the presence of 60 mM CTAB (*line b*). The experiments were carried out with 60 mM CTAB in 5 mM phosphate buffer, pH 7.4, at 3.7 K and 9.485501 GHz. Spectrometer conditions: gain 5×10^3, modulation amplitude 1.0 mT, microwave power 4 mW, time constant 10.24 ms and conversion time 81.92 ms

The minimal MP-9 and CTAB concentrations necessary for EPR measurements were those in which the micelle-to-MP-9 molar ratio was approximately 6.0. According to the EPR spectra shown in Fig. 1b, in a homogeneous medium (line a), at this high concentration, MP-9 was found at low and high spin states. The low spin state could be due to the coordination of the heme iron with the terminal N and the ε-amino groups of a neighboring MP-9 molecule. In the presence of CTAB (line b), only the high spin state of MP-9 was found; however, this heme iron spin state exhibits different EPR parameters compared with the high spin state observed in an aqueous medium. In CTAB micelles MP-9 heme iron exhibits axial symmetry with two g values ($g_{||} = 5.9600$ and $g_{\perp} = 2.0080$), and in a homogeneous medium, the prosthetic group exhibits a subtle rhombic distortion with three g values ($g_1 = 6.050$, $g_2 = 5.840$ and $g_3 = 2.009$). Moreover, the line widths are also different in the two media, suggesting different complex dynamics.

Catalytic cycle of MP-9/CTAB with *t*-BuOOH as an oxidizing and a reducing agent

The addition of 300 μM *t*-BuOOH to a medium containing 100 μM MP-9 associated to 60 mM CTAB (micelle-to-MP-9 molar ratio of 6.0) led to conversion of Fe(III)MP-9 to MP-9, compound II (oxoferryl form), in a time interval of 10 s (Fig. 2a). After 1.0 min of reaction, MP-9 compound II was completely converted to the native form, which characterized a typical peroxidase catalytic cycle (Fig. 2a). The peroxide cleavage process catalyzed by the MP-9/CTAB lipoenzyme was accompanied by oxygen consumption (not shown). Under this condition, the heme iron continuous-wave EPR spectra obtained at different reaction time intervals (Fig. 2b) showed that 2 s after *t*-BuOOH addition, the intensity of the heme iron EPR signal decreased, which was in accordance with the formation of compound II, an EPR-silent species [9]. After 17 s a partial regeneration of the Fe(III)MP-9 signal was observed, coherent with the occurrence of the peroxidase catalytic cycle.

A similar catalytic cycle was obtained at a micelle-to-MP-9 molar ratio of 66.0 (3 μM MP-9 and 20 mM CTAB, not shown) and was also accompanied by oxygen consumption (Fig. 3).

The conversion of Fe(III)MP-9 to the corresponding oxidized intermediate, compound II (oxoferryl species), could implicate the previous formation of compound I (oxoferryl π cation) that was no longer detected

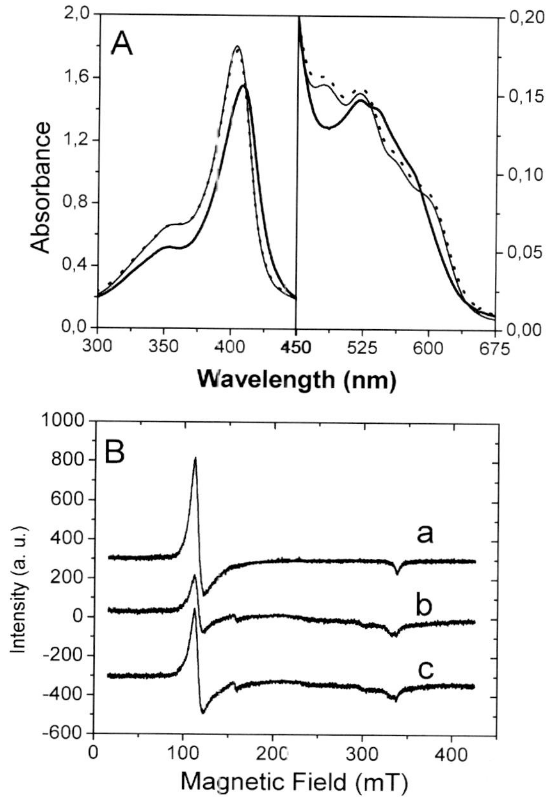

Fig. 2a MP-9 spectral changes during the peroxidase cycle. Fe(III)MP-9 associated to CTAB micelles (*thin solid line*); MP-9 compound II associated to CTAB micelles, formed 10 s after addition of 300 μM *t*-BuOOH (*thick solid line*); Fe(III)MP-9 associated to CTAB micelles recovered 1 min after addition of 300 μM *t*-BuOOH (*dotted line*). The experiments were carried out with 60 mM CTAB and 100 μM MP-9, in 5 mM phosphate buffer, pH 7.4, at 25 °C using a 0.1-cm optical path length cuvette. **b** Continuous-wave EPR MP-9/CTAB spectra during the peroxidase cycle. Fe(III)MP-9/CTAB EPR spectrum in the absence of peroxide (time zero) (*line a*); EPR spectrum 2 s after addition of 300 μM *t*-BuOOH (*line b*); EPR spectrum 17 s after addition of 300 μM *t*-BuOOH (*line c*). The experiments were carried out with 60 mM CTAB in 5 mM phosphate buffer, pH 7.4, at 3.7 K and 9.485501 GHz. Spectrometer conditions: gain 5×10^3, modulation amplitude 1.0 mT, microwave power 4 mW, time constant 10.24 ms and conversion time 81.92 ms

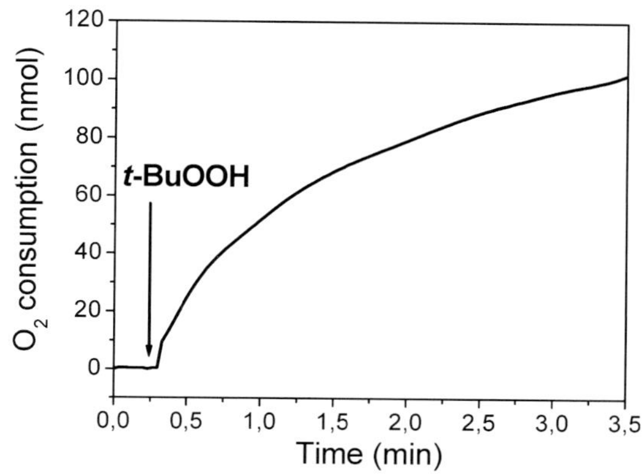

Fig. 3 Oxygen consumption during the MP-9/CTAB cycle. The experiment was carried out at 25 °C, in 5 mM phosphate buffer, pH 7.4, with 3 μM MP-9, 20 mM CTAB and 20 μM *t*-BuOOH added to the medium at the time indicated by the *arrow*

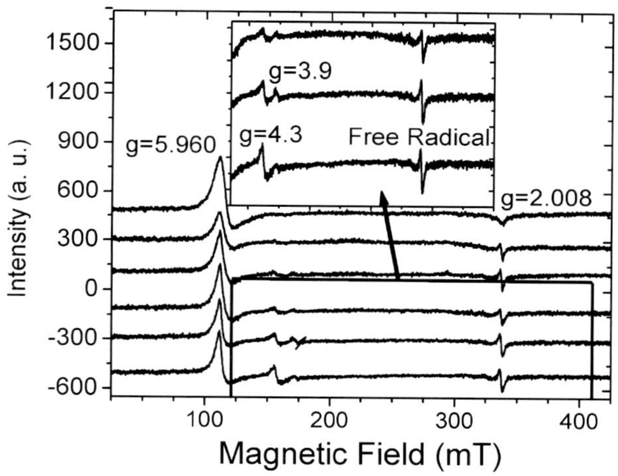

Fig. 4 Formation of MP-9 compound I (oxoferryl π cation) during the reaction with iodoso benzene. The spectra, from *top* to *bottom*, were obtained at the following times: 0 (absence of iodoso benzene), 4, 10, 15, 35 and 60 s after addition of 3 mM iodoso benzene. The *insert* shows a zoomed-in spectrum of the species with $g=3.9$ (compound I), its isoelectronic form, the species with $g=4.3$ and the iodoso benzene derived free radical. The experiments were carried out under the same experimental conditions described for Fig. 2b

(Fig. 2b). However, this intermediate ($g=3.9$) and its isoelectronic form ($g=4.3$) [10] were identified by EPR during the reaction of MP-9/CTAB with high concentrations of peroxide (over the range 0.2–1.2 mM) and with iodoso benzene (Fig. 4).

Discussion

The structure of the catalyst

MP-9 has a net negative charge owing to its heme propionate groups and the carboxy group of the C-terminal oligopeptide chain, providing electrostatic affinity for the cationic CTAB micelle surface. However, the hydrophobic groups present in the oligopeptide chain with the hydrophobic heme group must exhibit affinity for the micelle core. Hence, we propose a model for MP-9 associated with CTAB, where the heme propionate groups are associated to the micelle surface and the porphyrin ring and oligopeptide chain sides are partially inserted into the micelle (Scheme 1).

This model is in accordance with the UV–vis and EPR spectral changes observed when MP-9 is bound to CTAB

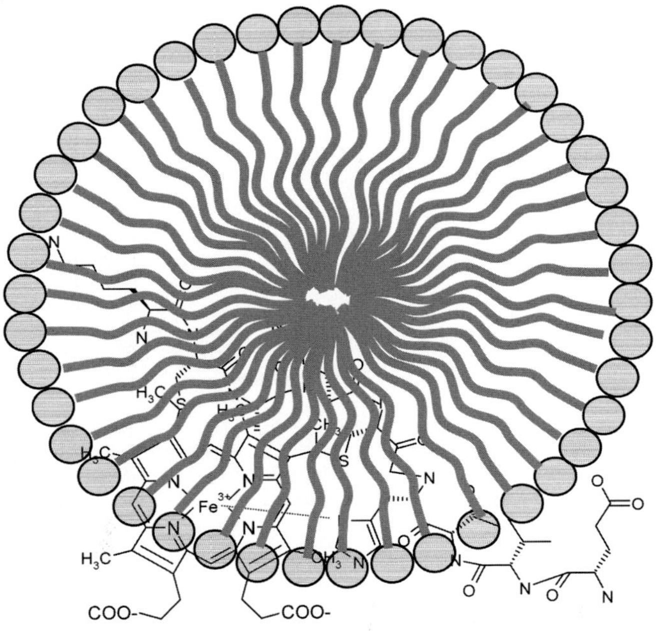

Scheme 1 Proposed model for the microperoxidase-9 (*MP-9*)/cetyltrimethylammonium bromide complex

micelles (Fig. 1). The bathochromic effect observed for the Soret band is typical of the heme group in a hydrophobic microenviroment [11]. Changes in the EPR parameters for the high spin state of MP-9 indicate the possibility that inside the micelle MP-9 is not coordinated with water but probably with OH^- or with the ε-amino group of its N-terminal lysine residue. Further experiments are necessary to characterize the ligand at the sixth coordination position of MP-9 associated to CTAB micelles.

In this model, the micelle constitutes an artificial enzymatic active site that exhibits higher affinity for hydrophobic substrates. In fact, we have previously observed that MP-9/CTAB and MP-8/CTAB complexes exhibit different affinity for *t*-BuOOH and HOOH. At pH 7.4, the Michaelis constant observed for *t*-BuOOH, $K_m = 0.17 \pm 0.01$ mM, was lower than that observed for HOOH, $K_m = 0.28 \pm 0.03$ mM. For this comparison it was necessary to use higher peroxide concentrations (over the range 0.2–1.2 mM) since conversion of microperoxidases to the oxoferryl form (compound II) was not detected at low HOOH concentrations (up to 0.05 mM), reinforcing the proposal of the catalyst structure.

MP-9/CTAB behaves as a true peroxidase

The occurrence of a peroxidase catalytic cycle during the reaction of MP-9/CTAB with *t*-BuOOH (Fig. 2a) shows that this heme peptide/micelle aggregate behaves as an artificial enzyme, here named lipoenzyme. The catalytic cycle observed when peroxide (in this case, *t*-BuOOH) was the only available substrate could implicate that the peroxide acts as an oxidizing (formation of compound I) and a reducing agent (formation of compound II and regeneration of the native form) of the prosthetic group. At this point an interesting question arises concerning the reaction mechanism: what is the reaction mechanism for the MP-9 enzymatic cycle in the absence of other reducing substrates? Certain peroxidases have been found to utilize HOOH to reduce compound I when no other substrate is available [12, 13]. In this case two pathways are proposed: (1) two-electron reduction of compound I by HOOH with production of O_2 [12, 13]; (2) one-electron reduction of compound I by HOOH generating O_2^- and compound II, which reacts with another HOOH molecule, producing H_2O and compound III. Compound III formation results in enzyme turnover in the absence of reducing substrates; however, progressive irreversible enzyme inactivation has also been identified under this condition [13]. However, Primus et al. [14] described the formation of MP-8 compound II in the presence of high HOOH concentrations (0.2–1.2 mM), with this intermediate being subsequently inactivated to species that exhibit Soret band bleaching. In the presence of CTAB micelles and high peroxide concentrations (0.2–1.2 mM), we have identified the occurrence of the same process for MP-8 and MP-9, suggesting that, both in a homogeneous medium and associated to CTAB micelles, MP-8 and MP-9 compound I are able to oxidize HOOH to produce compound II. Therefore, the reaction mechanism is in Scheme 2 is proposed:

It is known that protons in the distal heme active site influence the formation of the catalytic reactive species for both peroxidases and cytochrome P450 [15]. In horseradish peroxidase, the distal His42 ligand acts as an acid/base catalyst that favors the deprotonation of the hydroperoxide substrate at the enzyme active site (compound 0) and the subsequent heterolytic cleavage [16–18]. Therefore, the deprotonation of the hydroperoxide in compound 0 is a crucial step in the catalytic cycle of peroxidases and cytochrome P450. Recently, the competitive effect of protons has been demonstrated for the reaction of Fe(III)MP-8 or Mn(III)MP-8 with hydrogen peroxide, suggesting that compound 0 precedes the formation of a high valent intermediate of Fe(III)MP-8 or Mn(III)MP-8 [14]. In comparison with horseradish peroxidase, MP-8 in an aqueous medium does not exhibit a site pocket with a distal basic residue participating in the peroxide deprotonation and favoring the O–O peroxide bond cleavage. These characteristics could somehow be responsible for the low reactivity of MP-8 at low pH. In an aqueous medium,

Scheme 2 Proposed mechanism for the MP-9 cycle accompanied by oxygen consumption

$$\text{PorFe(III)-OH}_2 \rightleftharpoons \text{PorFe(III)-OH} + \text{H}^+ \quad \text{(favored by the micellar alkaline interface)} \quad (1)$$

$$\text{PorFe(III)-OH} + t\text{-BuOOH} \rightleftharpoons \text{PorFe(III)-OO}t\text{-Bu} + \text{HOH} \quad (2)$$

$$\text{PorFe(III)MP-9-OO}t\text{-Bu} \longrightarrow \text{Por}^{+}\text{Fe(IV)MP-9} + t\text{-BuOH (heterolytic cleavage)} \quad (3)$$

$$\text{PorFe(III)MP-9} + t\text{-BuOOH} \longrightarrow \text{PorFe(IV)MP-9} + t\text{-BuO}^{\bullet} \text{ (homolytic cleavage)} \quad (4)$$

$$\text{Por}^{+}\text{Fe(IV)MP-9} + t\text{-BuOOH} \longrightarrow \text{PorFe(IV)MP-9} + t\text{-BuOO}^{\bullet} \quad (5)$$

$$\text{PorFe(IV)MP-9} + t\text{-BuOOH} \longrightarrow \text{PorFe(III)MP-9} + t\text{-BuOO}^{\bullet} \quad (6)$$

$$t\text{-BuO}^{\bullet} \longrightarrow \text{CH}_3^{\bullet} + \text{acetone} \quad (7)$$

$$\text{CH}_3^{\bullet} + \text{O}_2 \longrightarrow \text{CH}_3\text{OO}^{\bullet} \quad (8)$$

alkaline pH leads to deprotonation of the MP-8 bound water, which then either assists concerted hydrogen peroxide deprotonation and coordination of the hydroperoxo group to the metal center or is directly oxidized by hydrogen peroxide to a metal-hydroperoxo MP-8 [14]. Since, in the presence of CTAB micelles, the enzymatic cycle occurs efficiently at pH 7.4, probably all MP-9 molecules associated to CTAB micelles are coordinated with deprotonated water, favoring the formation of compound 0. Considering that compound II had been detected spectroscopically and compound I has not been detected by EPR, a two-electron reduction of MP-9 compound I can be discarded. However, the mechanism to MP-9 compound II formation is uncertain. The micellar alkaline interface favors the deprotonation of the water molecule bound to heme iron and subsequent change for the substrate molecule with formation of MP-9 compound 0 (Eqs. 1, 2). After the formation of MP-9 compound 0, there are two possible mechanisms for the cleavage of *t*-BuOOH: a heterolytic cleavage generating *t*-BuOH and compound I (Scheme 2, Eq. 3), which reacts with another peroxide molecule generating compound II and *t*-BuOO$^{\cdot}$ (Scheme 2, Eq. 5) or a homolytic cleavage of *t*-BuOOH to form compound II directly (Scheme 2, Eq. 4). Whatever the mechanism for peroxide cleavage, native Fe(III)MP-9 can be regenerated after the reaction of compound II with another *t*-BuOOH molecule (Scheme 2, Eq. 6). The fact that compound I has not been detected during the MP-9/CTAB peroxidase cycle (Fig. 2b) could somehow induce us to accept the second possibility; however, the fact that EPR signal from a oxoferryl π cation could not be detected is not surprising considering the high reactivity of this peroxidase species. Moreover, the MP-9 compound I EPR spectrum has been detected in the presence of both *t*-BuOOH and HOOH at high concentration and in the presence of iodoso benzene (Fig. 4), showing that the MP-9/CTAB lipoenzyme can form this oxidized species.

Oxygen consumption during the reaction (Fig. 3) should come from the radical propagation as previously determined for the reaction of cytochrome c with *t*-BuOOH and depicted in Scheme 2, Eqs. (7) and (8) [19].

Conclusion

The MP-9/CTAB micelle complex behaves as a true peroxidase and is an interesting model to study peroxidase reaction mechanisms and for applications in nanotechnology. For the first time the enzymatic peroxidase cycle could be achieved for a microperoxidase molecule.

Acknowledgements We are grateful to Fundação de Amparo à Pesquisa do Estado de São Paulo and Conselho Nacional de Desenvolvimento Científico for financial support and to Mauren Luciana Estevam for technical assistance.

References

1. Carraway AD, Povlock SL, Houston ML, Johnston DS, Peterson J (1995) J Inorg Biochem 60:267
2. Veeger C (2002) J Inorg Biochem 91:35
3. Peterson J, Saleen MMM, Silver J, Wilson MTJ (1983) Inorg Biochem 19:165
4. Sharma VS, Schmidt MR, Ranney HM (1976) J Biol Chem 251:426
5. Kimura K, Peterson J, Wilson MT, Cookson DJ, Williams RPJ (1981) J Inorg Biochem 15:11
6. Tatsuma T, Watanabe T (1992) Anal Chem 64:143
7. Mazumbar S, Medhi OK, Mitra S (1991) Inorg Chem 30:700
8. Aron J, Baldwin DA, Marques HM, Pratt JM, Adams PA (1986) J Inorg Biochem 27:227
9. Dunford HB, Stillman JS (1975) Coord Chem Rev 19:187
10. Fujii H (2002) Coord Chem Rev 226:51
11. Makarska M, Radzki St., Legendziewicz JJ (2002) AlloyCompd 341:233
12. Claiborne A, Fridovich I (1979) J Biol Chem 254:4245
13. Hiner ANP, Ruiz JH, López JNR, Cánovas FG, Brisset NC, Smith AT, Arnao MB, Acosta M (2002) J Biol Chem 277:26879
14. Primus JL, Grunenwald S, Hagedoorn PL, Albrecht-Gary A-M, Mandon D, Veeger C (2002) J Am Chem Soc 124:1214
15. Mukai M, Nagano S, Tanaka M, Ishimori K, Morishima I, Ogura T, Watanabe Y, Kitagawa T (1997) J Am Chem Soc 119:1758
16. Ortiz de Montellano PR (1992) Annu Rev Pharmacol Toxicol 32:89
17. Savenkova MI, Kuo JM, Ortiz de Montellano PR (1998) Biochemistry 37:10828
18. Smith AT, Veitch N (1998) Curr Opin Chem Biol 2:269
19. Nantes IL, Faljoni-Alario A, Nascimento OR, Bandy B, Gatti R, Bechara EJH (2000) Free Radical Biol Med 28:786

Progr Colloid Polym Sci (2004) 128: 199–202
DOI 10.1007/b97072

Qing S. Mu
Ying H. Ma
Andrew L. Lewis
Steve P. Armes
Jian R. Lu

Complexation of DNA with biocompatible diblock copolymers

Q.S. Mu (✉) · J.R. Lu
Physics Department, University of Manchester Institute of Science and Technology, P.O. Box 88, Manchester, M60 1QD, UK
e-mail: q.mu@postgrad.umist.ac.uk

Y.H. Ma · S.P. Armes
School of Chemistry, Physics and Environmental Science, University of Sussex, Falmer, Brighton, BN1 9QJ, UK

A.L. Lewis
Biocompatibles Ltd., Farnham Business Park, Weydon Lane, Farnham, GU9 8QL, UK

Abstract A novel diblock copolymer, consisting of a biocompatible phosphorylcholine block (MPC) and a pH sensitive dimethylamino ethyl block (DMA), denoted as $MPC_{30}DMA_{60}$, shows interesting properties of surface adsorption and solution aggregation. These features have been exploited to condense DNA into nanoaggregates in bulk solution and to immobilise DNA onto a substrate surface. These molecular manipulations may find their applications in controlled release and bionanotechnology.

Keywords Surface adsorption · Nanoaggregates · DNA · Neutron reflection · Dynamic light scattering

Introduction

The delivery of DNA across a cell membrane represents a major limitation in nonviral gene delivery. The DNA molecule is negatively charged and the entire chain is rather stiff. Recent research has shown that complexation with an oppositely charged macromolecule helps to neutralise DNA and reduce the effective molecular size, resulting in significant improvement in gene transfection [1]. However, most synthetic polymers are cytotoxic and are inadequate as a nonviral gene delivery vehicle.

Phosphorylcholine (PC) is the lipid head group of the outer cell membrane and has the inherent biocompatibility to keep cells nonfouled in the blood stream. An extensive list of literature has demonstrated that synthetic polymers incorporating PC head groups offer haemo- and biocompatibility in vitro and in vivo [2]. Diblock copolymers are known to have attractive interfacial and aggregation properties that can be exploited for complexation with DNA. In this work, we show the basic colloidal behaviour of a diblock PC copolymer from a combined physical characterization of surface tension, neutron reflection and dynamic light scattering (DLS).

Materials and methods

Materials

The copolymer was made by reacting methacryloylphosphorylcholine (MPC) with 2-(dimethylamino) ethyl methacrylate (DMA) in a molar ratio of MPC to DMA of 1:2 [3] The copolymer, termed $MPC_{30}DMA_{60}$, has a methacrylate backbone pendent with PC groups on the MPC block and DMA groups on the DMA block. The molecular structure is shown schematically in Fig. 1. The molecular weight of $MPC_{30}DMA_{60}$ is 18,000 Da.

DNA was purchased from Sigma, UK (cat. no. D1501) and was highly polymerised from calf thymus. The pH of the solutions was adjusted using sodium hydroxide and hydrochloric acid. The total ionic strength was fixed at 0.02 M using NaCl. The large surface of the silicon block used in neutron reflection was polished using the procedures described previously. The freshly polished block was soaked in acid peroxide solution (600 ml 98%H_2SO_4 and 100 ml 25% H_2O_2) for 2 min at 90 °C, followed by rinsing with 5%

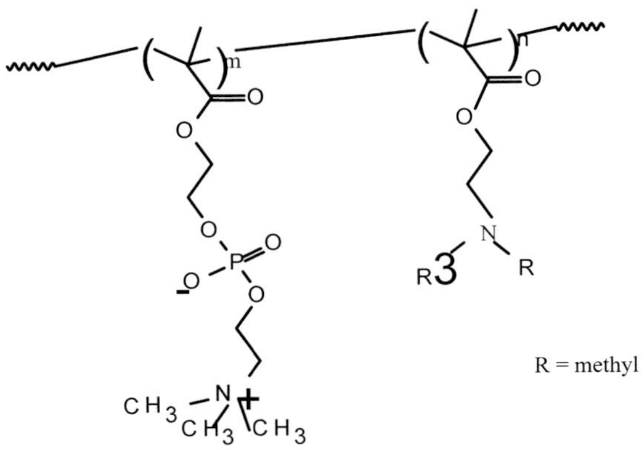

Fig. 1 Chemical structure of $MPC_{30}DMA_{60}$ copolymer

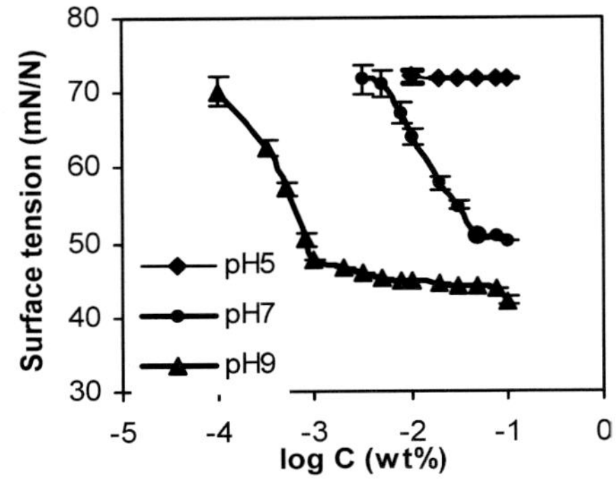

Fig. 2 Surface tension of $MPC_{30}DMA_{60}$ measured at 298 K

neutral Decon solution (Decon Laboratories, UK) and ultrapure water (Purelab UHQ, Elga, UK). This procedure was found to generate a surface with a reproducible hydrophilicity and thickness of the silicon oxide layer.

Physical characterization

The surface tension of the aqueous diblock solution was determined with a Krüss Tensiometer K11 (Krüss, Germany) using a Pt/Ir ring and was calibrated using the Harkins and Jordan method. The surface tension was found to show time-dependent variation over the first 20–30 min. All the measurements were made after the solution had been equilibrated for more than 30 min.

The neutron reflectivity (NR) measurements were carried out on the reflectometer SURF at the Rutherford Appleton Laboratory near Oxford, UK [4, 5]. The reflectivity, determined by the ratio of the reflected neutrons at each wavelength to the number incident on the sample, is measured as a function of momentum transfer Q [$Q=(4\pi\sin\theta)/\lambda$, where θ is the glancing angle of incidence and λ the wavelength of the neutrons]. The calibration and background treatment were handled as in Ref. [4].

DLS was performed on the polymer solution at pH 5, 7 and 9, using a Malvern Zeta 3000 spectrometer equipped with a 40-mW He–Ne laser. Measurements were made at an angle of 90° to the incident beam.

Results and discussion

Surface tension

The surface tension of the diblock copolymer, $MPC_{30}DMA_{60}$, from aqueous solution at 25 °C is shown in Fig. 2. It can be seen that at pH 7 and 9 the surface tension shows a break point, indicating the occurrence of a critical aggregation concentration (cac). At pH 5, the surface tension is close to that of pure water and shows no variation with bulk concentration, indicating no surface adsorption. The three surface tension plots together demonstrate a clear trend of surface activity against solution pH. They show that surface tension decreases with increasing solution pH.

The surface tension measurement serves as a useful indication of the colloidal behaviour of the copolymer. It shows that the copolymer tends to aggregate early when the solution pH is enhanced. The pK_a for the DMA group was reported to be around 6.5 [3] Thus at pH 7, the copolymer is likely to be partly charged. An increase in pH to 9 reduces the charge density on the DMA block and the DMA block becomes more hydrophobic, resulting in micellar aggregation at lower copolymer concentration. In contrast, lowering the pH below 6.5 results in increased protonation of the DMA block and the entire polymer becomes hydrophilic, resulting in a reduced driving force for surface adsorption or bulk aggregation.

Dynamic light scattering

DLS was first used to determine the size of the copolymer aggregates at pH 7. It was found that the copolymer itself gave a very weak scattering signal and the diameters of the aggregates were estimated to be around 8–10 nm. DNA solutions were subsequently titrated into the copolymer solutions and the scattering measurements performed accordingly. The distributions of the complex sizes determined under different charge ratios of the base pair of DNA to DMA at pH 7 are shown in Fig. 3. It can be seen that without any copolymer, the DNA is approximated by a very broad size distribution. With the addition of copolymer, the peak goes down drastically and the width of the distribution also narrows significantly. Further increase in the ratio of the polymer above 1:1 does not appear to cause any observable effect on the distribution of the complex sizes.

Similar experiments were performed at other pH values and the same trend of the condensing effect was observed. Note that below pH 7 the charge neutralisation may become important. However, within the experimental error, it is difficult to ascertain the difference arising from the pH effect, indicating that either the apparent charge effect is less significant than the hydrophobic interaction, or the light scattering is too coarse to reveal the detailed size distributions.

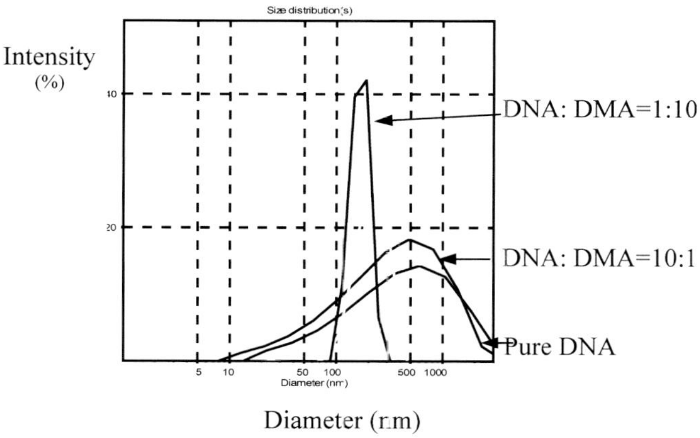

Fig. 3 Polymer complex size with different charge ratios of DNA to $MPC_{30}DMA_{60}$ at pH 7.1

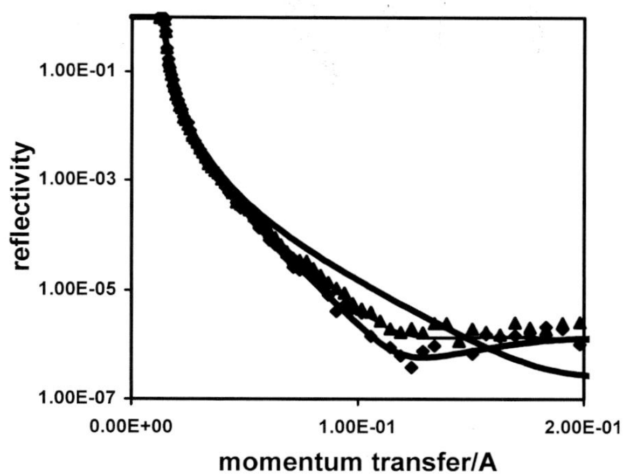

Fig. 4 Neutron reflectivity measured at the bare silicon oxide/D_2O interface (*top line*), after the copolymer adsorption (*middle line*) and the subsequent DNA binding (*bottom line*)

Specular neutron reflection

NR measured at the solid/water interface offers information about the adsorption of the polymer layer and any subsequent immobilisation of DNA onto the preadsorbed polymer. In this work, NR was first measured at the silicon oxide/D_2O interface to quantify the copolymer adsorption using 0.1% copolymer solution at pH 6.9. The difference between this reflectivity profile and that from pure D_2O indicates polymer adsorption. The continuous lines shown in Fig. 4 represent the best model fits calculated via the optical matrix method. The bare oxide layer was well represented by a uniform layer of 15 Å with scattering length density $\rho = 3.4\times10^{-6}$ Å^{-2}, indicating no cavity inside the oxide layer.

The modelling of the adsorbed copolymer layer was subsequently conducted assuming that the adsorption formed a uniform polymer layer on top of the silicon oxide. The best fit in the case of polymer adsorption gave a layer 30-Å thick and $\rho = 5.5\times10^{-6}$ Å^{-2}. The volume fraction of the copolymer inside the layer (χ_p) can be estimated from the following equation:

$$\rho = \chi_p \rho_p + \chi_w \rho_w, \tag{1}$$

where ρ_p and ρ_w are the scattering length densities of the polymer and solvent, respectively. The value of ρ_p was estimated to be around 0.68×10^{-6} Å^{-2} and that for pure D_2O to be 6.35×10^{-6} Å^{-2}; this gave $\chi_p = 0.15$. It should be noted that the uniform layer model did not reproduce the reflectivity over the low momentum transfer range well. It was found that the measured reflectivity was better modelled by adding a broad diffuse layer on the water side. This broad layer was some 40-Å thick with $\rho = 6.1\times10^{-6}$ Å^{-2}, giving $\chi_p = 0.04$. The best fit to the measured reflectivity with the adsorbed polymer film represented by a two-sub-layer distribution is shown in Fig. 4.

The adsorbed polymer layer was then rinsed with pure D_2O buffer and the reflectivity was found to be unchanged, as expected. This indicated the irreversible adsorption of the polymer layer on the SiO_2 surface. Subsequent reflectivity measurement was made when the buffer was replaced by 0.01 wt/vol % DNA solution with the charge ratio of DMA to the DNA base pair kept at 1:1, shown in Fig. 4. The resultant reflectivity is compared with the reference obtained from the bare silicon oxide/D_2O interface in Fig. 4. The increased gap between the two reflectivity profiles clearly indicates increased adsorption and hence DNA binding. The use of two sublayers was found to be adequate for modelling the mixed polymer/DNA layer distribution. The best fit shown in Fig. 4 represents an inner dense layer of 40 Å with $\rho = 5.5\times10^{-6}$ Å^{-2} and an outer layer of 65 Å with $\rho = 6\times10^{-6}$ Å^{-2}. This indicates that the binding of DNA has increased the overall layer thickness, but it does not appear to have altered the layer density or distribution.

Conclusions

The surface tension data show a strong pH-dependent adsorption and solution aggregation for the diblock copolymer. When the solution pH increases above the pK_a 6.5, the DMA block becomes less charged and more hydrophobic, and the copolymer is more surface-active. This is manifested in a reduced cac concurrent with a lowering of surface tension at the cac. When the solution pH decreases below 6.5, the DMA becomes more charged. The copolymer is then more hydrophilic and its tendency for surface adsorption and aggregation becomes weaker. It should also been noted that the aggregation behaviour of the copolymer is temperature-sensitive as well (data not shown here). The shift in the ratio of DMA to MPC should make it possible to optimise the copolymer to show a pH-sensitive aggregation around the human body temperature.

The preliminary DLS data serve to show the occurrence of complexation between the copolymer and DNA. More accurate measurements need to be carried out to ascertain the effect of charge ratio and solution pH. It is, however, clear that complexation of DNA with the copolymer reduces the size of the aggregates and narrows their range of distributions. The NR data are useful to demonstrate that the immobilisation of DNA onto the polymeric support can be monitored in situ. NR offers a useful means for the quantitative determination of the amount and distribution of the condensing copolymer and the subsequent DNA to be attached onto a support surface, making it feasible to link molecular details to effective surface functionalisation.

An important extension of this work in the near future will be to assess the cytotoxicity of the systems and verify the efficacy of gene transfection using bioactive DNA samples.

Acknowledgements We thank Biocompatibles UK Ltd., for financial support to Q.S.M. We also thank EPSRC for the support of the neutron experiments.

References

1. Labhasetwar V, Bonadio J, Goldstein S et al (1998) J Pharm Sci 87:1347–1350
2. Lobb EJ, Ma I, Billingham NC et al (2001) J Am Chem Soc 123:7913–7914
3. Ma YH, Tang YQ, Billingham NC et al (2003) Macromolecules 36:3475–3484
4. Penfold J, Ward RC, Williams WG (1987) J Phys E Sci Instrum 20:1411–1417
5. Lee EM, Thomas RK, Penfold J et al (1989) J Phys Chem-us 93:381–388

Progr Colloid Polym Sci (2004) 128: 203–206
DOI 10.1007/b97073

Ana Lúcia C.F. Souto
Erick F. Poletti
Clóvis R. Nakaie
Shirley Schreier

Fluorescence and circular dichroism study of the interaction between indolicidin, a tryptophan-rich antimicrobial peptide, and model membranes

A.L.C.F. Souto · S. Schreier (✉)
Department of Biochemistry,
Institute of Chemistry,
University of São Paulo, CP 26077,
05513-970 São Paulo, Brazil
e-mail: schreier@iq.usp.br
Tel: +55-11-30912179;
Fax: +55-11-30912179

E.F. Poletti · C.R. Nakaie
Department of Biophysics,
Federal University of São Paulo,
São Paulo, Brazil

Abstract We have examined the interaction of the antimicrobial peptide indolicidin (IND) with large unilamellar vesicles of the zwitterionic phospholipid 1-palmitoyl-2-oleoyl phosphatidylcholine (POPC) and of 2:1 mixtures of POPC and negatively charged 1-palmitoyl-2-oleoyl phosphatidylglycerol (POPG). We also investigated the interaction between IND and lysophosphatidylcholine (LPC) and LPC–lysophosphatidylglycerol (LPG) (2:1) micelles. The peptide's fluorescence intensity increased and the wavelength of maximum emission decreased upon binding to all model membrane systems. IND's binding to zwitterionic interfaces indicates the contribution of hydrophobic interactions. The stronger binding to POPC:POPG (2:1) than to pure POPC evinced the contribution of electrostatic interactions. Fluorescence quenching studies with 2,2,6, 6-tetramethylpiperidine-*N*-oxyl-4-amino-4-carboxylic acid (TOAC) suggest that IND resides in the water–bilayer interface region in vesicles and in LPC–LPG (2:1) micelles, being more deeply inserted in LPC micelles. IND's pH titration is drastically altered in the presence of lipid, probably due to p*K* shifts of the side chain residues, resulting from surface charge effects, and to binding-induced conformational changes. Circular dichroism spectra corroborate the conclusion that binding to model membranes induces peptide conformational changes.

Keywords Antimicrobial peptide · Fluorescence · Circular dichroism · Indolicidin · 2,2,6,6-Tetramethylpiperidine-*N*-oxyl-4-amino-4-carboxylic acid

Introduction

The increase of multi-drug-resistant bacteria has prompted the search for new antibiotics, leading to growing interest in naturally occurring antimicrobial peptides and their mechanism of action [1]. Antimicrobial cationic peptides are thought to be important in innate host defense against infectious agents. Indolicidin (ILPWKWPWWPWRR-NH_2, IND), a cationic antimicrobial peptide from cytoplasmic granules of bovine neutrophils [2], has a broad spectrum of antimicrobial activity. Its action is thought to involve interaction with membranes, leading to their disruption [3–5]. However, the detailed molecular mechanism of action is still unknown. The current study aims at providing further insight into this process.

Materials and methods

IND was synthesized according to the method in Ref. [6]. Peptide purity was checked by analytical high-performance liquid chromatography and mass spectrometry. Chloroform phospholipid solutions were evaporated with a N_2 stream and the sample was dried under vacuum for 2 h. The lipid films were suspended in 10 mM phosphate–borate–citrate buffer, pH 7.4. The suspensions were extruded to form 100-nm diameter large unilamellar vesicles (LUVs).

Lysophosphatidylcholine (LPC) and LPC:lysophosphatidylglycerol (LPG) (2:1) micellar stock solutions were prepared in 10 mM phosphate–borate–citrate buffer, pH 7.4. Fluorescence spectra were acquired using a Hitachi F-4500 fluorimeter. Circular dichroism (CD) spectra were obtained in a Jobin-Yvon CD6 spectrometer. All spectra were obtained at room temperature (22 ± 2 °C).

Results and discussion

IND's fluorescence intensity increased (Fig. 1) and the wavelength of maximum emission decreased from 348 nm in solution to 342 ± 1 nm in the presence of model membranes. The binding constants (K_a, per mol per liter), calculated from binding isotherms, were LPC 4.6×10^3, LPC:LPG (2:1) 4.2×10^3, 1-palmitoyl-2-oleoyl phosphatidylcholine (POPC) 16×10^3, and POPC–1-palmitoyl-2-oleoyl phosphatidylglycerol (POPG) (2:1) 24×10^3. The binding of IND to zwitterionic vesicles and micelles was previously observed [4, 7] and indicates that hydrophobic interactions contribute to binding. IND's stronger binding to negatively charged LUVs evinces that electrostatic interactions also play a role in this process.

The accessibility of the water-soluble quencher 2,2,6,6-tetramethylpiperidine-*N*-oxyl-4-amino-4-carboxylic acid (TOAC) to IND's tryptophan residues yielded information about the peptide insertion in the model membranes (Fig. 2). The order of fluorescence quenching was solution > POPC:POPG (2:1) γ√ POPC γ√ LPC:LPG (2:1) > > LPC, showing that IND is exposed to approximately the same extent in all model membranes, except for LPC. This indicates a more superficial peptide location (at the water–membrane interface) in the former systems and a deeper penetration in LPC. This latter result is probably due to a greater water exposure of the micelle hydrophobic region in view of its greater curvature, allowing the deeper peptide insertion. When comparing micelles, the greater accessibility in LPC:LPG (2:1) is very likely due to electrostatic interactions with the negatively charged head groups, which keep the peptide closer to the interface.

IND's pH titration (Fig. 3) in solution differs from the theoretical prediction based on the individual

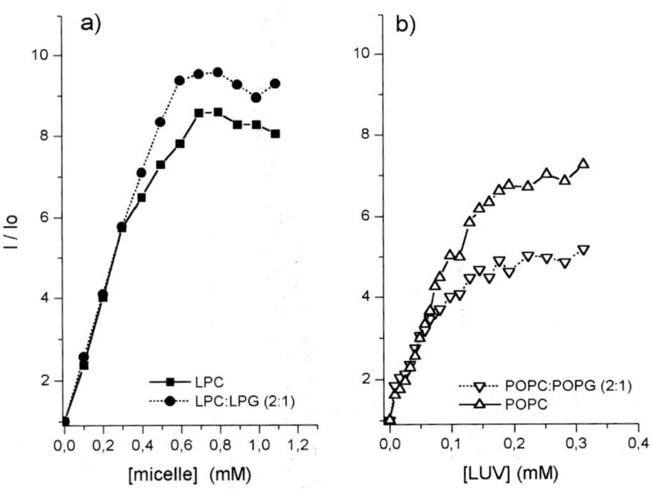

Fig. 1 Relative fluorescence of 20 μM indolicidin (*IND*) as a function of **a** micelle concentration and **b** large unilamellar vesicle (*LUV*) concentration. λ_{exc} = 290 nm, λ_{em} = 342 ± 1 nm, lysophosphatidylcholine (*LPC*), lysophosphatidylglycerol (*LPG*), 1-palmitoyl-2-oleoyl phosphatidylcholine (*POPC*), 1-palmitoyl-2-oleoyl phosphatidylglycerol (*POPG*)

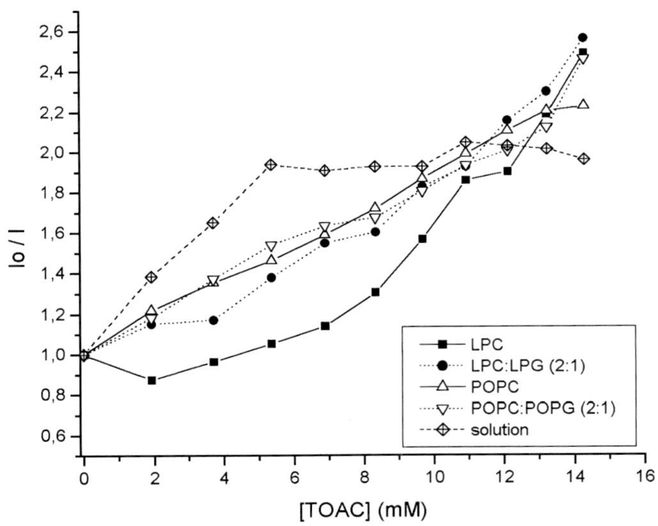

Fig. 2 Stern–Volmer plot for the quenching of 20 μM IND fluorescence by 2,2,6,6-tetramethylpiperidine-*N*-oxyl-4-amino-4-carboxylic acid (*TOAC*) in solution and in the presence of 0.8 mM micelles

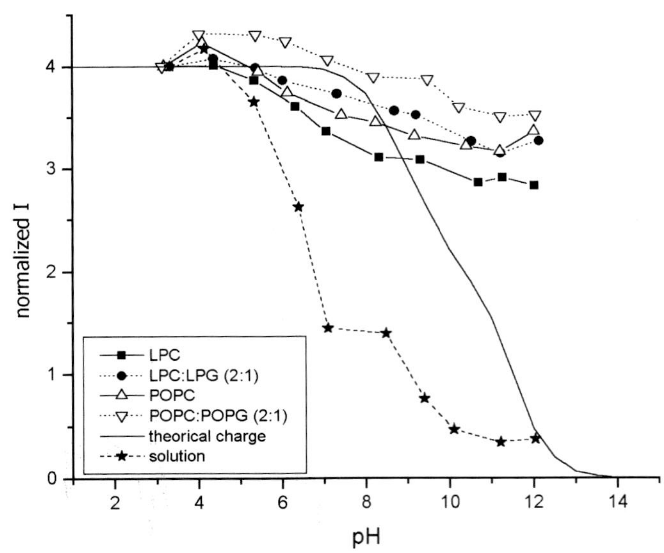

Fig. 3 IND normalized fluorescence as a function of pH in solution and in the presence of 0.8 mM micelles or 0.25 mM LUVs. λ_{exc} = 290 nm, λ_{em} = 343 ± 1 nm. The theoretical variation of IND's charge with pH is also shown. λ_{exc} = 290 nm, λ_{em} = 343 ± 1 nm [IND] = 20 μM

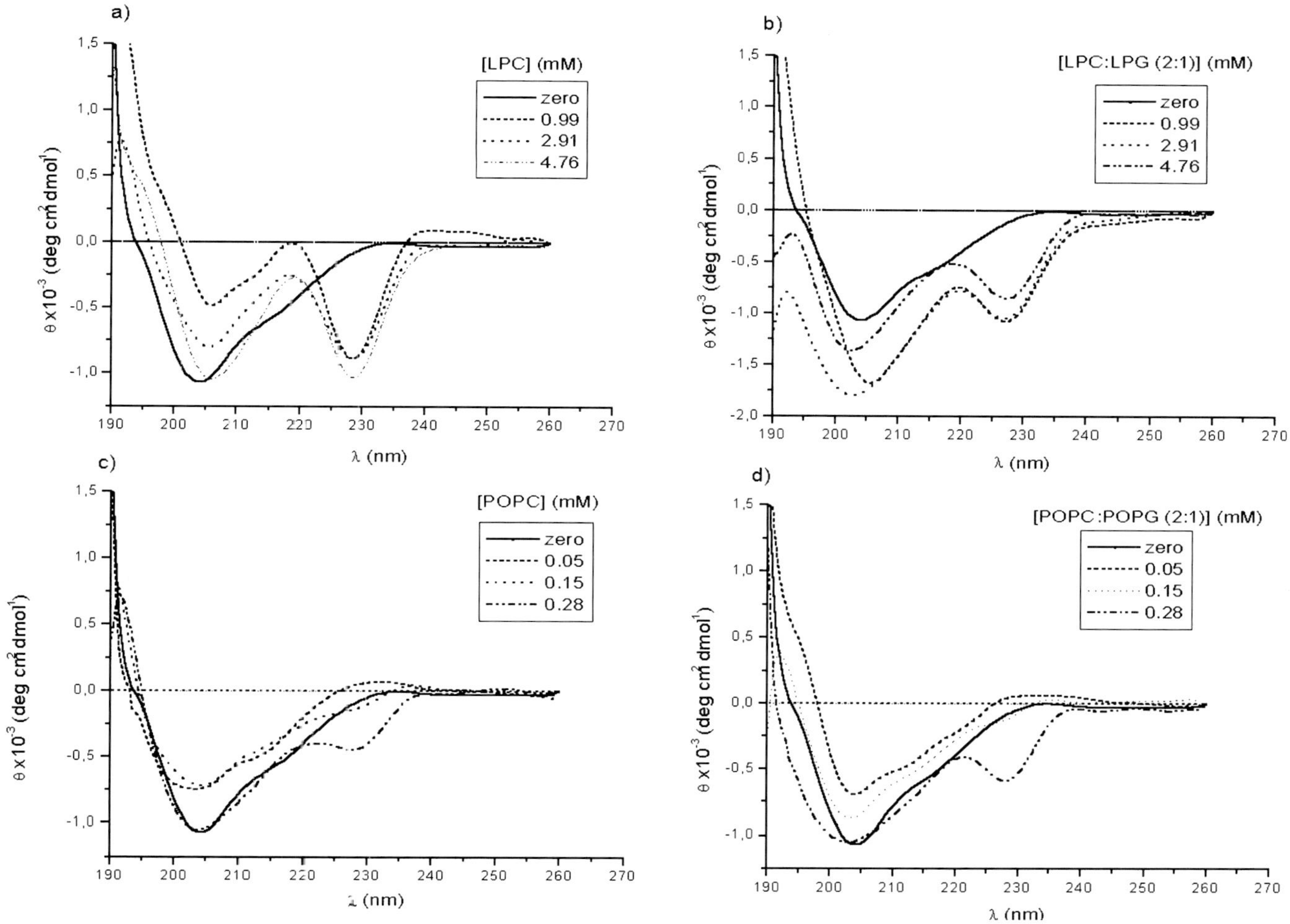

Fig. 4 Circular dichroism spectra of 100 μM IND in solution and in the presence of micelles of **a** LPC, **b** LPC:LPG (2:1), and of LUVs of **c** LUVs of POPC and **d** POPC:POPG (2:1)

amino acids p*K*s, probably because of the mutual effect of the charged residues upon each other's p*K*. Moreover, one would expect that binding to zwitterionic interfaces would cause a decrease in the p*K*s of positively charged groups, while binding to negatively charged interfaces would cause an increase [8]. Although striking alterations in the titration curves occurred in the presence of model membranes, this was not observed, strongly suggesting that, in addition to p*K* shifts, binding-induced conformational changes also contribute to the observed pH effects.

The occurrence of conformational changes was corroborated by CD spectroscopy. IND's CD spectrum in aqueous solution has a negative band with a minimum near 202 nm; a sharp negative band near 227 nm appears in the presence of model membranes [3, 7, 9–12]. Although IND's CD spectra in the peptide region are strongly influenced by the contribution of the five tryptophan residues, precluding reliable conformational analyses, the significant spectral changes in the presence of model membranes found in the present study (Fig. 4) clearly indicate that conformational changes occur.

In conclusion, the data show that IND binds to both zwitterionic and negatively charged interfaces and that both hydrophobic and electrostatic interactions play a role in binding. The quenching studies revealed the different accessibility of IND in the model membranes, the peptide being more deeply inserted in LPC micelles and more superficially located in the other systems. In membranes carrying negative charge, peptide–lipid electrostatic interactions play a role in maintaining the peptide at the water–membrane interface. The variable pH studies evinced p*K* shifts of the peptide's ionizable groups in solution. In the presence of membranes the effects were interpreted as reflecting, in addition to p*K* shifts, binding-induced conformational changes. This conclusion was corroborated by the changes in IND's CD spectra in the presence of membranes.

Acknowledgements This work was supported by grants from FAPESP and CNPq.

References

1. Schibli DJ, Epand RF, Vogel HJ, Epand RM (2002) Biochem Cell Biol 80:667
2. Selsted ME, Novotny MJ, Morris WL, Tang YQ, Smith W, Cullo JS (1992) J Biol Chem 267:4292
3. Falla T, Karunaratne DN, Hancock REW (1996) J Biol Chem 271:19298
4. Friedrich CL, Moyles D, Beveridge TJ, Hancock RE (2000) Antimicrob Agents Chemother 44:2086
5. Vogel HJ, Schibli DJ, Jing W, Lohmeier-Vogel EM, Epand RF, Epand RM (2002) Biochem Cell Biol 80:49
6. Barany G, Merrifield RB (1980) The peptides. Academic, New York
7. Ladokhin AS, Selsted ME, White SH (1997) Biophys J 72:794
8. Lee AG, Schreier S (1993) In: Gregoriadis G (ed) Liposome technology. Entrapment of drugs and other materials, vol II. CRC, Boca Raton, pp 1–25
9. Rozek A, Friedrich CL, Hancock REW (2000) Biochemistry 39:15765
10. Ladokhin AS, Selsted ME, White SH (1999) Biochemistry 38:12313
11. Friedrich CL, Rozek A, Patrzykat A, Hancock RE (2001) J Biol Chem 276:24015
12. Nagpal S, Kanwal JK, Deepti J, Salunke DM (2002) Protein Sci 11:2158

Progr Colloid Polym Sci (2004) 128: 207–211
DOI 10.1007/b97074

Daniel Kalnin
Modibo Ouattara
Michel Ollivon

A new method for the determination of the concentration of free and associated sodium caseinate in model emulsions

D. Kalnin (✉) · M. Ouattara · M. Ollivon
UMR CNRS 8612 Equipe Physico-Chimie des Systèmes Polyphasés,
92296 Châtenay-Malabry, France
e-mail: daniel.kalnin@web.de
Tel.: +33-1-46835465
Fax: +33-1-46835312

Abstract This work shows a close correlation between density, sound velocity and X-ray scattering measurements at very small angles (SAXS) for solutions of industrial grade sodium caseinate (SC). As SAXS measurements at synchrotron radiation sources are not applicable on a laboratory scale every day we prove a close correlation between the structural data derived from scattering techniques with combined density and sound velocity measurements. As a new method for the determination of SC concentration we propose density measurements which require minimal sample preparation, have a short measuring cycle of just a few minutes, and yield very accurate, reproducible results, which allows sample reuse after the measurements. Linear density changes on the order of $\Delta\rho^{1\ \mathrm{bar}}{}_{20\ ^\circ\mathrm{C}} = 0.0027 \pm 0.0003$ g cm^{-3} per weight percent of protein in the range of $0.99 \le \rho^{1\ \mathrm{bar}}{}_{20\ ^\circ\mathrm{C}} \le 1.04$ g cm^{-3} are addressed. Such density dependence enables us to accurately determine the SC concentration in the continuous phase of model emulsions. The droplet mean size $D[4,3]$ of emulsion droplets is measured by laser light diffraction using the Mie theory coupled with polarization intensity differential scattering. To determine the SC concentration in the serum in equilibrium with the surrounding emulsion particles dispersed and continuous phase are separated using centrifugation at 2400*g*. Coupling of density measurement that provides the SC concentration in the aqueous phase and the mean droplet size measurement allows us to evaluate the surface density on emulsion globules, which is essential for emulsion studies.

Keywords Density measurements · Small-angle scattering · Sodium caseinate · Emulsifier concentration · Emulsions

Introduction

Many dispersed multiphase-systems are thermodynamically unstable per se and thus require stabilization. In the food industry, stabilization of emulsions is frequently attained by the addition of milk proteins such as sodium caseinate (SC). The addition of such proteins has a double stabilizing effect since it provokes together with other added surfactants an increase of both the viscosity of the continuous phase and stability of the interface [1–3]. Then, multicomponent systems – with any component being susceptible to influence the physical chemistry of the others – render investigation of emulsions very difficult [4–6]. Numerous phases coexist in emulsions. Moreover, a supplementary phase is added to the system each time one of the components crystallizes or is at least partly crystalline [2]. The crystallization behaviour inside the emulsion droplets not only depends on the sample's time/temperature history but also on the interface around the fat globule [7].

For many years the existence of the casein submicelle has been discussed intensely. Holt et al. [8] brought a little more light in the discussion using small-angle X-ray scattering (SAXS) and contrast-variation small-angle neutron scattering. We will not focus on structural determination but on a simple way of concentration measurements for solutions of industrial grade SC [9] (for native caseinates see Refs. [1, 8, 10]) and SC-stabilized emulsions. Previous studies showed that small-angle scattering (SAS) measurements of SC-stabilized emulsions show an intense central scattering [9] which is varies with the SC concentration. Other studies have shown that crystallization in the emulsion is dependent on the surface coverage and the mean size of the emulsion droplets [7]. As the characterization of thermal and structural behaviour of fat in emulsion droplets, or more generally of any substance in finely dispersed systems, represents a main challenge in many fields, a simple technique determining the surface coverage of the interface in dispersed systems is needed. In this paper, we present a new method for the determination of particle surface coverage by SC based on density measurements. In this respect, easy-to-apply techniques are well suited to study colloids of biological, pharmaceutical, medical or industrial interest.

Material and methods

Binary SC/water system

SC was industrial food grade (Armor Protéines, Saint Brice en Cogles, France). The same batch of SC was used for binary water systems and emulsion stabilization. Samples for the study of the binary SC/water system were prepared by dispersing weighed amounts of SC in Millipore-quality water and after mixing they were weighed for concentration determination. Samples of about 20 cm^3 were prepared with and without potassium sorbate for preservation purposes. Some samples (2, 5, 10 wt %) were prepared for scattering experiments by adding 0.5 and 1 wt % of GMS, which was mainly composed of glycerol monostearate and glycerol distearate and was provided by Danisco Coulter (Denmark).

Oil-in-water model emulsions

Ingredients of industrial food grade were weighed before emulsion preparation. SC was first dissolved in preheated water and so hydrated at $T\approx60$ °C. Then, under mechanical mixing conditions, preheated fully hydrogenated vegetal fat (a palm oil/palm kernel oil mixture) was added to the preemulsion at $T\approx60$ °C. Model emulsions containing 20 wt % fat were prepared using a APV Lab 2000 two-stage homogenizer using a homogenization pressure of about 60 MPa. Different protein concentrations ranging from 1 to 4 wt % were used. The mean droplet size $D[4,3]$ was determined in all emulsions using a laser light diffraction technique using the Mie theory (Coulter LS 230, Miami, USA) coupled with polarization intensity differential scattering (PIDS). One has to note that those emulsions are stable for months at room temperature. They display the same fat globule size distribution after that time providing they are lightly agitated by shaking by hand before the actual measurement.

SAS measurements

X-ray diffraction was performed using the high-flux of synchrotron radiation at beamline D24 of LURE (DCI, Orsay, France) using a wavelength of 1.6 Å. Calibration was carried out using silver behenate [11] and tristearin (SSS) which perfectly agree. A gas-filled position-sensitive linear detector was used for data collection using a sample-to-detector distance of about 2.1 m. All X-ray diffraction patterns were recorded by transmission using glass capillaries (GLAS W. Müller, Berlin, Germany, diameter about 1.5 mm (hand calibrated), wall thickness 0.01 mm). Samples were prepared by filling these glass capillaries with about 0.02 cm^3 of sample using a special laboratory-made syringe. The samples were placed in a MICROCALIX calorimeter designed in our laboratory for measurements in the temperature range –30 to 120 °C. To ensure good dispersion of the lipid emulsifier in the binary SC/water system, temperature cycling was performed between 30 and 60 °C for 6 h. X-ray scattering was monitored at 20.0 °C for all the samples.

Density and ultrasound velocity measurements

Density and ultrasound velocity measurements were carried out about 24 h after preparation of the binary samples to ensure equilibration of protein solutions using 2 cm^3 of bubble-free sample. For emulsion samples (storage condition 4 °C) the creaming process leading to complete phase separation was accelerated by centrifugation (SIGMA 3K30) at 20 °C, 2,400*g* for 600 min. After centrifugation, differential scanning calorimetry was used to ensure the absence of lipids in the inferior phase in order to verify the effective separation of the dispersed lipid phase from the continuous aqueous phase (T. Cantat, D. Kalnin, M. Ollivon, unpublished work) [12]. The complete absence of exothermic events – during the cooling scan at 10 °C min^{-1} – warrants for the completeness of the separation. Observation of peaks means incomplete phase separation since crystallization of either pure or emulsified fat leads to exothermic events [7]. The density measurements along with the ultrasound velocity measurements were performed using a Paar Beer Analyzer 2 (Anton Paar, Graz, Austria). A volume of 2 cm^3 of the inferior phase is needed for a correct measurement at 20 °C. Milli Q-quality water was taken as a standard for calibration ($\rho^{1\,\mathrm{bar}}{}_{20\,°\mathrm{C}}=0.9983$ g cm^{-3}, $c^{1\,\mathrm{bar}}{}_{20\,°\mathrm{C}}=$ 1.482 km s^{-1}). Samples were conserved using 0.05 wt % potassium sorbate.

Results and discussion

The systems investigated have two origins. On the one hand, the binary system of SC/water was investigated and compared with the ternary system SC/water/GMS. On the other hand, the triglyceride emulsion was phase-separated and the aqueous phase analyzed. The results of measurements on the aqueous phase were then also compared with those for the binary system.

SAS of binary SC/water and ternary SC/water/GMS systems

Scattering from SC solutions is detectable using SAS at a sample-to-detector distance of approximately 1.2–1.5 m. However, quantitative measurements of scattering for

SC concentration determination require a longer sample-to-detector distance [9] such as 2.1 m at D24. At the low concentrations examined (2 and 5 wt %), each SAXS pattern, shown in Fig. 1, is the sum of three expositions of 10 min each. For higher amounts of SC (e.g., 10 wt %) a scattering peak is clearly visible in the log–log plot. The dependence of the peak maxima on SC concentration is similar to that described by Kalnin et al. [9]. Adding GMS to the binary SC/water system might give birth to a lamellar structure at $q \sim 0.10$ Å^{-1} ($d \sim 60$ Å). This diffraction peak is characteristic (Fig. 1 insert, higher-q region) of an excess of lipid emulsifier [13]. However, the presence of the added lipid emulsifier does not influence the position of the scattering peaks at about $q \sim 0.03$ Å^{-1}, indicating that GMS addition changes neither the protein nor the calcium phosphate structure of the micelle [8]. At low concentrations GMS phase-separates from the other components and does not affect the SC measurement.

Density measurements of the binary SC/water system

Solubility of SC in aqueous solutions is influenced by many factors, including temperature, pressure and pH [6]. Measurements were performed at pH 6.5, atmospheric pressure and $T = 20.000 \pm 0.001$ °C. The bottom part of Fig. 2 reflects then density measurements of the binary SC/water solution with and without potassium sorbate as a preservative The region of concentrations measured is above the concentration at which submicelles or clusters are formed. The measured densities in the concentration range from 0.06 to 19 wt % SC comprise between $0.998 \le \rho^{1\ \mathrm{bar}}20\ °\mathrm{C} = 1.043$ g cm^{-3}. A linear change of density is measurable as a function of the concentration in this range, which indicates that the SC concentration in the aqueous phase is always above the SC concentration of aggregation but below the SC concentration of gel formation. This change in density correlates with the change of the scattering maximum (q_{max}) (Fig. 2, top) [9]. As seen in Fig. 2 both density and SAXS at very small angles measurements allow us to monitor even slight decreases of the concentration.

Density measurements from the aqueous phase of SC-stabilized oil-in-water emulsions

It has been observed that emulsions similar to those examined here, and also stabilized with different amounts of SC but with no lipidic emulsifier, show changes in the central scattering depending on SC concentration as binary SC/water solutions [9]. Then also the concentration of SC in the aqueous phase should change. Emulsions containing 20% fat were stabilized with 1–4 wt % SC, which leads to a different size distribution using the same homogenization pressure. The total amount of SC $[\mathrm{SC}]_{tot}$ partitions into the amount of SC at the interface $[\mathrm{SC}]_{lip}$ and the free SC $[\mathrm{SC}]_{aq}$ according to

$$[\mathrm{SC}]_{tot} = [\mathrm{SC}]_{lip} + [\mathrm{SC}]_{aq}.$$

The mean droplet size $D[4,3]$ was monitored using the PIDS technique and laser light diffraction and is shown versus $[\mathrm{SC}]_{tot}$ in Fig. 3. For small amounts of $[\mathrm{SC}]_{tot}$ the

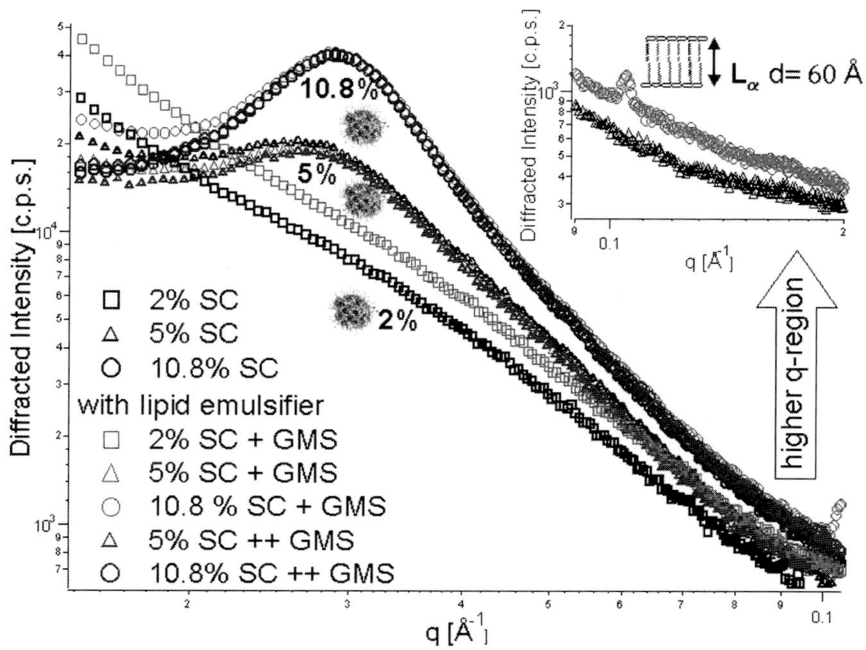

Fig. 1 Evolution of small-angle scattering (*SAS*) patterns of binary and sodium caseinate (*SC*)/water and ternary SC/water/glycerol monostearate (*GMS*) systems at a sample-to-detector distance of 2.1 m as a function of SC concentration in aqueous solutions with and without emulsifier. *Insert*: SAS pattern of a higher-q region shows the diffraction peak of a lamellar phase formed by excess lipid emulsifier GMS in the ternary SC/water/GMS system when present

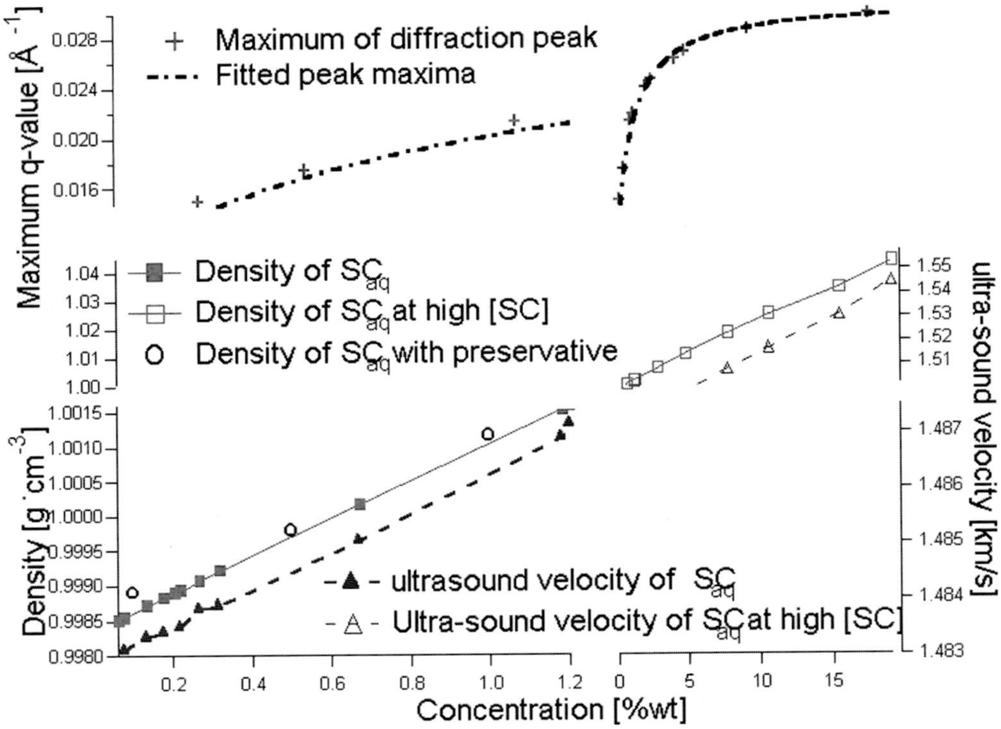

Fig. 2 Linear evolutions of density (*squares*) and ultrasound velocity (*triangles*). Measurements are plotted on two different scales, *left* 0–1.2 wt %, *right* 0–19 wt %, in order to reveal linearity for both concentrated and dilute samples of the binary SC/water system. SC/water solutions preserved with potassium sorbate (*circles*) are drawn for comparison. *Top*: For both scales, the corresponding maximum positions of the SAS peaks of the binary SC/water system, determined from the normalized SAS patterns such as in Fig. 1, are shown

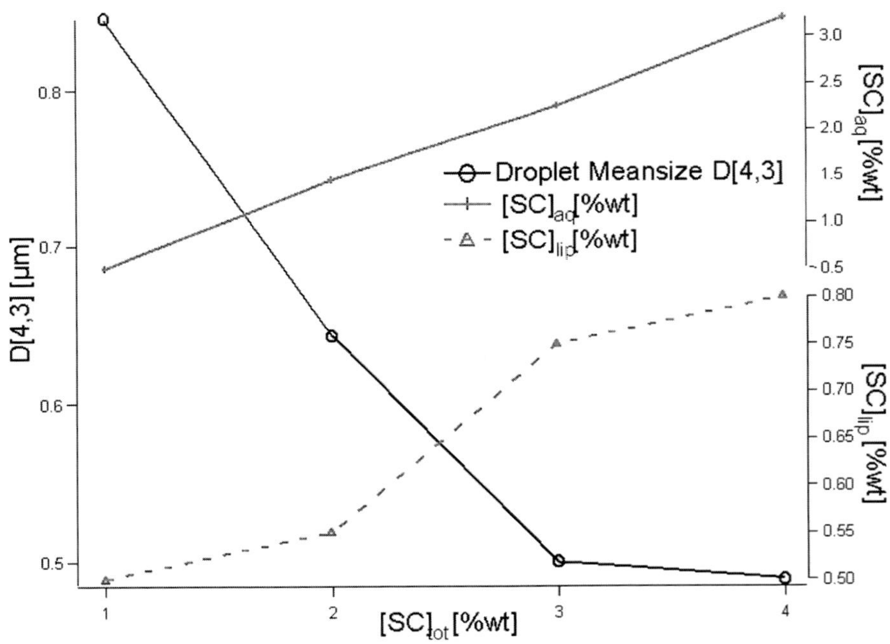

Fig. 3 Mean droplet-size $D[4,3]$ (*circles*) measured with laser light scattering as a function of the total protein concentration. $[SC]_{tot}$ is plotted against the free and bound $[SC]_{aq}$ (*vertical bars*) and $[SC]_{lip}$ (*triangles*) determined by density measurements of the aqueous phase resulting from the centrifuged emulsion

concentration seems to limit the fat globule size in the emulsion, whereas for higher concentrations the homogenization pressure seems to be the limiting factor. Since the globule size distribution does not change at $[SC]_{tot} > 3$ wt %. This means that there will be a partition $K = [SC]_{lip}[SC]_{aq}^{-1}$ between the aqueous phase and the interface depending on the surface area of the interface.

We suppose for the further interpretation that the gravitational force that we apply to reinforce the creaming procedure is too weak to deplete the fat globules of their $[SC]_{lip}$ and only separates the aqueous phase "as it is" without changing this partition. Thus, we can use density measurements to directly measure the density in the aqueous phase. The $[SC]_{lip}$ concentration of the interface can then be determined from the $[SC]_{aq}$ as shown in Fig. 3.

Additional 1 wt % in excess of the $[SC]_{tot}$ limit concentration of 3 wt % (then 4 wt % $[SC]_{tot}$) does not change the size distribution of the emulsion droplets then the SC in excess is expected to be located in the aqueous phase, which is observed. Those emulsions are stable for months without destabilization or flocculation even for high concentrations of $[SC]_{tot}$. When no destabilization occurs and in this respect the fat globule interface stays intact then both SAXS at very small angles and density measurements allow us to determine free and associated SC in complex emulsion systems providing the substances added do not interfere with scattering and the density of free SC.

Conclusion

Emulsions are complex systems in which the interface plays a critical role; however, this interface cannot be studied directly. Actually, the interface has an influence on the physical chemical properties of the emulsion. Providing that the partition coefficient K is not affected by the centrifugation step, our new method based on density measurements allows, as SAS experiments, the quantitative characterization of SC content in the aqueous phase of oil-in-water emulsions. Once K is known and the mean size of the emulsion droplets has been measured, the interfacial composition can be determined after measuring the "excess" concentration in the aqueous phase.

Acknowledgements The authors thank Nestlé and ANRT (Paris, France) for supporting this research. Beamline setup was performed by Claudie Bourgaux. Beamtime at LURE D24 workbench is acknowledged.

References

1. Farrer D, Lips A (1999) Int Dairy J 9:281
2. Kalnin D, Arnaud G, Amenitsch H, Ollivon M (2002) Food Sci Int 35:927
3. Dickinson E (1999) Int Dairy J 9:305
4. Cornec M, Wilde PJ, Gunning PA (1998) J Food Sci 63:39
5. Walstra P (1979) J Dairy Res 46:317
6. Gayot P, Lorient D (1998) Structures et technofonctions des protéines du lait. Tec et Doc Lavoisier, Paris
7. Lopez C, Bourgaux C, Bernadou S, Keller G, Lesieur P, Ollivon M (2002) J Colloid Interface Sci 254:64
8. Holt C, de Kruif CG, Tuinier R, Timmins PA (2003) Colloids Surf A 213:275
9. Kalnin D, Quennesson P, Artzner F, Schafer O, Narayanan T, Ollivon M () Prog Colloid Polym Sci (in press)
10. Storthart DG, Cebula DJ (1982) J Mol Biol 227:391
11. Huang TC, Toraya H, Blanton TN, Wu Y (1993) J Appl Crystallogr 26:180
12. Vanapalli AS, Coupland JN (2001) Food Hydrocolloids 15:507
13. Krog N (1997) In: Friberg SE, Larsson K (eds) Food emulsions, 3rd edn. Dekker, New York

Progr Colloid Polym Sci (2004) 128: 212–215
DOI 10.1007/b97075

Amane Jada
Hamidou Haidara

Effects of the solid surface and polyelectrolyte on the two-dimensional organisation of calcium carbonate particles

A. Jada (✉) · H. Haidara
Institut de Chimie des Surfaces et Interfaces, ICSI-CNRS,
Université de Haute Alsace,
15 rue Jean Starcky,
68057 Mulhouse, France
e-mail: a.jada@univ-mulhouse.fr
Tel.: +33-3-89608709
Fax: +33-3-89608799

Abstract Calcium carbonate ($CaCO_3$) particles were precipitated from supersaturated solutions. The effects of the poly(sodium 4-styrene sulfonate) used as an additive and of the silicon wafer surface present in the precipitation medium on the two-dimensional organisation of $CaCO_3$ particles were investigated. The nature of the solid substrate was varied, using a hydrophilic nonmodified silicon oxide containing hydroxylated surface groups, and a hydrophobic silicon oxide modified with *n*-hexadecyltrichlorosilane. The micrographs indicate that the surface ordering and densities of the $CaCO_3$ particles depend on the nature of the silicon wafer dipped during the precipitation experiments. Two-dimensional ordering of single and aggregated $CaCO_3$ particles were obtained, respectively, on the homogenous hydrophobic *n*-hexadecyltrichlorosilane and on the hydrophilic nonmodified silicon oxide surfaces. The results indicate that the surface coating can be controlled not only by the nature of the solid substrate and additives present in the precipitation medium but also by the coating mode.

Keywords Crystallisation · Polyelectrolyte · Hydrophilic–hydrophobic silicon oxide dipped surface

Introduction

Calcium carbonate, $CaCO_3$, in its various forms, is the most widely used mineral filler and pigment; however large-scale production of $CaCO_3$ particles with uniform morphology and size is still a challenge. Hence, the most practical uses of $CaCO_3$ particles are dependent on the particle size distribution and its morphology. Compounds such as surface-active additives [1] or enzymes [2] were used to control the $CaCO_3$ particle morphology and size. Further studies dealing with crystallisation of $CaCO_3$ on solid surface [3] or in solution [4] have shown that the morphology and the growth of the crystal may be affected in the presence of foreign compounds or additives. For instance, in biological mineralisation, calcium-containing salts are the major constituents of mollusc shells. In these mineralised tissues, precipitation of oriented aragonite and calcite $CaCO_3$ forms occurs within polymeric tissue matrices. Acidic proteins [5] which are located at the surfaces of the matrix interact from the solution with the growing $CaCO_3$ crystals and are responsible for both nucleation and inhibition of crystallisation. Others studies on morphology changes have included growing crystals, using an organic template, such as β-chitin, or an ammonium surfactant to form aragonite or vaterite [6, 7].

It should be emphasised that the $CaCO_3$ mineral exists in various polymorph phases that are, in order of decreasing solubility, calcium carbonate hexahydrate, calcium carbonate monohydrate, vaterite, aragonite, and calcite. The two first polymorphs are unstable and are converted into the most thermodynamically stable form, calcite. Vaterite is also unstable and is formed first in aqueous solution under conditions of spontaneous precipitation. This phase transforms rapidly into calcite, but may be stabilised in the crystallisation medium in the presence of polymeric additives.

Adsorption of particles from colloidal suspensions onto solid substrates and their organisation is of great interest to understand the mechanisms involved in the surface coating. The important dependence of the substrate nature on the orientation and organisation of adsorbed homogeneous particles from colloidal suspension has been demonstrated [8]. Accordingly the organisation and orientation of the deposited particles result from an electrostatic interaction between the solid surface and the polar colloidal particle. However, the mechanism by which the solid surface affects the organisation of colloidal particles may involve other solid surface properties and forces, such as wetting properties of the surface and the competition between particle–particle, particle–solvent and solvent–surface interaction forces. Further, it is known that positively or negatively charged colloidal particles adsorb and may form aggregates on hydrophobic or hydrophilic substrates, whatever the sign of the solid surface charge. The surface wettability and van der Waals and capillary forces mainly control such processes.

In previous works we investigated the effect of the polyelectrolyte on the modification of the structure and morphology of $CaCO_3$ particles [9, 10]. The new results that we show in the present study deal with the preparation of $CaCO_3$ particles from supersaturated solutions and their organisation on hydrophilic and hydrophobic silicon wafers that were dipped in the precipitation medium prior to the $CaCO_3$ nucleation.

Experimental

Materials

The polyelectrolyte used was the polystyrene sulfonate, which was purchased from Aldrich and has a molecular weight value of 7×10^4 g mol^{-1}. The anhydrous sodium carbonate, Na_2CO_3, and the calcium nitrate tetrahydrate, $Ca(NO_3)_2\ 4H_2O$, were purchased, respectively, from Fluka Chemie and from Prolabo. All the reagents were used as received without further purification.

Coverage of the silica surface by *n*-hexadecyltrichlorosilane molecules

n-Hexadecyltrichlorosilane (HTS) was deposited on silica oxide as described elsewhere [11]. This organic film forms a self-assembled monolayer which is densely packed and ordered. By depositing such an organic molecular film on silicon oxide, we achieve full coverage of the silica surface forming a hydrophobic layer. The hydrophobicity of the final HTS-modified silicon oxide surface was checked by measuring the water advancing contact angle ($\theta_a = 111 \pm 1.5°$) and the receding contact angle ($\theta_r = 101 \pm 1.5°$).

Preparation of calcium carbonate particles and coating of solid substrates

The $CaCO_3$ particles were prepared in the presence of the polyelectrolyte by addition of equal volumes of Na_2CO_3 to $Ca(NO_3)_2\ 4H_2O$ solutions [9, 10]. In these experiments the aqueous polymer solution was first added to calcium nitrate solution and the mixtures were stirred for about 15 min prior to the addition of sodium carbonate. Further, in each precipitation experiment, the silicon oxide whether it was modified or not was dipped and kept in contact with the solution for about 5 min prior to the addition of Na_2CO_3 solutions. The period of time of the $CaCO_3$ precipitation experiments was about 10 min and the experiments were made at ambient temperature in a water-jacketed Pyrex glass vessel of 300-ml capacity. At the end of the experiments, the silicon substrates covered with the $CaCO_3$ crystals were withdrawn from the crystallisation medium, washed with pure water, dried and examined by optical microscopy.

Results and discussion

Micrographs at various magnifications and representing the single spherical $CaCO_3$ particles randomly distributed on the hydrophobic HTS surface are given in Fig. 1.

Previous X-ray diffraction analyses have shown that such particles obtained in the presence of the polyelectrolyte are vaterite [10]. The resulting low surface density on the HTS surface and the absence of large aggregated structures result from the balance between the particle–particle, the particle–surface, the particle–solvent and the solvent–surface interactions. It should be emphasised that $CaCO_3$ aggregates do not form in solution and a narrow size distribution for single $CaCO_3$ particles with a mean size around 2 μm was observed in the presence of polyelectrolyte [10]. Further, the particles are negatively charged owing to the presence of polyelectrolyte on their surface. However, such aqueous $CaCO_3$ dispersions are not stable and we observed settling of $CaCO_3$ particles in the reactor a few minutes after the end of the precipitation experiments. Therefore, it is likely that the vaterite $CaCO_3$ particles have hydrophobic and hydrophilic surface sites owing to partial screening of the polyelectrolyte surface charge by the residual salt present in the precipitation medium. In fact, during the complexation of the polyelectrolyte by the calcium cations, an exchange process of the polyelectrolyte counterions, i.e. the sodium ions, by the calcium ions occurs and leads to a reduction of the mutual repulsion between the polyelectrolyte segments. In addition, when the carbonate salt is added to the precipitation medium, the $Na(NO_3)$ salt is formed. Several authors [12–14] reported the hydrophilic and hydrophobic behaviours of poly(sodium styrene sulfonate) as function of salt. In the present work it should be noted that the final $CaCO_3$ particle results from nanocrystallite aggregation that occurs at the isoelectrical point of the mineral (i.e. pH 10.3). In order to control the surface charge of $CaCO_3$ particles we measured their zeta potentials by microelectrophoresis. We found that the zeta potential decreases from –50 mV for $CaCO_3$ particles in their precipitated medium to –13 mV after recovery by filtration, washing with water and dispersion of the $CaCO_3$ particles in salt-free water. Thus, owing to their low surface charges, single $CaCO_3$ particles

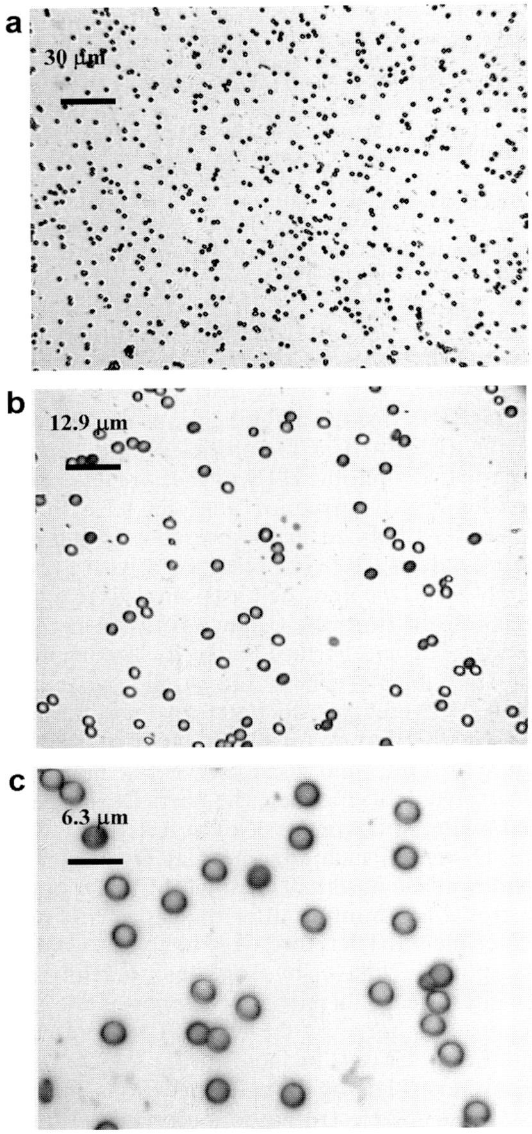

Fig. 1a–c Various magnifications of optical micrographs of $CaCO_3$ vaterite particles grown in the presence in solution of the polyelectrolyte at a concentration of 1.6 g L^{-1} and a dipped hydrophobic *n*-hexadecyltrichlorosilane (*HTS*) surface. $[Ca^{2+}]=[CO_3^{2-}]=0.007$ M

prepared in the presence of the polyelectrolyte may adsorb either on hydrophobic HTS or on the hydrophilic nonmodified SiO_2 surface. In this adsorption process, the interactions between the particles and the surfaces involve mainly the long-range London–van der Waals dispersion forces. Further, the absence of large aggregates on the HTS surface results from the nonwetting properties of the substrate.

Micrographs of the hydrophilic silicon covered with agglomerates of spherical $CaCO_3$ primary particles that were prepared in the presence of the polyelectrolyte are shown in Fig. 2.

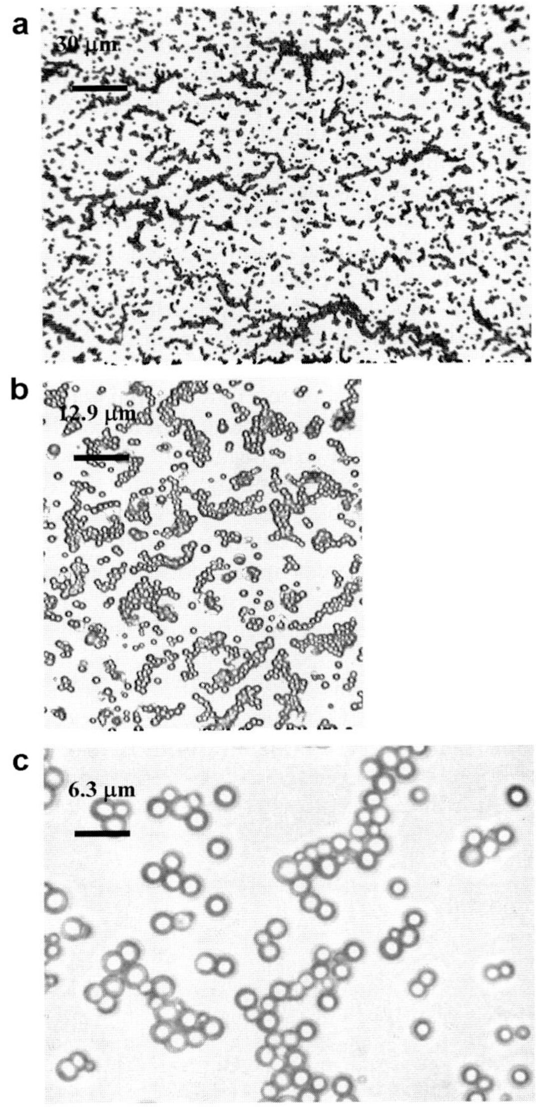

Fig. 2a–c Various magnifications of optical micrographs of $CaCO_3$ vaterite aggregates grown in the presence in solution of the polyelectrolyte at a concentration of 1.6 g $L^{-1}$and a dipped nonmodified hydrophilic SiO_2 surface. $[Ca^{2+}]=[CO_3^{2-}]=0.007$ M

The various magnifications indicate aggregates of larger size and higher surface coverage on the hydrophilic substrate compared with the HTS surface. In addition, the organisation of $CaCO_3$ aggregates on the hydrophilic surface seems to form an ordered two-dimensional network at the microscopic scales compared with the randomly distributed particles on the hydrophobic HTS surface. The higher density of the $CaCO_3$ aggregates on the hydrophilic surface results mainly from the spreading of water on this substrate, while the particle aggregation is driven by the capillary forces which become operative from a certain water film thickness, during the drying process. Hence the hydrophilic silicon surface is water-wet and after the withdrawal of the substrate from the precipitation medium and washing with pure water, the remaining water film on the surface allows the $CaCO_3$ particles to move freely [15]. Further, during the drying step the water film thickness decreases and reaches a critical value at which the attractive lateral component of the capillaries forces becomes significant, leading to aggregate formation and lateral ordering of the $CaCO_3$ particles on the hydrophilic surface. Besides the factors already mentioned and dealing with the surface nature and the polyelectrolyte effects, the surface-coating mode was also investigated. Thus, the surface organisation of $CaCO_3$ particles observed on a dipped solid surface during the precipitation experiments was different from those obtained from drying drops of aqueous dispersions on the same substrates. In fact, the $CaCO_3$ particles deposited on hydrophilic and hydrophobic substrates give, respectively, an interconnected network and isolated aggregates.

Conclusion

The results shown in this study indicate the effects of, respectively, the anionic polyelectrolyte present in the solution and the nature of the solid surface dipped in the precipitation medium on the size and the organisation of the $CaCO_3$ precipitated on the solid surface. Thus, high surface density of agglomerates made of tiny spherical $CaCO_3$ particles is observed on the hydrophilic silicon substrate, while dispersed single spherical $CaCO_3$ particles are observed on the hydrophobic silicon substrate. This study shows that depending on the surface composition of the substrates, selective crystallite adsorption occurs on the solid surface and the organisation of the $CaCO_3$ particles on the solid surface is mainly dependent on its wetting properties.

A silicon wafer having a given surface composition can be used as a tool to control the size of the calcium carbonate particles formed by precipitation for specific purposes. This precipitate can be removed from the precipitation medium by hydrophobic or hydrophilic silicon wafers dipped in the solution and its mineralogy remains unchanged for at least 1 year.

Further work is in progress to investigate the formation of the $CaCO_3$ particles on mixed hydrophilic–hydrophobic dipped silicon wafers. The aim of such work is to verify if the surface-coating process results from the adsorption of the precipitated colloidal particles from the bulk solution or from the particle nucleation on the solid surface.

References

1. Canselier JP (1993) J Dispersion Sci Technol 14:625
2. Sondi I, Matijevic E(2001) J Colloid Interface Sci 238:208
3. Mann S, Heywood BR, Rajam S, Birchall JD (1988) Nature 334:692
4. Cölfen H, Antonietti M (1998) Langmuir 14:582
5. Falini G, Alveck S, Weiner S, Addadi L (1996) Science 271:67
6. Walsh D, Mann S (1995) Nature 377:320
7. Litvin A, Samuelson LA, Charych DH, Spevak W, Kaplan DL, (1995) J Phys Chem 99:12065
8. Teixeira-Neto E, Galembeck F (2002) Colloids Surf 207:147
9. Jada A, Pefferkorn E (2000) J Mater Sci Lett 19:2077
10. Jada A, Verraes A (2003) Colloids Surf 219:7
11. Haidara H, Mougin K, Schultz J (2000) Langmuir 16:7773
12. Dobrynin A, Rubinstein M, Obukhov SP (1996) Macromolecules 29:2974
13. Essafi W, Lafuma F, Williams C (1995) J Phys 5:1269
14. Essafi W, Lafuma F, Williams C (1994) In: Schmitz KS (ed) Macroion characterisation from dilutes solutions to complex fluids. ACS symposium series 548. American Chemical Society, Washington, DC, chap 2
15. Mougin K, Haidara H (2002) Langmuir 18:9566

Progr Colloid Polym Sci (2004) 128: 216–220
DOI 10.1007/b97077

Saeed Farrokhpay
Gayle Morris
Daniel Fornasiero
Peter Self

Role of polymeric dispersant functional groups in the dispersion behaviour of titania pigment particles

S. Farrokhpay · G. Morris (✉)
D. Fornasiero
Ian Wark Research Institute,
University of South Australia,
Mawson Lakes, SA 5095, Australia
e-mail: gayle.morris@unisa.edu.au
Fax: +61-8-83023683

P. Self
Adelaide Microscopy,
University of Adelaide, Adelaide,
SA 5005, Australia

Abstract The adsorption of polymeric dispersants onto titania pigment particles has been investigated at pH 9.5 as a function of the type of polymer functional group, using polyacrylic acid and modified polyacrylamides. The polyacrylamides include homopolymer and copolymers with either hydroxyl groups or carboxylate and hydroxyl groups. The experimental methods used were adsorption isotherm, zeta potential, particle size measurements and rheology. The anionic dispersants increased the magnitude of the titania pigment particle zeta potential, while the nonionic dispersants had a negligible effect. Both polyacrylic acid and polyacrylamide copolymers altered the titania pigment dispersion behaviour, reducing the suspension yield stress and the mean pigment particle size, while the polyacrylamide homopolymer had a negligible effect on the pigment dispersion properties. Generally the optimum dispersion effect (minimum particle yield stress and particle size) coincided with the polymer adsorption density; however, for polyacrylic acid, the maximum dispersion effect occurred prior to maximum adsorption. Acrylamide polymer functional groups adsorbed onto the pigment surface but they were unable to influence the pigment particle dispersion properties; however, adsorption and dispersion properties were enhanced in the presence of hydroxyl groups.

Keywords Titania pigment · Polyacrylamide · Rheology · Zeta potential · Polymer functional groups

Introduction

Titania pigment dispersion properties are very important for consumer industries, such as paints, papermaking and plastics. If aggregates are present, the end-use properties, including gloss, opacity, tint strength, colour distribution and storage stability, will be highly affected. Dispersion properties can be improved by addition of a dispersant with the dispersant selection largely controlled by its functional group composition. The functional groups are required to interact with the pigment surface as well as to contribute to the repulsion between particles. The interaction between titania pigment and a variety of dispersants has been investigated extensively over the last few years [1–4].

Polyacrylic acid and polyacrylamide homopolymers and copolymers are used as dispersants in a wide range of industries, including mining, water treatment, ceramics, paints and pigments. The use of polyacrylic acid to disperse oxide minerals such as titania and alumina is well established and documented [1, 5, 6]. Adsorption of polyacrylamide on mineral oxides [7], anionic polyacrylamide onto titania [8–10] and also polyacrylamide copolymers onto talc particles [11] has been studied individually. The aim of this study is to investigate the interaction between polymeric dispersant functional groups and titania pigment particles, at pH and ionic strength conditions used in industry. Four polymers, polyacrylic acid and three modified polyacrylamides, were carefully chosen in this study. The functional groups are acrylamide, hydroxyl and carboxylate groups of varying density. The role of the functional groups on the polymeric dispersant adsorption onto titania pigment particles and the resultant electrochemical and rheological properties are subsequently reported.

Experimental

Materials

The titania pigment sample used throughout this study was an industrial pigment. It is an aluminate and zirconia-coated rutile, with a density of 4.1 g/cm^3. The Brunauer–Emmett–Teller surface area of the pigment particles measured by N_2 adsorption was 19.0 m^2/g. The mean diameter of the particles measured by transmission electron microscopy was 0.2 μm. The major surface elements determined by X-ray photoelectron spectroscopy were Al, Zr, Ti, O and C (C is present as adventitious carbon). The aluminate phase is the hydrated form, boehmite (AlOOH), as indicated by Auger parameters [12]. The pigment isoelectric point (iep) was 7.8 ± 0.3.

The polymers used were polyacrylic acid and polyacrylamide. The polyacrylic acid (polymer-A) was obtained from Aldrich Chemical Co. as the sodium salt. The polyacrylamide reagents (polymer-N, polymer-C and polymer-H) were obtained courtesy of Cytec Research & Developments, Stamford, USA. Polymer-N is a polyacrylamide homopolymer, Polymer-C is an anionic polyacrylamide with a low level (about 10%) of COOH and OH substitution and Polymer-H is a nonionic polyacrylamide with about 10% OH substitution. The molecular weights of the polymers, measured using size-exclusion chromatography relative to poly(ethylene glycol) standards, were 13,600, 12,600, 17,000 and 13,100 Da for polymer-A, polymer-N, polymer-C and polymer-H, respectively. More details of the polymer structure are presented else where [2].

High-purity water (Milli-Q) was produced by the following sequential treatment: reverse osmosis, two-stage mixed-bed ion exchange, two-stage activated carbon treatment and a final filtering step through a 0.22-μm filter. The conductivity was less than 0.5 μS/cm with a surface tension of 72.8 mN/m at 20 °C. All other reagents used were analytical grade (unless otherwise stated).

Adsorption isotherms

Adsorption isotherm studies were conducted in a glass-conditioning vessel at 25 °C at pH 9.5. In industry, dispersion of titania pigments is generally carried out at alkaline pH values; therefore, pH 9.5 was chosen for this study. Pigment suspensions (2.5 wt %) were conditioned by stirring continuously overnight in 0.01 M KCl solution at natural pH value. After pH adjustment to the desired value with a small amount of HCl or KOH solution, a known concentration of polymer stock solution was introduced to the vessel and stirred for 40 min, which was the time to reach adsorption equilibrium as indicated by adsorption kinetics. During the adsorption period, the pH was monitored and no considerable change was observed. A sample of the suspension was then removed and centrifuged with an Eppendorf 5416 centrifuge at 10,000 rpm for 10 min. The supernatant was analysed to determine the amount of polymer remaining in solution using the complexation procedure explained later. Individual adsorption measurements were undertaken for a range of polymer concentrations and it was assumed that the amount of polymer depleted from the solution had adsorbed onto the pigment surface.

Polyacrylic acid complexation involves the interaction of carboxylate groups with a Bio-Rad protein assay dye (Bio-Rad Chemical Division, USA) [13]. The polymer solutions were added to the dye solution and it was found that the absorbance at 585 nm increased linearly with increasing polymer concentration. The absorbance was measured using a Varian Cary1E UV/vis spectrophotometer.

For analysis of the polyacrylamide solutions, a sensitive method based on bromine oxidation of the amide group was used [14]. Following destruction of excess bromine with sodium formate, the amide oxidation product oxidizes iodide ion to iodine, which is then measured as a tri-iodide–starch complex by visible spectroscopy at 600 nm.

Zeta potential measurements

Zeta potential values of the pigment particles were determined from dynamic mobility measurements using an Acoustosizer II (Colloidal Dynamics, Australia) at 25 °C. Pigment suspensions (2.5 wt %) were conditioned by continuous stirring overnight in 0.01 M KCl solution at the natural pH of the pigment (pH 6.8). After pH adjustment to 9.5 with a small amount of 0.5 M KOH solution, the zeta potential of the titania pigment particles was measured as a function of polymer concentration. The particle mean size (D_{50}) was also measured simultaneously with the Acoustosizer II.

Rheology

Shear-stress values of the pigment suspensions were determined at 25 °C as a function of polymer concentration at pH 9.5 using a Haake CV20 couette-type rheometer fitted with a Mooney-Ewart (45-mm diameter, 0.5-mm gap) concentric cylinder sensor. Pigment suspensions were prepared at a solid content of 20 vol % (50 wt %) in 0.01 M KCl. After conditioning by continuous stirring overnight and then adjusting the pH, the titania suspensions were sheared at an increasing shear rate from 0 to 300 1/s and back to rest in 2 min with the torque sensor continuously determining the shear stress.

The yield stress (τ_B) was obtained using the Bingham model: $\tau = \tau_B + \eta_{pl} D$ [15], where η_{pl} is the plastic viscosity, D is the shear rate and τ is the shear stress.

Results and discussion

Adsorption isotherms of the polymers on titania pigment particles at pH 9.5 in 0.01 M KCl are presented in Fig. 1. The Langmuir adsorption model was applied using the Langmuir equation [16]. In all cases, the amount of adsorbed polymer increases with increasing concentration of polymer in the suspension, until a plateau is reached. It is also observed that the polymer functional group has a significant effect on the adsorption density.

As the zeta potential of the titania pigment is approximately –31 mV at pH 9.5, electrostatic repulsion between the negative carboxyl groups of the anionic polymers, polymer-A and polymer-C ($pK_a = 4.5$ [11]), and the negatively charged pigment surface is likely to hinder the adsorption process. If electrostatic interaction is considered as the only adsorption mechanism, one

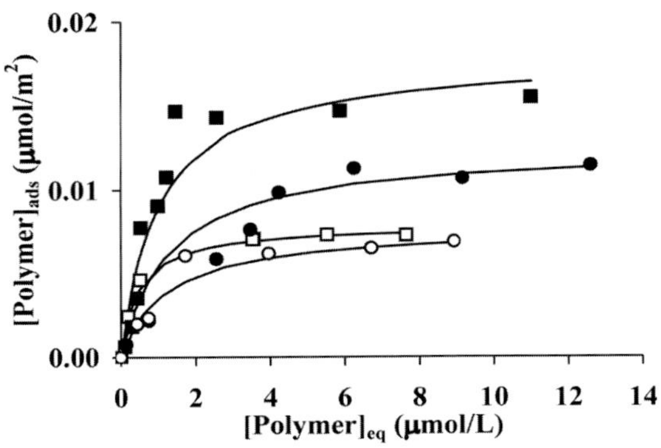

Fig. 1 Adsorption isotherms of polymer-A (*closed squares*), polymer-C (*open squares*), polymer-H (*closed circles*) and polymer-N (*open circles*) on titania pigment particles at pH 9.5 in 0.01 M KCl (pigment concentration 2.5 wt %). The *lines* were obtained from the Langmuir equation fit

would not expect any anionic polymer adsorption at pH values above the iep of the pigment, where both the polymer and pigment are negatively charged. Interestingly, polyacrylic acid (polymer-A) has a higher adsorption than all polyacrylamide polymers, even though it has a larger density of carboxyl groups than polymer-C; therefore, other driving forces must exist which become effective and cause the adsorption in the alkaline medium. When comparing the adsorbed amount of nonionic polymers, polymer-H and polymer-N, the higher adsorption density of polymer-H suggests that the presence of hydroxyl groups increases the polymer adsorption density, possibly through hydrogen bonding.

The free energy of adsorption values ($\Delta G_{ads} = -RT \ln K$, where R and T are the gas constant and the absolute temperature, respectively, and K is the Langmuir constant) are similar for all polymers, within experimental error: –10.6, –11.9, –10.4 and –10.1 kJ/mol for polymer-A, polymer-C, polymer-N and polymer-H, respectively. The negative sign of the adsorption free-energy values is due to a favourable adsorption of the polymers at the experimental conditions. The interaction between the polymers and pigment surface is, however, rather weak and unlikely to be of a chemical nature as the ΔG_{ads} values obtained are less than –25 kJ/mol [17].

The adsorbed polymer concentration, the zeta potential and the yield stress of titania pigment particles are shown in Fig. 2 as a function of added polymer concentration at pH 9.5 in 0.01 M KCl.

The adsorption density of the nonionic polymers (polymer-N and polymer-H) on the pigment surface increases with increasing polymer concentration until the maximum adsorption; however, the yield stress and zeta potential values are unaltered for polymer-N and are partially altered for polymer-H. The adsorption density of the anionic polymers (polymer-A and polymer-C) on the pigment surface increases up to 0.015 µmol/m^2 for polymer-A and 0.007 µmol/m^2 for polymer-C with a decrease in the yield stress values and an increase in the negative magnitude of the zeta potential. The maximum adsorption density occurs at an initial polymer concentration of 0.02 and 0.01 µmol/m^2 for polymer-A and polymer-C, respectively. The maximum particle zeta potential and maximum yield stress reduction is observed below the adsorption plateau for polymer-A and at the maximum adsorption density for polymer-C (Fig. 2). At higher polymer-A concentrations, approaching the maximum adsorption density, the yield stress and zeta potential values remain unaltered. It is observed that as the amount of anionic polymer is increased to about 0.02 µmol/m^2 for polymer-A and 0.01 µmol/m^2 for polymer-C, the corresponding zeta potential approaches a plateau value of approximately –70 and –45 mV,

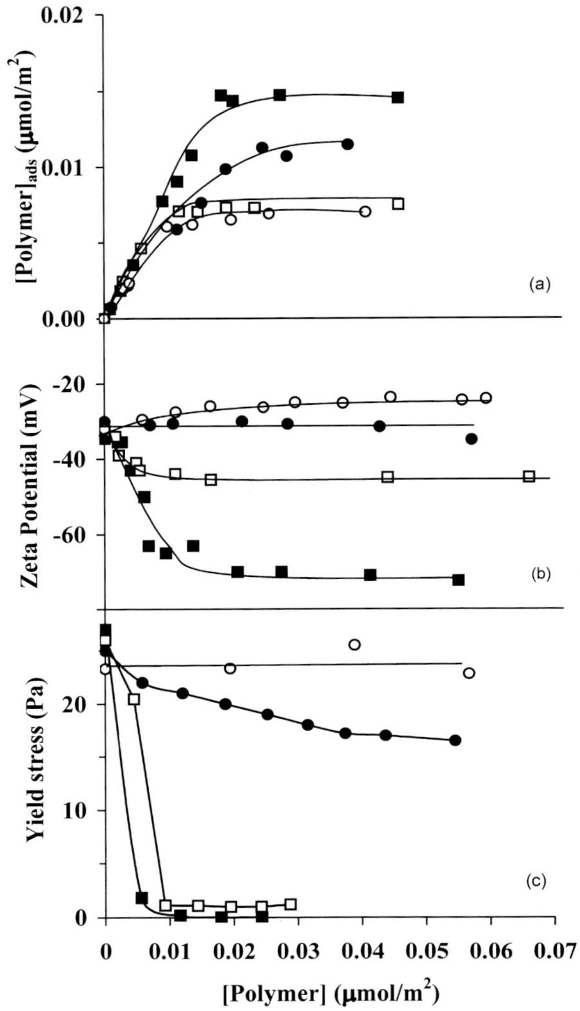

Fig. 2 Adsorbed polymeric dispersant concentration, **b** zeta potential and **c** yield stress of titania particles at pH 9.5 in 0.01 M KCl as a function of polymer-A (*closed squares*), polymer-C (*open squares*), polymer-H (*closed circles*) and polymer-N (*open circles*) concentration

respectively. Interestingly for polyacrylic acid (polymer-A), the optimum dispersion effect is observed at a polymer concentration (0.01 μmol/m^2) much lower than that required for maximum adsorption density (0.02 μmol/m^2), but this is not the case for the anionic polyacrylamide (polymer-C: both values are about 0.01 μmol/m^2) or for the hydroxyl-modified polyacrylamide (polymer-H: both values are about 0.025 μmol/m^2). Consequently, under these conditions, the optimum dispersion effect of polyacrylamide copolymers is obtained at a polymer concentration corresponding to the maximum adsorption density, as has been observed previously for low molecular weight polymers such as phosphate adsorbed onto oxides [4, 18]. For polymer-A, however, the maximum negative zeta potential value is observed to occur at the corresponding adsorption plateau (0.02 μmol/m^2), but the yield stress value reduces close to zero prior to this concentration. This suggests that the repulsion between the negative polymer-A carboxylate groups adsorbed onto the titania pigment particle surface is not the sole source of particle stabilization and the adsorbed polymer chains have a steric stabilization effect. In a similar study involving adsorption of low molecular weight polyacrylic acid (3,400 Da) on kaolin particles, the zeta potential and viscosity of the suspension approached a plateau at a polymer concentration lower than that needed for the corresponding adsorption plateau [19]. This correlation between maximum zeta potential and minimum viscosity in the presence of polyacrylic acid of such a low molecular weight indicates that the polymer is not able to provide a steric repulsion effect and the stabilization mechanism is likely to be only due to electrostatics. In the current case, owing to the higher molecular weight of polymer-A, the adsorbed polymer is able to provide a steric stabilization component.

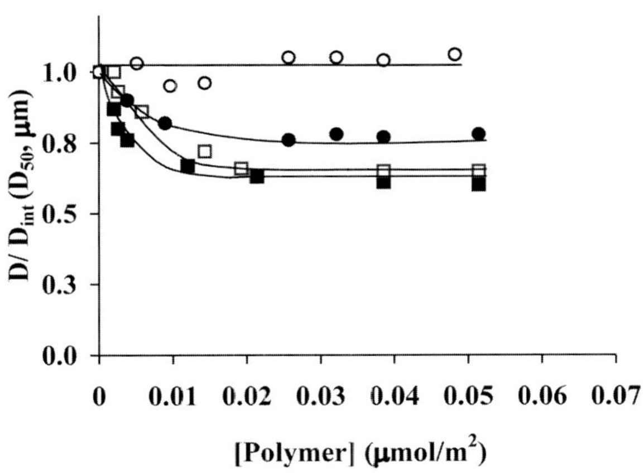

Fig. 3 Effect of polymer-A (*closed squares*), polymer-C (*open squares*), polymer-H (*closed circles*) and polymer-N (*open circles*) on the relative average titania pigment particle size (D_{50}) at pH 9.5 in 0.01 M KCl (D_{int} is the initial pigment particle size)

When comparing the maximum adsorption density of polymer-A and polymer-C, we can observe that although in the case of polymer-A both a higher adsorption density and a higher number of anionic groups cause higher negative zeta potential values for the pigment particles (–70 mV for polymer-A compared with –45 mV for polymer-C), a lower number of carboxylate groups is able to provide adequate repulsion between particles (Fig. 2).

The mean particle size (D_{50}) of the titania pigment particles is shown as a function of polymer concentration in Fig. 3. Clearly the anionic polymers (polymer-A and polymer-C) and also the hydroxyl-substituted polyacrylamide (polymer-H) have a significant effect on the pigment particle size, even at low concentrations where the homopolymer polyacrylamide (polymer-N) has minimal effect. A strong correlation between particle size results and the rheology data is observed.

Conclusions

The interaction of polymer dispersants with titania pigment particles at pH 9.5 has been shown to be strongly influenced by the type and the number of the polymer functional groups. The presence of hydroxyl groups (polymer-H) increases the polymer adsorption density compared with that of the homopolymer polyacrylamide (polymer-N), probably by hydrogen bonding.

The anionic polymers (polymer-A and polymer-C) increase the negative magnitude of the pigment zeta potential and therefore increase the repulsive force between particles. This concurrently affects the rheology of the particles, greatly decreasing the particle yield stress and mean particle size, resulting in a strong pigment dispersion. For the nonionic polyacrylamides, the homopolymer (polymer-N) does not alter the zeta potential and the yield stress of the pigment particles, while the hydroxyl-substituted copolymer (polymer-H) decreases the particle yield stress and particle size with a minimal effect on the zeta potential.

Comparing the effect of the different functional groups studied in this work on the dispersion of titania pigment particles, we conclude that while acrylamide groups have no effect on the dispersion process, polymer adsorption and dispersion properties are enhanced in the presence of hydroxyl groups. Although carboxylate groups are very effective in the pigment dispersion process, the high density of carboxylate groups in polyacrylic acid does not provide a better dispersion compared with the low-density carboxylate polyacrylamide.

Acknowledgements S.F. gratefully acknowledges the University of South Australia and Ian Wark Research Institute for the provision of a PhD research scholarship.

References

1. Boisvert JP, Presello J, Castaing JC, Cabane B (2001) Colloids Surf A 178:187–198
2. Farrokhpay S, Morris GE, Fornasiero D, Self PG (2004) J Colloid Interface Sci 274:33–40
3. Esumi K (2001) Colloids Surf A 176:25–34
4. Taylor ML, Morris GE, Smart RStC (2001) Colloids Surf A 190:285–294
5. Strauss H, Heegn H, Strienitz I (1993) Chem Eng Sci 48:323–332
6. Clayton J (1997) Surf Coat Int 9:414–420
7. Lee LT, Somasundaran P (1989) Langmuir 5:854–860
8. Chibowski S (1990) J Colloid Interface Sci 140:444–449
9. Girodo G, Lamarche JM, Foissy A (1988) J Colloid Interface Sci 121:265–272
10. Chong P, Curthoys G (1979) Int J Miner Process 5:335–347
11. Morris GE, Fornasiero D, Ralston J (2002) Int J Miner Process 67:211–227
12. Morris GE, Skinner WA, Self PG, Smart RStC (1999) Colloids Surf A 155:27–41
13. Vedula RR (1990) PhD thesis. Clemson University, USA
14. Scoggins MW, Miller JW (1979) Soc Pet Eng J 19:151–154
15. Hunter RJ (2001) Foundations of colloid science, 2nd edn. Oxford University Press, Oxford
16. Shaw DJ (1992) Introduction to colloid and surface chemistry, 4th edn. Butterworth-Heinemann, Oxford
17. Atkins P, de Paula J (2002) Physical chemistry, 7th edn. Oxford University Press, Oxford
18. Huynh L, Jenkins P (2001) Colloids Surf A 190:35–45
19. Zaman AA, Tsuchiya R, Moudgil BM (2002) J Colloid Interface Sci 256: 73–78

Progr Colloid Polym Sci (2004) 128: 221–226
DOI 10.1007/b97089

Pedro H.C. Camargo
Giovana G. Nunes
David J. Evans
G. Jeffery Leigh
Germano Tremiliosi-Filho
Eduardo L. Sá
Aldo J.G. Zarbin
Jaísa F. Soares

Single-source precursor and homometal approaches to the sol–gel synthesis of iron and titanium oxides

P.H.C. Camargo · G.G. Nunes
E.L. Sá · A.J.G. Zarbin · J.F. Soares (✉)
Departamento de Química,
Universidade Federal do Paraná,
Centro Politécnico, Jardim das Américas,
81531-990 Curitiba, PR, Brazil
e-mail: jaisa@quimica.ufpr.br
Tel.: +55-41-3613180
Fax: +55-41-3613186

D.J. Evans
Department of Biological Chemistry,
John Innes Centre, Norwich Research
Park, Colney, Norwich, NR4 7UH, UK

G.J. Leigh
Department of Chemistry, University of
Sussex, Brighton, BN1 9QJ, UK

G. Tremiliosi-Filho
Departamento de Físico-Química,
Instituto de Química de São Carlos,
Universidade de São Paulo, 13560-970
São Carlos, SP, Brazil

Abstract The recently described titanium and iron heterometal isopropoxide, $[FeCl\{Ti_2(OPr^i)_9\}]$ (complex 1), has been used as a single-source sol–gel precursor to produce mixed TiO_2/Fe_2TiO_5 materials. The products were dried and submitted to thermal treatment from 55 to 1,000 °C. Oxide composition and phase transitions were studied by X-ray powder diffractometry and Raman, electron paramagnetic resonance and Mössbauer spectroscopies. A mixture of nanometric-size TiO_2 (anatase, 4.3–6.1 nm) and amorphous iron(III) oxide was obtained up to 500 °C, while TiO_2 (rutile), α-Fe_2O_3 (hematite) and Fe_2TiO_5 (pseudobrookite) were found at 700 °C. At 1,000 °C, only rutile and pseudobrookite were observed. These results are very similar to those obtained by a homometallic approach from a 1:2 mixture of $FeCl_2$ and $[Ti(OPr^i)_4]$. Scanning electron microscopy/energy-dispersive X-ray spectroscopy results for the powder obtained from complex 1 at 1,000 °C showed no phase segregation at a submicrometre level. The average chemical composition of different grains confirms the complete conversion of **1** to the metal oxides. The solids obtained from the homometal precursors present significant phase segregation. This indicates that complex 1 can generate oxide–oxide (nano)composites with a degree of homogeneity that cannot be achieved by an analogous multicomponent route.

Keywords Sol–gel · Single-source precursor · Heterometal oxides · Titanium · Iron

Introduction

Multimetal oxides are an important class of advanced materials with a range of technological applications. Their sol–gel synthesis often employs mixtures of unimetal precursors [1, 2], although this may lead to inhomogeneous products because of differences in the hydrolysis rates of the starting materials [3]. This is a problem for applications with specific uniformity requirements. The utilisation of heterometal complexes as single source sol–gel precursors (SSP) has been a powerful alternative to the production of pure, highly homogeneous multimetal materials.

Heteronuclear alkoxides have been used as SSP for mixed-metal oxides in sol–gel and chemical vapour deposition processes. Several systems have been described involving combinations of second- and third-row transition metals with lanthanides and/or main-group metals [4–8], but there are no reports of preparations where the SSP contains only early, first-row transition metals.

Here we describe the sol–gel treatment of complex 1, $[FeCl\{Ti_2(OPr^i)_9\}]$, and of a mixture of unimetal $FeCl_2$ and $[Ti(OPr^i)_4]$ (1:2). Results of the SSP approach are compared with those from the multicomponent route. Complex 1, recently reported by our group [9], is the first structurally characterised isopropoxide of Fe^{II} and Ti^{IV}. Our interest in Fe/Ti oxides derives from their electrochromic behaviour in the UV–vis region [10] as well as their potential applications for electrodes, gas sensors and catalytic ammonia production [11, 12].

Experimental

Operations were carried out under air or N_2 as indicated. X-band electron paramagnetic resonance (EPR) data were obtained from solid samples at 77 K on a Bruker ESP-300E instrument. Raman spectra were obtained using a Renishaw Raman image spectrophotometer. A He–Ne laser (632.8 nm) with an incident power of 2 mW was used. Mössbauer data were recorded in zero field at 77 K using an ES-Technology MS105 spectrometer with a 875 MBq ^{57}Co source in a rhodium matrix. Spectra were referenced against iron foil at 298 K. X-ray diffraction (XRD) powder patterns for the oxides were obtained at room temperature with a Philips-Rigaku diffractometer, using Co Kα radiation ($\lambda = 1.790$ Å) in the 2θ range 10–60°. Powder Si reflections were used for 2θ calibrations. Crystallite diameters were determined by the use of the Scherrer equation following Gaussian deconvolution of the XRD peaks. Scanning electron microscopy (SEM) and energy-dispersive X-ray spectroscopy (EDS) analyses were carried out on a LEO 440 electron microscope equipped with an Oxford 7060 analyser. Complex 1 (Scheme 1) was prepared according to our previous report [9].

Sol–gel synthesis – SSP approach

An aqueous solution of HNO_3 (16 cm^3, 0.05 M) was added to a solution of complex 1 (0.82 g, 1.1 mmol) in 18 cm^3 of tetrahydrofuran. This mixture was stirred at 60 °C for 50 h under N_2. The resulting white sol was dried at 55 °C in air for 50 h, giving 0.27 g of a brown solid (SSP55). Portions of SSP55 were heated for 2 h at 300, 500, 700 and 1,000 °C, giving the solids SSP300, SSP500, SSP700 and SSP1000, respectively.

Sol–gel synthesis from homometal precursors

A solution of [Ti(OPr^i)$_4$] (1.3 g, 4.4 mmol) in toluene/propan-2-ol (3.5:1) was mixed with a previously prepared solution of $FeCl_2$ (0.28 g, 2.2 mmol) in 60 cm^3 of propan-2-ol. This mixture was then added to an aqueous solution of HNO_3 (36 cm^3, 0.12 M) at 60 °C. These steps were carried out under N_2. The reaction mixture was then stirred for 50 h in air, giving an orange-brownish suspension that was dried for 3 days at 55 °C to produce 0.52 g of a brown solid (H55). Portions of H55 were heated as described earlier, giving the solids H300, H500, H700 and H1000, respectively.

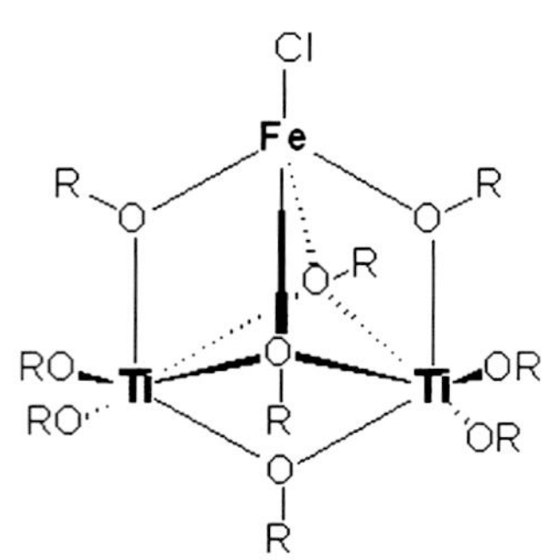

Scheme 1 (R = isopropyl) [9]

Results and discussion

Product characterisation by X-ray powder diffractometry and spectroscopic methods

Sol–gel treatment of complex 1

XRD powder data for SSP55–SSP500 (Fig. 1) show broad TiO_2 (anatase) reflections that indicate small crystallite sizes. For SSP55, SSP300 and SSP500, Scherrer's law calculations yielded crystallite diameters of 4.6, 4.3 and 6.1 nm, respectively. No diffraction peak was observed for the iron species in these materials, and this is probably due to the amorphous character of those species up to 500 °C. The presence of crystalline α-Fe_2O_3 (hematite) was detected in SSP700 (Fig. 1), as well as the occurrence of an anatase-to-rutile phase transition and the formation of Fe_2TiO_5 (pseudobrookite). Finally, SSP1000 contains only rutile and pseudobrookite. Raman spectral data support this interpretation, as absorption bands characteristic of anatase, hematite, rutile and pseudobrookite were observed in SSP55–SSP300 (anatase), SSP500 (anatase plus traces of hematite), SSP700 (hematite, rutile and pseudobrookite) and SSP1000 (rutile and pseudobrookite).

Two bands at $g = 2.0$ and 4.3 were observed in the EPR spectra for SSP55–SSP500 and were assigned to Fe^{III} sites in cubic (octahedral or tetrahedral) and distorted rhombic environments, respectively [13].

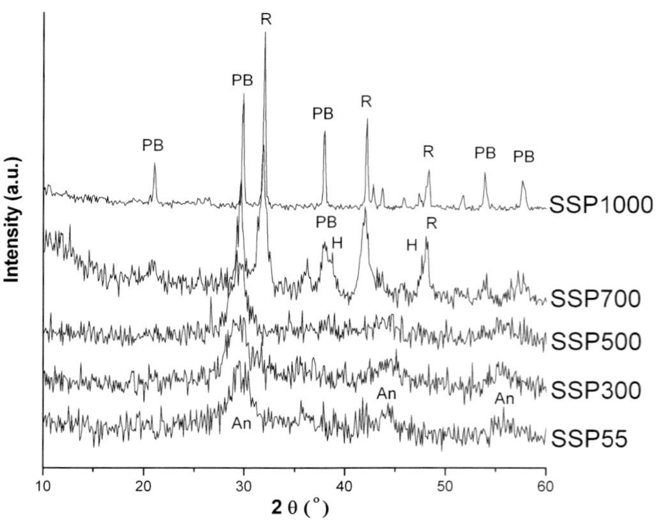

Fig. 1 X-ray diffraction powder analyses for *SSP55–SSP1000*. Anatase (*An*), rutile (*R*), hematite (*H*) and pseudobrookite (*PB*)

SSP700 and SSP1000 gave additional resonances ($g = 8.1$, 5.6, 3.4, 2.8 and 1.9) characteristic of Fe^{III} substitutionally incorporated into Ti^{IV} sites in the rutile lattice. This behaviour is accounted for by iron diffusion through open channels located along the rutile crystallographic *c*-axis, leading to the formation of pseudo-brookite [14].

The ^{57}Fe Mössbauer data recorded at 77 K for SSP55–SSP1000 are shown in Fig. 2. The results given by solids heat-treated at lower temperatures (55–500 °C) suggest iron-containing grains of nanometre size with isomer shifts and quadrupole splittings typical of amorphous Fe_2O_3 (Fig. 2, Table 1) [15, 16]. The spectrum of SSP55 shows a quadrupole-split sextet possibly given by a high proportion of surface-bound hydroxyl groups on very small particles of amorphous iron oxide. This generates an iron-oxo-hydroxo-like behaviour evidenced by the sextet signal [17–20]. The spectra of SSP300–SSP500 consist of a doublet arising from superparamagnetic α-Fe_2O_3 nanoparticles [21]. For SSP700, the spectrum contains both the main doublet and the sextet, and this is accounted for by a transition from superparamagnetic to ferromagnetic interactions on increasing the particle size [22]. Finally, SSP1000 contains only a doublet with isomer shift and quadrupole splitting values compatible with Fe_2TiO_5 [23].

Fig. 2 ^{57}Fe Mössbauer spectra of *SSP55–SSP1000* at 77 K

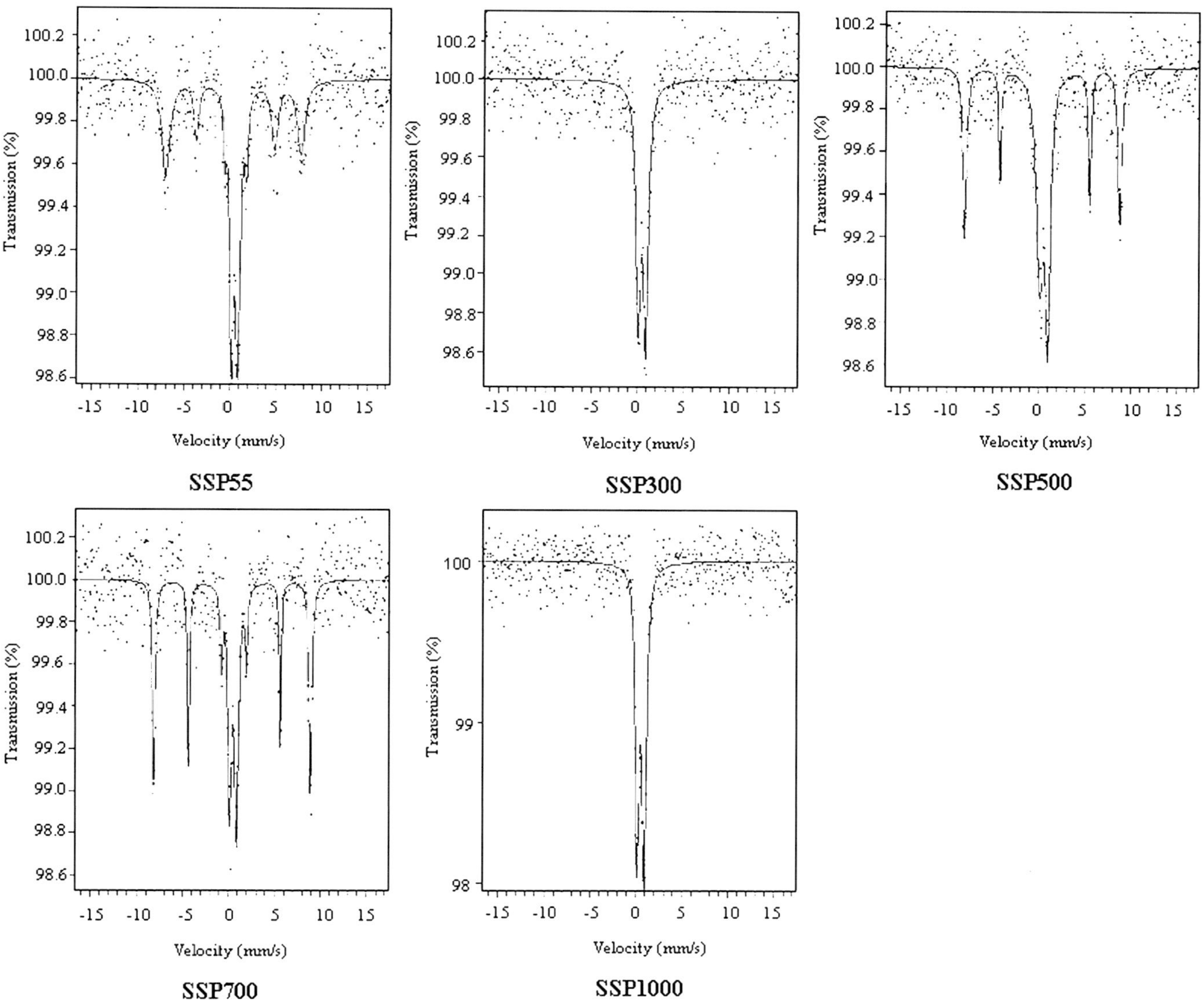

Homometal approach

The powder X-ray diffractograms recorded for H55–H1000 are very similar to the ones obtained for

Table 1 ^{57}Fe Mössbauer parameters obtained for SSP55–SSP1000 at 77 K. Numbers in parentheses correspond to the experimental error in the last significant figure

Samples	Mössbauer parameters	
	Isomer shift/ mm s^{-1}	Quadrupole splitting/mm s^{-1}
SSP55	0.48(2)	0.45(1)
SSP300	0.47(2)	0.49(1)
SSP500	0.45(1)	0.47(1)
SSP700	0.49(1)	0.53(2)
SSP1000	0.47(1)	0.53(2)

SSP55–SSP1000. This was also observed for the results of EPR, Raman and Mössbauer spectroscopy analyses and suggests no significant effect of the type of precursor and experimental sol–gel conditions on the chemical nature of the oxides. Morphological and semiquantitative microanalytical data were then obtained by scanning electron microscopy (SEM)/energy-dispersive x-ray spectroscopy (EDS) in order to try and distinguish the solids produced by the heterometal and the homometal approaches.

Morphological features and sample homogeneity

Scanning electron micrographs of both SSP1000 and H1000 show large grains (20–400 μm) consisting of a number of smaller particles (1–3 μm). Two distinct grain morphologies were observed: randomly packed, brittle aggregates with no definite shape (Fig. 3a) and a much more compact crystallite aggregation, with noticeable faces and edges (Fig. 3b). These differences may be related to higher or lower exposition of the solids to O_2 during the drying and/or heat-treatment steps. Closer inspection at the porous surfaces in Fig. 3b revealed the presence of well-packed submicrometric granules of globular shape.

EDS analyses performed on different grains in SSP1000 (Table 2) gave Ti-to-Fe-to-O ratios very close to the proportion 3(TiO_2):(Fe_2TiO_5), corresponding to the original 2:1 titanium-to-iron ratio in the molecular precursor. This confirms the stoichiometric conversion of complex 1 to the corresponding oxides. Random dot-mapping of different grains of SSP1000 showed no phase segregation at a submicrometre level (Fig. 4a). For H1000, however, EDS analyses (Table 2) and random dot-mapping (Fig. 4b) suggest significant segregation of iron and titanium oxides, as metal contents vary largely in distinct regions of the sample. Hence, despite the compositional and spectroscopic similarities, the materials obtained from the homometallic route are clearly inhomogeneous, while those synthesised from complex 1 are mixed oxides with remarkably good submicrometric phase dispersion.

Fig. 3 Scanning electron microscopy photographs of representative large grains in **a** SSP1000 and **b** H1000

Conclusions

The sol–gel processing of complex 1 produces a multiphase Fe_2TiO_5/TiO_2 system. The formation of TiO_2 is explained on stoichiometric grounds, as complex 1 does not contain the two metal ions in the required proportion to give pure pseudobrookite.

Materials with very similar chemical and spectroscopic features were produced by the sol–gel route starting from complex 1 and also from a 1:2 mixture of $FeCl_2$ and [Ti(OPr^i)$_4$]. In spite of this, the multiphase oxide system obtained from complex 1 is uniformly dispersed, while the mixture of homometallic precursors yielded inhomogeneous materials. Accordingly, com-

Table 2 Energy-dispersive X-ray spectroscopy experimental composition of different grains in SSP1000 and H1000

Sample	Grain	Atom percentage			Total Ti-to-Fe ratio	Total O-to-metal ratio	O-to-Ti ratio in TiO_x^a	Empirical composition[a]
		O	Ti	Fe				
SSP1000	1	62.80	25.01	12.19	2.1	1.7	1.7	$3(Ti_{1.11}O_{1.89})/(Fe_2TiO_5)$
	2	60.13	26.30	13.57	1.9	1.5	1.3	$3(Ti_{1.28}O_{1.72})/(Fe_2TiO_5)$
	3	63.16	22.14	14.70	1.5	1.7	1.8	$3(Ti_{1.08}O_{1.92})(Fe_2TiO_5)$
	4	61.74	26.46	11.80	2.2	1.6	1.6	$3(Ti_{1.17}O_{1.83})/(Fe_2TiO_5)$
	5	61.73	27.08	11.19	2.4	1.6	1.6	$3(Ti_{1.21}O_{1.90})/(Fe_2TiO_5)$
H1000	1	81.08	15.42	3.50	4.4	4.3	5.3	-[c]
	2	29.24	43.61	27.14	1.6	0.4	-[b]	-[c]
	3	76.31	20.04	3.64	5.6	3.2	3.7	-[c]
	4	50.69	48.15	1.16	41.5	1.0	1.0	-[c]
	5	41.18	50.54	8.28	6.1	0.7	0.4	-[c]

[a] Assuming complete conversion of iron to Fe_2TiO_5
[b] Not applicable
[c] Not determined

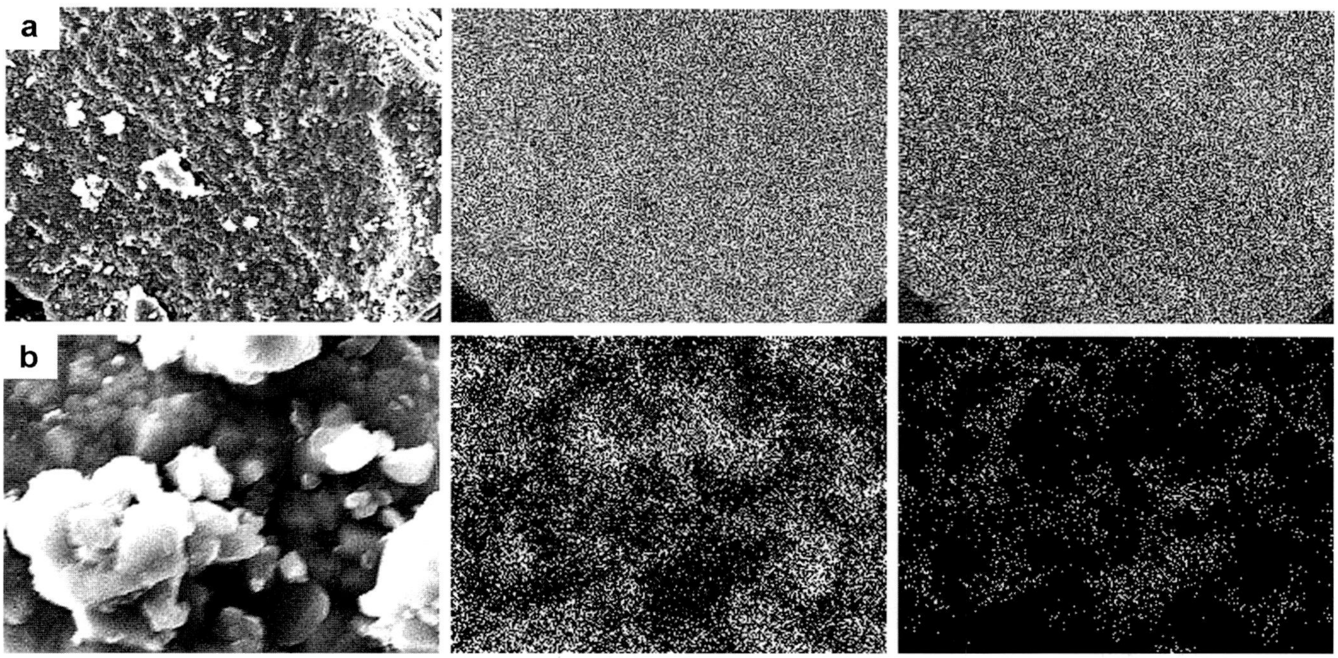

(1) (2) (3)

Fig. 4 Representative elemental dot maps for samples of **a** SSP1000 and **b** H1000: original image (*1*), titanium (*2*) and iron (*3*)

plex 1 is a potential precursor for iron–titanium oxide (nano)composites with a degree of homogeneity not given by homometallic precursors.

Acknowledgements We are grateful to Marcela M. Oliveira and the Laboratório de Espectroscopia Molecular (Universidade de São Paulo, *campus* São Paulo, Brazil) for the Raman analyses, to Dayane M. Reis for the help with the preparation of complex 1, to Irineu Mazzaro and Marcelo G. Honnicke for their attention during the powder X-ray analyses, and to Geraldo R. Friedermann and Antonio S. Mangrich for the EPR spectra. This work was supported by the Brazilian PRONEX Program, the Brazilian Research Council (CNPq), Coordenação de Aperfeiçoamento de Pessoal de Nível Superior, Fundação Araucária, Fundação de Amparo à Pesquisa do Estado de São Paulo, Financiadora de Estudos e Projetos (MCT) and Universidade Federal do Paraná. D.J.E. thanks the Biotechnology and Biological Sciences Research Council (UK) for financial support.

References

1. Hernandez BA, Chang K–S, Fisher ER, Dorhout PK (2002) Chem Mater 14:480
2. da Costa E, Avellaneda CO, Pawlicka A (2001) J Mater Sci 36:1407
3. Hubert-Pfalzgraf LG (2003) Inorg Chem Commun 6:102
4. Veith M, Mathur S, Shen H, Hufner S, Jilavi M (2002) Chem Mater 14:568
5. (a) Daniele S, Tcheboukov D, Hubert-Pfalzgraf LG (2002) J Mater Chem 12:2519; (b) Daniele S, Tcheboukov D, Hubert-Pfalzgraf LG (2001) J Phys IV 11:283; (c) Hubert-Pfalzgraf LG (1998) Polyhedron 17:889
6. (a) Werndrup P, Kessler VG (2002) Transition Met Chem 27:541; (b) Kessler, VG (2003) Chem Commun 1213
7. Starikova ZA, Yanovsky AI, Kotova NM, Yanovskaya MI, Turova NY, Benlian D (1997) Polyhedron 16:4347
8. (a) Veith M, Mathur S, Lecerf N, Bartz K, Heintz M, Huch V (2000) Chem Mater 12:271; (b) Lecerf N, Mathur S, Shen H, Veith M, Hüfner S (2001) Scripta Mater 44:2157; (c) Veith M (2002) J Chem Soc Dalton Trans 2405
9. Nunes GG, Reis DM, Amorim PT, Sá EL, Mangrich AS, Evans DJ, Hitchcock PB, Leigh GJ, Nunes FS, Soares JF (2002) New J Chem 26:519
10. Macek M, Orel B, Meden, T (1997) J Sol-Gel Sci Technol 8:771
11. Phani AR, Santucci S (2001) Mater Lett 50:240
12. (a) Schrauzer GN, Guth TD (1977) J Am Chem Soc 99:7189; (b) Edwards JG, Davies JA, Boucher DL, Mennad A (1992) Angew Chem Int Ed Engl 31:480; (c) Davies JA, Edwards JG (1993) Angew Chem In Ed Engl 32:552; (d) Rusina O, Linnik O, Eremenko A, Kisch, H (2003) Chem Eur J 9:561
13. (a) Prakash C, Husain S, Singh RJ, Mollah S (2001) J Alloys Compd 326:47; (b) Ennas G, Musinu A, Piccaluga G, Zedda D, Gatteschi D, Sangregorio C, Stanger JL, Concas G, Spano G (1998) Chem Mater 10:495
14. Amorelli A, Evans JC, Rowlands CC (1987) J Chem Soc Faraday Trans I 83:3541
15. Zboril R, Mashlan M, Petridis D (2002) Chem Mater 14:969
16. Howard DG, Nussbaum RH, (1980) Surf Sci 93:L105
17. Conforto E, Rechenberg HR, Jafelicci MJ (1986) J Phys Chem Solids 47:1179
18. Benz M, van der Kraan AM, Prins R (1998) Appl Catal A 172:149
19. Rodmacq BJ (1984) J Phys Chem Solids 45:1119
20. Morup S, Madsen MB, Franck J, Villadsen J, Koch CJW (1983) J Magn Magn Mater 40:163
21. Rancourt DG, Julian SR, Daniels JM (1985) J Magn Magn Mater 49:305
22. Kündig W, Bömmel H, Constabaris G, Lindquist RH (1966) Phys Rev 142:327
23. Greenwood NN, Gibb TC (1971) Mössbauer spectroscopy. Chapman and Hall, London

Progr Colloid Polym Sci (2004) 128: 227–231
DOI 10.1007/b97091

Gabriel L.T. Nascimento
Luciana M. Seara
Bernardo R.A. Neves
Nelcy D.S. Mohallem

Textural characterization of porous silica films prepared by the sol–gel process

G.L.T. Nascimento · L.M. Seara
B.R.A. Neves · N.D.S. Mohallem (✉)
Laboratório de Materiais,
Departamento de Química, ICEx,
UFMG Cx. 702, CEP 31270–901,
Belo Horizonte, MG, Brazil
e-mail: nelcy@dedalus.lcc.ufmg.br

Abstract Porous thin silica films were prepared on glass plates using the dip-coating process from colloidal solutions of silicon alkoxide dissolved in ethyl alcohol. The influence of solution parameters such as viscosity, pH, dipping velocity and thermal treatment temperature on the film formation was determined. The films were adherent, transparent and free of cracks. The porosity was estimated by the Lorentz–Lorenz equation using the refractive index measured by UV–vis spectroscopy and compared with values obtained by the nitrogen adsorption method. For a certain thickness and porosity, it was possible to prepare antireflective layers, which were identified by the transmission spectra in the visible region.

Keywords Porous thin silica films · Sol–gel process · Antireflective layers

Introduction

Interest in the preparation of thin films using the sol–gel process by dip-coating has increased in the last few years. Brinker and Scherer [1] reviewed studies made until 1990. Other authors [2, 3] have reviewed relevant results published after that.

Single-component and multicomponent oxide coatings can be obtained by the dip-coating process on a large scale at a cost lower than by other methods. The dry films obtained by this process can be porous, a characteristic that can be used to prepare antireflective coatings. The colloidal solution used as a precursor must have the appropriate physical-chemical properties to produce vitreous or crystalline, porous or dense films with a large variation in their characteristics. Porous films exhibit low density, thermal conductivity, a dielectric constant and a refractive index applicable to electronic and optical devices. Depending on the substrate used, it is possible to choose the type of oxide film and the refractive index suitable to obtain antireflective surfaces [1–4] or an intermetal dielectric [5].

Most commercial glasses have a refractive index of approximately 1.51, which means a reflection of 4% per interference. To increase the efficiency of transmitted radiation devices it is necessary, if not to eliminate this reflection, at least to minimize it. Antireflective thin layers can minimize surface reflection loses.

When the environment is air, an antireflective film on a glass substratum with a refraction index of 1.51 should have a postcoating refractive index of approximately 1.23. Since the refractive index of a material is related to its density, it is possible to obtain values on this order with porous oxide thin films such as SiO_2, Al_2O_3 and TiO_2 or a mixture of them. Even so, it is necessary to have a homogeneous pore distribution with a pore size smaller than the wavelength of light [1] so that the layer does not interfere with the transmission of radiation.

In the literature [1, 4, 5], the porosity of the layers has always been calculated using the Lorentz–Lorenz equation and the measured refractive index values. In this work, porous SiO_2 thin films prepared from silica colloidal solution on glass plates with a refractive index of 1.51 were characterized using UV–vis spectroscopy and the Lorentz–Lorenz equation for comparison with results obtained by the nitrogen adsorption method and atomic force microscopy.

Experimental

Porous SiO_2 coatings were prepared by the dip-coating process from starting colloidal solutions containing stoichiometric amounts of tetraethyl orthosilicate dissolved in a mixture of ethyl alcohol and 8 M HCl. The solution with a SiO_2 concentration of 0.1 kg L^{-1} was maintained under agitation at 25 °C for 2 h and then was aged. The coatings were deposited on the two sides of a glass plate substrate with a refractive index of 1.51 to obtain layers with antireflective properties. Thermal treatment of the gel films was done in air at temperatures between 100 and 700 °C for 30 min. Different viscosities and withdrawal speeds were used in the preparation of the films to determine the variation of the thickness with these parameters. The optical transmittance of the films was measured with a UV–vis spectrometer (U3010 – Hitachi). Thicknesses were measured using a profilometer (Alphastep 100, Tencor Instruments, USA). Refractive indices were measured using the transmittance spectrum taking into consideration the layers deposited onto the two sides of the substrate [6]. These values were used to evaluate the film porosity by the Lorentz–Lorenz equation.

Specific surface area, porosity and pore size distribution of the samples were evaluated through nitrogen gas adsorption (Autosorb – Quantachrome Nova 1200) at liquid nitrogen temperature and the application of the Brunauer–Emmett–Teller equation [7] and the Barrett–Joyner–Halend method [8]. Samples were outgassed for 3 h at 300 °C before each analysis. In this experiment, the cell was calibrated with pieces of uncoated substrate inside it. The measurements were made with the same substrate pieces coated with the porous silica films.

Some samples were analyzed by atomic force microscopy (AFM) with a Nanoscope III and a Dimension 3000 equipped with an extender modulus for phase imaging, all from Digital Instruments, to observe the film porosity and the topography and to verify possible coating defects such as cracks and peeling. The images were generated through the intermittent contact mode between the microscope tip and the sample.

Results and discussion

In order to obtain SiO_2 films with optical quality, the solution pH was kept between 2 and 4, the viscosity was kept between 2 and 5 cP and the relative humidity of the air was kept below 40%, since the films are opaque and not adherent for other pH, viscosity and relative humidity values. The thermal treatment temperature of the gel films was chosen to be 500 °C for 30 min, because this heating did not densify the films completely, but it is enough for the fixation of the film to the substrate through chemical reactions.

The SiO_2 films obtained using these best parameter values were transparent, adherent and with thicknesses ranging from 100 to 600 nm. The film thickness increased with increasing withdrawal speed, keeping a constant viscosity of the sol. Likewise, the film thickness increased when the solution viscosity increased and the withdrawal speed was kept constant (Fig. 1).

According to the Landau and Levich expression [9], the film thickness is linearly related to the square root of the withdrawal rate and the solution viscosity; this relationship originated from various works related to the dip-coating process. Some experimental works report that the film thickness is linearly related to exponential values of the withdrawal speed between 0.1 and 1 [10–15] in similar systems but with different composition. Yang et al. [16] derived the following expression relating the thickness of a polymer film prepared by a simple withdrawal method with parameters such as solution viscosity (η), solvent viscosity (η_s), withdrawal speed of

Fig. 1 Thickness of SiO_2 films as a function of withdrawal velocity and sol viscosity

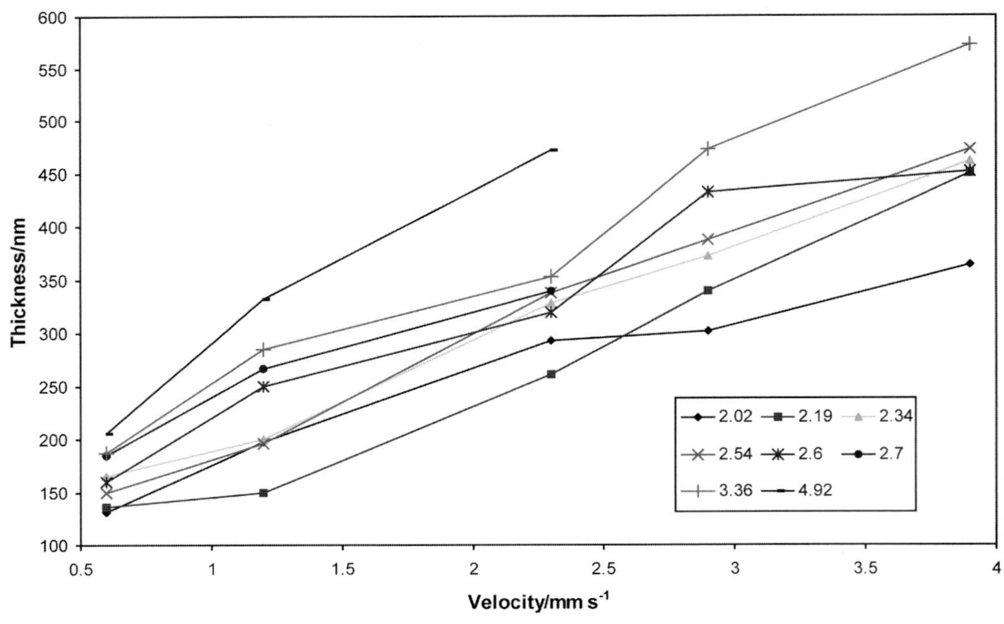

the substrate (U), polymer density (ρ_p) and solvent density(ρ_s):

$$t_p = J\xi\left(\frac{\eta - \eta_s}{\eta_o}\right)^{0.84}\left(\frac{\eta U}{g\rho_s}\right)^{1/2},$$

where t_p is the film thickness measured after thermal treatment, J the dimensionless fluid flow whose value (0.29) was determined experimentally, η_o is a constant related to the solution concentration, g is the gravitational acceleration and $\xi = \rho_s/\rho_p$.

Guglielmi and Zenezini [17] obtained a similar expression for a densified silica thin film prepared by the sol–gel process by modifying an expression derived by Strawbridge and James:

$$t_p = 10^{-3}\left(\frac{C_p}{\rho_p}\right)J\left(\frac{\eta U}{\rho g}\right)^{1/2},$$

where C_p is the solution concentration and the other parameters are similar to those in the expression of Yang et al. Both groups of authors obtained a good fit of the experimental results using their derived expressions.

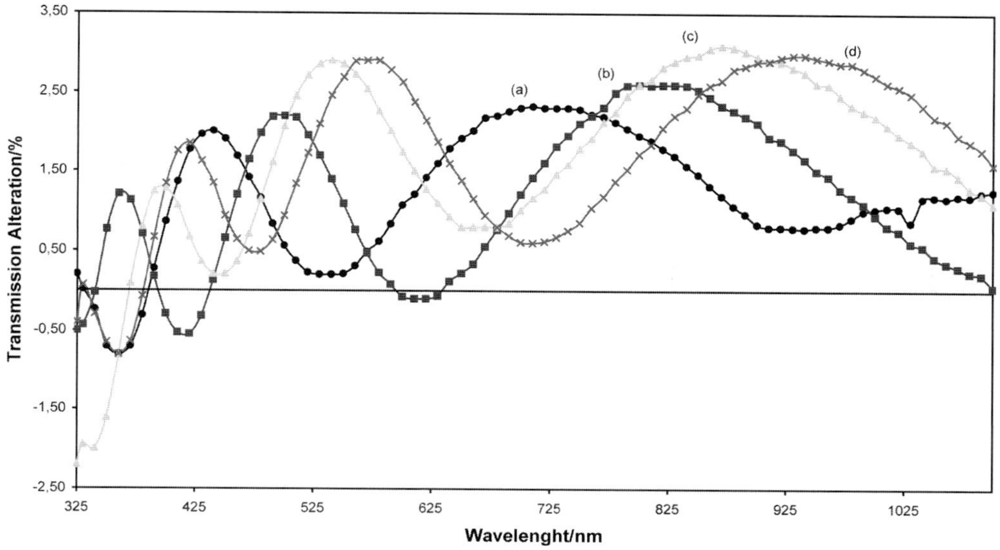

Fig. 2 UV–vis transmittance spectra of SiO_2 films prepared at a velocity of 2.9 mm s^{-1} and a viscosity of 4 cP with thickness of 339 nm (*a*), 388 nm (*b*), 432 nm (*c*) and 472 nm (*d*)

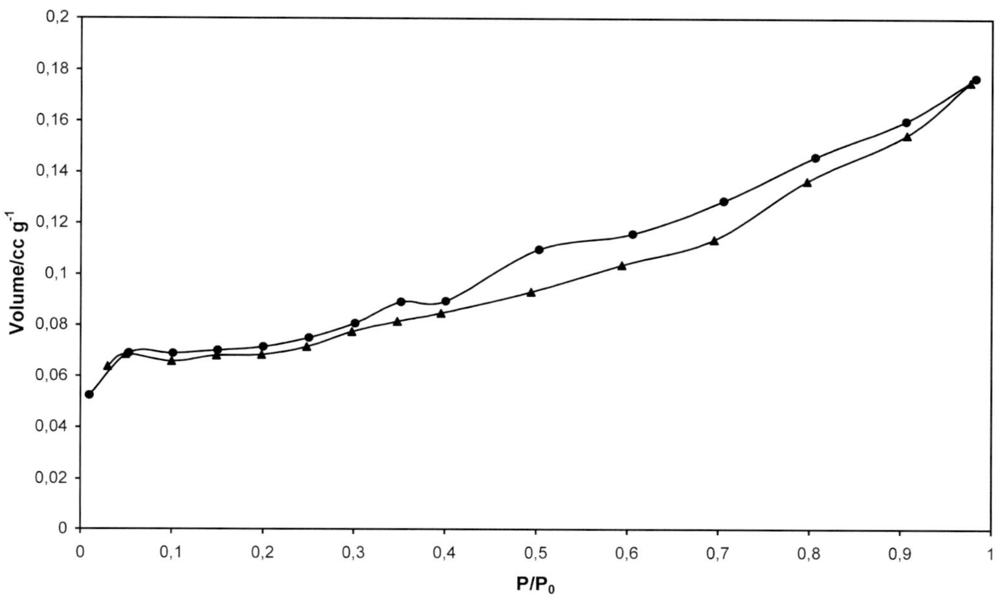

Fig. 3 Adsorption–desorption isotherm of SiO_2 films heated at 500 °C for 30 min

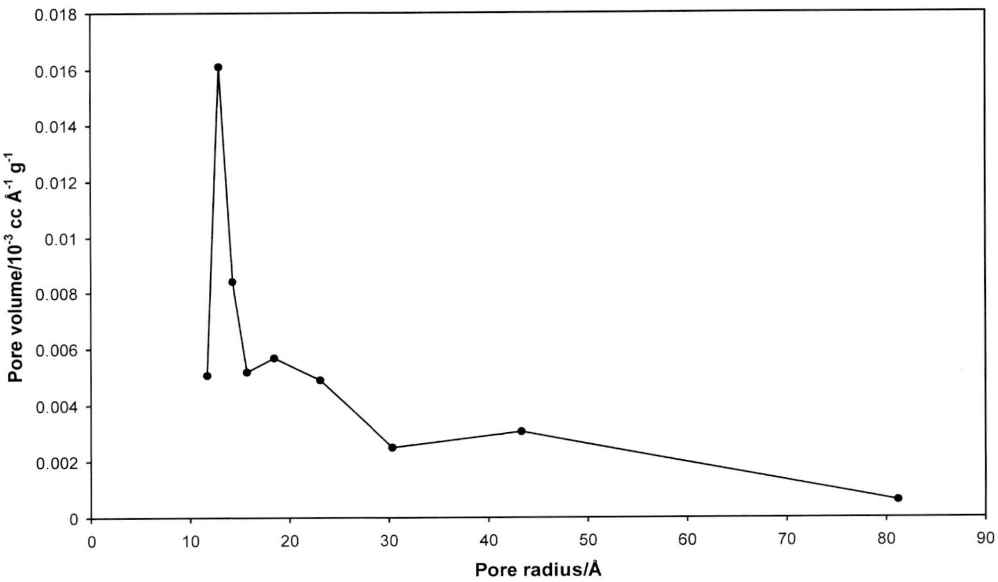

Fig. 4 Pore size distribution by the Barrett–Joyner–Halend method of SiO_2 films treated at 500 °C for 30 min

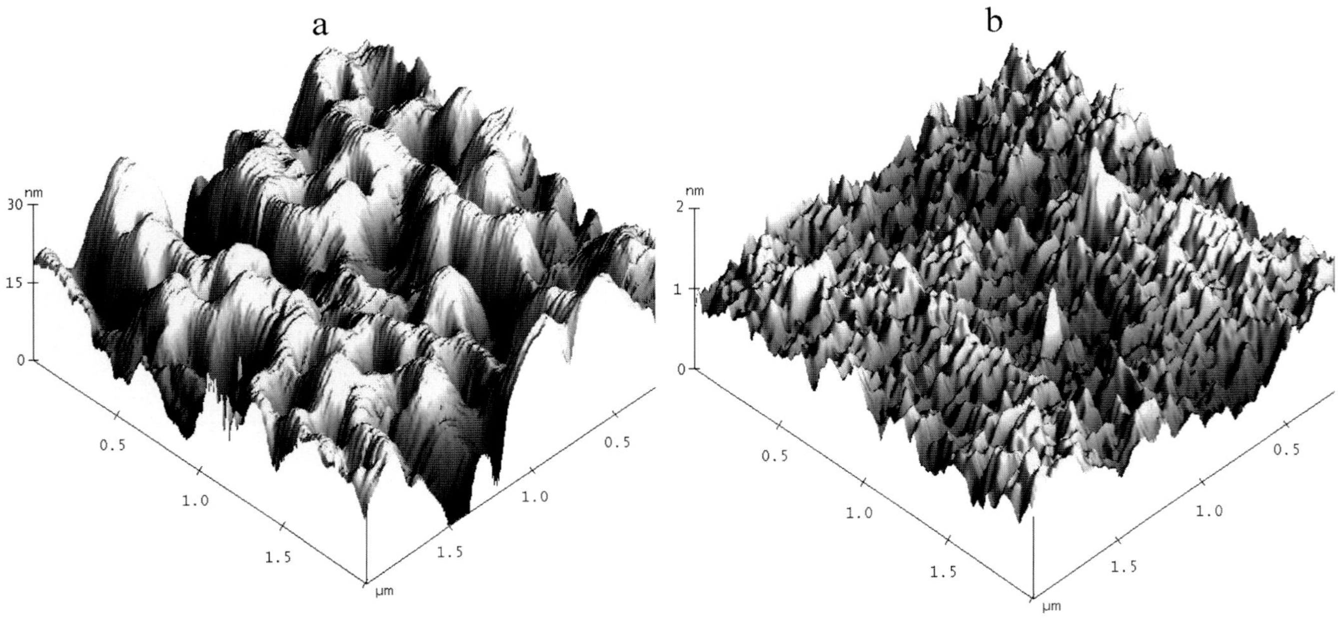

Fig. 5 Atomic force microscope image of SiO_2 film surface: **a** film thickness 570 nm; **b** film thickness 190 nm

In our system, the measured thickness is linearly related to the square root of the withdrawal rate and the solution viscosity, in accordance with the results of Yang et al. and Guglielmi et al. This fact facilitates the reproducibly of the films obtained.

Transmission spectra of SiO_2 films with thicknesses ranging from 339 to 472 nm deposited onto glass with glass as a baseline are shown in Fig. 2. Refractive indices calculated from these spectra range from 1.34 to 1.31 ($\lambda = 550$ nm), evidencing the porosity of the material on the order of 29–36% calculated by the Lorentz–Lorenz equation:

$$\frac{n_c^2 - 1}{n_c^2 + 2} = (1 - P)\frac{n_o^2 - 1}{n_o^2 + 2},$$

where P is the porosity, n_c is the refractive index of the coating and n_o is the refractive index of the silica thin films densified at 700 °C, which reached 1.458. Films with thickness less than 330 nm have refractive indices larger than 1.34, porosity less than 29%, and consequently they do not have antireflection behavior.

As expected, gas adsorption analyses for the substrate without a film resulted in zero porosity. The results of the analyses for samples with a film show a high surface area, $(1,900 \pm 130)$ m^2 g^{-1}, and porosity of $(35 \pm 3)\%$. An adsorption–desorption isotherm for films of SiO_2 typical of porous materials with cylindrical pores, according to deBoer [7], is shown in Fig. 3. The pore size distribution of films with a medium radius of 22 Å is shown in Fig. 4.

Microscopically, the films present a regular topography with no flaws, microcracks or any other defect in the coating. The substrate defects, like bubbles, are eliminated with the deposition of the SiO_2 film.

Typical AFM images of the porous film surface with two different thicknesses are shown in Fig. 5. These films were prepared from solution with a viscosity of about 3.9 cP and withdrawal speeds of 3.9 (Fig. 5a) and 0.6 mm s^{-1} (Fig. 5b), respectively. Films with thickness larger than 300 nm have more rugosity (about 4.5 nm) than the thinner films (about 0.2 nm), and present antireflection behavior.

This analysis confirms the results obtained by UV–vis spectroscopy and gas adsorption analysis.

Conclusion

Colloidal silica sol-based porous films produced by the sol–gel process using dip-coating are transparent and have a regular topography, without flaws or microcracks and with good adhesion to the substrate. The porosity of the coatings determined by gas adsorption analyses was in agreement with the results estimated by the Lorentz–Lorenz equation. The AFM images confirm these results. The porosity of the SiO_2 films implied a reduction of the refractive index of the material and an increase in the intensity of energy transmission of 3%. Thicker films were more porous and, therefore, presented antireflection behavior.

Acknowledgements The authors thank and Flávio Plentz for the thickness measurements. This work was supported by CNPq and FAPEMIG.

References

1. Brinker CJ, Scherer GW (1990) Sol–gel science – the physics and chemistry of sol–gel processing. Academic, San Diego
2. Uhlmann DR, Suratwala T, Davidson K, Boulton JM, Teowee G (1997) J Non-Cryst Solids 218:113
3. Mezza P, Phalippou J, Sempere R (1999) J Non-Cryst Solids 243:75
4. Yoldas BE (1980) Appl Opt 19:1425
5. Jo MH, Park HH, Kim DJ, Hyun SH, Choi SY (1997) J Appl Phys 82:1299
6. Bennett HE (1979) J Opt Soc Am 68:1426
7. Lowell S, Shields JE (1996) Powder surface area and porosity. Wiley, New York
8. Webb PA Orr C (1997) Analytical methods in fine particle technology. Micromeritics, Norcross, GA
9. Landau L, Levich B (1942) Physicochem 17:42
10. Duran A, Fernandez Navarro JM, Casariego P, Joglar A (1986) J Non-Cryst Solids 82:391
11. Strawbridge I, James PF (1986) J Non-Cryst Solids 86:381
12. Orgaz F, Capel F (1988) J Mater Sci 100:531
13. Beier W, Goktas AA, Frischat H (1986) J Non-Cryst Solids 82:391
14. Sakka S, Kamiya K, Makita K, Yamamoto Y (1984) J Non-Cryst Solids 63:223
15. San Vicente G, Morales A, Gutierrez MT (2001) Thin Solid Films 391:133
16. Yang C, Josefowicz JY, Alexandru L (1980) Thin Solid Films 74:117
17. Guglielmi M, Zenezini S (1990) J Non-Cryst Solids 121:303–309

Progr Colloid Polym Sci (2004) 128: 232–238
DOI 10.1007/b97092

Paula S. Haddad
Evandro L. Duarte
Mauricio S. Baptista
Gerardo F. Goya
Carlos A.P. Leite
Rosangela Itri

Synthesis and characterization of silica-coated magnetic nanoparticles

Abstract The present work reports on the synthesis and properties of magnetic nanoparticles based on magnetite coated by a silica layer. A series of conditions was tested to allow the synthesis of the nanoparticles, usually performed in an aqueous medium, and could be efficiently coupled with the silica sol–gel process. The resulting particles were characterized by means of X-ray diffraction, transmission electron microscopy associated with electron energy loss spectra and electron spectroscopy imaging, Fourier transform IR, Mössbauer and magnetization measurements. Initially, magnetite particles with an average crystalline grain size of about 100 Å and a polydispersion of 30%, as revealed by X-ray diffraction and transmission electron microscopy analyses, respectively, were obtained with a mixture of $FeCl_2$ and $FeCl_3$ in aqueous and acid solution. The magnetization measurements, at room temperature, show that the particles are in the superparamagnetic regime. Magnetite was also synthesized in acid solutions in an alcoholic environment (25% methanol-to-base ratio), a medium that allows us to proceed with silica-coating by the sol–gel process in a one pot reaction. The resulting particles present size dispersion ranging from around 15 Å to about 200–300 Å as evidenced by electron micrographs. The superparamagnetic behavior is preserved, although its saturation magnetization value decreases from about 92 to about 50 emu g^{-1}, probably owing to the contribution of the smallest particles as well as to the surface spin disorder induced by addition of methanol to the synthesis medium. For higher values of the alcohol-to-base ratio, the resulting particles are amorphous, becoming crystalline under thermal treatment. When tetraethylorthosilicate is added to a solution containing 25% of methanol to base, iron oxides are SiO_2-coated at room temperature, as evidenced by electron spectroscopy imaging and Fourier transform IR spectroscopy. The magnetization results are dependent on the Si-to-Fe volume ratio, in such a way that the values decrease as the SiO_2 amount increases, reflecting the nanoparticle coating.

Keywords Magnetic nanoparticles · Sol–Gel synthesis · Silica coating · Superparamagnetism · Electron spectroscopy imaging

C.A.P. Leite
Instituto de Química, Universidade Estadual de Campinas, UNICAMP, CP 6154, 13083-970, Campinas, SP, Brazil

P.S. Haddad · M.S. Baptista
Departamento de Bioquímica, Instituto de Química, Universidade de São Paulo, Av. Lineu Prestes, 748, 05508-900 São Paulo, SP, Brazil

E.L. Duarte · G.F. Goya · R. Itri (✉)
Instituto de Física, Universidade de São Paulo, CP 66318, 05315-970 São Paulo, SP, Brazil
e-mail: itri@if.usp.br
Tel.: +55-11-30917012

Introduction

Magnetic nanocomposites have been the subject of extensive investigation owing to their important properties and applications [1], such as information storage [2], color imaging [3], and magnetic refrigeration [4].

The preparation methods of ferrite-like nanocomposites (Fe_3O_4 or γ-Fe_2O_3) are very important owing to the fundamental and technological perspectives. Small particles exhibit very peculiar properties compared with those presented in bulk materials, and these peculiar properties can provide new ways of using magnetic structures also in biotechnology [5]. A challenging issue in the nanomagnetic field is to develop strategies to functionalize these particles. A new biocompatible magnetic nanoparticle covered by drugs would have potential application in medicine [6].

Several strategies are being tested to functionalize these particles, including reaction with dimercaptosuccinic acid [6], which has been used to graft a variety of molecules and biomolecules to magnetic nanoparticles [6–8]. However, this route usually provides relatively labile bridges with the magnetic nanoparticle [6–8]. Another possible route to graft these particles, which has received special attention in recent years, involves the use of an inorganic material such as silica [9–17]. This special interest arises from the fact that the silica grafting somehow stabilizes the particles in the ferromagnetic γ-Fe_2O_3 form, i.e., it decreases the transition efficiency to the antiferromagnetic α-Fe_2O_3 form with an increase in temperature [12, 15, 16]. It also provides a transparent layer that can be used to stabilize the particles in suspension, providing a route to easily further graft it with organic molecules [9, 10, 12, 18, 19].

There are basically two ways to obtain the γ-Fe_2O_3/SiO_2 composites:

1. Disperse the previously synthesized iron oxide particles in different sol–gel matrices. This is usually performed by coprecipitating the particles in aqueous solution, filtering and redispersing the precipitate in ethanol solution to perform the silica coating by silane hydrolysis and polymerization [9, 10, 11].
2. In situ precipitation of the magnetic nanoparticles during the matrix formation [12, 15–17], which does not involve the step of filtration and redispersion. However, this method can have relatively long gelation times, sometimes longer than 9 days [17, 20], and the solution has to be heated to relatively high temperatures (above 700 °C) [12, 16, 17, 20].

In the current work, an effort has been made to find an alternative route to prepare magnetic nanoparticles and to coat them with silica. Ferrite-like nanoparticles were prepared by mixing $FeCl_2$ and $FeCl_3$ acidic solutions (to avoid the formation of iron hydroxides) in an alcoholic environment and by adding tetraethylorthosilicate (TEOS) in an one-pot reaction scheme. With this procedure we would not need the filtration and redispersion step characteristic of method 1 nor the long gelation time and thermal treatment of method 2.

A variety of structural and spectroscopic methods were used to characterize these particles. Structural determination of nanoparticles was performed by means of X-ray diffraction (XRD). Fourier transform IR (FTIR) spectroscopy was used to identify the stretching vibration of Si–O–Fe. Transmission electron microscopy (TEM) with an energy filter was also carried out to observe the nanoparticle size and shape. Electron spectroscopy imaging (ESI) was used to probe the elemental distribution. Mössbauer and vibrating sample magnetometer (VSM) techniques, at room temperature, were used to characterize the magnetic properties of the nanoparticles.

Experimental

Materials

All reagents were obtained from commercial suppliers and were used without any further purification. Water was distilled in all-glass apparatus, filtered and deionized (Milli-Q water). All experiments were performed at room temperature of 23 ± 1 °C.

Synthesis

Three different kind of samples were studied. The first one was synthesized in an acidic medium according to the well-established protocol in the literature [21]. A 1.0-mL aliquot of $FeCl_2$ (Merck) (1.0 M) and 4.0 mL $FeCl_3$ (Merck) (0.5 M) prepared in 1.0 M HCl (Aldrich) were vigorously stirred, while 50.0 mL NH_3 (0.7 M) (Aldrich) was slowly dropped into the solution, resulting immediately in a black precipitate. At this stage, the precipitate was decanted, washed thoroughly with acetone and dried in vacuo. The second one involved the same iron mixture of acid solution in an alcoholic medium containing methanol (Merck) and base in a volume ratio of 25%. After the same washing and drying procedures as described for the fist sample, a black powder was also obtained. Both samples were crystalline at room temperature (see later). It is worthy of mention that solutions containing higher ratios of methanol to base were also investigated, but the nanoparticles obtained were amorphous at room temperature, becoming crystalline under thermal treatment at 500 °C for 2 h in a nitrogen atmosphere.

In order to obtain silica-coated magnetic nanoparticles at room temperature, TEOS was used as the Si source and was added to the 25% methanol-to-base solution just after the black precipitate had been formed. The amount of TEOS was changed depending on the desired Si-to-Fe volume ratio (10%, 20%, 30%, 50% and 70%). A brown gel was immediately obtained, and was washed with acetone. The nanoparticles were isolated by decantation.

Methods

XRD: Structural identification was performed by using a Rigaku-Denki powder diffractometer with a conventional X-ray generator (Cu Kα radiation $\lambda = 1.5418$ Å and a graphite monochromator) coupled to a scintillation detector. The angular scanning performed on all samples ranged from 20° up to 70° with 0.05° step width. The average size of the nanoparticles was calculated from the full width at half maximum of the (311) reflection (spinel structure) using Scherrer's equation [22].

TEM: A Carl Zeiss CEM 902 transmission electron microscope equipped with a Castaing Henry–Ottensmeyer filter spectrometer within the column was used. One drop of the particle dispersion was deposited onto carbon-coated parlodion films supported in 300-mesh gold grids. The images were obtained using zero loss energy, low loss energies between 20 and 30 eV and specific elemental energies; they were recorded with a slow-scan CCD camera (Proscan) and processed in the AnalySis 3.0 system. Electron energy loss spectra in the range 70–150 and 680–760 eV for Si and Fe were acquired for use as a reference to ESI, in the CEM-902 transmission electron microscope, by using the parallel acquisition method [23].

FTIR: The IR spectra were recorded using a Nicolet IMPACT 400 spectrophotometer in a wavenumber range from 4,000 to 400 cm^{-1}, at a resolution of 2 cm^{-1}, using KBr pellets.

VSM: A VSM was used to obtain the magnetization versus magnetic field loop (M versus H) at room temperature up to H = 20 kOe. The apparatus was calibrated with a Ni pattern. The magnetization measurements were carried out on a known quantity of sample powder, slightly pressed and conditioned in cylindrical holders of Lucite.

Mössbauer: The measurements were performed with a conventional constant-acceleration spectrometer in a transmission geometry with a source of about 50 mCi ^{57}Co in a Rh matrix between 78 and 296 K. Hyperfine parameters, such as the distribution of the hyperfine magnetic field (B), the isomer shift (IS) and the quadrupole shift (QS), were determined by the NORMOS program.

Results and discussion

Structural and morphological evaluations

X-ray diffraction data, typical of a spinel structure [24], obtained from samples prepared in acid solution in the absence (sample 1) and in the presence of an alcoholic medium (no TEOS, sample 2) are shown in Fig. 1. In fact, such a structure corresponds to magnetite, as demonstrated by the Mössbauer data presented later. The analysis of the main X-ray reflection (311) peak width and position by using the Debye–Scherrer equation led to a mean grain size of 100 Å for sample 1 and 170 Å for sample 2 (without TEOS). Then, the presence of methanol in the synthesis environment seems to induce the formation of larger magnetic particles.

At the same time, the TEM micrograph (Fig. 2a) reveals that sample 1 is constituted by particles with well-defined spherical shape contours with a mean dimension of 60 Å and a polydispersity of about 30%; therefore, the X-ray reflection width is predominantly weighted by the largest particles. However, lower-quality micrographs were obtained from sample 2 synthesized in an alcoholic medium. Firstly, most of the images did not contain isolated particles, making it difficult to define their morphology and size. Secondly, the images show

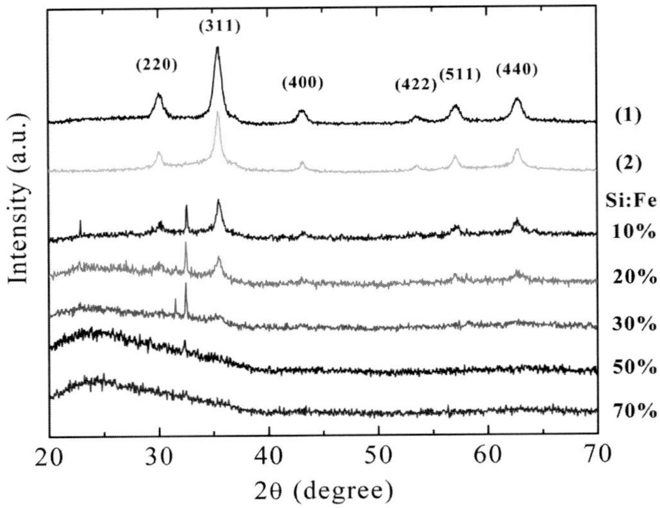

Fig. 1 X-ray powder diffraction from samples synthesized in an acidic medium (*1*), and in a methanolic medium (25% methanol-to-base ratio) with increasing Si-to-Fe volume ratio, as indicated in the figure, ranging from 0% (*2*) to 70%, at room temperature

particles with a very broad size distribution, including small particles of average sizes of 15 Å (Fig. 2b) and large particles with dimensions of about 200–300 Å (Fig. 2c). Therefore, the presence of methanol in the synthesis medium also seems to amplify the polydispersity of the system.

It should be remarked that the magnetic nanoparticles prepared by the usual coprecipitation method in aqueous solution [25] have the same features, revealed by XRD and TEM, as those obtained in acidic solution (sample 1).

For samples containing a Si-to-Fe volume ratio of 10% and 20% the main X-ray reflections (Fig. 1) still indicate the spinel-like crystalline structure and the nanoparticles keep their average grain size of about 170 Å. The extra peak at $2\theta \sim 32.5°$ is identified as the SiO_2 reflection, evidencing its formation during the synthesis. Further, a large band is observed in the angular range from 20° to 40° that increases with addition of silica to the system. This behavior indicates the formation of an amorphous silica phase, in such a way that for large silica content (30, 50 and 70% of Si to Fe) there is a predominance of the amorphous phase attributed to amorphous SiO_2 in the X-ray spectra. It should be emphasized, however, that dark-field electron micrographs (data not shown) revealed the preservation of some crystalline phase.

The FTIR spectra obtained from samples containing 10–70% of Si to Fe are shown in Fig. 3. The frequencies of the most important bands are depicted in Fig. 3. FTIR spectroscopy is a helpful technique to identify the stretching vibrations of Si–O, Si–O–Fe and Fe–O–Fe bonds [26–30]. The FTIR spectra of all the samples

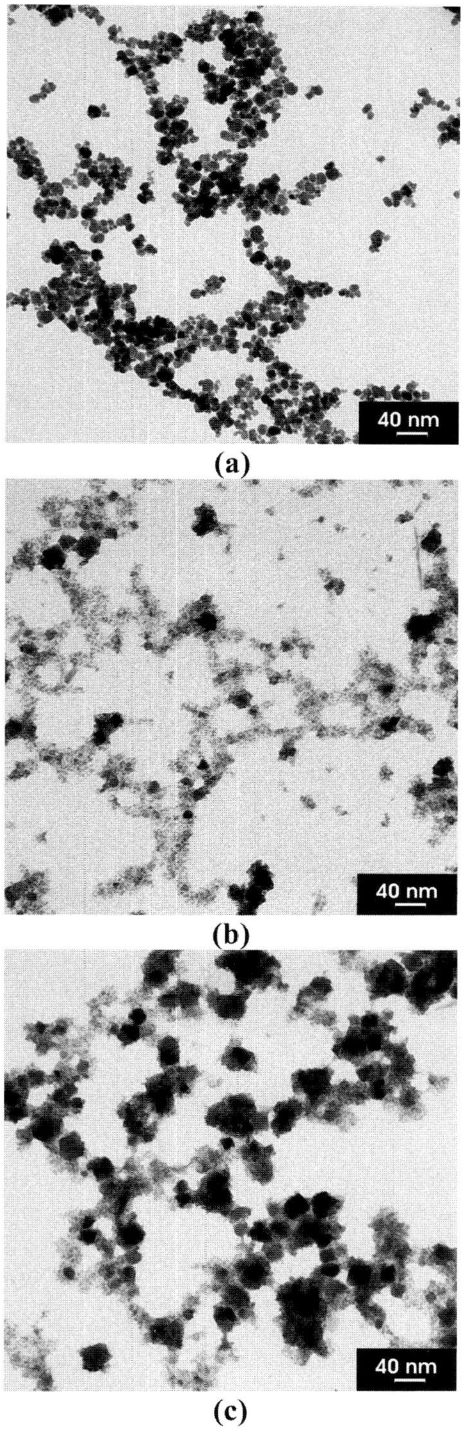

Fig. 2 Transmission electron microscopy micrographs recorded from samples prepared in **a** an acidic environment (sample 1), and a methanolic environment (25% methanol-to-base ratio) (sample 2) representing **b** smaller nanoparticles and **c** larger ones

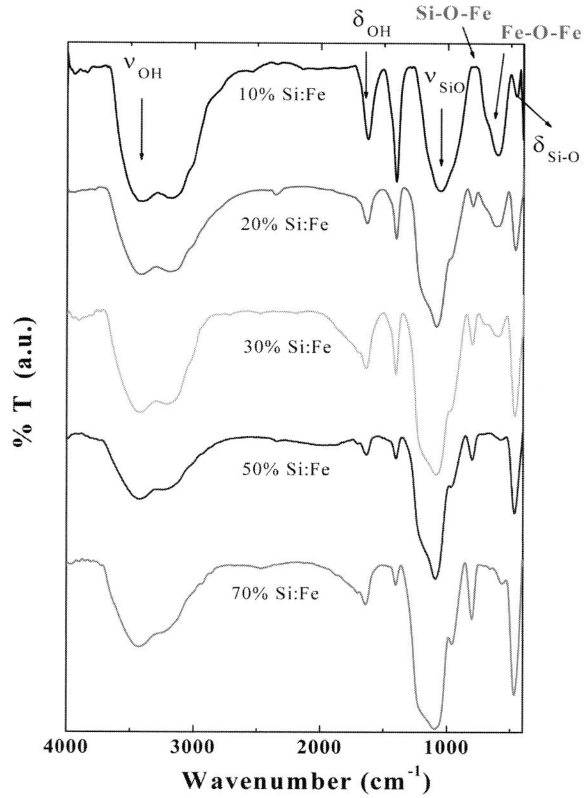

Fig. 3 Fourier transform IR spectra from nanoparticles synthesized in a methanolic medium (25% methanol-to-base ratio) with increasing Si-to-Fe volume ratio from 10% to 70%, at room temperature

(Fig. 3) show bands at 3,440 and 1,640 cm^{-1}, assigned to ν_{OH} and δ_{OH}, respectively. The bands around 1,100, 1,400 and 470 cm^{-1} are attributed to the Si–O bond. The δSi–O increases as the amount of silica increases. The sharp band around 800 cm^{-1} suggests the presence of Si–O–Fe bonds. The bands at 600 cm^{-1} must be evidence of a Fe–O–Fe bond [31–33]. The increase in the silica content implies the decrease of such a band, whereas the Si–O–Fe band increases. The bands are not shifted significantly; therefore, the FTIR analysis suggests the binding of silica to the iron oxide nanoparticles at the surface.

In order to certify that these samples were actually silica-coated, we made use of the electron energy loss spectra and ESI of the particle surface. The Si and Fe elemental maps for 10% and 50% of Si-to-Fe samples are shown in Fig. 4. As one can see, Fig. 4a evidences (for the sample prepared with 10% of Si to Fe) the elemental Si map (i.e., the bright region corresponds to high elemental Si concentration), whereas Fig. 4b represents the elemental Fe distribution for the same sample. Considering that the particles are coated by silica, one can interpret the ESI results by comparing both images. Note that in the same region of the elemental Si map (clear region in Fig. 4a) there is an underlying Fe

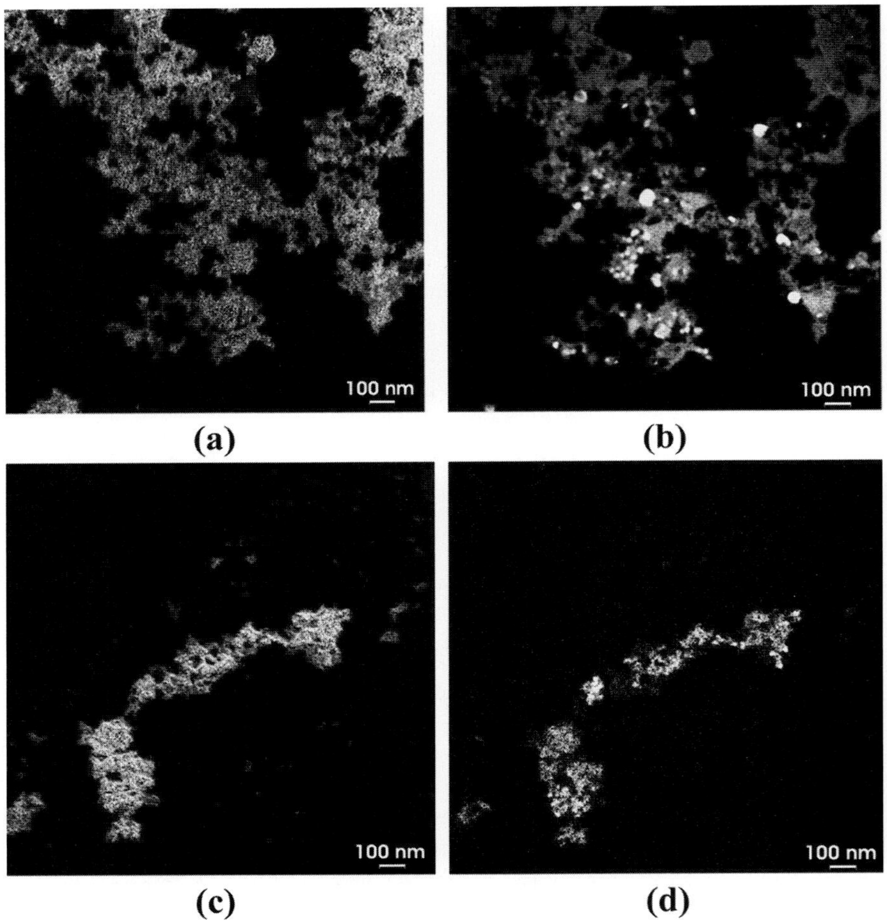

Fig. 4 Electron spectroscopy imaging micrographs of **a** Si and **b** elemental Fe maps from a sample with a Si-to-Fe ratio of 10% and **c** Si and **d** elemental Fe maps from a sample with a Si-to-Fe ratio of 50% (see text for details)

component (bright region in Fig. 4b). This suggests the superposition of a silica layer over iron oxide particles. Similar behavior is observed for 50% of Si to Fe (Fig. 4c, d). These findings corroborate those obtained from FTIR results, where a sharp band at 804 cm^{-1} suggests the presence of a Si–O–Fe bond [12].

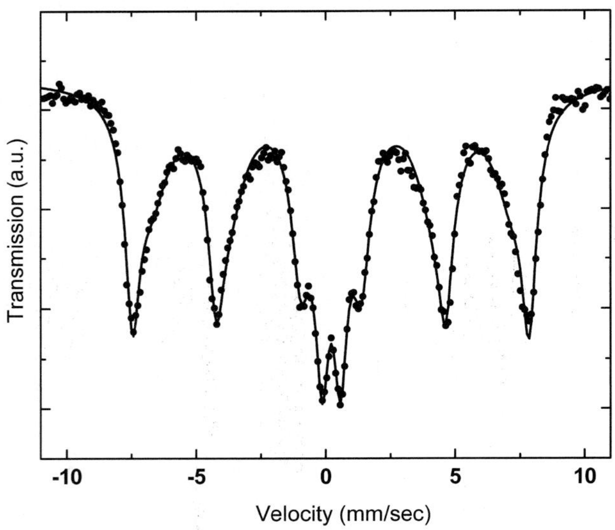

Fig. 5 Mössbauer spectra (*circles*) of the magnetite sample 2 at room temperature. The *solid line* is the best fit to the experimental data

Magnetic properties

To assess the phase composition of the nanoparticles, we performed Mössbauer measurements at room temperature. For sample 2 (0% Si), the spectrum (Fig. 5) displays a mixture of magnetically blocked and superparamagnetic contributions. The hyperfine parameters of the blocked sextets are $B_1 = 47$ T and $B_2 = 43$ T, $QS_1 = 0.00$ mm s^{-1} and $QS_2 = 0.00$ mm s^{-1} and $IS_1 = 0.31$ mm s^{-1} and $IS_2 = 0.33$ mm s^{-1}, which correspond to A and B sites of the magnetite phase [34–36]. For samples with higher silica contents, the spectra showed only a superparamagnetic doublet, and thus no information about the phase can be extracted. Low-temperature spectra (80 K) were obtained, but the superparamagnetic state still persists at this temperature; therefore, data at 4.2 K are being currently obtained to identify the magnetic phase in the blocked state.

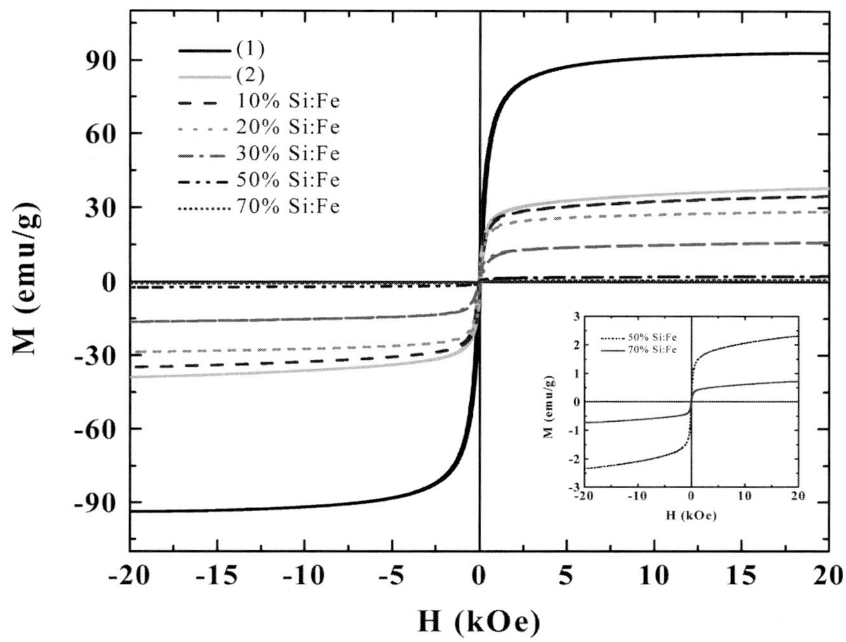

Fig. 6 Magnetization hysteresis curves at room temperature from samples prepared in: an acid medium (*1*), and in a methanolic medium (25% methanol-to-base ratio) with increasing Si-to-Fe volume ratio, as indicated in the figure, ranging from 0% (*2*) to 70%, at room temperature. The *inset* amplifies the magnetic behavior from samples containing the highest amount of silica

Typical magnetization curves obtained by VSM analysis are shown in Fig. 6. The samples are found to be superparamagnetic from the complete reversibility of the M–H curve recorded at room temperature [34]. The saturation magnetization (M_s) of sample 1 prepared in an acid medium (also in water, data not shown) was $M_s = 92(2)$ emu g^{-1}, which is coincident with the value for bulk magnetite [37], reinforcing the Mössbauer results. On the other hand, the strong decrease of M_s (about 38 emu g^{-1}) observed on addition of methanol (sample 2) may be due to the increase in the particle size polydispersity, as evidenced by XRD (Fig. 1) and TEM images (Fig. 2). The smaller particles usually yield surface spin disorder and thus lower magnetization values [35, 38]. With addition of TEOS to the system, M_S decreases until about 1 emu g^{-1} for a 70% Si-to-Fe volume ratio (Fig. 5). This behavior is, indeed, expected once a nonmagnetic material such as SiO was added to this system $_2$ [33].

Conclusions

We have presented an alternative method to synthesize silica-coated magnetic nanoparticles, by mixing $FeCl_2$ and $FeCl_3$ acid solutions in an alcoholic medium, followed by a sol–gel process using TEOS as the Si source, that produced a gel phase, immediately. ESI and FTIR measurements give strong support to the conclusion that the iron oxide particles are silica-coated. Moreover, Mössbauer data and XRD reflections are consistent with the magnetite structure for nanoparticles prepared in a solution containing Si-to-Fe volume ratios of 10% and 20%. The particles have a mean crystalline grain size of about 170 Å, in agreement with that obtained in the absence of TEOS. The magnetic properties of these particles are preserved, in such a way that superparamagnetic behavior is still observed, although their saturation magnetization values decrease slightly, probably owing to an increase of the surface spin disorder and nonmagnetic SiO_2 coating. Therefore, we are now able to proceed towards the synthesis of a new magnetic particle that contains photoactive molecules attached to its surface, by functionalizing the silica layer with photoactive molecules [18].

Acknowledgements The authors thank CNPq, CAPES/PROCAD, FAPESP and Millenium Institute/IQ-UNICAMP.

References

1. Beecroft LL, Ober CK (1997) Chem Mater 9:1302
2. Awschalom DD, DiVicenzo DP (1995) Phys Today 43
3. Ziolo RF, Giannelis EP, Weinstein BA, O' Horo MP, Ganguly BN, Mehrotra V, Russel MW, Huffman DR (1992) Science 257:219
4. McMicheal RD, Shull RD, Swartzendruber LJ, Bennett LH, Watson RE (1992) J Magn Mater 11:29
5. Häfeli U, Schüt W, Teller J, Zborowski M, (1997) Scientific and clinical applications of magnetic carriers, 1st edn. Plenum, New York
6. Roger J, Pons JN, Massart R, Halbreich A, Braci JC (1999) Eur Phys J A 5:321
7. Fauconnier N, Pons JN, Roger J, Bee A (1997) J Colloid Interface Sci 194:427
8. Safarik I, Safarikova M (1999) J Chromatogr B 722:33
9. Klotz M, Ayral A, Guizard C, Ménaer C, Cabuil V (1999) J Colloid Interface Sci 220:357
10. Ma M, Zhang Y, Yu W, Shenm HY, Zhang HQ, Gu N (2003) Colloids Surf A 212:219
11. Morales MP, Munoz-Aguado MJ, Garcia-Palacio JL, Lazaro FJ, Serna CJ (1998) J Magn Magn Mater 183:232
12. Moreno EM, Zayat M, Morales MP, Serna CJ, Roig A, Levy D (2002) Langmuir 18:4972
13. Tartaj P, González-Carreño T, Sena CJ (2003) J Phys Chem 107:20
14. Del Barco E, Asenjo J, Zhang XX, Pieczynsky R, Juliá A, Tejada J, Ziolo RF, Fiorani D, Testa AM (2001) Chem Mater 13:219
15. Cannas C, Casula MF, Concas G, Corrias A, Gatteschi D, Falqui A, Musinu A, Sangregorio C, Spano G (2001) J Mater Chem 1:3180
16. Cannas C, Gatteschi D, Musinu A, Piccaluga G, Sangregorio C (1998) J Phys Chem B 102:7721
17. Ennas G, Musinu A, Piccaluga G, Zedda D, Gatteschi D, Sangregorio C, Stanger JL, Concas G, Spano G (1998) Chem Mater 10:495
18. Rodrigues MA, Tada DB, Politi MJ, Baptista MS (2002) J Non-Cryst Solids 304:116
19. Rodrigues MA, Politi MJ, Miranda MTM, Bemquerer MP, Brochsztain S, Baptista MS (2002) Chem Lett 6:604
20. Savii C, Popovici M, Enache C, Subrt J, Niznansky D, Bakardzieva S, Caizer C, Hrianca I (2002) Solid State Ionics 151:219
21. Berger P, Adelman NB, Beckamn KJ, Campbell DJ, Ellis AB, Lisensky GC (1999) J Chem Educ 76:943
22. Klug HP, Alexander LE (1974) X-ray diffraction procedures for polycrystalline and amorphous materials, 2nd edn. Wiley-Interscience, New York
23. Costa CAR, Leite CAP, Galembeck F (2003) J Phys Chem B 107:4747
24. Itri R, Depeyrot J, Tourinho FA, Sousa MH (2001) Eur Phys J E 4:201
25. Massart R (1983) IEEE Trans Magn 17:1247
26. Nogami M, Asuha N (1993) J Mater Sci Lett 12:1705
27. Chaneac C, Tronc E, Jolivet JP (1996) J Mater Chem 6:1905
28. Ying JY, Benziger JB, Navrostsky A (1993) J Am Ceram Soc 76:2571
30. Ayers MR, Song XY, Hunt AJ () J Mater Sci 31:6251
31. Nyquist RA, Kagel RO (1971) Infrared spectra of inorganic compounds: 3,800–45 cm^{-1}, 1st edn. Academic, New York
32. McDevitt NT, Baun WL (1964) Spectrochim Acta 20:799
33. Del Monte F, Morales MP, Levy D, Fernandez A, Ocaña M, Mollins E, ÓGray K, Serna CJ (1997) Langmuir 13:3627
34. (a) Biddlecombe GB, Gun'ko YK, Kelly JM, Pillai SC, Coey JMD, Venkatesan M, Douvalis AP (2001) J Mater Chem 11:2937; (b) Rajendran M, Pullar RC, Bhattacharya AK, Das D, Chintalapudi SN, Majumdar CK (2001) J Magn Magn Mater 232:71
35. Goya GF, Berquó TS, Fonseca FC, Morales MP (2003) J Appl Phys 94:3520
36. Voogt FC, Hibma T, Smulders P, Niesen L (1997) J Cryst Growth 174:440
37. Cullity BD (1972) Introduction to magnetic material, 1st edn. Addison-Wesley, London
38. Dormann L, Fiorani D, Tronc E (1997) In: Prigogine I, Rice SA (eds) Advances in chemical physics. New York

Progr Colloid Polym Sci (2004) 128: 239–243
DOI 10.1007/b97093

Juliana Flor
Sergio Antonio Marques de Lima
Marian Rosaly Davolos

Effect of reaction time on the particle size of ZnO and ZnO:Ce obtained by a sol–gel method

J. Flor · S.A. Marques de Lima ·
M.R. Davolos (✉)
Instituto de Química, Universidade
Estadual Paulista-UNESP, P.O. Box 355,
CEP 14801-970, Araraquara, SP, Brazil
e-mail: davolos@iq.unesp.br
Tel.: +55-16-2016634
Fax: +55-16-2227932

Abstract In this work particles of ZnO of size range 33–56 Å were prepared by a sol–gel method. The effect of reaction time on the particle size of ZnO or ZnO:Ce was investigated by transmission electron microscopy measurements, UV–vis absorption and luminescence spectroscopy. A linear increase of the mean particle size is observed as a function of reaction time. The cerium-doped particles are bigger than the pure ZnO ones obtained at the same reaction time. A shift to lower energy at the maximum of the bands is observed in all absorption, emission and excitation spectra as a function of particle growth. From the absorption spectra the optical energy gap values (Eg) for these particles were determined. In the quantum size regime, Eg was found to decrease with particle growth.

Keywords Cerium · Zinc oxide · Nanoparticle · Luminescence

Introduction

Solid zinc oxide compounds are used in numerous applications [1]. It is well known that ZnO presents unique physicochemical properties, rendering it a suitable compound for widespread applications, whether as ceramic material in electro-optical devices and varistors, or powder or emulsion in pharmaceutical and cosmetic products [2]. For these applications the control of the particle diameter is fundamental and the particle diameter may change the spectroscopic properties. In cosmetics, for example, ZnO is used in sun block because it has the very opportune property of being transparent to visible light, but opaque to UV radiation [3]. However this behavior depends on particles or cluster size, which must be smaller than wavelengths in the visible region. If not, the sun block will absorb visible light and this is undesirable. In optical devices, the particle size is also important because both excitation and emission spectra can be modulated by the well-known quantum effect of size. This effect was studied in ZnO by van Dijken et al. [4, 5], who have made important contributions in order to understand the transitions involved in the emission phenomenon.

The sol–gel process is widely used to obtain nanoparticles with well-defined size. It is fundamental, however, that all the parameters are strictly controlled, since a small change in one of the parameters, such as temperature, reaction time or concentration, can significantly influence the results [6]. Not only these parameters, but also the introduction of doping can change morphological and spectroscopic characteristics of the sample.

Rare earth (RE) ions are widely used in solids because their intrashell transitions (*f–f*) allow them to be used as structural probes, since the surrounding atoms weakly shift those transitions. RE ions are used as monochromatic light emitters due to their narrow emission lines. The *f–d* transitions and charge transfer transitions play important roles in the absorption of RE ions in the UV range, which is also desirable for sunscreen applications because a higher performance is achieved [7]. Attempts to measure the photoluminescence or electroluminescence of ZnO doped with various RE ions produced unsatisfactory results, since only a weak energy transfer was observed for just a few RE ions [8–11].

The aim of this work is to investigate the effect of some experimental parameters such as condensation time and doping on the particle size and spectroscopic properties of the ZnO obtained.

Experimental

Suspensions of nanocrystalline ZnO particles were prepared in 2-propanol (P.A. – Synth). A solution of 0.06 g $Zn(C_2H_5O_2)_2{\cdot}2H_2O$ (Qeel) in 80 mL of 2-propanol was made at 50 °C. After dilution of the solution to 230 mL with 2-propanol, it was cooled to 0 °C. At this temperature 0.016 g NaOH (Mallinckrodt) was added while the mixture was stirred. After aging in an ultrasound bath at 65 °C for 10, 60, 120, 180, 240 and 300 min, a transparent, colorless suspension of ZnO particles was obtained. ZnO:Ce samples were prepared with 3% of Ce using $CeCl_3$ as precursor that was added at the beginning of the reaction.

Each sample of ZnO or ZnO:Ce in 2-propanol was collected after the times mentioned above, cooled and then immediately characterized by luminescence spectroscopy (Fluorog SPEX F212I spectrofluorimeter equipped with a 450-W xenon lamp and a Hamamatsu R928 photomultiplier) using a front-face acquisition mode and absorption spectroscopy (PerkinElmer Lambda 14P) using 1-cm-thick cell. The morphology and the particle size were examined with a Phillips CM200 transmission electron microscope with an accelerating voltage of 135 kV. For this measurement, a carbon-coated grid was dipped into the solution and then dried.

Results and discussion

Since the particle size can shift the absorption feature when the quantum size regime is achieved and the particle size depends on the reaction time, an absorption redshift as a function of reaction time is expected. In the quantum size regime the energy gap (Eg) is modulated as a function of particle radius. Thus, changes in the experimental parameters, which change the particle size, will shift the Eg. Since the Eg is the energy difference between the valence band (VB) and the conduction band (CB), a shift in the observed bands is expected. This effect is observed in absorption spectra for pure ZnO and ZnO:Ce (see Figs. 1, 2, respectively). The observed bands attributed to an electronic transition from the VB to the CB shift to longer wavelengths as a function of reaction time.

In 1972, Wood and Tauc [12] proposed a method to determine the optical Eg of a semiconductor from its absorption spectra, which is widely described in the literature. A plot of the experimental optical Eg versus reaction time for ZnO and ZnO:Ce is shown in Fig. 3. For both, the optical Eg decreases as a function of reaction time, suggesting that the particle size increases with an increase in reaction time. Lower Eg values are observed for all cerium-doped samples when compared with the pure samples, indicating that the presence of cerium ion as a dopant increases the particle size. When calculating the slope of the curves in Fig. 4, an interesting result was found. For pure ZnO, we found a slope of $1.7{\times}10^{-4}$ and for ZnO:Ce a slope of $2.5{\times}10^{-4}$, suggesting that particles grow faster when cerium ions are introduced into the solution (the time of 10 min was not taken into account for these calculations).

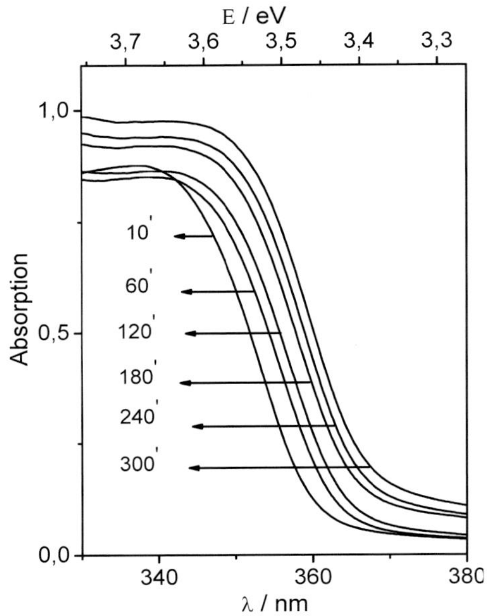

Fig. 1 UV–vis absorbance spectra of ZnO samples with different reaction times

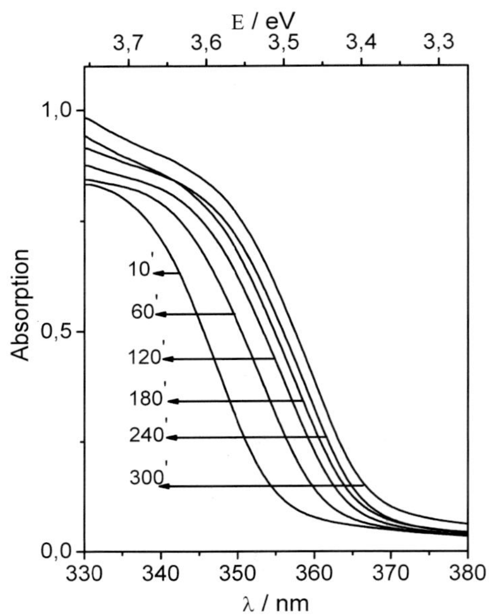

Fig. 2 UV–vis absorbance spectra of ZnO:Ce samples with different reaction times

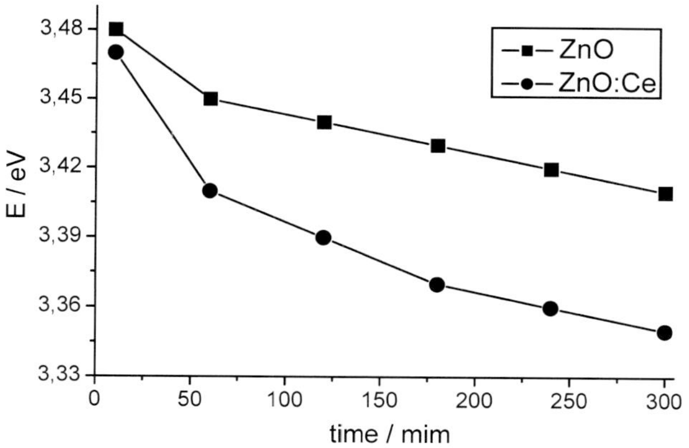

Fig. 3 Optical energy gap of ZnO and ZnO:Ce samples, determined by the Tauc method from absorption spectra (Figs. 1, 2) versus reaction time

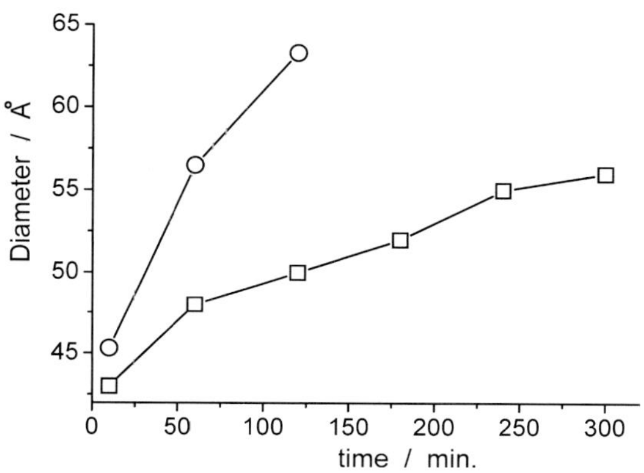

Fig. 4 Dependence of the particle size on the reaction time of ZnO and ZnO:Ce. Particle diameters from Table 1 determined were the relationship between the energy gap and the size from Ref. [13]. *Squares* represent ZnO and *circles* represent ZnO:Ce. *Squares* represent ZnO and *Circles* represent ZnO:Ce

Using an experimental relationship between the absorption and the particle radius described by Meulenkamp [13], it was possible to determine the particle sizes as, shown in Table 1. This method has the limitation that it only allows the diameter to be determined in the range 25–65 Å. For particles larger than 65 Å the Eg value varies only slightly with particle size and the error in the calculated value for the mean radius of the ZnO particles is fairly large. The particle size determined are between 43 and 63 Å. From transmission electron microscopy (TEM) it was also possible to determine the mean particle size of the samples of pure ZnO (10, 120 and 300 min) and ZnO:Ce (120 min). The particle sizes determined are between 33 and 50 Å. These data are also shown in Table 1; the difference between the particle size determined by optical measurements and by TEM can be due to the error in the determination of the band gap energy from the absorption spectra. The results indicate that the mean particle size increases with increasing reaction time. The ZnO:Ce sample (60 min) has the same particle size as the ZnO sample (300 min), 57 Å, determined from the absorption spectra, and the ZnO:Ce sample (120 min) has larger particles, 65 Å, than the ZnO sample (300 min), 56 Å. These data confirm that the smaller values of the optical Eg from the ZnO:Ce samples, when compared with the ZnO samples, are related to larger particle size.

Table 1 Particle diameters determined by transmission electron microscopy (*TEM*) photomicrographs and by using an experimental relationship between the absorption and the particle radius described by Meulenkamp [13]

Sample/time (min)	Diameter (Å)	
	From TEM	From absorption
ZnO/10	33 (±3.5)	43
ZnO/60	40 (±4.7)	48
ZnO/120		50
ZnO/180		52
ZnO/240		55
ZnO/300	45 (±5.6)	56
ZnO:Ce/10		42
ZnO:Ce/60		53
ZnO:Ce/120	50 (±6.1)	65

The plot of the particle size, calculated from the absorption spectra, versus the reaction time is presented in Fig. 4. According to Van Dijken et al. [4], immediately after addition of the NaOH a rapid formation of extremely small ZnO particles followed by a relatively slow growth of the particles occurs. From Fig. 4 it is possible to show that the dependence of the particle growth on the reaction time, which is linear at a constant temperature of 65 °C. It is observed, from the two curves in Fig. 4, that particles grow faster up to 60 min and then the growth rate decreases to give a linear dependence on reaction time.

Figure 5 shows that the excitation spectra for the pure ZnO samples at different reaction times have the same features as the absorption spectra. One excitation band is observed, with the maximum around 350 nm, attributed to transitions from the VB to the CB. This band shifts to longer wavelengths as the reaction time increases. This suggests an increase in the particle size as a function of time, since particle growth results in a decrease in the Eg. The excitation spectra of ZnO:Ce samples are similar to the excitation spectra of pure ZnO samples (data not shown).

By comparing the absorption and excitation spectra of the ZnO samples, one can notice that although they have the same features, the observed intensities follow the opposite tendency with increasing particle size. To understand this inversion, light scattering must be considered. In absorption measurements, if the light is

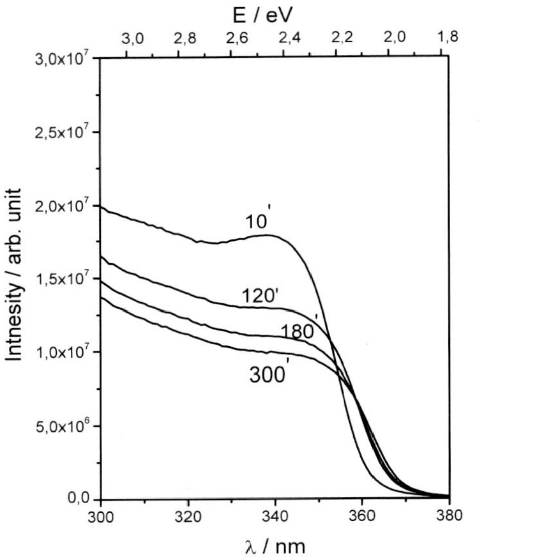

Fig. 5 Excitation spectra of ZnO samples at room temperature, $\lambda_{em} = 575$ nm, with different reaction times

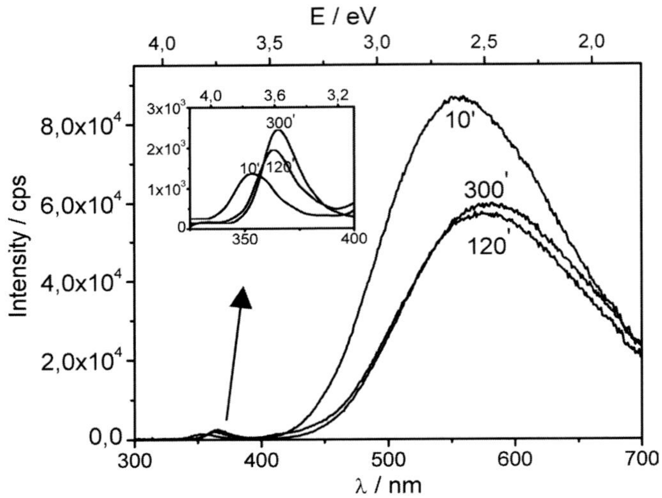

Fig. 6 Emission spectra of ZnO samples at room temperature, $\lambda_{ex} = 345$ nm, with different reaction times

scattered by the particles in solution the intensity of the transmitted light collected by the detector will be lower. In this work, the absorption results indicate an increase in the absorption intensity with an increase in reaction time. This effect can be related to the increase of the particle size, which increases not only the absorbed light but also the scattered light. However, in photoluminescence spectroscopy, more scattered light means less emitted light collected by the detector at an angle of 22.5° to the excitation beam. This can explain the inversion in the intensity of absorbed light (Figs. 1, 2) and the emitted light (Fig. 5).

We can observe that the emission spectra from pure ZnO samples exhibit the same features as the ZnO:Ce spectra (data not shown) in Fig. 6. Two emission bands are observed, one in the UV region attributed to exciton emission, and the other in the visible region attributed to a transition from the CB (or a shallow level close to this band) to a trap level approximately 2 eV below the CB edge [5]. Both bands shift to longer wavelengths with an increase in particle size; however, the excitonic emission shifts more than the visible one. According to van Dijken et al. [5] this behavior is expected. They found a slope of approximately 0.6 from the plot of the visible versus UV maximum emission bands. From our results, slopes of 0.74 for ZnO and 0.70 for ZnO:Ce were found.

Conclusions

This work has demonstrated that from simple analyses of optical Eg values calculated by the Tauc method, it is possible to predict changes in particle size within the quantum size regime, with lower values of the optical Eg indicating an increase in particle size and vice versa. The optical change can be observed in both excitation and absorption spectra. The observations from optical Eg values were confirmed by the particle size calculated from TEM micrographs and were also determined from the absorption spectra using an empirical relationship. At a constant temperature of 65 °C, an increase of the reaction time and the presence of cerium ions as doping agents cause an increase in the particle size of ZnO obtained by a sol–gel method.

Acknowledgements Financial support by FAPESP is gratefully acknowledged. We are grateful to I.R.P. Maniette for TEM measurements. J.F. thanks CAPES and S.A.M.L. thanks FAPESP for scholarships.

References

1. Castellano M, Matijevic E (1989) Chem Mater 1:78
2. Andrés-Vergés M, Martinez-Gallego M (1992) J Mater Chem 27:3756
3. Mitchnick MA (1992) Cosmet Toilet 107:111
4. van Dijken A, Meulenkamp EA, Vanmalkelbergh D, Meijerink A (2000) J Lumin 87–89:454
5. van Dijken A, Meulenkamp EA, Vanmalkelbergh D, Meijerink A (2000) J Lumin 90:123
6. Brinker CJ, Scherer GW (1990) Sol gel science: the physics and chemistry of sol gel processing. Academic, Boston
7. Yabe S, Sato T (2003) J Solid State Chem 171:7
8. Bacher S, Azuma K, Kossanyi J, Valat P, Ronfard-Haret JC (1997) J Lumin 75:35
9. Hayashi Y, Narahara H, Uchida T, Nuguchi T, Ibuki S (1995) J Appl Phys 34:1878
10. Bacher S, Sandouly C, Kossanyi J, Ronfard-Haret JC (1996) J Phys Chem Solids 57:1869
11. Lima SAM, Sigoli FA, Davolos MR (2003) J Solid State Chem 171:287
12. Wood DL, Tauc J (1972) Phys Rev B 5:3144
13. Meulenkamp EA (1998) J Phys Chem B 102:5566

Progr Colloid Polym Sci (2004) 128 : 244–248
DOI 10.1007/b97095

Ferenc Csempesz
Katalin F. Csáki
Miklós Nagy

Thermodynamic properties and interfacial behaviour of uncharged polymers and binary polymer mixtures

F. Csempesz (✉) · K.F. Csáki · M. Nagy
Department of Colloid Chemistry,
Eötvös University, P.O. Box 32,
1518 Budapest 112, Hungary
e-mail: csf@ludens.elte.hu
Tel.: +36-1-2090594
Fax: +36-1-2090602

Abstract The thermodynamic properties of uncharged polymers in aqueous solution and the interfacial behaviour of single polymer and binary polymer mixtures, respectively, on colloidal dispersions of negatively charged polystyrene latex were studied. The water activities in binary and ternary solutions of chemically different polymers were determined by a gel-deswelling method and the relevant solvent–segment and segment–segment pair interaction parameters were calculated. In aqueous dispersions of polystyrene latex, with and without adsorbed polymer, respectively, the osmotic pressures were directly measured by a block osmometer and the corresponding solvent activities were estimated. Independent adsorption isotherms were determined for each polymer, both individually and under simultaneous, competitive adsorption conditions. As a measure of the preferential affinity for particle surfaces of chemically different macromolecules in simultaneous competitive adsorption, preferential adsorption parameters for the polymers in pairs were used. The thermodynamic characteristics proved to be relevant indicators for the interfacial behaviour of the polymers. Correlation was found between the thermodynamic activity and the affinity for polystyrene surfaces of the polymeric adsorbates.

Keywords Uncharged polymers · Solvent activity · Osmotic pressure · Gel deswelling · Adsorption preference

Introduction

Adsorption of polymers plays a fundamental role in controlling interfacial processes in a variety of disperse systems. The effects observed result mostly from the change of interparticle interactions due to the macromolecules adsorbed at the particle–solution interfaces, but in certain cases, the presence of free polymer in the solution may play a role [1–3]. Surface modification by macromolecules has also attracted much attention because of its widespread (biological, medical, pharmaceutical, etc.) applications, including improvement of the wettability, biocompatibility or adhesion of the surfaces [4].

The interfacial behaviour of macromolecules is closely related to the magnitude of relevant (solvent–segment, segment–surface and solvent–surface) interaction energies. The thermodynamic quality of a solvent is, therefore, a key factor that substantially affects both the chain dimensions in solution and the properties of a polymer layer at an interface. The classical mean-field theory derived by Flory and Huggins [5] is still widely used to represent the thermodynamic properties of polymer solutions. The quality of a solvent is characterized through the interaction parameter χ. For a binary solution, χ_{12} measures the solvent (1)–segment (2) interaction and theoretically ranges from below zero for a very good solvent to 0.5 or greater. A ternary solution that consists of two polymeric components and a solvent can also be described on the basis of extensions of the Flory–Huggins equations. Besides the two solvent–segment interaction parameters (χ_{12}, χ_{13}), information on the thermodynamic interaction between the different macromolecular species (χ_{23}) should also be obtained.

For a full understanding of the interfacial behaviour of polymeric substances, the extent of adsorption, the chain conformations and the segmental adsorption (free) energy ($\chi_S\ kT$) should certainly be established for each macromolecular adsorbate [6]. A reasonable displacement model to estimate χ_S parameter for a polymer has already been proposed [7]. This method pays attention to the critical point of the multipoint adsorption at which desorption of the last segments is finished. The detailed mechanism of the desorption process in solvent-displacer mixtures has not been discussed. The fact that synthetic polymers are, in general, polydisperse also has considerable influence on the displacement [8].

The adsorption from multicomponent polymer solutions differs in many aspects from the individual adsorption of macromolecules [3, 9–11]. Chemically different polymers usually have different affinities for the surface, which may lead to greatly different adsorption characteristics for the various components. In general, adsorption from a polymer mixture leads to preferential adsorption of the polymer with the highest adsorption energy [1, 3, 11] and displacement of the weakly attached molecules from the interfaces may take place. Considerable difficulties arise, however, in obtaining segmental adsorption energies even for two polymers in simultaneous adsorption from solution, since six different interactions should be taken into account. No general method has yet been developed for determination of the various adsorption energies in systems where different polymeric species compete for surface sites.

This paper is focused on the main factors controlling the interfacial behaviour of uncharged polymers and polymer mixtures at particle–solution interfaces. In this study, thermodynamic properties of uncharged polymers in binary and ternary solutions, respectively, and their adsorption from these solutions on polystyrene latex in aqueous dispersions were investigated. The solvent activity of the dissolved polymers together with that of adsorbed macromolecules were determined, and a possible link of the thermodynamic variables to the adsorption characteristics of the polymers is discussed.

Theoretical background to the thermodynamic investigations

For an exact characterization of the thermodynamic properties of a solution, one of the most important quantities is the chemical potential of the solvent (μ_1). A basic experimental measurement which can be used to determine the relative thermodynamic activity of the solvent (a_1) in a solution is the osmotic pressure [$\Pi(=-RT\ln a_1/V_1)$, where R is the gas constant, T the absolute temperature and V_1 the partial molar volume of the solvent]. Macromolecular solutions exhibit deviation from ideal behaviour, even at low concentrations. The concentration dependence of the osmotic pressure for a dilute solution allows the evaluation of the second and third virial coefficients, as well, which are basic measures of the interactions between polymer coils.

According to the theory of Flory and Huggins [5], the change in the chemical potential of the solvent (μ_1) in a solution of an uncharged polymer is $\Delta\mu_1 = \mathrm{RT}[\ln(1-\mathrm{v}_2) + (1-1/\mathrm{x}_n) + \chi_{12}\mathrm{v}_2^2]$. Here, v_2 is the volume fraction of the polymer and x_n is the ratio of the number-average molar volume of a polydisperse polymer and the molar volume of the solvent. From the volume-fraction dependence of the solvent activity, the χ interaction parameter can be calculated. Once χ_{12} is known, one may evaluate the chemical potential ($\Delta\mu_2$) of the macromolecular solute:

$$\Delta\mu_2 = \mathrm{RT}\left[\ln(\mathrm{v}_2) - (\mathrm{x}_n - 1)(1 - \mathrm{v}_2) + \chi_{12}\mathrm{x}_n(1 - \mathrm{v}_2)^2\right].$$

Osmotic pressure may be directly related to the thermodynamic and statistical mechanical properties of colloidal dispersions. In dispersions with electrically charged particles a major contribution to the total osmotic pressure may originate from interparticle forces. The variation in osmotic pressure as a function of the volume fraction of dispersed particles and electrolyte concentration may be used to obtain information about these colloidal interactions in (moderately concentrated) dispersions [12].

Experimental

Aqueous dispersions of polystyrene latices (PS-74 and PS-133) with negatively charged monodisperse particles were prepared according to the Vanderhoff method [13]. The required ion concentration in the dispersion was adjusted with NaCl. The mean particle size and the size distribution were determined by photon-correlation spectroscopy and controlled by scanning electron microscopy or sedimentation measurements [10, 17, 20].

Water-soluble uncharged polymers prepared from commercial products of Tylose MH50 methylcellulose (MC), hydrolysed Poval 420 poly(vinyl alcohol) (PVA), Fluka K-60 (PVP) and GAF K-30 (PVP-1), K-60 (PVP-2), K-90 (PVP-3) poly(vinylpyrrolidone) were used. To sharpen the molar mass distribution of the samples, fractions of the highest and lowest molar mass were removed from the aqueous solutions by suitable precipitation procedures. For measurements of solvent activity, the samples were intensively dialysed.

The number-average and/or the viscosity-average molecular mass of the polymers and the volume-average mean diameter of the latex particles together with the polydispersity index and the surface charge density are given in Table 1.

Table 1 Characteristics of the polymers and the dispersions

	Polymer samples		Dispersions		
	$M_n \times 10^3$	$M_\eta \times 10^3$	Particle size (nm)	Polydispersity index	Surface charge density ($\mu C\ cm^{-2}$)
PS-74	–	–	74	0.025 ± 0.005	5.6
PS-133	–	–	133	0.030 ± 0.005	3.2
MC	32	63			
PVA	74	113			
PVP	78	254			
PVP-1		32			
PVP-2		153			
PVP-3		890			

Table 2 Water activities in polymer solutions, the activity of macromolecules, and the solvent activity in dispersions of PS-74 latex with and without adsorbed polymer

	Polymer concentration ($g\ dm^{-3}$)	Dissolved polymer		PS-74 latex + polymer
		$\ln a_1$	$\ln a_2$	$\ln a_1$
	–	–	–	-1.4×10^{-6} (without polymer)
MC	4.6	-6.7×10^{-6}	−770	-3.7×10^{-6}
PVA	3.2	-1.7×10^{-6}	−1700	-2.4×10^{-6}
PVP	3.7	-1.8×10^{-6}	−1760	-1.7×10^{-6}

Solvent activity measurements

Two sets of measurements at 298 ± 0.1 K were carried out. At low volume fractions of the solutes ($v_2 \leq 1.5 \times 10^{-2}$), the osmotic pressure in solutions of uncharged macromolecules was measured by a block membrane osmometer using an acetylated Visking cellulose membrane (Medicell, UK) [14]. The detectable limit, with adequate accuracy of osmotic pressure, was ± 10 Pa ($\ln a_1$: -7×10^{-8}), and the lowest measurable value for $\ln a_1$ was about -3×10^{-7}. Equilibrium was reached within 48–72 h. The osmotic pressure of the latex dispersions, with and without adsorbed polymer, respectively, at various volume fractions of polystyrene particles was determined in the same way.

In more concentrated solutions ($1 \times 10^{-2} < v_2 < 1.2 \times 10^{-1}$) of the polymers and the binary polymer mixtures, the solvent activities were determined using a gel-deswelling method described by Nagy [15]. Poly(vinyl alcohol-*co*-vinyl acetate) gels with 8 mol % acetate content were used for these measurements. Prior to use, the gels were calibrated using PVP solutions. The time taken to reach equilibrium increased in these measurements up to 10 days.

Polymer adsorption

A 24-h time period was generally allowed for the adsorption in a thermostat at 298 ± 0.1 K. After removing the solid particles from the polymer solution by ultracentrifugation, the concentration of the polymer in the supernatant was measured spectrophotometrically. In ternary solutions, the concentration of both polymers was independently established by adding the appropriate reagents separately to the solutions [9, 10]. A method developed to explore precisely the preferential affinity of uncharged polymers for adsorbent surfaces in simultaneous competitive adsorption has been reported elsewhere [10]. This method is based on the determination of preferential adsorption parameters (f^p) for the competing polymers. They can be calculated from the "characteristic adsorption isotherms", the isotherm pairs corresponding to zero adsorbent concentration [16].

Results and discussion

Thermodynamic investigations

For a theoretical approach to the interfacial behaviour of polymers, the thermodynamic activity of water in polymer solutions was determined at various volume fractions by osmotic pressure and gel-deswelling measurements, respectively, and the solvent–segment interaction parameters (χ_{12}) were calculated. The volume-fraction dependence of the χ_{12} parameters fitted well to a straight line, which had a small positive slope for each of the polymers. Extrapolating these functions to zero polymer volume fraction, we can obtain the χ parameters at lower v_2 and the intrinsic χ values. Full knowledge of the χ_{12} interaction parameters in a concentration range allows the evaluation of the thermodynamic activity (chemical potential) of the macromolecular solute.

The osmotic pressures of the latex dispersions in the absence and the presence, respectively, of adsorbed polymer were also measured and the corresponding water activities calculated. Relevant results of the thermodynamic investigations are collected in Table 2. The water activities in solutions of single polymers ($\Delta\mu_1/RT = \ln a_1$), the activity of the dissolved macromolecules ($\Delta\mu_2/RT = \ln a_2$), as well as the solvent activity in dispersions of "bare" and polymer-covered PS-74 latices at particle concentration of 0.0502 g/g are shown together.

The values of $\ln a_1$ and $\ln a_2$ for each polymer are given at the concentration which corresponds to whole amount of the polymer adsorbed on the particles. These results show that all the polymers reduce, although to different extents, the thermodynamic activity of water. By comparing the water activity of the polymer-free and polymer-covered latices, a definite reduction in the solvent activity due to the adsorbed polymer layers can be seen. In other words, adsorption of the macromolecules onto particle-solution interfaces is accompanied by a decrease in their solvent activity, as well.

Polymer adsorption

Typical adsorption isotherms of high-affinity character were obtained for each of the polymers in both individual and competitive adsorption on polystyrene latices [9, 10]. For illustrating the differences in the adsorbability of polymers on PS-74 particles, the adsorbed amounts corresponding to the onset of a (semi-) plateau region of the individual adsorption isotherm for the polymer are given in the following. The adsorbed amounts for the single polymers on PS-74 latex of 5% (w/w) are as follows, Γ/mg m^{-2}: MC = 1.19 ± 0.05, PVA = 0.84 ± 0.02, PVP = 0.95 ± 0.03. In order to compare the adsorption results with those of the thermodynamic investigations, solvent activity measurements with these polymer amounts were also carried out.

Adsorption isotherms determined for a polymer mixture in the usual way lead only to the order of preferential adsorption of the macromolecules [6, 17]. In preceding papers we showed [10, 14] a thermodynamic approach developed to characterize the adsorption preference in simultaneous competitive adsorption. This method has recently been evaluated and used [3, 18] for clarifying the reversibility of the adsorption of polymers.

On the basis of the characteristic adsorption isotherms, preferential adsorption parameters for competing polymers have been established. For polymer 1 in competition with polymer 2, f^{P}_1 is given by $f^{P}_1 = \Gamma_1/(\Gamma_1 + \Gamma_2)$, where Γ_1 and Γ_2 are the adsorbed amounts per unit area of surface for polymer 1 and polymer 2, respectively. The higher f^{P} is for a given polymer (from 0 to 1), the more pronounced is its adsorption preference in competitive adsorption at a solid–solution interface. The preferential adsorption parameters obtained for some pairs of the polymers in competitive adsorption on PS-133 are collected in Table 3.

These results demonstrate a strong adsorption preference for MC in combination with either PVA or PVP samples. When the two vinyl polymers compete for polystyrene surfaces, the molecules of PVP-3 are adsorbed preferentially over the more hydrophilic PVA. Changes in the preferential adsorption of macromolecules, due to the chain length effect, are also well reflected in the values of f^{P}. The order of preferential adsorption of these polymers on the latex particles is MC > PVPs > PVA. The f^{P} values obtained with different polystyrene latices exhibited no definite dependence of these parameters on the particle size [10]. Accordingly, these parameters can be really regarded as adequate characteristics of the preferential affinity for surface sites of polymer molecules in competitive adsorption. Moreover, these results provide evidence that preferential adsorption of a polymer can mainly be accounted for by the higher adsorption energy of the molecules in question.

To compare the results of the thermodynamic measurements with those of adsorption investigations, the change in the chemical potential of the dissolved macromolecules ($\Delta\mu_2$) due to their adsorption on particle surfaces was estimated. It must be noted, however, that estimation of the chemical potential of the macromolecules in the adsorbed state ($\Delta\mu_2^*$) on the theoretical basis of the Flory–Huggins approximation requires much caution. Fractionation of polymer samples at the interfaces may occur, and also adsorbed macromolecules can influence the interparticle van der Waals attraction, the electrical (double-layer) potential and/or give rise to steric repulsion between charged particles. Taking all these effects into account is very difficult but, under certain conditions, interfacial fractionation and the effect of macromolecules on the double-layer potential can be minimized and/or eliminated [10, 19, 20]. By calculating the $\Gamma(\Delta\mu_2 - \Delta\mu_2^*)$ product for the polymers, information on the affinity for surface sites ("surface activity") of the adsorbed macromolecules can be obtained. These estimated quantities for the polymers are as follows. MC: –9–12 mJ·m^{-2}, PVA: –2–3 mJ·m^{-2}, PVP: –5–8 mJ·m^{-2}.

A reasonable reduction in the free energy of particle–solution interfaces due to the adsorption of polymer molecules is indicated by these results. These data might be interpreted, in one way, as if the solvent for the adsorbed macromolecules had become "poorer". As to the qualitative trends, some correlation both with the individual and with the preferential adsorption of the polymers can be revealed. For a quantitative evaluation, the procedure suggested for the determination of preferential adsorption parameters in simultaneous competitive adsorption should be applied in the thermodynamic investigations as well.

Table 3 Preferential adsorption parameters for polymer pairs in competitive adsorption on PS-133 particles

Polymer 1	f^{P}_1				
	MC	PVA	PVP-1	PVP-2	PVP-3
MC	–	0.21	0.32	0.38	0.41
PVA	0.79	–			0.88
PVP-1	0.68		–		
PVP-2	0.62			–	
PVP-3	0.59	0.12			–

Conclusions

The thermodynamic characteristics determined for uncharged polymers in solution and at particle–solution interfaces, respectively, proved to be relevant indicators for the interfacial behaviour of the polymers. Correlation was found between the thermodynamic activity and the affinity for polystyrene surfaces of the polymeric adsorbates.

Acknowledgement The Hungarian Science Foundation under grant OTKA T 034929 supported this work.

References

1. Fleer GJ, Cohen Stuart MA, Scheutjens JMHM, Cosgrove T, Vincent B (1993) In: Polymers at interfaces. Chapman and Hall, London, pp 234–297
2. Napper DH (1983) Polymeric stabilization of colloidal dispersions. Academic, London
3. Lipatov YS (2002) In: Hubbard AT (ed) Encyclopedia of surface and colloid science A. Dekker, New York, pp 558–572
4. Decher G, Eckle M, Schmitt J, Struth B (1998) Curr Opin Colloid Interface Sci 3:32
5. (a) Flory PJ (1942) J Chem Phys 10:51; (b) Huggins ML (1942) J Phys Chem 46:151; (c) Huggins ML (1942) J Am Chem Soc 64:1712
6. Kawaguchi M (1990) Adv Colloid Interface Sci 32:1
7. Cohen Stuart MA, Fleer GJ, Scheutjens JHHM (1984) J Colloid Interface Sci 97:515
8. Chen M-H, Furusawa K (1999) Colloids Surf A 151:65
9. (a) Csempesz F, Rohrsetzer S (1984) Colloids Surf 11:173; (b) Csempesz F, Rohrsetzer S, Kovács P (1987) Colloid Surf 24:101
10. Csempesz F, Csáki K, Kovács P, Nagy M (1995) Colloids Surf A 101:113
11. Kawaguchi M (1990) Adv Colloid Interface Sci 32:1
12. Ottewill R H, Parentich A, Richardson RA (2000) Colloids Surf A 161:231
13. Van Den Hul HJ, Vanderhoff JW (1972) J Electroanal Chem 37:161
14. Nagy M (2003) Mag Kem Foly (in press)
15. Nagy M (2000) Phys Chem Chem Phys 2:2613
16. Nagy M (1991) Langmuir 7:334
17. (a) Csempesz F, F-Csáki K (2000) Langmuir 16:5917; (b) F-Csáki K, Csempesz F (2001) Colloids Surf A 190:3
18. Lipatov Y, Todosijchuk T, Chornaja V, Menzheres G (2002) J Colloid Interface Sci 246:29
19. (a) Cohen Stuart MA, Waajen FHWH, Dukhin SS (1984) Colloid Polym Sci 262:423; (b) Cohen Stuart MA, Mulder JW (1985) Colloids Surf 15:49
20. Csempesz F (2000) Chem Eng J 80:43

Progr Colloid Polym Sci (2004) 128: 249–254
DOI 10.1007/b97097

Rosane Costa Beber
Clifford Bunton
Gianfranco Savelli
Faruk Nome

Incorporation and differential reactivity of anions in zwitterionic sulfobetaine micelles

R.C. Beber · F. Nome (✉)
Departamento de Química,
Universidade Federal de Santa Catarina,
88040-900 Florianópolis, SC, Brazil
e-mail: faruk@qmc.ufsc.br

C. Bunton
Department of Chemistry and
Biochemistry, University of California,
Santa Barbara, California, USA

G. Savelli
CEMIN, Dipartimento di Chimica,
Università di Perugia, 06100 Perugia, Italy

Abstract Zwitterionic surfactants, such as sulfobetaines and carboxybetaines, are formally neutral, but their micelles incorporate anions owing to the higher charge density of the cationic ammonium inner surface, compared with that of the anionic sulfonate, or carboxylate, surface and to anion-specific interactions. In this work, we used the following sulfobetaines: *N*-decyl-, *N*-tetradecyl- and *N*-hexadecyl-*N*,*N*-dimethylammonio-1-propanesulfonate, *N*-tetradecyl-*N*,*N*-diethyl-, *N*-tetradecyl-*N*,*N*-dipropyl- and *N*-tetradecyl-*N*,*N*-dibutylammonio-1-propanesulfonate and *N*-tetradecyl-*N*,*N*-dimethyl-, *N*-tetradecyl-*N*,*N*-diethyl- and *N*-tetradecyl-*N*,*N*-dipropylammonio-1-butanesulfonate. We have determined the partitioning of bromide and perchlorate ions between water and micelles in terms of anions incorporated in the interfacial region of the micelles as a function of concentrations of added ions in the bulk solution. The concentrations of free and micellar-bound ions were determined by using ion-selective electrodes, in the presence and absence of sulfobetaines, at pH 9.0 (borate buffer). The incorporation of anions into zwitterionic micelles fits Langmuir adsorption isotherms, and the results are related to changes in the micellar surface potential. The incorporation of perchlorate is in the range 15–30% of the concentration of micellized surfactant for the different sulfobetaines, and is greater than that for bromide ion (about 10%). These results fit differences in anionic reactivities and in the NMR line widths of micellar incorporated anions.

Keywords Sulfobetaines · Perchlorate ion · Bromide ion · Potentiometry NMR spectroscopy

Introduction

Interactions of aqueous ionic association colloids (micelles, oil-in-water microemulsions) with counterions are readily understandable in terms of a combination of electrostatic effects, which can be treated theoretically, plus ion-specific effects which follow the Hofmeister series of the anions [1–3]. This series describes the specificity of interactions of ions with a variety of environments. Zwitterionic micelles are formally neutral, but their interactions with ions depend on electrostatic and ion-specific effects [2, 4]. In this paper, we consider only sulfobetaine micelles and their ability to bind anions, which depends significantly on the anionic charge density and also on the structures of the sulfobetaines, as shown for linear alkyl derivatives.

$$C_nH_{2n+1}N^+R_2(CH_2)_mSO_3^-$$

For simplicity we designate the derivatives where R is methyl as SB*m-n*, and the *N*-alkyl derivatives ethyl, propyl and butyl are written as SBE*m-n*, SBP*m-n* and SBB*m-n*, respectively. Most work has been on the C_{10}–C_{16} derivatives with a trimethylene tether.

Evidence for anion binding comes from both kinetic and physical work [5], plus theoretical consideration of the interactions of these micelles with anions [2–4]. These results are briefly summarized and more recent evidence is considered in some detail, with comparison of conclusions from various methods.

Early evidence for the interaction of sulfobetaines and carboxybetaines with anions [2, 6–9] was the observation that spontaneous anionic dephosphorylations and decarboxylations are accelerated by betaine micelles which, in this respect, behave like cationic micelles in incorporating the substrate into a region which is less polar than water, thereby speeding up the reaction [5, 10–15].

Most bimolecular spontaneous hydrolyses are, as expected in terms of polarities and water content, inhibited by ionic micelles, but inhibition is greater for anionic than for cationic and sulfobetaine micelles [9, 11, 12, 16]. These effects are insensitive to addition of hydrophilic, non-nucleophilic anions, but added ClO_4^- readily enters the interfacial micellar region, rendering it anionic and strongly inhibiting hydrolysis [11]. Many bimolecular reactions of nucleophilic or basic anions with organic substrates are accelerated by cationic micelles that concentrate reactants into the interfacial region. Sulfobetaine micelles mildly inhibit reactions of hydrophilic nucleophiles, for example, OH^-, but assist reactions of less hydrophilic anions, for example, Br^-, and this acceleration is eliminated by an anion such as ClO_4^- which drives Br^-out of the interfacial region [5, 11].

These kinetic results show that betaine micelles can incorporate anions and the salt-order of binding, which follows the Hofmeister series, is that observed with cationic micelles and has the same origin. However, this evidence is qualitative and we need both physical evidence of anion affinities for sulfobetaine micelles and theoretical models of the binding.

Zwitterionic micelles are formally uncharged, but for sulfobetaine micelles with an extended trimethylene sulfonate head group it is obvious that for an approximately spherical micelle the positive charge density at the surface encompassing the quaternary ammonium group is slightly higher than that of the sulfonate anions [6, 9]. Thus, there should be a favorable electrostatic component for interaction with anions, but not cations [17, 18]. Indeed, interactions of ions with foams generated by amphiphilic ions follow the pattern observed with association colloids [2, 19]. Examination of distributions of ions between water and ionic or zwitterionic micelles is a powerful method for studying the sources of ionic interactions with interfacial regions. This interaction between very hydrophilic anions and zwitterionic micelles has been treated quantitatively by considering only the electrostatic term [2], but Iso and Okada [17] and Masudo and Okada [18] included ion-specific interactions with less hydrophilic anions in their treatment.

There is compelling evidence of specific interactions of anions with cationic micelles, depending on anionic polarizability and charge density, which also control interactions of anions with sulfobetaine micelles. Thus experimental methods of investigation should be similar for both, cationic and zwitterionic, micellar systems [1, 3, 17, 18].

The incorporation of anions into cationic, and sulfobetaine, micelles is examined on the assumption that water and micelles behave as distinct regions, i.e., as pseudophases [1, 3]. Therefore an ion-selective electrode (ISE) should sense ions in the bulk water rather than those bound to the micelle. These electrodes have been developed for many ions and this method is generally applicable. Electrolytic conductance is a general and conceptually simple method of quantifying relative numbers of free and micellar-bound ions. Another useful method is NMR spectroscopy [5], because the chemical shift of an ion should be different in water, typically H_2O–D_2O, and in the interfacial region. Some ions with quadrupole nuclei have very broad signals, which complicates estimating chemical shifts, but incorporation of such an ion into the interfacial region decreases its mobility and changes its local environment, and the consequent increase in line width provides evidence for the binding. At the same time interaction of an anion with quaternary ammonium ^{14}N affects its line width [5, 19]. These methods provide evidence for the partitioning of anions between water and sulfobetaine micelles.

There is evidence, some of it qualitative, for the binding of anions and dediazonization trapping is being used to estimate local anionic concentrations in the micellar pseudophase of sulfobetaines [20]. In this work we describe the binding of anions to sulfobetaine micelles, which results from contributions of nonspecific, electrostatic, and anion-specific interactions, and we show how these interactions can be treated quantitatively.

Experimental

Materials

We used the following sulfobetaines: *N*-decyl-, *N*-tetradecyl- and *N*-hexadecyl-*N*,*N*-dimethylammonio-1-propanesulfonate (SB3-10, SB3-14 and SB3-16), from Sigma, and *N*-tetradecyl-*N*,*N*-diethyl-, *N*-tetradecyl-*N*,*N*-dipropyl- and *N*-tetradecyl-*N*,*N*-dibutylammonio-1-propanesulfonate (SBE3-14, SBP3-14 and SBB3-14) and *N*-tetradecyl-*N*,*N*-dimethyl-, *N*-tetradecyl-*N*,*N*-diethyl- and *N*- tetradecyl-*N*,*N*-dipropylammonio-1-butanesulfonate (SB4-14, SBE4-14 and SBP4-14) were prepared as described in Ref. [5].

Methyl tricapryl ammonium chloride (Aliquat 336-Cl, Aldrich) was used as received. Sodium perchlorate (Riedel) and bromide (Vetec) were dried before use. All other reagents and solvents were of analytical grade and were used without further purification. Deionized water was used in the preparation of all solutions.

Methods

Potentiometric measurements

Salts of Aliquat 336-Cl were dissolved in chloroform and converted into the perchlorate form by repeated shaking with aqueous sodium perchlorate (about 1 M) [22]. The electrode membrane was prepared by dissolving polyvinyl chloride in tetrahydrofuran and this mixture was shaken gently to facilitate dissolution. Then, Aliquat–ClO_4 and dibutyl phthalate were added. The ISEs were prepared on graphite rod supports by immersing them five times in the solution just described, with 1 min between each immersion cycle. Then, the electrodes were air-dried for 24 h, with slow evaporation of the solvent. Finally, the electrodes were wrapped tightly with an insulating film, maintaining a constant contact area. All perchlorate electrodes were soaked for 24 h in 0.1 M $NaClO_4$ before use, stored in air between experiments and reconditioned immediately before use by soaking them for 15 min in the perchlorate solution just described.

The pH of the solutions was maintained with borate buffer (0.01 M). In general, solutions were prepared immediately before use, and all potentiometric and pH measurements were carried out with a Metrohm, model 713, pH meter or a Hanna, model 200, digital pH meter calibrated with standard buffers, pH 7.00 and 9.00 (Carlo Erba). In general, potentiometric measurements with ISEs were done by using a standard calomel electrode as a reference, and the response time of the electrodes was usually 1 min. Perchlorate ion standards were prepared in borate buffer (0.01 M), and the response varied linearly with the logarithm of anion activity for solutions in the range 10^{-1}–10^{-4} M, with a slope of 58 ± 3 mV. Borate ions do not perturb the response of the electrodes.

Solutions of the different sulfobetaines (generally 0.05 or 0.1 M) contained 0.01 M borate buffer, and perchlorate ion concentrations were monitored in the concentration range of the standard.

Viscosity measurements

The viscosities of sulfobetaine solutions in the presence of 0–1 M sodium perchlorate were determined using a Brookfield Reomether LV DV-III (spindle geometry Cp-42) at 25.0 ± 0.1 °C.

NMR spectroscopy

The NMR spectroscopic and kinetic measurements were previously described [5]. In general, ^{23}Na, ^{35}Cl, ^{14}N and ^{81}Br NMR spectra were recorded and a Bruker AC 300 P spectrometer operating at 79.391, 29.407, 21.688 and 81.053 MHz, respectively. Typical acquisition conditions were ^{23}Na, 90° pulses, 0.2–0.6 s acquisition, 2,000 Hz sweep width; ^{35}Cl, 90° pulses, 0.02–0.07 s acquisition, 1,000 Hz sweep width; ^{14}N, 90° pulses, 0.02–0.06 s acquisition, 80,000 Hz sweep width; T_1 measurements for ^{23}Na were made with the standard inversion recovery pulse sequence. All measurements were made at 25 °C in H_2O/D_2O in a ratio of 1:4.

Results and discussion

Quantitative aspects

The qualitative evidence cited earlier shows that incorporation of anions into sulfobetaine micelles is very sensitive to anionic charge density. Several methods have been used to estimate partitioning. Conductance is useful, provided that allowance can be made for the contribution due to developing negative charge on the micelles and that there is significant counterion incorporation. For the sulfobetaine systems, ISE measurements distinguish between ions in the aqueous and micellar pseudophases and are well adapted for the examination of anions such as ClO_4^- and Br^-, which transfer readily from water into the micelles (Figs. 1, 2). Observed differences between the experimental plots in the presence and absence of surfactant are larger for perchlorate than for bromide ion.

The distribution of $Br^-$and ClO_4^- between water and sulfobetaine micelles of SB3-10, SB3-12, SB-14 and

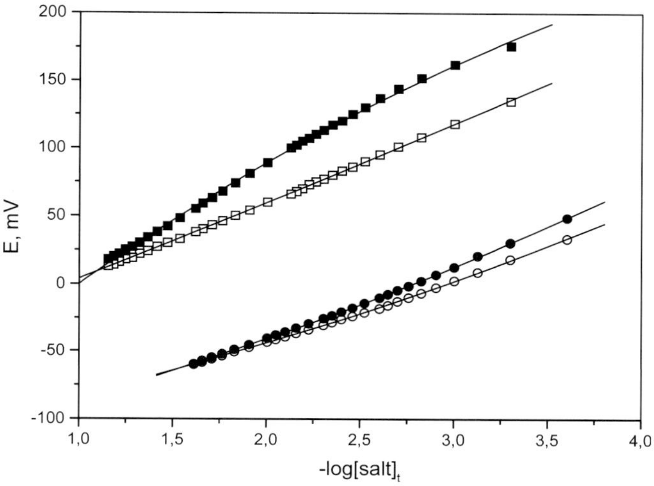

Fig. 1 Electrode potentials of perchlorate ion in the presence (*closed squares*) and absence (*open squares*) of 0.1 M SB3-16 and of bromide ion in the presence (*closed circles*) and absence (*open circles*) of 0.1 M SB3-16, pH 9.0, 25 °C

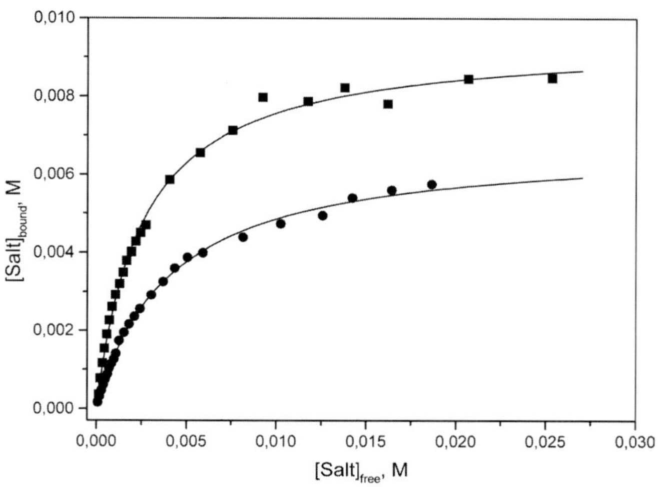

Fig. 2 Langmuir plot for free versus bound ClO_4^- (*squares*) and Br^- (*circles*) in 0.1 M SB3-16, pH 9.0 and 25 °C

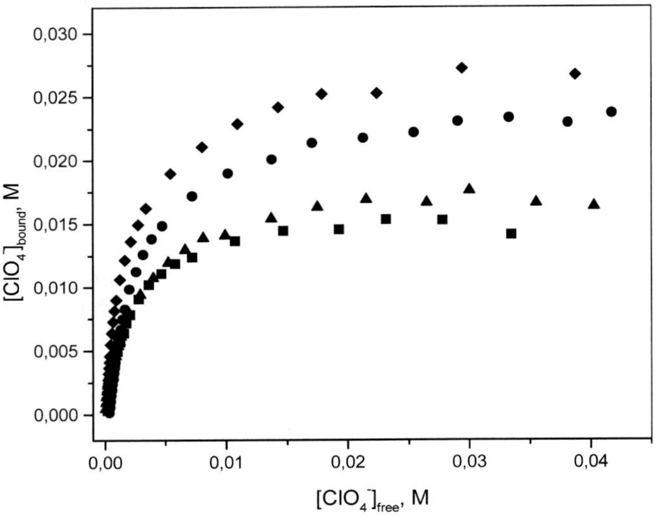

Fig. 3 Langmuir plot for free versus bound ClO_4^- in 0.1 M SB3-14 (*triangles*), SBE3-14 (*squares*), SBP3-14 (*circles*) and SBB3-14 (*diamonds*), pH 9.0 and 25 °C

SB-16 and related alkyl derivatives, estimated by use of an ISE from data similar to those in Fig. 2, are shown in Fig. 3 for some selected examples. In all cases the concentrations of the anions incorporated into the micellar surface follow, as a function of free ion concentration, the saturation behavior typical of a Langmuir isotherm, Eq. (1) [23]:

$$[ClO_4^-]_{bound} = \frac{\theta_{max} K_L [ClO_4^-]_{free}}{1 + K_L [ClO_4^-]_{free}},$$

where θ_{max} indicates the percentage of binding sites (defined as the ratio of bound counterions per head group of micellized surfactant) in the micelle which can take up the anion, K_L is the Langmuir association constant and concentrations are molarities in terms of the total solution volume. The constants for surfactants with different head groups and tether lengths are given in Table 1.

Table 1 Percentages of binding sites, θ_{max}, and Langmuir association constants, K_L, for perchlorate and bromide ions with sulfobetaine micelles

Surfactants	θ_{max} (%)		K_L (M^{-1})	
	ClO_4^-	Br^-	ClO_4^-	Br^-
SB3-10	20.5±4	5.8	184	60
SB3-14	19.2±2	4.6	388	171
SB3-16	20.5±5	6.6	548	256
SBE3-14	17	–	–	–
SBP3-14	25	–	272	–
SBB3-14	27	–	470	–
SB4-14	26	–	164	–
SBE4-14	24	–	252	–
SBP4-14	29	–	833	–

These results demonstrate that there is very strong anion specificity, as noted earlier, and that the capacities of these micellized surfactants for anions are very much lower than those of cationic micelles where moderately hydrophilic counterions neutralize 70–80% of the micellar charge. There is an evident balance between anionic uptake and development of anionic character in the micelles which suppresses further anion uptake. As a result, anions such as Br^-, and especially Cl^-, are weakly bound and for very hydrophilic anions saturation of the micelles is not reached at realistic concentrations.

Iso and Okada [17] and Masudo and Okada [18] used capillary electrophoresis to estimate the zeta potential of salt-modified sulfobetaine micelles as an indicator of anion uptake. They noted that the anion order for uptake is that for the free energy of anion transfer between water and dipolar aprotic solvents such as dimethylformamide and dimethyl sulfoxide, indicating a role for partial ionic desolvation.

This anion selectivity is evident in the results from the use of the ISE, which show that ClO_4^- is much more strongly bound than Br^-. The kinetic data are consistent with this anionic sensitivity, although they provide only qualitative evidence [5, 6, 10, 11, 21]. For reactions with a variety of substrates the overall rate constants for reactions in sulfobetaine solutions, relative to those in water, follow the sequence $Br^- > Cl^- > F^- > OH^-$ [1, 2, 24]. Some spontaneous, water-catalyzed reactions are significantly inhibited by anionic micelles of sodium dodecyl sulfate, while sulfobetaine micelles have only slight rate effects, but addition of $NaClO_4$ increases their anionic character and the rates then decrease sharply [11]. Consistently bimolecular nucleophilic reactions of Br^- with moderately hydrophobic substrates are accelerated by sulfobetaine micelles and these reactions are strongly inhibited by ClO_4^-, which expels Br^- from the micelles. The salt-order for competition between anions, which can be deduced from the kinetic effects, is that seen for competition in cationic micelles and, as is general for ion specificity, follows the Hofmeister series [1, 2, 5].

NMR spectroscopy

The NMR spectra of some ions are sensitive to their transfer from water into micelles, which is sensed by changes in chemical shift or signal line width, and interactions of anions with an ammonium ionic center change the symmetry about the ^{14}N nucleus and therefore its NMR line width. This general method has been used extensively to monitor interactions of $^{35}ClO_4$, ^{79}Br and ^{81}Br with sulfobetaine and cationic micelles [5, 19].

The line width of ^{81}Br increases approximately linearly to 5,170 Hz with up to 0.1 M SB3-14 [5]. The corresponding line width of ^{81}Br in cationic micelles [5, 19, 25, 26] is in the range of 15,100 Hz and, on the reasonable assumption that this value can be applied to SB3-14 micelles, we estimate an approximate value of 7%, relative to SB3-14, for binding of Br^-, which fits the value estimated by using Eq. (1) and the parameters in Table 1. There is only a modest increase in ^{23}Na T_1 values although the micelle is developing a negative charge [5], consistent with other evidence that hydrophilic cations do not bind to sulfobetaine micelles, even when they develop anionic character, probably because Na^+ and the SO_3^- residue are strongly hydrated. As expected from the measurements with the ISE, the line widths of ^{14}N in the presence of ClO_4^- should increase very sharply, reflecting the transfer of ClO_4^- from water into micelles of SB3-14, and then remain constant. However, the line width of the quaternary ^{14}N signal increases monotonically as ClO_4^- perturbs the environment and the symmetry around the ^{14}N (Fig. 4). These changes in the ^{14}N line width are fitted quantitatively by assuming that the initial increase is due to the partitioning of ClO_4^- into the zwitterionic micelle (first and second terms in Eq. 2) and that all changes in the line width of ^{14}N with ClO_4^- concentrations greater than 0.1 M (third term in Eq. 2) are due to the observed increase in viscosity (Fig. 4). Thus, the variation of the ^{14}N line widths is fitted by Eq. (2) and the parameters given in Table 1:

$$\mathrm{LW}^{14}\mathrm{N} = \frac{\mathrm{K_L LW}^{14}N_{SAT}\left[\mathrm{ClO_4^-}\right]_t}{1+\mathrm{K_L}\left[\mathrm{ClO_4^-}\right]_t} + \frac{\mathrm{LW}^{14}N_0}{1+\mathrm{K_L}\left[\mathrm{ClO_4^-}\right]_t} + 0.5[\mathrm{ClO_4^-}]_t$$

where ($LW^{14}N$) is the experimental ^{14}N line width, $[ClO_4^-]_t$ is the total concentration, ($LW^{14}N_{SAT}$) is the ^{14}N line width of the ClO_4^- saturated surfactant, ($LW^{14}N_0$) is the ^{14}N line width in the absence of ClO_4^-, and K_L was previously defined.

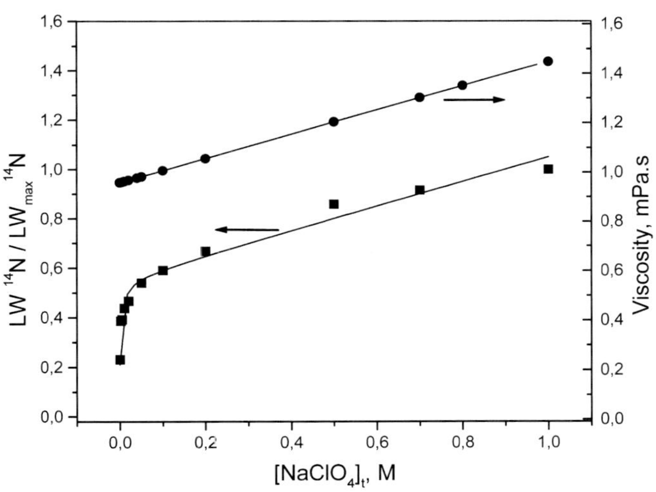

Fig. 4 NMR line widths of ^{14}N and of the viscosity of aqueous solutions of 0.055 M SB3-14, as a function of perchlorate ion concentration, LW_{max} $^{14}N = 57$ Hz. The *lines* were calculated by using Eq. (2)

Forces that control anion binding.

The uptake of anions into sulfobetaine micelles depends on nonspecific, electrostatic interactions, which can be treated classically [2, 17, 18], and also on anion-specific interactions. With dilute electrolytes these interactions are additive in terms of free energy, reinforcing ion binding, but as anions enter the interfacial region, rendering it anionoid, the anion-specific interaction becomes a very important factor against ion binding. This balance of forces means that the interfacial region becomes anion-saturated at modest occupancy, unlike the situation for ionic micelles where, depending on the charge density of the counterion, approximately 70% of the head-group charge is neutralized. However, the occupancy should increase modestly with high electrolyte concentration as entropy drives ions into the interfacial region, so that in the limit of very high electrolyte concentration there should be no ionic gradient between water and micelles. It is generally impossible to reach this situation experimentally. This generalization will be even more applicable to sulfobetaine micelles where the anion occupancy, θ_{max}, reaches limiting values of 20–30% with 0.10 M $NaClO_4$ and is not very sensitive to structures of the surfactant head groups. The situation is qualitatively similar to that for ionic micelles where, except with very high electrolyte concentrations, the values of the fractional micellar charge are nearly constant, typically 0.3.

Effect of sulfobetaine structure

The binding of both Br^- and ClO_4^- to sulfobetaine micelles is sensitive to the length of the long alkyl chain, the length of the tether between the ammonium and sulfonate centers and the size of the dialkyl groups on nitrogen. These structural changes have only modest effects upon values of θ_{max} (which increase with increasing hydrophobicity of the surfactant), but they affect those of the Langmuir constants, K_L, which are sensitive to head group structure in sulfobetaine micelles (Table 1). Indeed, the values of K_L depend on the size of the tether, the chain length of the sulfobetaine and the *R* group in the cationic N center.

An increase in the length of the alkyl group from $C_{10}H_{21}$ to $C_{14}H_{29}$ and $C_{16}H_{33}$ increases the Langmuir constants for both Br^- and ClO_4^- (Table 1). The constants for binding of ClO_4^- to the C_{14} sulfobetaines increase with increases in tether length from $(CH_2)_3$ to $(CH_2)_4$ and for the *N*-alkyl groups in the sequence Me < Et < *n*-Pr < *n*-Bu.

Partitioning of anions between micelles and water is measured rather than the local anion concentrations in the micellar interfacial region, although they can, in principle, be related to the partitioning parameters in terms of the molar volume of the interfacial region. This question has been discussed in detail elsewhere [1, 3, 27].

It appears that an increase in the volume of the interfacial region owing to changes in surfactant geometry increases the extent of partitioning from water into micelles. An increase in the size of the *N*-alkyl groups will displace water from this region and, following Iso and Okada [17] and Masudo and Okada [18], we assume that it will increase the transfer of polarizable anions, such as ClO_4^-, from water to the region. In considering these differences we note questions regarding conformations of the head groups. Weers et al. [28] have pointed out that the tether groups may not be extended, but they are eclipsed owing to intermolecular or intramolecular ion-pairing between the ammonium and sulfonate centers [28].

Conclusions

Several structural effects have to be considered. As noted earlier kinetic data provide qualitative evidence for incorporation of anions into sulfobetaine micelles. Several caveats in the interpretation of this evidence have to be considered: the rates of bimolecular reactions in micelles, or other association colloids, depend on local concentrations and on second-order rate constants which may differ in water and micelles. Kinetic data on spontaneous, unimolecular reactions indicate that interfacial regions of sulfobetaine micelles are similar to those of ionic micelles in being somewhat less polar and less aqueous than water. Structural effects of the micellized sulfobetaines upon the rate constants of bimolecular S_N2 reactions of Br^- at an alkyl group are similar qualitatively to those for binding of Br^- or ClO_4^- to sulfobetaine micelles (Table 1) [27].

The reported results are consistent with the observation that cationic micelles grow sharply on addition of anions such as Br^- or arene carboxylate and sulfonate ions, which interact strongly with the head groups. Solutions become viscous as growth induces a sphere-to-rod transition, whereas changes in viscosities are smaller for sulfobetaine micelles. Thus salt effects are, as expected, much less evident for sulfobetaine than for cationic micelles, as expected from the weaker anion–micelle interactions. However, addition of sodium perchlorate increases the viscosities of sulfobetaine solutions and modestly increases aggregation numbers. Interactions of ClO_4^- with cationic micelles cannot be examined because quaternary ammonium perchlorates are insoluble in water [5]. Clearly, there are limits to the incorporation of anions into sulfobetaine micelles and it seems that the extents of incorporation of perchlorate and bromide ions are in the range 20–30% and 5–7%, respectively. This result is important in evaluating kinetic effects of micellized sulfobetaines and the physical properties of these systems.

Acknowledgements We are grateful to PRONEX, CNPq (Conselho Nacional de Desenvolvimento Científico e Tecnológico, Brazil), MIUR, Rome, Italy and the US Army Office of Research for support of this work.

References

1. Bunton CA, Savelli G (1986) Adv Phys Org Chem 22:213
2. Baptista MS, Cuccovia I, Chaimovich H, Politi MJ (1992) J Phys Chem 96:6442
3. Bunton CA, Nome F, Quina FH, Romsted LS (1991) Acc Chem Res 24:357
4. Brochsztain S, Filho PB, Toscano VG, Chaimovich H (1990) J Phys Chem 94:6781
5. Profio PD, Germani R, Savelli G, Cerichelli G, Chiarini M, Mancini G, Bunton CA, Gillitt ND (1998) Langmuir 14:2662
6. Kamenka N, Chorro M, Chevalier Y, Levy H, Zana R (1995) Langmuir 11:4234
7. Chevalier Y, Perchec PL (1990) J Phys Chem 94:1774
8. Lee B, Nome F (2000) Langmuir 16:10131
9. Bunton CA, Mhala MM, Moffatt JR (1987) J Org Chem 52:3832
10. Profio PD, Brinchi L, Germani R, Sevelli G, Cerichelli G, Bunton CA (2000) J Chem Soc Perkin Trans 2: 2162
11. Brinchi L, Profio PD, Germani R, Sevelli G, Spreti N, Bunton CA (1998) J Chem Soc Perkin Trans 2: 361
12. Brinchi L, Profio PD, Germani R, Savelli G, Bunton C (1998) Colloids Surf A 132:303
13. Brinchi L, Dionigi C, Profio PD, Germani R, Sevelli G, Bunton CA (1999) J Colloid Interface Sci 211:179
14. Brinchi L, Profio PD, Germani R, Savelli G, Gillitt ND, Bunton CA (2001) J Colloid Interface Sci 236:85
15. Brinchi L, Profio PD, Germani R, Marte L, Sevelli G, Bunton CA (2001) J Colloid Interface Sci 243:469
16. Bunton CA, Mhala MM, Moffatt JR (1989) J Phys Chem 93:854
17. Iso K, Okada T (2000) Langmuir 16:9199
18. Masudo T, Okada T (1999) Phys Chem Chem Phys 1:3577
19. Troganis AN, Tsanaktsidis C, Gerothanassis IP (2003) J Magn Reson (in press)
20. Cuccovia IM, Romsted LS, Chaimovich H (1999) J Colloid Interface Sci 220:96
21. Bertoncini C, Nome F, Cerichelli G, Bunton C (1990) J Phys Chem 94:5875
22. Rohm T, Guilbault G (1974) Anal Chem 46:590
23. Altin O, Ozbelge H, Dogu T (1998) J Colloid Interface Sci 198:130
24. Baptista MS, Politi MJ (1991) J Phys Chem 95:5936
25. Fabre H, Kamenka N, Khan A, Lindbiom G, Lindman B, Tiddy GJT (1980) J Phys Chem 84:3428
26. Bacaloglu R, Bunton C, Cerichelli G, Ortega F (1989) J Phys Chem 93:1490
27. Sevelli S, Germani RLB (2001) In: Texter J (ed) Reactions and synthesis in surfactant systems. New York, chap 8
28. Weers J, Rathman J, Axe F, Crichlow C, Foland L, Scheuing D, Wiersema R, Zielske A (1991) Langmuir 7:854

Progr Colloid Polym Sci (2004) 128: 255–263
DOI 10.1007/b97099

Yasar Yilmaz
Önder Pekcan
Demet Kaya

Sol–gel transition in polyacrylamide gels by using a fluorescence technique

Y. Yilmaz · Ö. Pekcan · D. Kaya (✉)
Department of Physics,
Istanbul Technical University,
34469 Maslak, Istanbul, Turkey
e-mail: demet@itu.edu.tr
Tel.: +90-212-2853213
Fax: +90-212-2856386

Abstract Pyranine (8-hydroxypyrene-1, 3,6-trisulfonic acid, trisodium salt) was used as a floroprobe for monitoring the solution free-radical cross-linking polymerization of acrylamide (AAm) to test the universality class of the sol–gel transition. The spectra of the pyranines shift to shorter wavelengths during the polymerization of AAm. The intensity of the shifted spectra monitors the sol–gel transition, without disturbing the system mechanically. Observations around the critical point show that there are three regimes of AAm concentration in which the gel fraction exponent and the weight-average degree of polymerization exponent differ drastically. They agree best with the static percolation results for higher AAm concentration, but cross over from percolation to mean-field values when AAm concentration is lower than 2 M.

Keywords Fluorescence technique · Sol–gel transition · Critical exponents

Introduction

The predictions of the theories [1–4] about the critical exponents for the sol–gel transition are different from the point of the universality. Consider, for example, the exponents γ and β for the weight-average degree of polymerization, DP_w, and the gel fraction, G (average cluster size, S_{av}, and the strength of the infinite network, P_∞, in percolation language) near the gel point, which are defined as

$$DP_w \propto (p_c - p)^{-\gamma}, p \to p_c^-$$

and

$$G \propto (p - p_c)^{\beta}, p \to p_c^+,$$

where the Flory–Stockmayer theory (the so-called classical or mean-field theory) gives $\beta = \gamma = 1$, independent of the dimensionality d, while the percolation studies based on computer simulations give γ and β around 1.7 and 0.43 in three dimensions [3–8].

These two universality classes for the gelation problem are separated by a Ginzburg criterion [9] that depends upon the chain length N between the branch points as well as the concentration of the nonreacting solvent. The vulcanization of long linear polymer chains (large N) belongs to the mean-field class. Critical percolation (small N) describes the polymerization of small multifunctional monomers [3–6].

No real experiment measuring directly the critical exponents γ and β together with great sensitivity and accuracy has been performed so far, to our knowledge, at the sol–gel transition owing to experimental difficulties. Therefore, the results of the classical and percolation theories could not be tested adequately with real experiments.

The experiments directly measuring some physical properties in the course of the polymerization reaction for monitoring the sol–gel transition must be very sensitive to the structural changes, and should not disturb the system mechanically. The fluorescence technique is particularly useful for elucidation of detailed structural aspects of the gels. This technique is based on the interpretation of the change in anisotropy, emission and/or excitation spectra, emission intensity, and viewing the lifetimes of injected aromatic molecules to monitor the change in their microenvironment [10–13].

Fluorescence probes can be used in two ways for the studies on polymerization and gelation. First, one can add a luminescent dye as a probe to the system (extrinsic fluoroprobe). By using this fluoroprobe it is possible to measure some physical parameters of the polymerizing system, like polarity [14, 15], viscosity [16–19] and hydrophobicity [20]. In the second approach, the florophore is covalently attached to the polymer, and serves as a polymer-bond label (intrinsic fluoroprobe) [21], where the polymer floroprobe association depends on some factors including Coulombic interactions and the hydrophobicity of the polymer–florophore pair.

These techniques have been successfully used to perform experiments on polymerization [22, 23], chemical gel formation [24–26], swelling of the gels [27], slow release of the probe molecules from a gel [28, 29], metal ion detection via a metal-ion templated polymeric gel [30], the affinity of the gels to the target molecules [31], and examination of the collapsed-state phases and the volume phase transition of the polymeric gels [32]. These are all well-established methods.

Recent studies, using pyrene as an extrinsic fluoroprobe, showed that the glass transition both for the linear bulk polymer [33] and for gels [34] could be described by percolation exponents. In these studies, the fluoroprobe monitors the change occurring in the rigidity of the medium near the glass transition. Therefore, this method seems particularly suitable for studying the abrupt changes occurring during the glass transition of poly (methyl methacrylate) polymerization and/or gelation. As for the sol–gel transition, for either the solution polymerization or the bulk polymerization, there are some difficulties to be overcome. For example, the probe can move randomly in the gel and be quenched by the sol molecules if the polymeric system is not in the glass state; therefore, the intensity of the floroprobe will not be a direct function of the monomer conversion.

We surmounted this difficulty and studied the free-radical cross-linking polymerization of acrylamide (AAm) using pyranine (POH, a derivative of the pyrene molecule, which has three functional groups on it [32]). The excited POH, (POH)* may decay to its ground state by fluorescence at 445 nm, or it may dissociate to $(PO^-)^*$ upon deprotonation, depending on the pH of the solution. The excited-state anion $(PO^-)^*$ may dissipate its energy by fluorescence emission at 510 nm. At pH above 5 the hydrogen ion concentration is insufficient to reprotonate $(PO^-)^*$ within its lifetime. The observable net result of POH excitation by a fast laser pulse is, therefore, the formation of PO^- and H^+, when the pH is between the pK_a values of the ground and excited state of POH [35].

The POH added to the prepolymerization solution in a very small amount shows a spectral shift to shorter wavelengths upon initiation of polymerization; but the pH of the samples does not change considerably during the polymerization process: it remains about 7. When the gels were washed many times with pure water, the POH molecules still existed in the gel and had short-wavelength spectra afterwards. It is concluded that this spectral shift is due to the binding of POH to the polymer chains during the AAm polymerization. Thus, the total fluorescence intensity of the POH bonded to the strands of the polymers allowed us to measure directly the average degree of polymerization and the gel fraction near the sol–gel transition, and thus the corresponding critical exponents γ and β.

Experimental

Materials

Gels were prepared by using various amounts of AAm (Merck) and various amounts of cross-linker [N′-methylenebis(acrylamide), BIS] (Merck) by dissolving them in 25 cm^3 water in which 10 μl tetramethylethylenediamine (TEMED) was added as an accelerator. The initiator, ammonium persulfate (APS, Merck), was recrystallized twice from methanol. The initiator and POH concentrations were kept constant at 7×10^{-3} and 4×10^{-4} M, respectively, for all experiments. All samples were deoxygenated by bubbling nitrogen through them for 10 min, just before the polymerization process.

Gelation processes

The experiments with varying amounts of AAm, where the number of cross-linker BIS molecules per AAm monomer was fixed (Table 1), were performed at room temperature.

The fluorescence intensity measurements were carried out using a PerkinElmer model LS-50 spectrometer, equipped with a temperature controller. All measurements were made at the 90° position and the slit widths were kept at 5 mm. POH was excited at 340 nm during in situ experiments and variation in the fluorescence spectra and emission intensity of the POH were monitored as a function of polymerization time.

Results and discussion

The emission spectra of POH dissolved in pure water as a function of pH are shown in Fig. 1. At considerably high pH, in addition to the main peak at 508 nm, a small peak (like a shoulder) appears at about 445 nm. After pH 6.5, this shoulder disappears and the 508-mm peak shifts to 512 nm. Below pH 1, the 508-nm peak completely disappears and only the 445-nm peak remains in the emission spectra of POH. This figure indicates that the 445-nm peak corresponds to the POHs neutralized by H^+ ions

Table 1 Experimentally measured parameters for various acrylamide (*AAm*) contents

AAm (M)	N′-Methylenebis (acrylamide) (M)	t_{gel} (s)	c^+/c^-	β	γ
5	0.032	179 ± 5	2.7	0.47 ± 0.02	1.80 ± 0.05
			3.5	0.50 ± 0.02	
			4.3	0.50 ± 0.02	
			10	0.47 ± 0.02	
			1.0	1.00 ± 0.02	
3	0.019	305 ± 5	2.7	0.50 ± 0.02	1.80 ± 0.05
			3.5	0.50 ± 0.02	
			4.3	0.49 ± 0.02	
			10	0.46 ± 0.02	
			1.0	0.70 ± 0.02	
2	0.013	310 ± 5	2.7	0.52 ± 0.02	1.70 ± 0.05
			3.5	0.52 ± 0.02	
			4.3	0.55 ± 0.02	
			10	0.50 ± 0.02	
			1.0	0.92 ± 0.02	
1	0.0065	322 ± 5	2.7	0.94 ± 0.02	1.00 ± 0.05
			3.5	0.94 ± 0.02	
			4.3	0.95 ± 0.02	
			10	1.00 ± 0.02	
			1.0	0.94 ± 0.02	
0.75	0.0048	330 ± 5	2.7	1.00 ± 0.02	1.00 ± 0.05
			3.5	0.92 ± 0.02	
			4.3	0.92 ± 0.02	
			10	0.92 ± 0.02	
			1.0	1.10 ± 0.02	
0.65	0.0042	375 ± 5	2.7	0.92 ± 0.02	0.92 ± 0.05
			3.5	0.92 ± 0.02	
			4.3	0.85 ± 0.02	
			10	0.94 ± 0.02	
			1.0	1.03 ± 0.02	
0.55	0.0035	872 ± 5	2.7	0.30 ± 0.02	–
			3.5	0.29 ± 0.02	–
			4.3	0.29 ± 0.02	–
			10	0.25 ± 0.02	
			1.0	1.00 ± 0.02	–
0.50	0.0032	1,650 ± 5	2.7	0.33 ± 0.02	–
			3.5	0.32 ± 0.02	–
			4.3	0.25 ± 0.02	–
			10	0.08 ± 0.02	
			1.0	1.20 ± 0.02	–
0.30	0.00190	No gel	–	–	–
0.10	0.00065	No gel	–	–	–
0.01	0.0	No gel	–	–	–

We conducted some experiments for measuring the pH of the prepolymerization solution. The pH of the prepolymerization solution, water–AAm–BIS–POH, was 6; therefore, a weak 445-nm peak (the shoulder) appeared. Addition of TEMED (accelerator) to this led to an increase in the pH of the solution, from 6 to 8, and the shoulder disappeared completely, and only the 512-nm peak remained. Addition of the initiator APS to the solution including TEMED did not lead to an extra change in the pH of the solution: the pH remained at 8.

No shift or change in the emission spectra of POH had been observed before the polymerization was initiated with APS. Upon the initiation of the polymerization the intensity of the 512-nm peak started to decrease and a new peak appeared around 410 nm. It should be noted that this new peak is different from the peak at 445 nm, which is a direct indication of the acidity of the solution (Fig. 1).

Typical fluorescence spectra for POH at different stages of the AAm copolymerization are shown in Fig. 2. At the beginning of the reaction, only the 512-nm peak existed, then the intensity of the new (short-wavelength) peak started to increase as the intensity of the 512-nm peak (long-wavelength peak) decreased during the course of AAm polymerization.

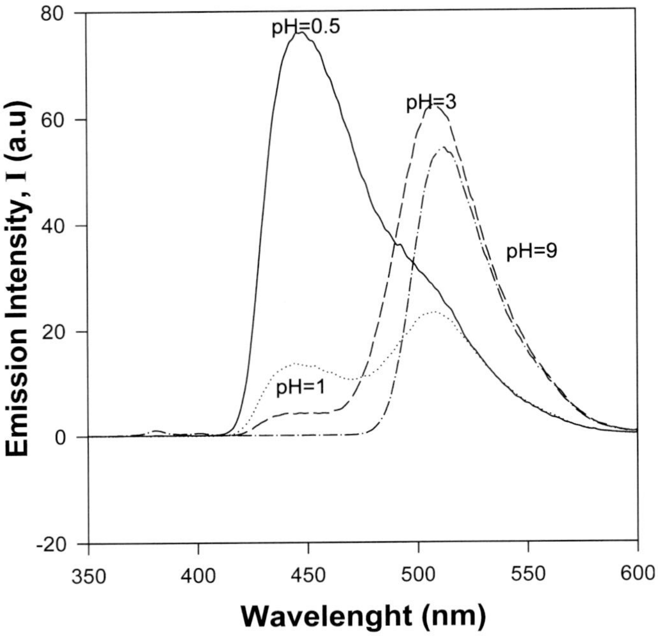

Fig. 1 Emission spectra of pyranine dissolved in pure water as a function of pH

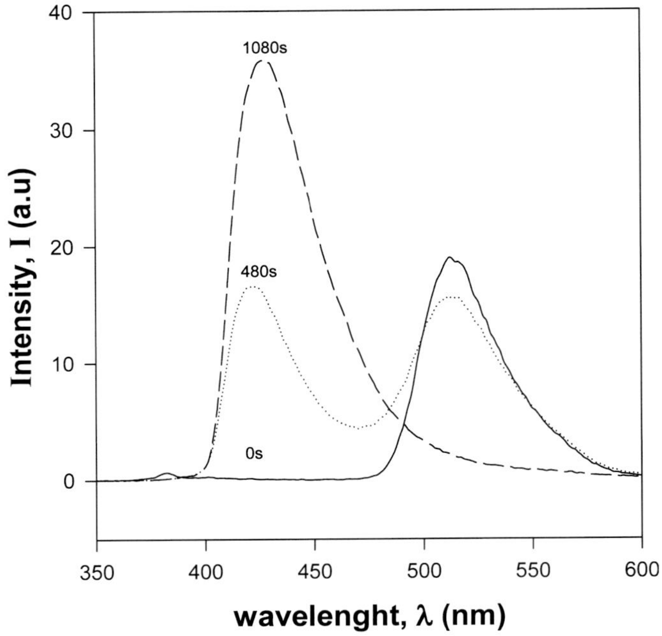

Fig. 2 Typical fluorescence spectra of pyranine at different stages during the acrylamide (*AAm*)–*N*′-methylenebis(acrylamide) polymerization. The *numbers* on the spectra show corresponding reaction times

A shift to short wavelengths and an increase in the intensity of the short-wavelength peak during the polymerization should be due to the neutralization of SO_3^- groups of some of the POHs in the reaction mixture. The disappearance of the 512-nm peak at the end of the polymerization reaction, as shown in Fig. 2, can only be attributed to the binding of the POHs to the polyacrylamide (PAAm) gel as previously observed by Yilmaz [32] for methacrylic acid-*co*-dimethyl(acrylamide) gels, where when POH molecules were bonded to the gel via methacryl aminopropyltrimethylammonium chloride molecules, they emit light at around 410 nm. The observation that the pH of the reaction did not change considerably during the polymerization, it remained around 7, indicates that the shift in the POH spectra cannot be due to the pH effect.

To show that POH is really bonded to the gel, we conducted some swelling experiments. The final gels were brought to the collapsed state and then put into pure water to swell. This process was repeated with fresh water a few times over a few weeks, and fluorescence measurements were performed, from time to time, on the gels during the swelling process. We observed only the short-wavelength peak from the washed gels, as shown in Fig. 3. This observation shows clearly that POH is bonded to the gel during the polymerization process, since POHs still existed in the gel and had short-wavelength spectra after the washing process had been completed.

A recent study showed that the charged and neutralized POH has two distinct spectra in PAAm gels or linear polymers; the charged one corresponds to the 512-nm peak at pH 8 (and 508 nm below pH 6.5) and the neutralized one corresponds to the peak around 410 nm [36]. The neutralization of POH is due to the binding to the PAAm gel during polymerization. POH was also used previously in protein binding of a fluorescence dye [37].

We conducted experiments for different AAm concentrations. For determining the gel points, each experiment was repeated at the same experimental conditions, and the gel points were determined by a dilatometric technique [38]. A steel sphere of 4.8-mm diameter was moved in the sample up and down slowly by means of a magnet applied to the outer face of the sample cell. The time at which the motion of the sphere was stopped was evaluated from the onset of the gel point, t_c. The t_c values are summarized in Table 1, together with the other parameters.

The steel sphere in the samples of higher polymer concentration cannot be moved after the polymerization is complete, but it can partly be vibrated around its equilibrium position for the loosely formed gels, by moving the magnet up and down on the surface of the sample cell. Another observation was that below 0.5 M polymer concentration, the polymeric system did not turn into a gel; it remained as a viscous liquid.

The same behaviour for the POH spectra was observed even for extremely low (below 0.2 M) AAm concentrations, i.e., the 512-nm peak shifted to about 410 nm during the polymerization although the pH of the polymer

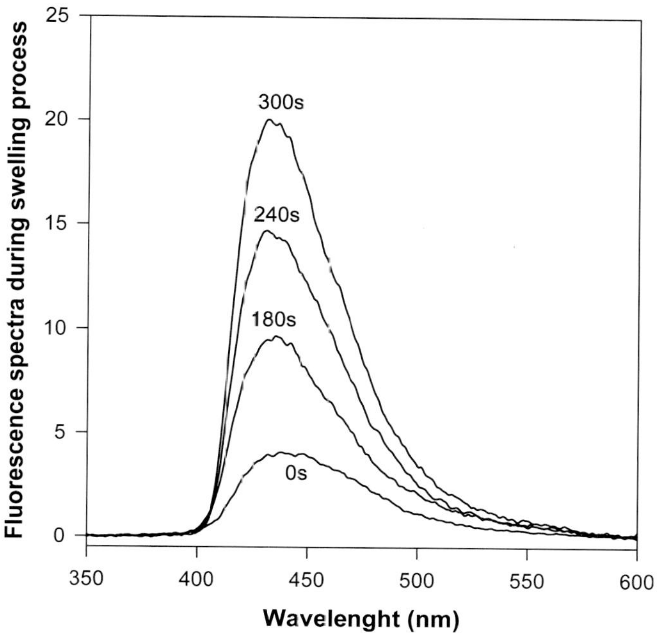

Fig. 3 Change in the fluorescence spectra of pyranine during the swelling of the AAm gels. The *numbers* on the curves represent corresponding reaction times

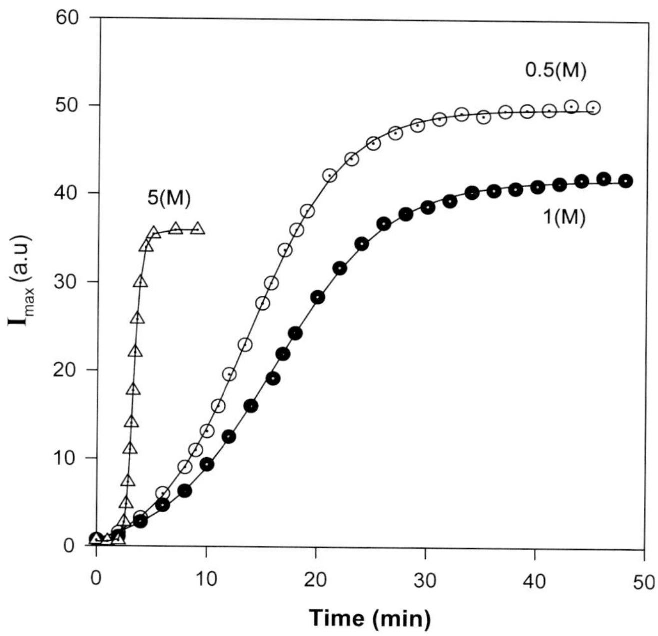

Fig. 4 Typical fluorescence intensity of pyranine bonded to the gel for different AAm concentrations, versus reaction time. The *numbers* on the curves indicate the corresponding AAm concentrations. The *symbols* and *dashed lines* represent the maximum intensity of the spectra and the best-fit lines to the experimental data, respectively

solutions did not change considerably from 7 during the reaction time. This point clearly shows that in our case a shift in the spectra from excited-state dissociation (about 510 nm) to the acid species (about 445 nm) is not due to water-restricted media as previously observed for reversed-micelle solubilized water pools in isooctane [35]. The probe molecule fluoresces at about 410 nm only when it is incorporated into the strand of the PAAm gels or AAm linear polymer chains.

One other observation is that the maxima of the short-wavelength peak (corresponding to the bonded POHs, in Fig. 2) shifted to relatively higher wavelengths, from 400 to 427 nm, during the polymerization process. The time elapsed for the shift from 400 to 427 nm depends on the polymer concentration. For example, it takes 240 s for 5 M, 720 s for 2 M, 1,200 s for 1 M, and 1,620 s for 0.1 M (at which gelation does not occur, a viscous solution) polymer concentrations.

The shift in wavelength from 427 to 400 nm (in contrast to the transition from 400 to 427 nm observed during polymerization) was observed during the swelling of the gels as can be seen from Fig. 3. These observations also shows that the degree of the restriction of the water cannot determine alone the fluorescence emission wavelength, from 445 nm or from 512 nm. The pH is an important parameter rather than the degree of restriction of water in our system.

POH has three functional groups by which it can be bonded to the polymeric network, branched or linear polymers. The probability that POH is bonded to the system over more than one functional group may increase with increasing polymer concentration, and also with the reaction time. As the polymerization progresses POH can have a chance to bind the polymeric system over two or three functional groups. The shift in the short-wavelength peak spectra from 400 to 427 nm may be due to the increasing number of neutralized SO_3^- groups of POH during the polymerization. The shift from 427 nm to shorter wavelengths during the swelling may be interpreted by the decreasing number of neutralized SO_3^- groups; some of the SO_3^- bonds may be broken upon swelling.

The fluorescence intensities from the bonded POH are shown in Fig. 4 against the reaction time for some different polymer concentrations. Since the maxima of the spectra (corresponding to bonded POH) shift from 400 to 427 nm as the polymerization progresses, one does not have a chance to monitor the intensity in the time-drive mode of the spectrometer (ten possible data in 1 s). Therefore, we monitored the fluorescence spectra over relatively large periods of time and plotted the intensity corresponding to the maxima of the spectra as a function of time, in Fig. 4. No considerable deviation is observed when the areas under the spectra are calculated instead of the maximum intensities. Therefore, we had a chance to get relatively more data in a certain time just by monitoring a small part (including the maxima) of the spectra.

We then used these data to evaluate the critical-point behaviour of the sol–gel transition. One can also use the continuous curves calculated by a best-fitting procedure to evaluate the critical-point behaviour of the sol–gel transition.

Here we would like to argue that the total fluorescence intensity from the bonded POHs monitors the weight-average degree of polymerization and the growing gel fraction below and above the gel point, respectively. This proportionality can easily be proven by using a Stauffer-type argument [4] as follows, under the assumption that the monomers occupy the sites of an imaginary periodic lattice.

The probability that a site belongs to a cluster of size s is given by $n_s s$, where n_s is the number of s-clusters (number of clusters including s sites) per lattice site. The probability that an arbitrary site belongs to any cluster is p; this is simply the probability that an arbitrary site is occupied. Thus, the probability w_s that the cluster to which an arbitrarily occupied site belongs contains exactly s sites is

$$w_s = \frac{n_s s}{\sum_s n_s s}$$

and thus the average cluster size S is calculated by the following relation [3–6]:

$$S = \sum_s w_s s = \frac{\sum n_s s^2}{\sum n_s s},$$

The definition of the average cluster size is the same for all dimensions, although n_s cannot be calculated exactly in higher dimensions [4]. This definition also holds for the bond percolation.

Now, let N_p be the number of POH molecules and N_m the number of other molecules (AAm, BIS, water, and APS) in the lattice; thus, the total lattice site N will be equal to $N_p + N_m$. The probability, P_p, that an arbitrary site is occupied by a POH molecule is simply N_p/N. The probability, P_y, that an arbitrary site both is a POH and belongs to the s-cluster can be calculated as a product of w_s and P_p:

$$P_y = P_p w_s = \frac{P_p n_s s}{\sum n_s s}.$$

Thus, $P_y s$ will be the total number of POH molecules in each cluster including s sites. The total fluorescence intensity, I, which is proportional to the total number of POHs trapped in the finite clusters, can be calculated as a summation over all s-clusters:

$$I \sim \sum_s P_y s = \sum_s \frac{P_p n_s s}{\sum n_s s} s = \frac{\sum P_p n_s s^2}{\sum n_s s},$$

where P_p can be taken out of the summation since the concentration of the POH is kept fixed for our work; thus,

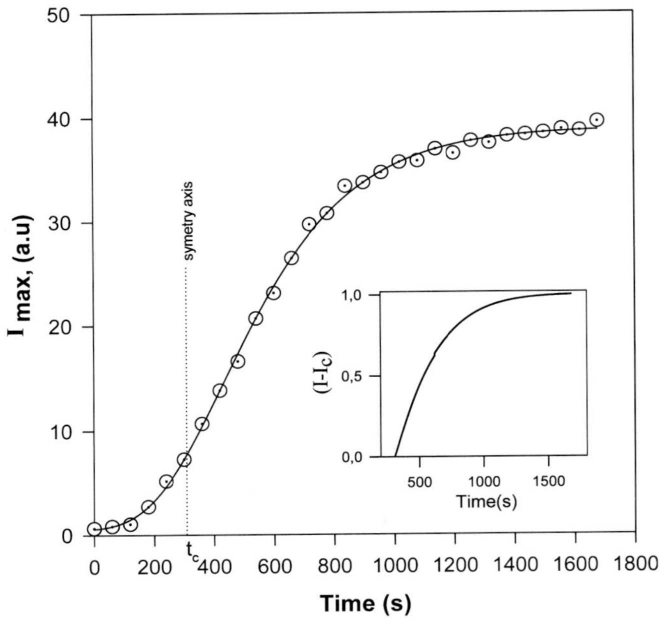

Fig. 5 A typical intensity–time curve for polymerization of AAm. The curve depicted by *dots* represents the mirror symmetry I_{ms} of the intensity according to the axis perpendicular to time axis at $t = t_c$. The intensity from the clusters above the gel point is calculated as $I_{ct} = \frac{c^-}{c^+} I_{ms}$. Thus, I–I_{ct} monitors the growing gel fraction for $t > t_c$. The *inset* represents the gel fraction for $\frac{c^-}{c^+} = 1$, as an example. The intensity from the bottom of the symmetry axis monitors the average cluster size for $t < t_c$

$$I \sim P_p \frac{\sum n_s s^2}{\sum n_s s} = P_p S.$$

The last expression shows that the total normalized fluorescence intensity, I, is proportional to the average cluster size, S. Note that the proportionality factor, P_p, is simply the fraction of POH molecules in the sample cell. The intensity will be linearly proportional to the average cluster size, provided that the POH concentration is not so high to quench the fluorescence intensity by an excitation-transfer effect.

Gelation theory often makes the assumption that the conversion factor, p, alone determines the behaviour of the gelation process, though p may depend on temperature, the concentration of monomers, and time [3, 4]. If the temperature and concentration are kept fixed, then p will be directly proportional to the reaction time, t. This proportionality is not linear over the whole range of reaction time, but it can be assumed that in the critical region, i.e., around the critical point, p–p_c is linearly proportional to t–t_c [33, 34, 39].

Therefore, below the gel point, i.e., for $t < t_c$, the fluorescence intensity, I, measures the weight-average degree of polymerization (or average cluster size). Above t_c, if the intensity I_{ct} from finite clusters distributed through the infinite network is subtracted from the total intensity, then the corrected intensity I–I_{ct} measures solely the gel fraction G, the fraction of monomers that belong to the macroscopic network. In summary, we have the following relations:

$$I \propto DP_w = c^+(t_c - t)^{-\gamma}, \quad t \to t_c^-,$$

$$I_{ct} \propto DP_w = c^-(t_c - t)^{-\gamma'}, \quad t \to t_c^+,$$

$$I - I_{ct} \propto G = B(t - t_c)^{\beta}, \quad t \to t_c^+,$$

where c^+, c^- and B are the critical amplitudes.

It is well known that the average cluster size of the finite clusters (distributed through the infinite network) above the gel point decreases with the same, but negative, slope of the increasing cluster size before the gel point [3, 6]. This means that the exponents γ and γ', defined for the cluster sizes below and above the gel point, have the same values. But, the critical amplitudes for the average cluster size defined below (c^- in Eq. 8a) and above (c^- in Eq. 8b) the gel point are different, and there is a universal value for the ratio c^+/c^-. This ratio is different for mean-field versus percolation as discussed by Aharony [7] and Stauffer et al. [3]. The estimated values for c^+/c^- is given in Ref. [7].

To determine the intensity I_{ct} in Eqs. (8b) and (9), we first choose the parts of the intensity–time curves up to the gel points, then the mirror symmetry I_{ms} of this parts according to the axis perpendicular to time axis at the gel point were multiplied by the ratio c^-/c^+, so that $I_{ct} = \frac{c^-}{c^+} I_{ms}$. Thus, I_{ct} measures solely the intensity from the cluster above the gel point, and I–I_{ct} measures the intensity from the gel fraction. This process is clarified explicitly in Fig. 5.

Using Eqs. (8) and (9) and the values for t_c summarized in Table 1 we calculated γ and β exponents as a function of AAm concentration, BIS concentration, and temperature. Figures 6 and 7 represent the log–log plots of some typical intensity–time data above and below the gel point, where the slopes of the straight lines, close to the gel points, give β and γ exponents, respectively.

The concentration and c^-/c^+ dependence of the exponents are summarized in Table 1, where three distinct concentration regimes appear. In the lowest concentration regime, the exponents β and γ obey neither

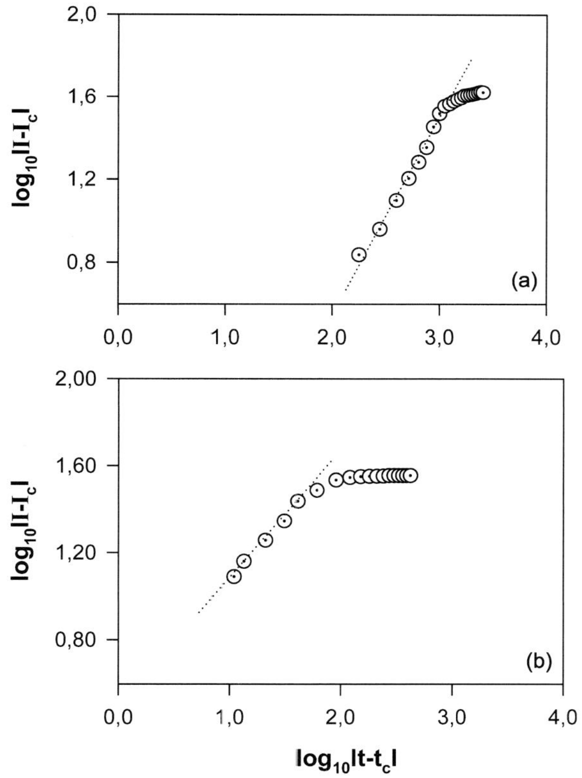

Fig. 6a, b Double logarithmic plot of the intensity I–I_{ct} versus time above t_c for 1 and 5 M AAm concentrations. The β exponents were determined from the slope of the straight lines near the gel points. As seen, the scaling region comes closer to the critical point as the AAm concentration is increased

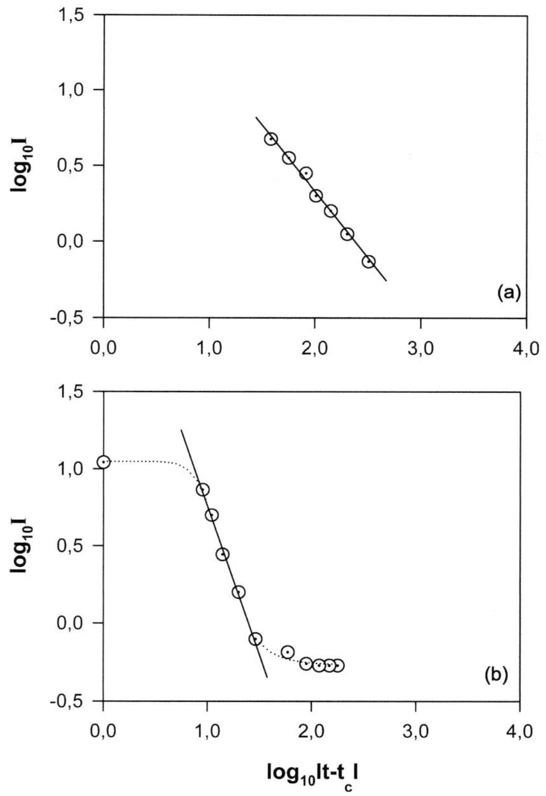

Fig. 7a, b Double logarithmic plot of the intensity versus time for $t < t_c$. The γ exponents were determined from the slope of the straight lines near the gel points. As seen, the scaling region comes closer to the critical point as the AAm concentration is increased, as in the case for β

the percolation nor the classical theory. In the intermediate regime, they have values close to unity, and agree with the classical theory independent of c^{-}/c^{+}. Finally, for the highest concentration regime they obey the percolation results in the experimental range of errors when c^{-}/c^{+} has the percolation values. The observation of three regimes of behaviour was also previously observed by Rubinstein and Colby [40] for modulus G_0 of gels and maximum swelling Q, as a function of the proximity to the gel point ε.

Conclusion

The application of fluorogenic probe molecules for in situ monitoring of free-radical cross-linking polymerization was illustrated by measuring the polymerization of AAm. The probe molecules fluoresce at different wavelengths when they are incorporated into growing polymer chains. The fluorescence intensity from the bonded probes is therefore a measure of the degree of polymerization and gel fraction below and above the gel point. In this study, we reported on a novel method to examine the percolation picture of gelation in an unequivocal way. We are, thus, able to measure directly polymerization kinetics without disturbing the system mechanically, and to test the universality of the sol–gel transition as a function of parameters like polymer concentration, cross-linker concentration and temperature.

The gel fraction exponent β and the average cluster size exponent γ agree best with the Flory–Stockmayer theory in the concentration range 0.65–1 M independent of the ratio c^{-}/c^{+}, and agree with percolation theory for higher AAm concentrations above 1 M when the ratio c^{-}/c^{+} takes the values given in the percolation theory.

It seems that this crossover from mean-field to percolation theory is related to the fact that we are able only to approach the critical point to $\left|\frac{p-p_c}{p_c}\right| \leq 0.1$ for concentrations below 2 M AAm, as can be calculated easily from Figs. 6 and 7. Above 1 M AAm, it seems we are able to approach the true critical region where $\left|\frac{p-p_c}{p_c}\right| \leq 0.01$.

An experimental system with intermediate N will exhibit a crossover between mean-field behaviour far from the gel point and critical behaviour, close to the percolation threshold, in a manner similar to other continuous phase transitions [9]. For large N the critical region in which the exponents are calculated is quite small. Thus, the entire range of experimentally accessible $|p-p_c|$ will be in the mean-field class and modelled by the Flory–Stockmayer theory [1, 2].

It seems that 0.65 and 2 M concentrations are crossover concentrations for γ and β exponents; 0.65 M from nonuniversal to mean-field, and 2 M from mean-field to percolation values for the AAm system. The Ginzburg criterion [9] is true for 2 M AAm. The crossover chain lengths in our systems for 0.65 and 2 M AAm are about 1,300 and 435. The nonuniversal behaviour for extremely low concentrations may be due to the fact that we are extremely far away from the critical region for these experimental conditions.

Acknowledgement Y.Y. was supported by the Turkish Academy of Sciences (TÜBA) in the framework of the Young Scientist Award Program (EA-TÜBA-GEBIP/2001-1-1).

References

1. (a) Flory PJ (1941) J Am Chem Soc 63:3083; (b) Flory PJ (1941) J Am Chem Soc 63:3091; (c) Flory PJ (1941) J Am Chem Soc 63:3096
2. (a) Stockmayer W (1943) J Chem Phys 11:45; (b) Stockmayer W (1944) 12:125
3. Stauffer D, Coniglio A, Adam M (1982) Adv Polym Sci 44:193
4. Stauffer D, Aharony A (1994) Introduction to percolation theory, 2nd edn. Taylor and Francis, London
5. Sahimi M (1994) Application of percolation theory. Taylor and Francis, London
6. de Gennes PG (1988) Scaling concepts in polymer physics. Cornell University Press, Ithaca, p 54
7. Aharony A (1980) Phys Rev B 22:400
8. Colby RG, Rubinstein M (1993) Phys Rev E 48:3712
9. Lusignan CP, Mourey TH, Wilson JC, Colby RH (1999) Phys Rev E 60:5657
10. Barrow GM (1962) Introduction to molecular spectroscopy. McGraw-Hill, New York
11. Birks JB (1965) Photophysics of aromatic molecules. Wiley-Interscience, London
12. Herculus DM (1965) Fluorescence and phosphorescence analysis. Wiley-Interscience, New York
13. Galanin MD (1995) Luminescence of molecules and crystals. Cambridge International Science, Cambridge
14. Jager WF, Volkers AA, Neckers DC (1995) Macromolecules 28:8153
15. Schaeken TC, Warman JM (1995) J Phys Chem 99:6145
16. Royal JS, Torkelson JM (1993) Macromolecules 26:5331
17. Miller KE, Krueger RH, Torkelson JM (1995) J Polym Sci B 33:2343
18. Vatanparast R, Li S, Hakala K, Lemmetyinen H (2000) Macromolecules 33:438
19. Warman JM, Abellon RD, Verhey HJ, Verhoeven JW, Hofstraat JW (1997) J Phys Chem B 101:4913
20. Erçelen S, Klymchenko AS, Demchenko AP (2002) Anal Chim Acta 464:273
21. Jager WF, Sarker AM, Neckers DC (1999) Macromolecules 32:8791
22. Pekcan Ö, Yilmaz Y, Okay O (1997) Polymer 38:1693
23. Yilmaz Y, Yagci Y, Pekcan Ö (2001) J Macromol Sci Pure Appl Chem A 38:741
24. Pekcan Ö, Yilmaz Y, Okay O (1994) Chem Phys Lett 229:537

25. Serrano B, Levenfeld B, Bravo J, Baselga J (1996) Polym Eng Sci 36:175
26. Okay O, Yilmaz Y, Kaya D, Keskinel M, Pekcan Ö (1999) Polym Bull 43:425
27. Pekcan Ö, Yilmaz Y (1997) J Appl Polym Sci 63:1777
28. Pekcan Ö, Yilmaz Y, Ugur S (1997) Polym Int 44:474
29. Yilmaz Y, Pekcan Ö (1998) Polymer 39:5351
30. Guney O, Yilmaz Y, Pekcan Ö (2002) Sens Actuators B 85:86
31. Oya T, Enoki T, Grosberg AY, Masamune S, Sakiyama T, Takeoka Y, Tanaka K, Wang G, Yilmaz Y, Feld MS, Dasari R, Tanaka T (1999) Science 286:1543
32. Yilmaz Y (2002) Phy Rev E 66:5801
33. Yilmaz Y, Erzan A, Pekcan Ö (2002) Eur Phys J E 9:135
34. Yilmaz Y, Erzan A, Pekcan Ö (1998) Phys Rev E 58:6
35. Politi M.J, Brandt O, Fendler J.H (1985) J Phys Chem 89:2345
36. .Kaya D, Pekcan Ö, Yilmaz Y (2004) Phys Rev E 69:16117
37. Pagnoni A, Kligman AM, Stoudemayer T (1998) J Cosmet Sci 49:33
38. Okay O, Kaya D, Pekcan Ö (1999) Polymer 40:6179
39. Tuzel E, Özmetin M, Yilmaz Y, Pekcan Ö (2000) Eur Polym J 36:727
40. Rubinstein M, Colby RH (1994) Macromolecules 27:3184

Progr Colloid Polym Sci (2004) 128 : 264–269
DOI 10.1007/b97104

Sandro Rogério de Sousa
Beatriz V. Kilikian
Miguel Jafelicci Jr.
C. Laluce

Flotation of biocolloids suspended in liquid phases of different compositions and pH values

S.R. de Sousa · B.V. Kilikian
Programa de Pós-Graduação Interunidades em Biotecnologia USP/IPT/I, Butantan, São Paulo, SP, Brazil

M. Jafelicci Jr. · C. Laluce (✉)
Instituto de Química, Universidade Estadual Paulista, UNESP, Araraquara, SP, Brazil
e-mail: claluce@iq.unesp.br
Tel.: +55-16-2016673
Fax: 55-16-2227932

Abstract The flotation capacity was determined for cells of yeasts strains belonging to the genera *Hansenula*, *Candida* and *Saccharomyces*. A heterogeneous group of yeasts, comprising strains from the three genera, was identified as showing high flotation capacities (degrees of flotation above 50%), which were practically not affected by variations in medium pH in both the synthetic medium and 2% molasses. Thus, the flotation capacity of the cells in this yeast group seemed strongly dependent on the liquid phase properties and/or growth medium composition, more than on the simple variation in pH of the cell suspensions. A second group of strains, belonging to the *Saccharomyces* genus, including also brewing yeast strains, was identified as having lower flotation capacities (degrees of flotation below 50% at pH 1.5), which showed no alterations or variations significantly affected by the medium pH. Foam volumes obtained with *Saccharomyces* strains were greater in synthetic media than in molasses owing to the higher air flow rates required for flotation in molasses. The flotation efficiency decreased in molasses in all cases as well as the foam volume, except in the case of *Hansenula* cells, which showed an increased foam volume. This was probably due to variations in product excretion by the different yeasts and/or differences in cell wall composition.

Keywords Froth flotation · pH effects · *Saccharomyces cerevisiae* · *Hansenula polymorpha* · *Candida albicans*

Introduction

Flotation process occurs when air is bubbled into a liquid phase containing colloidal particles, which in the present work are cells, and these cells adhering to surfaces of the air bubbles form a particle-enriched foam that is withdrawn from the system at the top of the liquid phase. The flotation efficiency has been defined as the capacity of the process to concentrate cells in the foam [1, 2].

The adhesion of particulate matter is an important step in the cell flotation process. As the sizes of the microbial cells can be less than 1 μm, the phenomenon can be described in terms of colloidal chemistry involving the Derjaguin–Landau–Verwey–Overbeek (DLVO) theory. However, the involvement of hydrophobic and steric interactions has also been described in the literature [3]. As in the case of an ordinary colloidal particle, cell surfaces also show a tendency to acquire an electrical charge owing to their preferential adsorption of ions as well as to the dissociation of charged groups on their surfaces [3]. In addition, other adhesion phenomena are found in microbial processes, such as flocculation and aggregation. The flocculation process is based on reversible cell-to-cell interaction mainly due to the calcium bridges between groups on cell surfaces [4]. The aggregation phenomenon is based on irreversible cell-to-cell interactions, possibly due to interactions between glycoprotein groups of cell surfaces. Nevertheless, these two phenomena are mainly dependent on the chemical composition and properties of the microbial cell surface [5]. However, little is known concerning the types of interactive reactions between surface groups or the free-energy requirements to display these processes.

The yeast cell envelope has a complex composition, comprising polymers, such as phosphomannans and chitin [6], in proportions dependent on the yeast strain and growth conditions, such as medium composition, pH and growth temperature [5]. Ionizing groups such as phosphate groups linked to mannans (glucose epimers) forming the phosphomannans are also found within the cell surfaces [7]. The other additional groups are the carboxylic groups from amino acids, such as aspartic and glutamic acids, and amino groups, such as the ε-amino group of lysine, and they ionize as a function of the environmental pH [8, 9].

At neutral pH values, most of the microbial cells possess a negative net charge [10]; however, the surface charge of the microbial cells is greatly dependent on the type of ionizing groups of the cell wall. In this case, the adsorption of ions onto the microbial surface particles, mainly through electrostatic interactions, creates an electrical double layer around the surface that causes repulsion when two double layers become near enough to interact [10]. Adhesion phenomena among microorganisms are also influenced by the microbial production of the collectors acting at concentrations below 5×10^{-2} mol L^{-1} or depressants such as lipids and amino compounds [11]. In a previous work [12], the initial pH of the media only showed a small effect on the degree of flotation, which was more dependent on the bioparticle/liquid phase interface than on the pH of the cell suspension. This previous work was carried out at a fixed pH and indicated that cell flotation is more dependent on the global growth medium dispersed phase composition than on the properties of the cell surface such as hydrophobicity. It seems that cells are susceptible to the dispersion medium of the colloidal system and the froth formation is a more complex phenomena that cannot be simply explained by DLVO theory, which is based on electrostatic and van de Waals interactions, and it may be extended taking into account other types of interactions such as the hydrophobic and steric interactions, and/or ion bridges [13, 14]. The aim of present work is to correlate the variations in flotation parameters with the pH of suspensions having yeast cells that resulted from the growth of yeasts belonging to different genera and species in a synthetic medium and molasses, which is an industrial medium.

Materials and methods

Yeast strains

The yeast strains used were strain FLT-01 of *Saccharomyces cerevisiae* having both high cell-wall hydrophobicity and high flotation capacity [2], strain FLT-01/4b resulting from sporulation of strain FLT-01 and showing a lower degree of flotation [2], strain CBS 4732 of *Hansenula polymorpha* having a high flotation ability [2, 15], strain DSM 2155 of *S. cerevisiae* showing high flotation and flocculation capacities as described by other authors [15], baker's yeast strain LTU, which was obtained by plating samples from the commercial baker's yeast supplied by Fleischmann and Royal, Brazil, and strain FCFAR-250, which was an isolate of *Candida albicans,* kindly donated by M.J.M. Giannini from the Faculdade de Ciências Farmacêuticas de Araraquara SP, Brazil. The top strain L27 and the bottom brewing yeast strain A30 were cultures originally donated by Labatt Breweries culture collection (LCC culture collection; London, ON, Canada). The bottom strains belong to the *Saccharomyces uvarum* species, and they are also known as *Saccharomyces carlsbergensis*. All the stock cultures were maintained in a medium of 1% yeast extract, 2% peptone and 2% glucose at 4 °C.

Culture media and cultivation conditions

A synthetic medium containing glucose, ammonium sulfate, vitamins and minor amounts of inorganic salts as described in the literature [1] was used for cell growth. The pH was adjusted to 4.5 before sterilization. Each Erlenmeyer flask (125 ml) containing 25 ml of sterilized medium was incubated for cell propagation as also described in the literature [1]. The stationary phase cells used in all experiments were submitted to flotation after a 24-h growth period. Molasses was diluted with distilled water to a 2% total reducing sugar concentration (w/v) and the diluted samples (25 ml) were sterilized at 120 °C and 1 atm for 20 min before inoculation.

Bench flotation system

A flotation column (17-cm high × 2-cm inner diameter) was used for the flotation assays [1] and the pH of the samples was adjusted by adding sodium hydroxide and hydrochloridric acid before flotation as required. The final pH was measured after 30-min agitation in a rotary shaker operating at 100 rpm at room temperature. The parameters obtained were [1, 15]

$$\mathrm{FLT}(\%) = (C_p - C_r) \times C_p^{-1} \times 100 \quad (1)$$

and

$$\mathrm{FLT_{Eff}}(\%) = \left[(C_p - C_r) \times C_p^{-1} \times 100\right] \times \left(V_r \times V_p^{-1}\right), \quad (2)$$

where FLT is the flotation, $\mathrm{FLT_{Eff}}$ is the flotation efficiency, C_p is the initial cell concentration, C_r is the cell concentration in the remaining medium in the column after flotation, V_p is the volume of medium added to the flotation column and V_r is the volume of the medium remaining in the column after flotation. The experimental setup was performed according to the method described in the literature [1] and the parameters were $C_p = 0.5$ mg L^{-1}, $V_p = 11$ ml, flotation time 60 s, and airflow rate 8 ml s^{-1}. When molasses was used, the flotation conditions were changed as follows: $V_p = 12$ ml, flotation time 120 s, airflow rate 6 ml s^{-1}. Wherever dilutions were required before flotation assays, supernatants of the cultures were used instead of water so that the supernatant composition was not altered. The airflow rate was adjusted by using a flow-meter controller from Cole-Parmer

Table 1 Effects of adjusting the pH values to 1.5 and 4.5 in a synthetic medium and molasses on the flotation parameters of cell suspensions of yeasts from different genera and showing degrees of flotation higher than 50%

Yeast	Medium	pH 1.5			pH 4.5		
		FLT (%)	FLT_{Eff} (%)	V_s (ml)	FLT (%)	FLT_{Eff} (%)	V_s (ml)
S. cerevisiae (FLT-01)	Synthetic medium	93.1 ± 3.4	51.6 ± 2.5	4.9	97.4 ± 2.7	63.8 ± 1.9	3.8
	Molasses	58.9 ± 4.5	47.1 ± 3.8	2.5	59.6 ± 4.1	45.8 ± 2.7	2.9
H. polymorpha (CBS 4732)	Synthetic medium	87.5 ± 1.9	75.6 ± 1.7	1.5	89.7 ± 1.0	81.5 ± 0.8	1.0
	Molasses	71.2 ± 1.2	26.2 ± 1.1	7.9	58.8 ± 3.5	25.9 ± 2.3	7.0
S. cerevisiae (DSM 2155)	Synthetic medium	77.5 ± 7.1	44.2 ± 1.9	6.3	76.3 ± 1.5	57.2 ± 1.6	8.2
C. albicans (FCFAR-250)	Synthetic medium	65.4 ± 1.5	56.2 ± 1.3	1.5	58.2 ± 2.7	53.2 ± 2.3	0.8

(catalogue no. H-01919–98) and the flotation assays were carried out in triplicate.

Biomass determination

Samples of cell suspensions were washed in a 0.65-μm filter membrane from Millipore and dried to a constant weight for 16 h at 106 °C. The cell biomasses were expressed as milligrams per millilitre.

Total reducing sugar

The amount of total reducing sugar was determined by using the method based on the reduction of 3,5 dinitrosalicylic acid [16] using glucose as a standard. Molasses samples were hydrolysed in 2 M HCl for 10 min at 60 °C prior to assays.

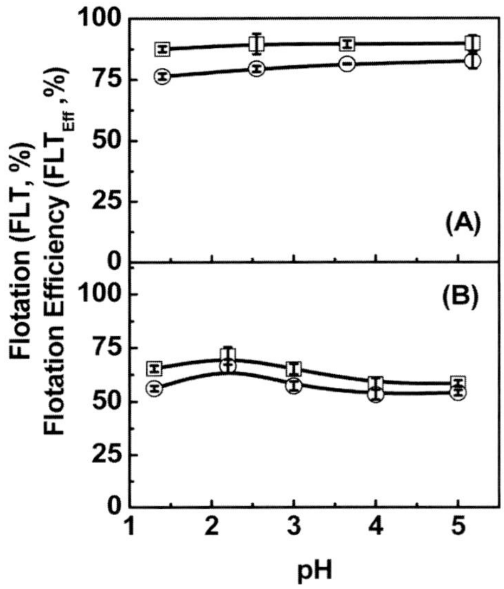

Fig. 1 Effect of increasing pH values (adjusted before flotation) on the degrees of flotation (*FLT*, *squares*) and the flotation efficiency (*FLT_{Eff}*, *circles*) obtained with cells of **A** strain CBS4732 of *Hansenula polymorpha* and **B** strain FCFAR-250 of *Candida albicans* grown in a synthetic medium

Results

The correlations established between variations in pH and the degrees of flotation resulted in the separation of yeasts strains into two groups. The separation was based on the values of degrees of flotation in a synthetic medium as well as on the variations that occurred when the medium pH was adjusted before flotation. A group of strains showing degrees of flotation greater than 50% for strains FLT-01 and DSM 2155 of *S. cerevisiae*, strain CBS 4732 of *H. polymorpha* and strain FCFAR-250 of *C. albicans* is described in Table 1. The degrees of flotation shown by these strains in a synthetic medium were high. On the other hand, decreases for both the degrees of flotation and the volume of the foam were observed in all cases after growth in molasses, excepted in the case of cells of *Hansenula*, which showed an increased foam volume. The flotation parameters (FLT and FLT_{Eff}) for strains of *Hansenula* and *Candida* are described in Fig. 1, while the flotation parameters for two distinct *Saccharomyces* strains are described in Fig. 2. The degrees of flotation described by these two figures were practically not affected by variations in the medium pH and the degrees of flotation were independent of the yeast genera. However, increases were observed in the flotation efficiency (FLT_{Eff}) when the medium pH was raised, particularly in the case of strain FLT-01 of *S. cerevisiae* (Fig. 2). In addition, the foam volumes mentioned in Table 1 decreased or increased with the rises in medium pH and these variations were strain-dependent.

A second group of yeast strains, showing smaller flotation capacities (below 50% at pH 1.5) in a synthetic medium as well as in molasses is listed in Table 2. Brewing strains A30 and L27, baker's yeast strain LTU and the strain FLT-01/4b of *Saccharomyces* are described in this group. The degrees of flotation obtained

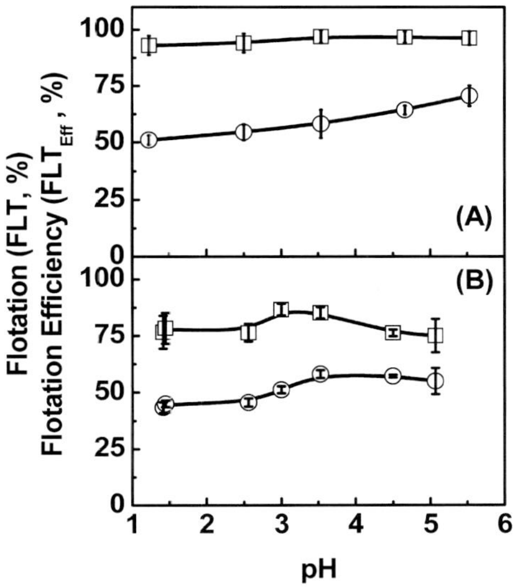

Fig. 2 Effect of increasing pH values (adjusted before flotation) on the degrees of flotation (*squares*) and the flotation efficiency (*circles*) obtained with cells of **A** strain FLT-01 and **B** strain DSM 2155 of *Saccharomyces cerevisiae* grown in a synthetic medium

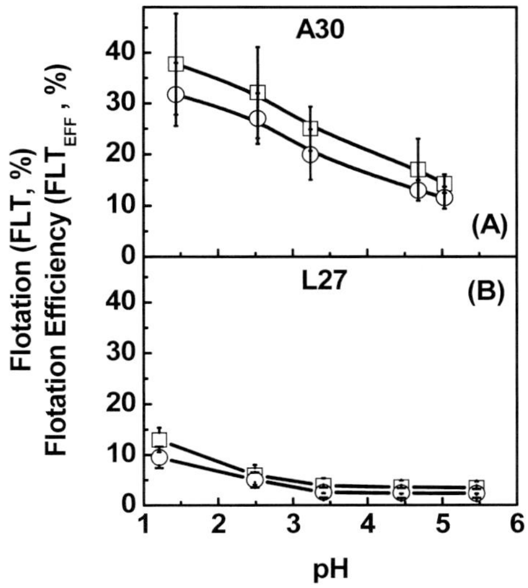

Fig. 3 Effect of increasing pH values (adjusted before flotation) on the degrees of flotation (*squares*) and the flotation efficiency (*circles*) obtained with cells of **A** the bottom strain *A30* and **B** the top strain *L27* of brewing yeasts grown in a synthetic medium

for cells belonging to this second group of strains were not equally affected by variations in medium pH, adjusted before flotation, as shown in Figs. 3 and 4. When the medium pH was raised (Fig. 3), decreases in the degrees of flotation were observed for brewing strains A30 and L27, while increases were observed with strain FLT-01/4b as shown in Fig. 4. Actually, the variations in pH before flotation affected both the flotation and flotation efficiencies of cells belonging to this second group while the variations in foam volumes were strain dependent. As the flotation efficiency is inversely correlated to foam volume, it decreased with the increases in foam volumes. Lastly, the degrees of flotation were not determined for strains DSM2155 and A30 (bottom brewing yeast) in molasses because strong flocculation occurred in this medium. Thus, flocculation is also a medium-dependent process.

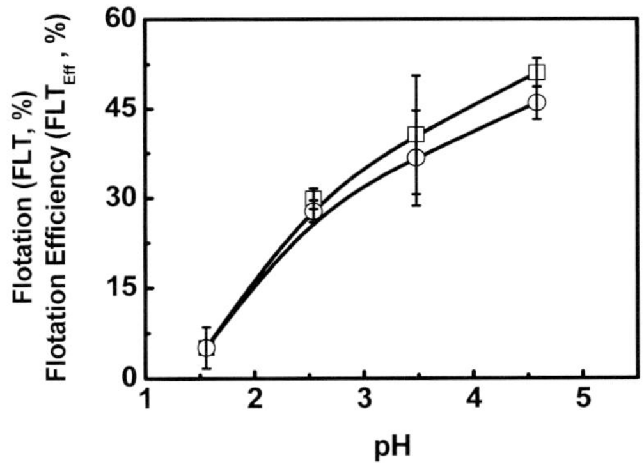

Fig. 4 Effect of increasing pH values (adjusted before flotation) on the degrees of flotation (*squares*) and the flotation efficiency (*circles*) obtained with cells of strain FLT-01/4b (resulting from the sporulation of strain FLT-01) of *S. cerevisiae* grown in a synthetic medium

Table 2 Effects of adjusting the pH values to 1.5 and 4.5, in a synthetic medium and 2% molasses, on the flotation parameters of the cell suspensions of *Saccharomyces* strains showing degrees of flotation lower than 50%. *ND* indicates not determined

Yeast *Saccharomyces* strains	Medium	pH 1.5			pH 4.5		
		FLT (%)	FLT_{Eff} (%)	V_s (ml)	FLT (%)	FLT_{Eff} (%)	V_s (ml)
A30	Synthetic medium	37.7 ± 4.2	31.9 ± 3.5	1.7	14.3 ± 1.8	11.0 ± 1.8	2.5
LTU	Synthetic medium	31.5 ± 3.2	15.8 ± 2.1	5.5	33.4 ± 2.5	19.1 ± 2.0	4.7
	Molasses	20.4 ± 3.0	15.2 ± 2.8	3.2	18.2 ± 4.7	13.0 ± 2.9	3.6
L27	Synthetic medium	12.9 ± 2.4	2.0 ± 0.7	2.7	3.4 ± 1.2	2.2 ± 1.2	3.8
	Molasses	1.5 ± 1.0	ND	ND	0.5 ± 0.3	ND	ND
FLT-01/4b	Synthetic medium	5.1 ± 3.4	5.1 ± 3.4	0.0	51.1 ± 2.4	46.0 ± 2.8	1.1

Discussion and conclusions

Synthetic media are routinely used in several microbiological studies owing to their simple and well-established composition. Using this type of medium, cells showing more closely related wall compositions as well as physiological behaviours are obtained in different trials. On the other hand, industrial media such as molasses, which is used as a raw material for bioethanol production, and other media containing by-products from the agroindustry usually show undefined mixtures of proteins, sugars and/or peptides. Actually, the raw material composition changes during the crop cycles and, for this reason, the synthetic medium allows more reliable comparisons among data from different laboratories. In addition, bacterial cell surfaces show highly dynamic architectures, which immediately respond to the environmental changes mainly through the adsorption of ions or other components of the medium onto their surfaces [10].

Yeast strains assayed in the present work were divided into groups. One group of yeasts (Table 1, Figs. 1, 2) was characterized by showing high degrees of flotation, which were slightly or not disturbed by variations in medium pH values of the cell suspensions. These characteristics seem related to strong interactions between the negative net charge of cell surfaces and the air bubble surfaces [17]. On the other hand, the second group of yeasts, described in Table 2, was characterized by lower degrees of flotation, probably owing to electrostatic repulsions between double layers. The cells in this second group of yeasts, showing lower flotation capacities, seem more susceptible to variations in pH, possibly owing to stronger effects of van der Waals interactions avoiding cell adhesion to air bubbles. Studies based on the bacteria adhesion to flat pieces of Teflon indicated the importance of van der Waals interactions and steric interactions between surfaces of both molecules and solids [18]. The ionic strength also plays an important role in adhesion [14]. Bacterial adhesions to Teflon were dominated by steric interactions in solutions having high ionic strength, while the interactions based on the DLVO theory dominated at low ionic strength [14].

Concerning molasses, the flotation process became more difficult in this complex medium and it was apparently less affected by changes in pH due to the formation of less stable foams. In addition, increases in the foam volume occurred in all cases except in the case of *Hansenula*. This apparently is also due to the complex molasses composition and its greater viscosity or secretion of surface-active agents by the yeast strains. However, this problem can be solved by the simple adjustment of the flotation conditions (to be published elsewhere).

The variations in flotation efficiency (FLT_{Eff}) are inversely proportional to variations in foam volume (Eq. 2). In addition, colloidal particles, such as the cells in this work, can act as foam stabilizers or destabilizes, depending on their properties [19]. In the present work, the degrees of flotation varied with the strains but the flotation efficiency (FLT_{Eff}) was more affected by variations in pH than the flotation itself. Thus, variations in pH can better control foam formation than cell adhesion to air bubbles in the flotation assays carried out in a synthetic medium. Strong interactions between the charged air bubble surfaces and the double layers surrounding cell surfaces, usually involved with adhesion in solutions, seems less affected by variations in the pH in a synthetic medium than the interactions between the medium/air bubbles, which are usually more involved with foam formation.

Despite the importance of the microbial adhesions to environmental phenomena and technological processes, their mechanisms are not clearly understood [3]. Physicochemical approaches related to free-energy requirements have frequently been employed in modelling microbial adhesive interactions to predict microbial adhesion to substrata. In addition, attempts have been made to embed the physicochemical approaches into current biochemical and microbiological knowledge [3].

Acknowledgement This work was generously supported by grants from FAPESP (project nos. 98/04299-9 and 98/16224-3).

References

1. deSousa SR, Laluce C (2000) Biotechnol Lett 22:753
2. Palmieri MC, Greenhalf W, Laluce C (1996) Biotechnol Bioeng 50:248
3. Oliveira R (1997) Exp Therm Fluid Sci 14:316
4. Guinard JX, Lewis MJ (1993) J Inst Brew 99:487
5. Walker GM (1998) Yeast physiology and biotechnology. Wiley, New York, p 17
6. Cabib E, Roberts R, Bowers B (1982) Annu Rev Biochem 51:763
7. Cawley TM, Letters R, Harrington MG (1972) Biochem J 129:711
8. Mozes N, Schinckus LL, Ghommidh C, Navarro JM, Rouxhet PG (1994) Colloid Surf B 3:63
9. Mozes N, Leonard AJ, Rouxhet PG (1988) Biochim Biophys Acta 945:324
10. Poortinga AT, Bos R, Norde W, Busscher HJ (2002) Surf Sci Rep 47:1
11. Hornsby D, Leja J (1982) Surface Colloid Sci 10:217
12. deSousa SR, Oleira KF, Souza CS, Kilikian BV, Laluce C (2003) Colloid Surf B 29:309
13. Bos R, vanderMei HC, Busscher HJ (1999) FEMS Microbiol Rev 23:179

14. Rijnaarts HHM, Norde W, Bouwer EJ, Lyklema J, Zehnder AJB (1995) Colloid Surf B 4:5
15. Gehle R, Sie T, Kramer T, Schugerl K (1991) J Biotechnol 17:147
16. Miller GL (1959) Anal Chem 31:426
17. Harvey RW (1991) In: Hust CJ (ed), Modeling the environmental fat of microorganisms. American Society for Microbiology, Washington, DC, p 89
18. Rijnards, HHM, Norde, W, Bouwer, EJ, Lyklema, J, Zehnder, AJB (1993) Environ Microbiol 59:3255
19. Ross S, Morriso ID (1988) Colloidal systems and interfaces. Wiley, New York

Progr Colloid Polym Sci (2004) 128: 270–272
DOI 10.1007/b97105

Encarnación Jurado
Vicente Bravo
Rafael Bailón
Josefa Núñez-Olea
Deisi Altmajer-Vaz
Miguel García-Román
Alejandro Fernández-Arteaga

Application of the statistical design of experiments and the Bath–Substrate–Flow method in formulating detergents for hard surfaces

E. Jurado (✉) · V. Bravo · R. Bailón
J. Núñez-Olea · D. Altmajer-Vaz
M. García-Román · A. Fernández-Arteaga
Departamento de Ingeniería Química,
Facultad de Ciencias, Universidad de
Granada, 18071 Granada, Spain
e-mail: ejurado@ugr.es

Abstract The present work studies the effect that both process and formulation variables exert over the detersive process on hard surfaces. The statistical design of experiments together with the Bath–Substrate–Flow method for evaluating detergency enabled us to develop detergent formulas for hard surfaces. The effectiveness of the detergent depends on the total concentration of the components in the washing bath and on the calcic hardness tested, which determines the optimal proportion of the components of the detergent.

Keywords Detergency · Detergent formulas · Statistical design · Bath–Substrate–Flow method · Oily soil

Introduction

Detergent formulas are complex systems composed normally of more than eight components [1]. Apart from the detergent composition, the variables of the washing process (temperature, pH, hydrodynamic conditions, etc.) are also key in detersive effectiveness, making it extremely difficult to choose the optimum detergent formula [2]. In the present work, a statistical design is applied together with the Bath–Substrate–Flow (BSF) method [3] for evaluating detergency, enabling the optimization of a detergent formula, both with regard to the components (surfactants and sequestering agents) as well as the effect of the washing conditions on detersive effectiveness.

Materials and methods

As detergent components, the commercial surfactants used were Glucopon 650 (alkylpolyglucoside) and Findet 1214 N/23 (fatty alcohol ethoxylate), supplied by Henkel and Kao Corporation, respectively, in addition to the sequestering agents Cellesh 100 (homopolymer of acrylic acid) and Cellesh 200 (copolymer of acrylic and maleic acids) both from the Kao Corporation.

The semisolid soiling agent tested consisted of a 72:28 mixture of oleic acid (PRS, Panreac) and commercial Edenor LLSM GS from Henkel (48.4% stearic acid and 50% palmitic acid). The washing tests were made at 45 °C and an initial bath pH of 8.0 in the BSF device [3].

The analysis of the soiling agent eliminated from the surface of the substrate (0.3-cm glass beads) 3 min after the beginning of the experiment was carried out using the acidity index [4]. The detergency (De) was calculated by

$$De(\%) = \left(\frac{V_{exp} - V_{cont}}{V_{theor}}\right) \times 100,$$

"where V_{exp} is the volume of 0.1 N KOH in ethanol used for titrating the sample, V_{theor} is the theorical volume needed for the titration of all the oily soil placed in the system, and V_{cont} is the volume used in the titration of the control."

To assess the washing process and to determine the best detergent formula for each set of "values of the process" variables, an experimental matrix was designed on the basis of four formulation factors (weight fractions of Glucopon, Findet, Cellesh 100 and Cellesh 200) and two process factors (calcic hardness of the bath and total detergent concentration). The lowest and highest values for the variables assayed are presented in Table 1.

The experimental design was prepared with the help of the statistical program MODDE 6.0 (Umetrics, Sweden), which at the same time enabled the fitting of the results. The *D*-optimum design implies the performance of 32 experiments, four of these being replicates of the same assay. The fitting used to model the detergency was a second-order polynomial model that takes into account the interactions between variables [5]. The goodness of the fit was evaluated by the determination coefficient R^2 and the goodness of the prediction Q^2, parameters between 0 and 1. For a good prediction with the equation, the values of both parameters should be near 1 and the differences between them lower than 0.3.

Table 1 Experimental domain of the design

Factors		Range
Weight fractions	Glucopon 650	0.25–0.50
	Findet 1214 N/23	0.25–0.50
	Cellesh 100	0.25–0.50
	Cellesh 200	0.00–0.25
Total detergent concentration (g/L)		1.0–2.0
Calcic hardness ($CaCO_3$) (g/L)		0.30–0.60

Results and discussion

The experimental results for the detergency were fit to a second-degree polynomial function

$$\begin{aligned} De =& 80.2402 - 0.7585C - 0.0384H_{Ca} \\ &+ 1.1913X_G + 2.3596X_F + 0.0454X_{C2} \\ &+ 1.8382X_{C1} + 0.7750X_F^2 + 1.03281CH_{Ca} \\ &+ 0.5799CX_G - 1.9316CX_F \\ &+ 1.8077CX_{C2} + 0.9905CX_{C1} + 0.7833H_{Ca}X_G \\ &- 0.1102H_{Ca}X_F \\ &+ 0.3573H_{Ca}X_{C2} - 0.7605H_{Ca}X_{C1}, \end{aligned}$$

where H_{Ca} is the calcic hardness, C is the total detergent concentration, X_G is the weight fraction of Glucopon, X_F is the weight fraction of Findet, X_{C1} is the weight fraction of Cellesh 100 and X_{C2} is the weight fraction of Cellesh 200.

The statistical analysis performed implies that the equation chosen represents satisfactorily the experimental data, with an R^2 of 0.949 and a Q^2 of 0.643. The highest detergency values predicted by the model were registered with the formula composed of the weight fractions of Glucopon, Findet and Cellesh 100 of 0.25, 0.50 and 0.25, respectively.

A comparison of the experimental and observed detergency values is presented in Fig. 1. The highest experimental detergency value recorded (95.86%) was reached at a total concentration of detergent in the bath of 1 g/L for a calcic hardness of 0.3 g/L and weight fractions of Glucopon 650, Findet 1214 N/23, Cellesh 100 and Cellesh 200 of 0.25, 0.50, 0.25 and 0, respectively. The detergency values obtained in the absence of Cellesh 200 at a total detergent concentration of 1 g/L and a calcic hardness of 0.3 g/L are shown in Fig. 2. With respect to the variables of the washing process, the detergency diminished with an increase in the total concentration of the components in the detergent mixture, and remained almost constant with greater calcic hardness.

The optimal composition of the formula differed depending on the hardness and the total concentration of the detergent. For hardness values between 0.30 and 0.45 g/L and a total detergent concentration between 1 and 1.5 g/L, the increase in the proportion of Findet in the medium improved the detergency. For high hardness values (0.60 g/L) it was found that, as the detergent concentration rose, the proportion of Glucopon should be greater in the detergent mixture in order to achieve the best washing results. In the case of C=2 g/L, for any

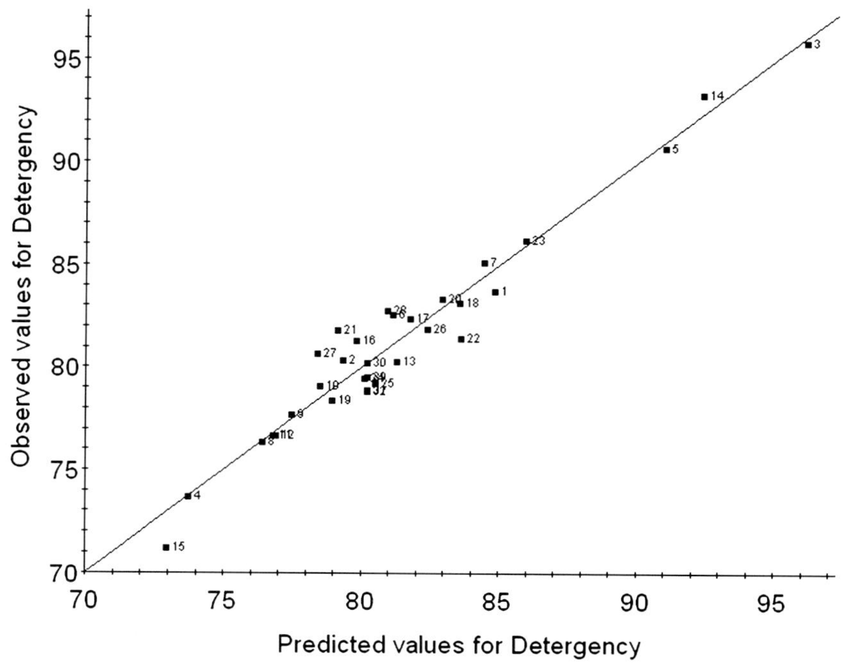

Fig. 1 Comparison between experimental and calculated detergency values

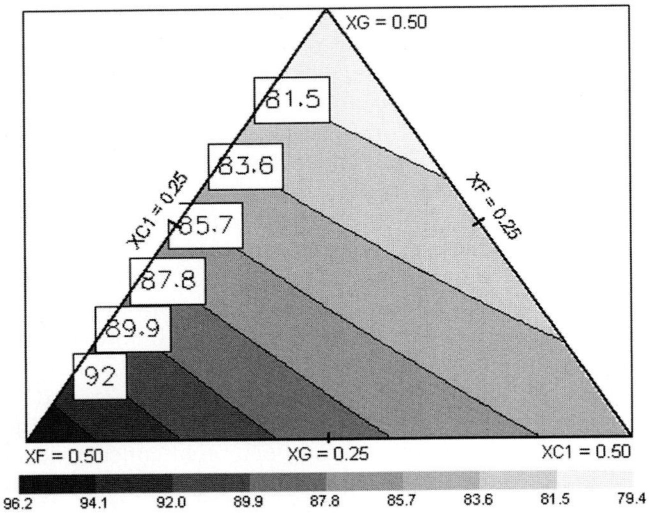

Fig. 2 Mixture contour plot

calcic hardness value, the detergency increased with a higher proportion of Glucopon and with a lower proportion of Findet.

Under all the conditions assayed, the presence of the sequestering agent Cellesh 200 lowered the detersive effectiveness and therefore Cellesh 200 should not be used in the detergent formula.

Conclusions

The experimental design together with the BSF method for evaluating detergency enabled us to develop detergent formulas for hard surfaces.

The detergent formulas developed depend on the total concentration of the components in the washing bath and on the calcic hardness assayed.

For low total concentrations of detergent and low hardness values, the detergent formula should contain a greater proportion of the surfactant fatty alcohol ethoxylate Findet 1214 N/23. For high concentrations of detergent and high calcic hardness values, the proportion of alkylpolyglucoside Glucopon 650 should be greater in order to achieve the greatest effectiveness in washing fatty soils.

Acknowledgements The Ministerio de Ciencia y Tecnología supported projects 1FD97–0931 and PB1998–1293, and the Agencia Española de Cooperación Internacional awarded study grants to D.A.-V.

References

1. Erikson L, Joahnson E, Wikstrom C (1998) Chemometr Intell Lab Syst 43:24
2. Cutler WG, Kissa E (eds) (1987) Detergency: theory and technology. Dekker, New York
3. Jurado Alameda E, Bravo Rodríguez V, Bailón-Moreno R, Núñez-Olea J, Altmajer-Vaz D (2003) Ind Eng Chem Res 42:19
4. IUPAC standard methods for the analysis of oils, fats and soaps 1964 II. D.1
5. Box GEP, Hunter WG, Hunter JS (1978) Statistics for experimenters. An introduction to design, data analysis and model building. Wiley, New York

Progr Colloid Polym Sci (2004) 128: 273–277
DOI 10.1007/b97107

Amós Luciano Carneiro
Maria Helena Andrade Santana

Production of liposomes in a multitubular system useful for scaling up of processes

Abstract This work describes the production of liposomes by hydration of a lipid film formed in a multitubular system. The process is a modified approach of the classical Bangham method, in which a thin lipid film is adsorbed in the inner surface of glass tubes. Initially, continuous adsorption was achieved by feeding a lipid dispersion in ethanol at the bottom of the tubes through a peristaltic pump, and circulating the dispersion for 30 min at 60 °C. In a second step, the excess of lipids was drained and the film was dried under vacuum. The dried film was hydrated using a buffered solution in order to produce liposomes. The adsorption of lipids was characterized through their isotherm and kinetics profiles. The hydration kinetics was monitored by the absorbance of the solutions, the mean diameter and the phospholipid concentrations at different initial concentrations of the lipid dispersion. A heterogeneous population of multilamellar liposomes was produced with a mean diameter of approximately 1,500 nm, as determined by dynamic laser light scattering. The tuberculostatic pyrazinamide was entrapped in liposomes and the incorporation efficiency was higher when the drug was added in the hydration solution (14.7%) than in the phospholipid film (6.7%). The stability profile in the presence of surfactants and the gradual release of the drug confirm the membrane properties of the liposomes produced. The results show the potentialities of the process for continuous production of liposomes and applications in the pharmaceutical industry.

Keywords Liposomes · Scaling up · Multitubular system

A.L. Carneiro · M.H.A. Santana (✉)
School of Chemical Engineering,
State University of Campinas,
CP 6066, 13083-970 Campinas, SP, Brazil
e-mail: lena@feq.unicamp.br
Tel.: +55-19-37883921

Introduction

Liposomes are phospholipid colloids useful as vehicles for controlled delivery of drugs in successful medical and pharmaceutical applications. Although various techniques have been developed for the preparation of liposomes on a laboratory scale, most of them are not suitable to large-scale liposome manufacturing. The scale-up, sterility and stability, the triple "S" problem, constitute the main limitation for industrial production of liposomes. Injection of organic solutions of lipids in an aqueous medium with continuous removal of solvent, microemulsification, and microfluidization are the conventional methods for scale-up of the liposome production [1]. Recently, spray drying has been studied for the massive production of liposomes [2]. Different aspects of the liposome manufacturing have been addressed and a number of different solutions to the problem of scale-up have been proposed. However, the poor control of production parameters which allow the reproducible process as well as low yields and productivities is still an unsolved problem. Furthermore, all known processes of

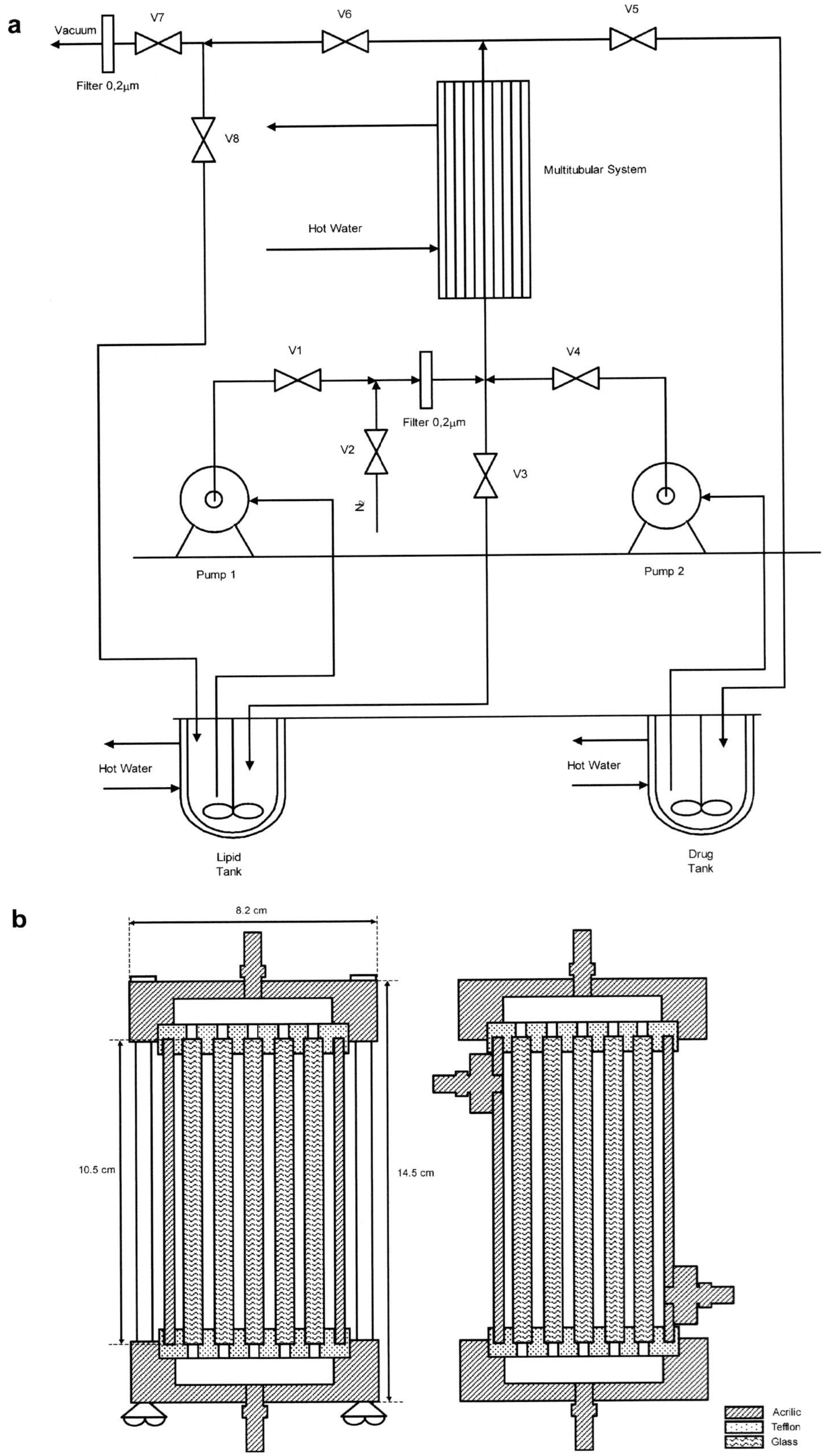

Fig. 1 **a** Flow sheet of the process for liposome production in a multitubular system. The dispersion from the lipid tank was delivered to the multitubular system through *pump 1*. The lipid film within the glass tubes was dried under vacuum and the liposomes were produced by feeding a hydration solution from the *drug tank* to the multitubular module. **b** Frontal and lateral views of the multitubular module

pilot or industrial scale, typically, are linked to small batches. As a consequence, these processes are too costly for large-scale commercial production.

Tournier et al. [3], based on the Bangham method [4], proposed that the lipid films obtained by evaporation of the solvent were formed in a parallel tube column, producing liposomes by hydration of the thin films. Furthermore, the lipid deposit may be expanded into a three dimensional structure with low density (approximately 0.01 g/cm^3) and high superficial area depending on the number and dimensions of the tubes. High yields of liposomes of desired size may be produced by the method, which is particularly easy to scale up and control. In addition, the method provides further advantages, which include reduced processing time, use of inexpensive materials and finally use of the same reactor for film formation, solvent evaporation, hydration of the film and sterilization of the liposomes. Except for the patent of Tournier et al., there are no studies in the literature related to this method.

The present work describes the production of liposomes in a multitubular setup based on the Tournier method. The adsorption of lipids in the tubes was characterized and the generation of liposomes analyzed.

Material and methods

Continuous adsorption of phospholipids

Hydrogenated natural soy lecitin, Epikuron 200SH (Degussa Bioactives, Hamburg, Germany), was used as a structural phospholipid for liposome production. The adsorption was carried out in a multitubular modulus system containing 21 glass tubes (inner diameter 4.0 mm) or 37 tubes (inner diameter 1.6 mm) of 11 cm length. The lipids were dispersed in ethanol at various concentrations and flowed inside the tubes for 30 min at 35 mL/min at 60 °C. The kinetic behavior of the adsorption was monitored by analysis of the phospholipid content in samples collected from the circuit during the flowing time. The amount of adsorbed phospholipids was determined by subtracting the phospholipid concentrations in the initial and flowing solutions. After the adsorption, the lipid solution was drained and the film was dried under a vacuum of 200 mbar at 80 °C for 60 min. For the adsorption isotherm, the phospholipid concentration at the equilibrium was analyzed in the drained solution. The adsorbed mass was determined from the phospholipid content after hydration (solubilization) of the lipid film as described next.

Hydration of the dried lipid film

The dried lipid film was hydrated at 60 °C by circulating a buffered solution [*N*-(2-hydroxyethyl)piperazine-*N'*-ethanesulfonic acid, 10 mm,pH 7.4] at a flow rate of 640 mL/min for 1 h. Samples were collected with time and assayed. The kinetics profiles for absorbance of the solutions at 340 nm, the mean diameter and the phospholipid concentrations were determined. The influence of the initial lipid concentration on the kinetic behavior of the hydration was analyzed.

Liposome characterization

The liposomes produced were characterized by their phospholipid contents [5], the mean diameter and the size distribution using laser light scattering (Malvern, Autosizer 4700). The membrane properties were characterized by the stability profile of liposomes in the presence of the nonionic surfactant polyoxyethylene 5 lauryl ether, and by its capacity for drug entrapment and gradual release. The stability profiles were determined by turbidity measurements of the solutions at 340 nm, after 2 h incubation in the presence of various surfactant concentrations. The drug release was determined by placing the liposome dispersion in a dialysis membrane and monitoring the drug concentration of the outside solution with time.

Results and discussion

The multitubular system was planned and constructed, with some modifications to the Tournier's process. A flow sheet of the process for liposome production and the frontal and lateral views of the multitubular module are shown in Fig. 1. The system is easy to

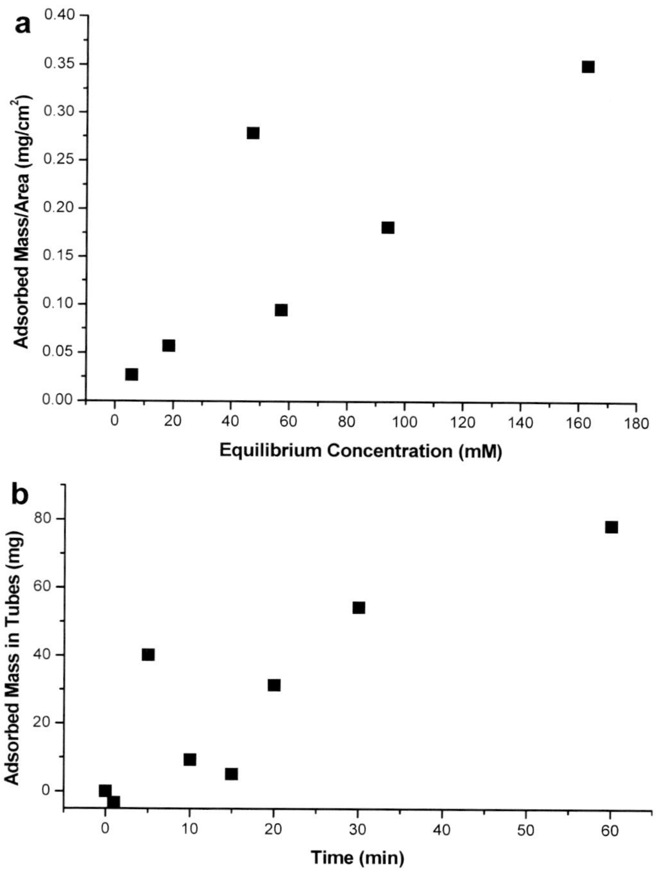

Fig. 2 a Isotherm and **b** kinetics profiles for adsorption of lipids on the inner walls of tubes

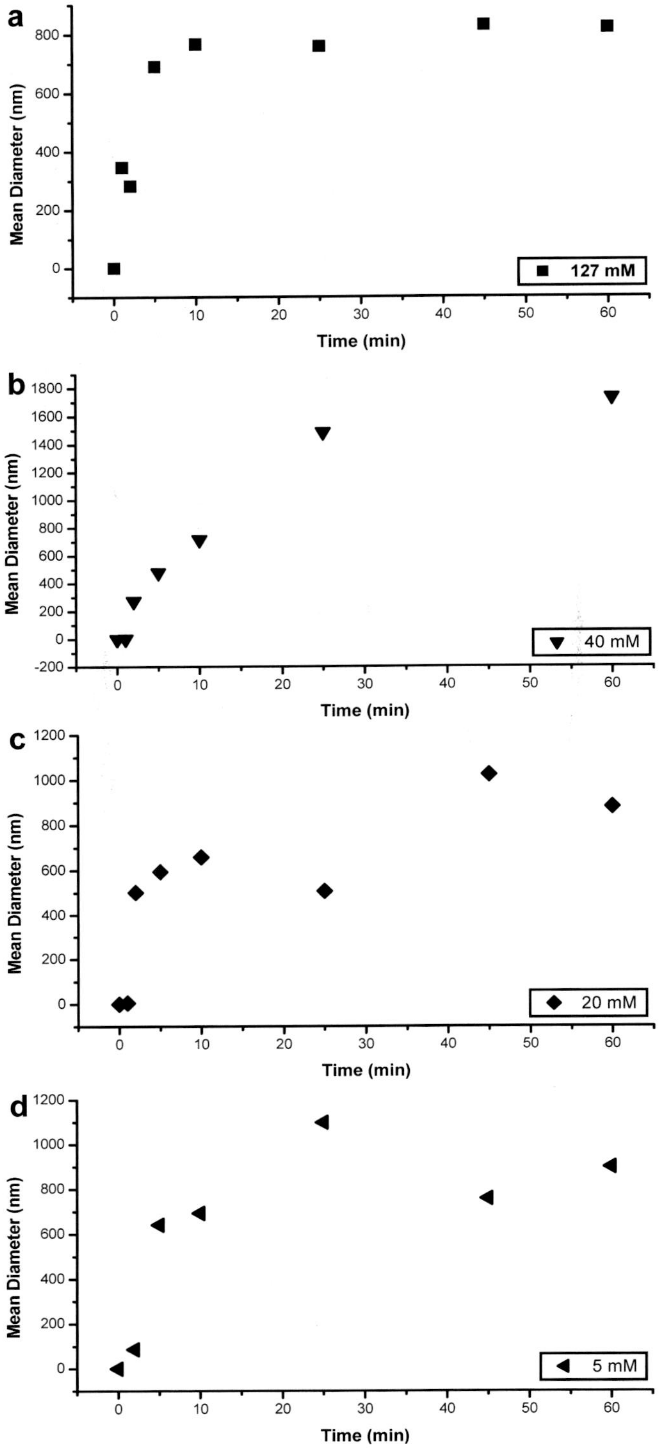

Fig. 3 Evolution of the mean diameter of liposomes with time at different concentrations of the initial lipid dispersion used for adsorption: **a** 127 mM, **b** 40 mM, **c** 20 mM and **d** 5 mM

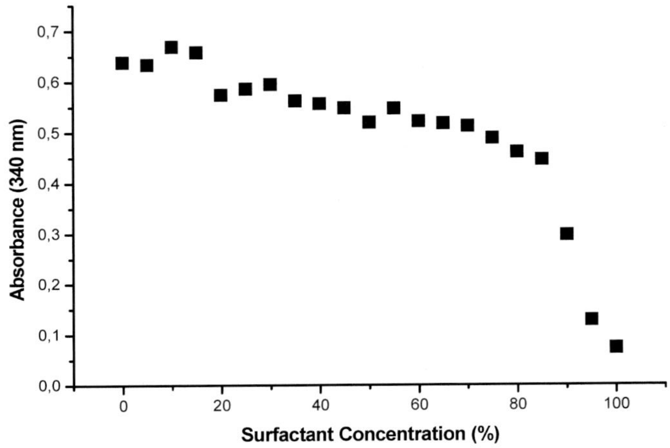

Fig. 4 Stability profile of the liposomes in the presence of the nonionic surfactant polyoxyethylene 5 lauryl ether

operate and control. The scale-up of the process may be done by increasing the length or the number of tubes in the module.

The kinetics profile and the isotherm shown in Fig. 2 reveal the adsorption of lipids on the inner walls of the tubes. In both curves, the dispersion of data in some regions is due to errors in the lipid concentration as a consequence of ethanol evaporation.

The generation of liposomes by hydration was monitored with time through the evolution of the profiles for absorbance of the solutions, the mean diameter and the phospholipid concentration. The mean diameter profiles for different concentrations of the initial lipid dispersion used for adsorption are shown in Fig. 3. The mean diameter increases with time, indicating the aggregation of lipids and a vesiculation phenomenon. After the initial time the profiles reached a maximum, characterizing the equilibrium of lipids in both forms, free and liposomal. At 60 min, the mean diameters of the liposomes produced were 819, 1,727, 879 and 898 nm, when the initial lipid concentrations were 127, 40, 20 and 5 mM, respectively. After the initial 5 min two main populations around 200- and 900-nm diameter were observed. The percentages of the larger population increased with time, demonstrating the aggregation of lipids.

In a typical stability profile for liposomes the turbidity of the solutions increases as a consequence of the surfactant penetration in the membrane, and decreases after the rupture of the liposomes in micelles [6]. The stability profile for the liposomes prepared is shown in Fig. 4. No significant changes in absorbance were observed up to 82% surfactant concentration, indicating poor penetration of the surfactant as a consequence of the dense packing of lipids in the membrane. The efficiency for pyrazinamide encapsulation was 14.7% and 6.7% when the drug was added to the hydration solution or to the lipid film, respectively. The gradual release of the drug from the liposomes was observed for 45 min. These results evidence the membrane properties of the liposomes.

Conclusions

In this study, the feasibility of liposome production in a multitubular system was demonstrated. The concentration and the mean diameter of the liposomes produced may be controlled by the initial conditions of the process. The simplicity and the efficiency of the system open unique possibilities for the production of liposomes on a large scale.

References

1. Lasic DD (1993) Liposomes from physics to applications. Elsevier, Amsterdam
2. Kikuchi H, Yamaguchi H, Hirota S (1991) Chem Pharm Bull 39:1522
3. Tournier H, Schneider M, Guillot C (1999) US Patent 5,980,937
4. Bangham AD, Standish MM, Watkins JC (1965) J Mol Biol 13:238
5. Chen P, Tonibara TY, Warner H (1956) Anal Chem 28:1756
6. Edwards K, Almgren M (1992) Langmuir 8:824

Progr Colloid Polym Sci (2004) 128: 278–282
DOI 10.1007/b97109

Carlos Alberto Policiano Almeida
Clodoaldo Machado
Nito Angelo Debacher

Adsorption of methylene blue as a model for the use of Barro Branco as an alternative adsorbent for color removal

C.A.P. Almeida · N.A. Debacher (✉)
Departamento de Química,
Universidade Federal de Santa Catarina,
88040-900 Florianópolis, SC, Brazil
e-mail: debacher@qmc.ufsc.br

C. Machado
Departamento de Química,
Universidade Regional de Blumenau,
CEP 89010-971, SC, Brazil

Abstract The application of the solid–liquid adsorption technique with low-cost adsorbents has been offering promising results in the treatment of textile wastewaters. In this study, we used as an adsorbent Barro Branco (aluminum-silicate – Al 16.2%, Si 21.5% and O 44.77%), one of the main by-products of the extraction of mineral coal. The samples were obtained at the Esperança mine, Santa Catarina, Brazil. The experiments were carried out in an adsorber vessel with controlled temperature, pH and a mechanical stirrer. A 1.0 g sample of Barro Branco (210–297 μm) was added to each 50 cm^3 of methylene blue dye solution. The results show that Barro Branco is a promising adsorbent for color removal, where the adsorption kinetics fit better to the pseudo-first-order rate law. The values of the rate constant of the adsorption decreased with the increase of the concentration. The isotherm constants agree well with the Langmuir equation and the Freundlich equation, with correlation coefficients close to 1.00 in both cases.

Keywords Methylene blue · Adsorption · Barro Branco · Langmuir isotherm · Freundlich isotherms

Introduction

The textile industry has nowadays a great number of chemical products which are used as dyes in the dyeing processes; however, many of these products are toxic, and thus demand a rigorous treatment of their effluents. Biological systems of treatment, although largely utilized in the removal of color, are very inefficient, mainly for synthetic dyes of complex structures [1, 2]. Therefore, many more effective methods have been developed, such as those based on solid–liquid adsorption treatment processes.

Adsorption has recently gained favor for its simplicity, low cost, and great efficiency in the removal of difficultly degradable dyes, when compared with conventional removal methods [3]. Furthermore, in some cases, it is possible to recover the dye through desorption [4]. Nowadays, diverse vegetable [1, 2] or mineral materials [3, 5, 6] are being investigated as possible low-cost adsorbents [7]. The adsorption technique has another great advantage that is related to the reuse of the large amounts of water employed by textile industries.

This work presents the results of solid–liquid adsorption of methylene blue as a model for the use of Barro Branco as an alternative adsorbent. Many adsorption parameters were investigated, such as the initial concentration of the methylene blue, the pH of the solution, temperature, and the rates of adsorption and desorption.

Adsorption isotherms

The Langmuir isotherm

$$q_e = \frac{K_L C_e}{1 + a_L C_e}, \tag{1}$$

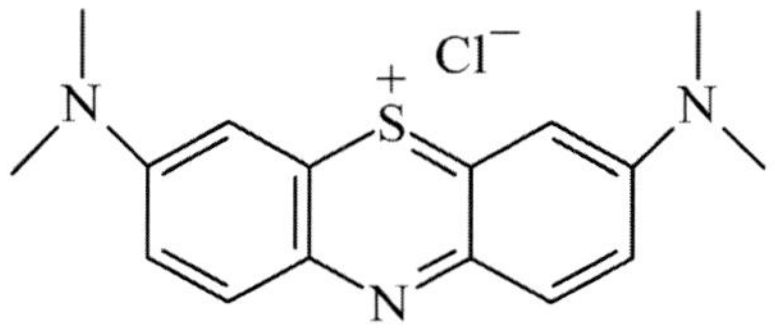

Scheme 1 Methylene blue

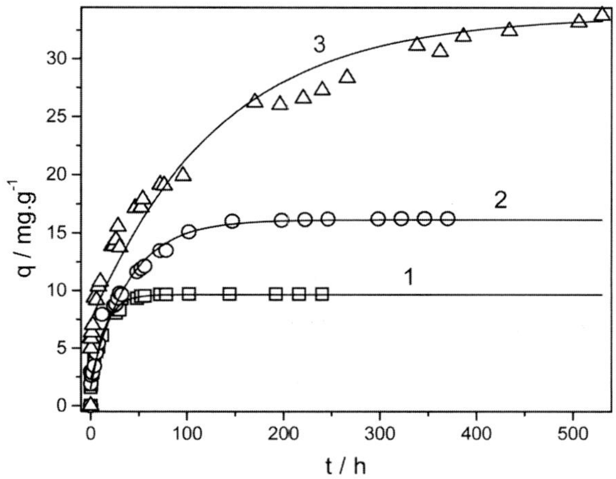

Fig. 4 Effect of initial concentration in the adsorption of methylene blue on Barro Branco at 45 °C, using a particle size of 210–297 μm, initial concentrations of the solutions of 194.42 mg dm^{-3} (*squares*), 325.79 mg dm^{-3} (*circles*), and 836.62 mg dm^{-3} (*triangles*), stirring at 250 rpm, a pH range of 5.5–7.0, and 1.0 g Barro Branco

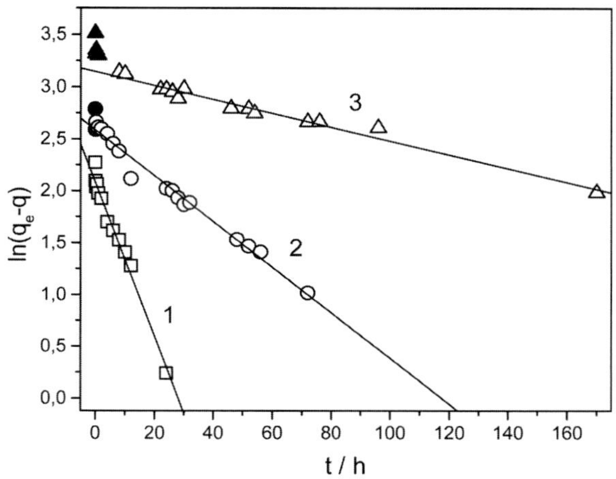

Fig. 5 Pseudo-first-order kinetics of adsorption for methylene blue on Barro Branco at 45 °C using a particle size of 210–297 μm, initial concentrations of the solutions of 194.42 mg dm^{-3} (*squares*), 325.79 mg dm^{-3} (*circles*), and 836.62 mg dm^{-3} (*triangles*), stirring at 250 rpm, a pH range of 5.5–7.0, and 1.0 g Barro Branco

predominant at the surface and favor the adsorption of positively charged substances.

Table 3 Parameters of the Lagergren equation (Eq. 5) in the adsorption of methylene blue

C_0 (mg dm^{-3})	q_e (mg g^{-1})	$10^2 k_{ads}$ (h^{-1})	$t_{0.5}$ (h)	Correlation coefficient
194.42	9.65	7.42	9.34	0.995
236.57	11.8	5.44	12.74	0.992
268.48	13.37	4.57	15.17	0.997
298.41	14.87	3.05	22.73	0.993
317.52	15.78	2.41	28.76	0.992
325.79	16.24	2.20	31.51	0.992
398.51	19.78	1.72	40.30	0.989
447.19	22.18	1.28	54.15	0.986
500.50	24.77	0.91	76.17	0.996
538.01	25.96	0.74	93.67	0,996
636.43	30.9	0.73	94.95	0.991
743.23	33.2	0.65	106.64	0.992
836.62	33.5	0.66	105.02	0.987

Effects of initial concentration of the dye

The adsorption of methylene blue in Barro Branco for increasing initial concentrations of the solutions is shown in Fig. 4. The plot of $\ln(q_e - q)$ versus t, Eq. (5), for the examples from Fig. 4, with the rate constants k_{ad}, is shown in Fig. 5. All the values of the constants k_{ad} obtained according to the example in Fig. 5, Eq. (5), are presented in Table 3. As can be seen in Fig. 5, with the increase of the initial sorbate concentration above 300 mg dm^{-3}, two distinct stages appear, and the first one has a greater k_{ad} constant. The first stage tends to disappear with the decrease of the initial concentration of methylene blue, suggesting a change in the adsorption mechanism directly related to the initial sorbate concentration.

The surface area value of the adsorbent was 9.0 m^2 g^{-1}; this is a low value, with low porosity when compared with activated carbon [10]. In this way, adsorption through the pores might not be significant. Therefore, it is possible that with concentrations lower than 300 mg dm^{-3}, in which there is only one stage in the kinetics of adsorption, the process occurs only on the available surface sites covering most of the surface. In higher concentrations, in order to adsorb, the adsorbate molecules need to reorganize, owing to the availability of active surface sites, decreasing the adsorption rate and triggering the second stage of the adsorption kinetics. These results corroborate those reported by Sarkar and Chattoraj [12] in their study on the adsorption of proteins in silica. Another suggestion for the change in the mechanism can be observed in Fig. 6.

Figure 6 shows the effect of the initial concentration of methylene blue on the adsorption rate constant k_{ad}. The values of the rate constants decrease showing three distinct slopes. The first one refers to lower concentrations, below 300 mg dm^{-3}. The second one is between

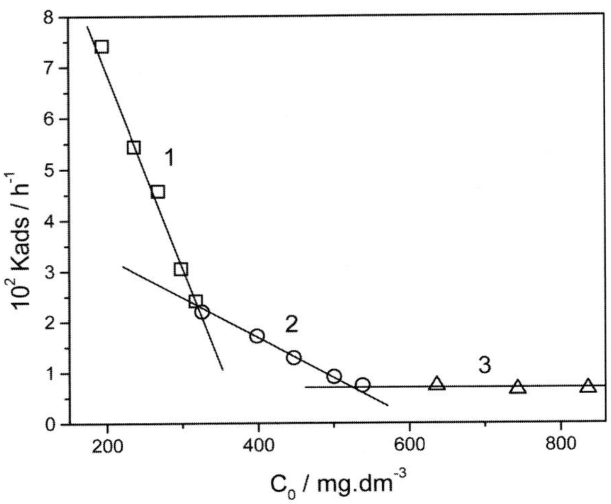

Fig. 6 Variation in the pseudo-first-order rate constant with the initial concentration of methylene blue for adsorption on Barro Branco at 45 ° C using a particle size of 210–297 μm, stirring at 250 rpm, a pH range of 5.5–7.0, and 1.0 g Barro Branco

300 and 600 mg dm^{-3}. Above this concentration the adsorption rate remains constant; this was expected, since this concentration is the same as the maximum in the adsorption isotherm, Table 3 (600 mg dm^{-3}, $q = 30.9$).

In the adsorption processes, in solid–liquid systems, the main stages are [13] (1) transfer of the adsorbate from the solution to the surface, (2) transfer of the adsorbate from the surface film, by diffusion, and (3) diffusion of the adsorbate from the surface pores onto the active adsorption sites. The most relevant factor in the control of the adsorption rate in solid–liquid systems is the diffusion of the adsorbate onto the surface. The fact that the adsorption rate changes linearly with the concentration suggests that adsorption is controlled by diffusion through the surface films. Figure 6 shows a linear decrease in the adsorption rate with the increase of the initial concentration. These results indicate that the adsorption rate of Barro Branco–methylene blue is controlled by the adsorbate diffusion onto the surface film.

The first slope, shown in Fig. 6, corresponds to one single kinetics stage in Fig. 5. The second slope corresponds to the rate constants of the second stage in Fig. 5. In the first stage in the kinetics of adsorption, the process occurs only on the available surface sites covering most of the surface. In higher concentrations, in order to adsorb, the adsorbate molecules need to reorganize, owing to the availability of active surface sites, decreasing the adsorption rate and triggering the second stage of adsorption kinetics. In high concentrations, the adsorption rate no longer depends on the initial concentration and becomes constant. This stage begins at the point at which the adsorbate surface reaches the maximum adsorption capacity, Fig. 3.

Conclusion

This study showed that Barro Branco is an effective adsorbent for the removal of methylene blue. This adsorbent is selective, allows the regeneration and concentration of the adsorbate, and, moreover, it may be used as a model for adsorption studies. The maximum amount adsorbed is significant, 33.5 mg g^{-1}, although the surface area is low when compared with that of activated carbon. Adsorption follows Langmuir, Freundlich, and pseudo-first-order models. In low concentrations, close to 300 mg dm^{-3}, the adsorption kinetics shows only one stage. With concentrations higher than 300 mg dm^{-3}, two adsorption rate stages were found. The adsorption rate depends on the initial concentration of the sorbate until the surface of the adsorbent reaches the maximum adsorption capacity, close to 600 mg dm^{-3}. After that, the adsorption rate no longer depends on the initial concentration and becomes constant.

Acknowledgements Carlos Alberto Policiano Almeida thanks Universidade Estadual do Centro–Oeste for the leave of absence during the PhD Course. He also thanks CAPES (PICDT) for a graduate fellowship.

References

1. Nigam P, Armour G, Banat IM, Singh D, Marchant R, McHale AP, McMullan G (2000) Bioresour Technol 72:219–226
2. Robinson T, Chandran B, Naidu, GS, Nigam P (2002) Bioresour Technol 85:43–49
3. Meshko V, Markovska L, Mincheva M, Rodrigues AE (2001) Water Res 35:3357–3366
4. Sivaraj R, Namasivayam C, Kadirvelu K (2001) Waste Manage 21:105–110
5. Penha FG, Spier VC, Debacher NA (2001) Quim Nova 5:612–615
6. Sun Q, Yang L (2003) Water Res 37:1535–1544
7. Choy KKH, McKay G, Porter JF (1999) Resour Conserv Recycl 27:57–71
8. Malik PK (2003) Dyes Pigm 56:239–249
9. Harris RG, Wells JD, Johnson BB (2000) Colloids Surf A 180:131–140
10. Pelekani C, Snoeyink VL (2000) Carbon 38:1423–1436
11. Tamai H, Yoschida T, Sasaki M, Yasuda H (1999) Carbon 37:983–989
12. Sarkar D, Chattoraj DK (1993) J Colloid Interface Sci 157:219–226
13. Slejko FL (1985) Adsorption technology, adsorption step by step approach to process evaluation and application. Dekker, New York

Progr Colloid Polym Sci (2004) 128: 283–287
DOI 10.1007/b97114

Márcia dos Anjos de Sousa
Geiza Esperandio de Oliveira
Elizabete Fernandes Lucas
Gaspar González

The onset of precipitation of asphaltenes in solvents of different solubility parameters

M. dos Anjos de Sousa
G. Esperandio de Oliveira
E. Fernandes Lucas
Institute of Macromolecules, Universidade Federal do Rio de Janeiro, Rio de Janeiro, RJ, Brazil

G. González (✉)
CENPES, PETROBRAS,
Rio de Janeiro, RJ, Brazil
e-mail: gaspar@cenpes.petrobras.com.br

Abstract The values of the solubility parameter prevailing in the medium at the onset of precipitation were estimated for asphaltenes dissolved in different solvents or for solvent mixtures presenting solubility parameters in the range 18.2–16.5 $(MPa)^{1/2}$. In all cases the onset of precipitation was observed at a particular concentration of the nonsolvent and this value was nearly independent of the origin and concentration of the asphaltenes. These results indicate that the precipitation effectively occurs at a particular value of the solubility parameter of the solvent medium. When ethyl alcohol was used to obtain the asphaltene separation, the solubility parameter calculated at the onset was 20.4 $(MPa)^{1/2}$; this value may be considered to represent the upper solubility limit for the asphaltene fraction.

Keywords Asphaltenes · Onset of precipitation · Petroleum fractions Crude oil · Solubility parameter

Introduction

Crude oil asphaltenes have been under investigation for more than half a century [1, 2]; nonetheless the chemical structure and the physicochemical properties of this crude oil fraction are still not well understood. There are two physical models to represent the structure of asphaltenes. In one of these, asphaltenes are considered to be intrinsically insoluble compounds that are kept dispersed by the peptizing action of the resin fraction [3]. Alternatively, asphaltenes are considered to be macromolecules dissolved in the oil as a thermodynamically stable solution [4]. Within these models, asphaltene precipitation by the addition of *n*-heptane is interpreted as the result of resin desorption from the particle surface or as a consequence of a reduction of the oil solubility parameter to a critical value. For crude oil it is not easy to decide which of the two alternatives effectively describes the asphaltene stability and precipitation because, although it is possible to determine the resin-to-asphaltene ratio, the overall solubility parameter of the solvent medium is unknown. In this work, solutions prepared using asphaltenes from different origins were used as a model system to assess the effect on the onset of precipitation with the intention to correlate this process with the solubility parameter prevailing in the solution at the flocculation point.

Experimental

Asphaltenes were separated from two samples of crude oil – Campos Basin, Brazil – (asphaltene A, extracted from an oil sample presenting an API gravity [5] of 19.2°, and 2.6% asphaltenes, and asphaltene B from an oil sample of API gravity 20.3° and 7.8% asphaltenes) and from an asphaltic residue obtained from a local refinery (asphaltene C). For the oil samples, the separation procedure was the IFP 143 standard procedure [6]. In the case of the asphaltic residue, the solid sample was extracted with *n*-heptane. Subsequently, the heptane insoluble fraction was dissolved in toluene, filtered and dried to constant weight in a rotary evaporators. All the asphaltene fractions were characterized by Fourier transform IR spectroscopy (PerkinElmer 1720X) by casting using toluene as the solvent. The solvents were obtained from VETEC Química Fina, Rio de Janeiro, Brazil, and were 99% pure.

Four solvent mixtures (SM1–SM4) were prepared. The composition and solubility parameters (δ) of these mixtures are presented in Table 1. In all cases the solubility parameters of the mixtures were calculated according to the following equation [7]:

$$\delta = \sum \delta_i \phi_i,$$

where δ is the solubility parameter of the solvents mixture, δ_i is the solubility parameter of each component and ϕ_i is the volume fraction of each component, assuming that the molar volumes of the various solvents are equal and that there are not specific interactions among them.

The onset of precipitation was identified as the dilution value, expressed as milliliters of *n*-heptane per milliliter of asphaltene solution (mLC7/mLAS) at the minimum in the absorbance versus dilution curve. The onset of precipitation of the asphaltene solution corresponds to the minimum in the absorbance. Such a minimum is preceded by an absorbance decrease due to the dilution by the solvent. Besides the minimum, almost all the curves show maxima or an increase in absorbance which may be due to an increase in the asphaltene particle size or to an increase in the number of asphaltene particles. The absorbance measurements were carried out at 850 nm using a Shimadzu 3101PC UV–vis spectrophotometer with 0.1-cm optical path quartz cells. For a few samples the onsets were also determined using a near-IR spectrophotometer (Quantum 1800, L.T. Industries). In this case, the measurements were carried out at 1,600 nm using an external optical fiber probe.

Results

The Fourier transform IR spectra for the three asphaltene samples were very similar. The same groups were identified by bands observed at the following wave numbers (reciprocal centimeters): 2,930 (aromatic ring CH); 2,923 and 2,853 (aliphatic CH, CH_2 and CH_3); 1,605 (C=C and C=O); 1,457 and 1,376 (C–CH_3 asymmetric and symmetric deformations); 1,032 ($C_2S=O$; sulfoxide functional groups [8]); 868, 812 and 752 (aromatic CH). Only minor differences in the intensities, revealing differences in elemental composition, were observed.

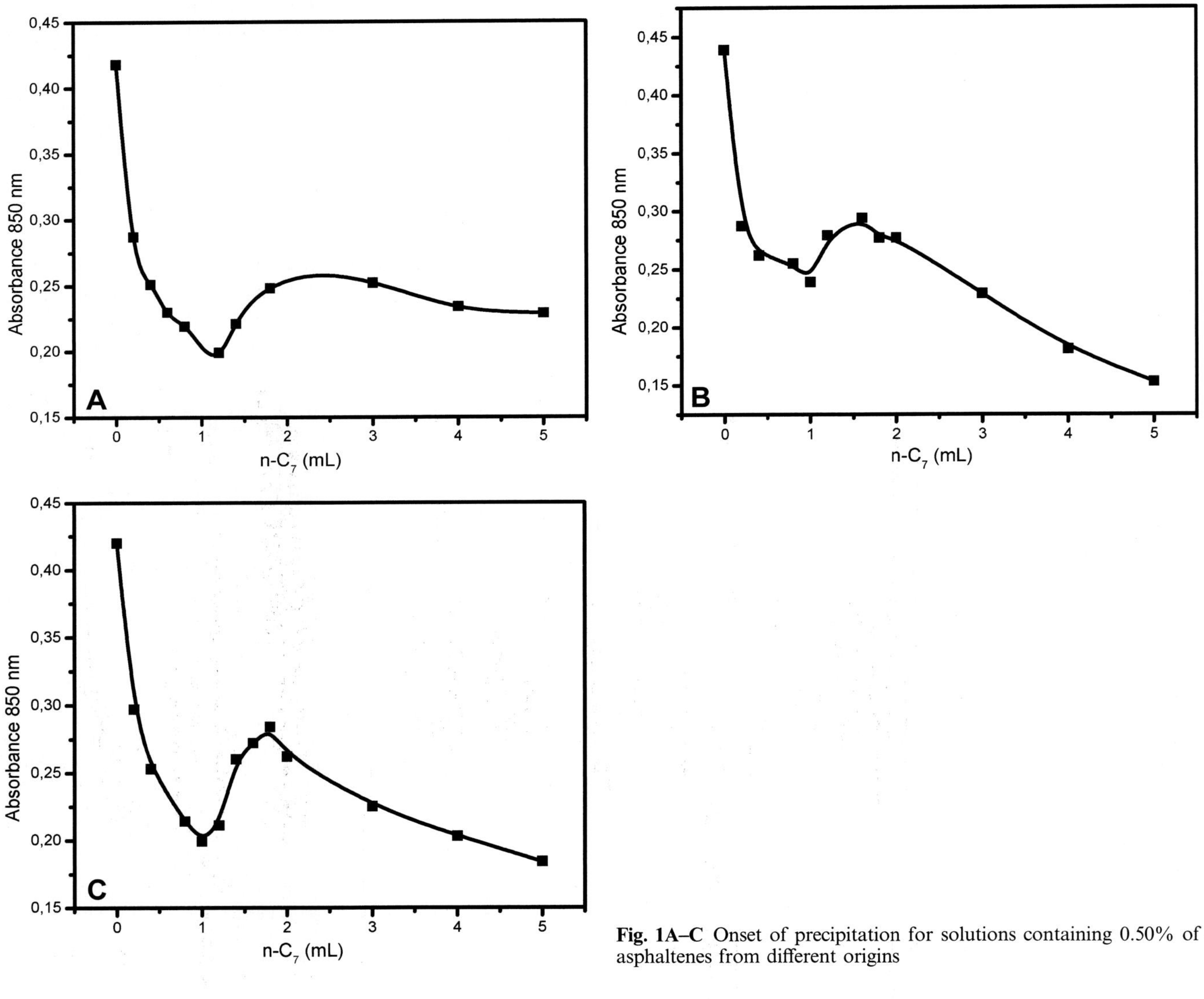

Fig. 1A–C Onset of precipitation for solutions containing 0.50% of asphaltenes from different origins

The influence of the origin of the asphaltenes on the onset of precipitation was evaluated, using solutions containing 0.5% asphaltene in toluene. The graphs for the determination of the onset are shown in Fig. 1. For asphaltene A, the onset of precipitation was obtained at 1.2 mLC7/mLAS solution (approximately 55% of heptane). For asphaltenes B and C, the onset was approximately 1.0 mLC7/mLAS (about 50%). The values are very close and seem to indicate that the origin of the asphaltenes has a rather limited influence on the onset of precipitation. The solubility parameters calculated at the onset of precipitation (i.e. including the contribution of the *n*-heptane) are 16.6 $(MPa)^{1/2}$ for asphaltene A and 16.7 $(MPa)^{1/2}$ for asphaltenes B and C. These values show good agreement with results obtained previously for asphaltenes isolated from another type of crude oil [9].

The effect of the asphaltene concentration on the precipitation onset was studied for solutions containing 0.25, 0.5 and 1.0% of asphaltene B. The graphs of the absorbance as a function of dilution with heptane are shown in Fig. 2. For the solutions with 0.25 and 0.5% of asphaltene, the onsets were around 1.0 mLC7/mLAS. For the solution containing 1.0% of asphaltene, the onset was around 0.6 mLC7/mLAS. These results may indicate a reduction of the onset at higher asphaltene concentrations; however, it must also be considered that, under these conditions, the number of particles necessary to produce a detectable scattering may be attained at lower heptane concentrations. The effect of concentration has been studied previously by other authors and the results are still not clear [10].

In order to assess the effect of the solvent medium, solutions containing 0.25% of asphaltenes were prepared using the solvent mixtures described in Table 1.The graphs of absorbance as a function of the dilution of these solutions with heptane, obtained using the near-IR spectrometer, are shown in Fig. 3. The onset of asphal-

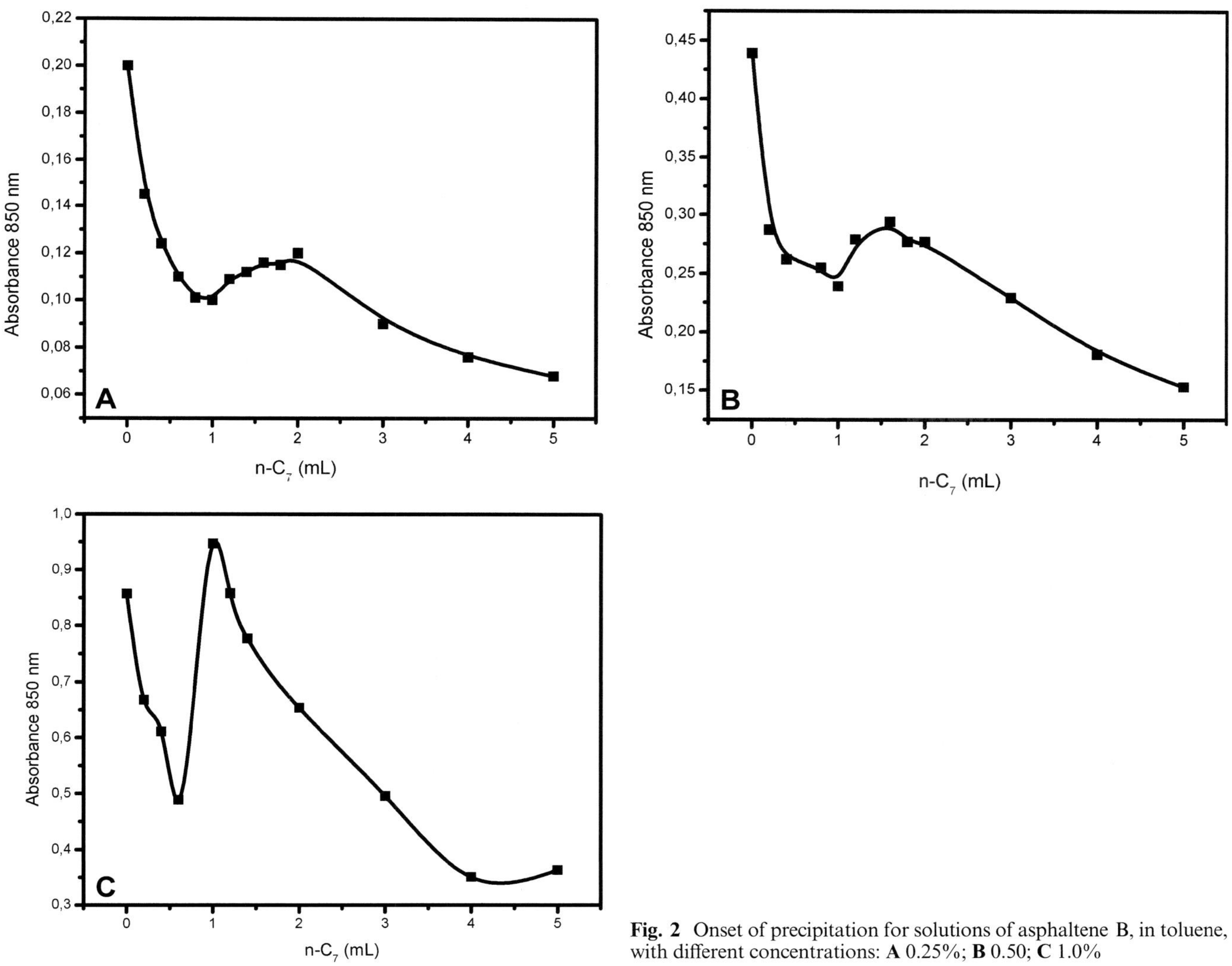

Fig. 2 Onset of precipitation for solutions of asphaltene B, in toluene, with different concentrations: **A** 0.25%; **B** 0.50; **C** 1.0%

Table 1 Composition and solubility parameters of the various solvents mixtures

Solvent	Decane	Toluene	Cyclohexane	Benzene	Dodecane	δ [(MPa)$^{1/2}$]
SM1	50	30	20	–	–	16.5
SM2	30	30	20	20	–	17.1
SM3	–	30	20	–	50	16.7
SM4	–	30	20	20	30	17.2

tene precipitation is evident only for the SM2 and SM4 solvent mixtures, at dilutions of 0.29 and 0.34 mLC7/mLAS, respectively. The solubility parameter values calculated at the onset were the same for these two mixtures, $\delta = 16.6$ (MPa)$^{1/2}$. The results for the SM1 and SM3 solvent mixtures show no evidence of asphaltene precipitation. Considering that the solubility parameters in these two solvent mixtures, 16.5 and 16.7 (MPa)$^{1/2}$, respectively, are very close to 16.6 (MPa)$^{1/2}$, it may be inferred that particles may be already present in the solvent medium prior to the addition of heptane. Therefore, the onset of precipitation would be in the range 16.5–16.7 (MPa)$^{1/2}$, independent of the composition of the solvent medium.

The onset of precipitation for solutions containing 0.2 and 0.5% of asphaltene A in toluene was determined using ethyl alcohol as a nonsolvent. The results are shown in Fig. 4. The onset of these solutions was about 0.4 mL ethyl alcohol per milliliter of asphaltene toluene solution, independent of the concentration. The solubility parameter calculated at the onset, in this case, was 20.4 (MPa)$^{1/2}$. Disregarding any possible specific interaction between the crude oil polar fractions and the alcohol, this value may be considered to represent the upper solubility limit for the asphaltenes. Similarly, the values of 16.5 and 16.7 (MPa)$^{1/2}$ obtained for *n*-heptane

Fig. 3 Onset of precipitation for solutions containing 0.25% of asphaltene B in different solvents mixtures: **A** SM1; **B** SM2; **C** SM3; **D** SM4

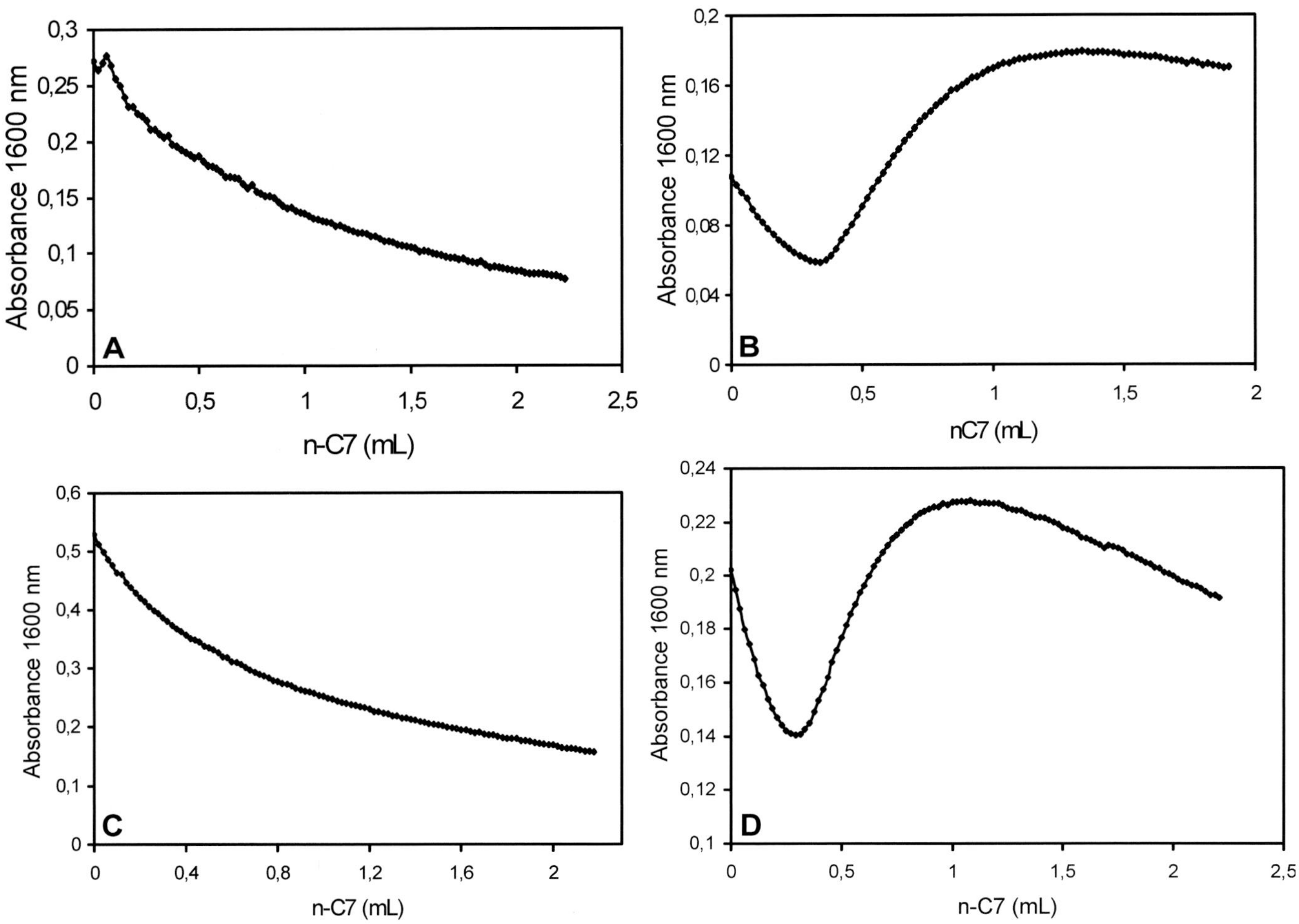

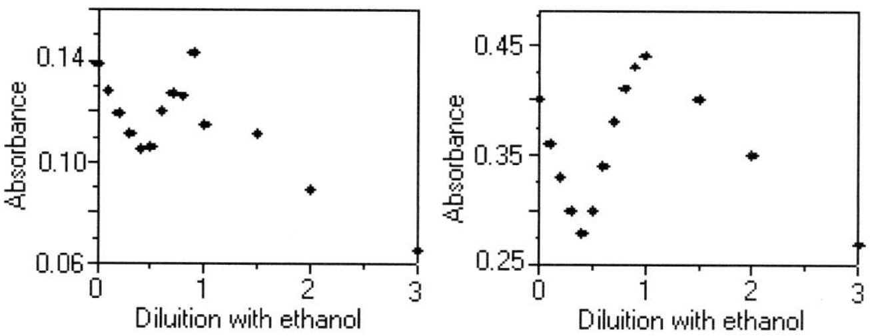

Fig. 4 Onset of precipitation for solutions containing 0.2% and 0.5% of asphaltene A in toluene, using ethyl alcohol as a nonsolvent

may be considered the lower solubility limit for this fraction. The reported solubility parameters [7] for asphaltenes are around 19 $(MPa)^{1/2}$, which is intermediate between these two values, showing good correlation with our results.

Conclusions

Some preliminary conclusions may be gathered from these preliminary results. The origin of the asphaltenes used in this work does not have an important influence on their onset of precipitation. The stability of this fraction seems to be determined mainly by the properties of the solvent medium. In bad solvents, presenting a solubility parameter close to 16.7 $(MPa)^{1/2}$, the asphaltenes tend to separate, showing a rather low onset of precipitation. Solutions presenting a solubility parameter close to 19 $(MPa)^{1/2}$ are stabler and present higher onsets. If these considerations were extrapolated to crude oil, it could be concluded that the stability of petroleum in relation to asphaltene deposition problems is determined by the solubility properties of the oil and not by any particular property of the asphaltene fraction.

References

1. Preckshot GW et al (1943) Trans AIME 151:188
2. Katz DL, Beu KE (1945) Ind Eng. Chem 37:195
3. Leontaritis KJ, Mansoori G (1987) Annu Proc Int Symp Oilfield Chem SPE 16258:149
4. Hirschberg A, De Jong LNJ, Schiper BA, Meijers JG (1984) SPE J June 283
5. Kinghorn RRF (1993) An introduction to the physics and chemistry of petroleum. Wiley, New York
6. The Institute of Petroleum (1965) IP standard method for petroleum and its products, part I, section 2, standard 143:596. The Institute of Petroleum, London
7. Barton AFM (1990) Handbook of polymer–liquid interaction parameters and solubility parameters. CRC, Boca Raton
8. Andersen SI (1997) Petrol Sci Technol 15:185
9. Anjos M (2000) Dissertação de Mestrado. IMA-UFRJ, Rio de Janeiro, Brazil
10. Buckley JS (1999) Energy Fuels 13:328

Progr Colloid Polym Sci (2004) 128: 288–292
DOI 10.1007/b97117

Márcia C.K. Oliveira
Elizabete F. Lucas
Gaspar González
José F. Oliveira

Heavy oil fraction removal from sand using hydrotropes containing oil-in-water microemulsions

M.C.K. Oliveira · J.F. Oliveira (✉)
Universidade Federal do Rio de Janeiro, COPPE, Rio de Janeiro, RJ, Brazil
e-mail: oliveira@metalmat.ufrj.br

E.F. Lucas
Universidade Federal do Rio de Janeiro, IMA, Rio de Janeiro, RJ, Brazil

G. González
Petrobras Research Center, CENPES, Rio de Janeiro, RJ, Brazil

Abstract Oil-in-water microemulsions were prepared with a nonionic surfactant and different cosurfactants using as the oil phase a hydrocarbon mixture of linear, cyclic and aromatic hydrocarbons. This organic mixture ensures the solubilization of the crude oil heavy fraction, such as asphaltenes and resins, which are insoluble in normal aliphatic solvents. The results showed that it was possible to obtain microemulsions with the mixture of hydrocarbons used. The microemulsion region inside the pseudo-ternary phase diagram was slightly increased using different hydrotropic compounds. The results indicate that the addition of hydrotropes causes a reduction in the minimum amount of surfactant and cosurfactant necessary for the solubilization of the oil phase. When samples of sand contaminated with asphaltene residue were treated with these microemulsions, it was observed that the oil fraction was incorporated in the organic phase of the dispersion and that the contaminant was almost totally removed. These studies are being extended aiming at the development alternative techniques to remove oil from soils contaminated with organic compounds.

Keywords Microemulsion · Hydrotropes · Crude oil heavy fraction

Introduction

Soils contaminated with aromatic hydrocarbons may pose serious risks to human health and the environment. Owing to the hydrophobic nature of these compounds, conventional remediation techniques are considered as being of limited practical value. Therefore, significant efforts are being devoted to the development of more efficacious approaches for the remediation of these contaminated sites [1]. Soil washing with surfactant solutions is an efficient treatment technology for the remediation of contaminated soils. Although this technique has shown significant potential for application in environmental remediation practices, a major disadvantage is the potential formation of unstable macroemulsions in the porous media.

The use of microemulsions seems to be attractive as an alternative process to remove oil from contaminated soils owing to their high stability and solubilization capacity, the small size distribution of the disperse phase and their low viscosity. Microemulsions are formed by surfactants, cosurfactants, water and oils. They have the ability to concentrate both oil-soluble and water-soluble materials within the same isotropic medium owing to their ultralow interfacial tension. There are two main mechanisms for oil recovery: mobilization and enhanced solubilization. Mobilization is carried out by decreasing the interfacial tension, while solubilization is accomplished by incorporating the contaminant into the micelles [2].

The application of microemulsions as cleaning systems has been studied using model hydrocarbons to represent typical oils [3, 4, 5, 6] and gasoline [7].

However, to our knowledge, the use of microemulsions to remove crude oil heavy fraction, such as asphaltenes and resins, from soil has not been investigated so far. Asphaltenes and resins consist of condensed aromatic rings, bound by alkyl chains and various functional groups. The composition of asphaltenes depends upon the origin of the crude oil and its extraction method.

This study aims to contribute to the development of new microemulsion systems containing hydrotrope compounds that can be used as an alternative technique to remove heavy oil from contaminated soils. Hydrotropes are highly aqueous soluble organic molecules, which show substantial ability to solubilize lipophilic compounds.

Experimental

This work was carried out in two steps, namely, the study of the phase behavior of microemulsions and the removal of the crude oil heavy fraction. First, a surfactant and cosurfactant which when added together with an oil phase and an aqueous phase form a microemulsion were selected. Then, the effects of three cosurfactants and the cosurfactant-to-surfactant (C/S) ratio were investigated. In order to increase the microemulsion region in the pseudo-ternary diagram, different hydrotropic compounds were incorporated into the system.

In this study the removal of the heavy oil fraction from soil was investigated including the effects of the sand-to-microemulsion ratio, the extraction velocity and the microemulsion solubility capacity.

The following components were used in the microemulsion systems: surfactant polyethoxylene-10 lauryl ether ($C_{12}E_{10}$) from Oxiteno Co., Brazil, cosurfactants isopropyl, butyl and isoamyl alcohols P.A. from Vetec Co., Brazil, hydrotropes sodium *p*-toluene sulfonate and sodium cumene sulfonate from BASF Co., and triethylene glycol ether from Oxiteno Co. The oil phase, named DTC, was a mixture of decane, toluene and cyclohexane, all obtained from Sigma Co. This mixture, which contains linear, cyclic and aromatic hydrocarbons, was selected as representative of the hydrocarbons normally found in gasoline and crude oils.

The pseudo-ternary phase diagrams were prepared by titration of a previously weighed volume of oil/surfactant/cosurfactant mixture with an aqueous solution containing a proper amount of hydrotrope. The experiments were carried out under constant stirring at 296 K, until the solution became cloudy. The mixture obtained was weighed at this point, to determine the amount of aqueous solution added to the system. In order to obtain the pseudo-ternary phase diagram, it was considered that the C/S ratio remained constant in the superior vertex of the triangular diagram. The aim was to obtain the maximum monophasic region, especially at the aqueous end.

The model soil sample used was quartz sand contaminated with asphaltene residue (rasf, 2900 mg/kg), whose composition was 20% asphaltene, 21% resins, 56% aromatic and 2.9% aliphatic. To assess the oil from the soil-contaminated sample, 2.5 g soil were contacted with the microemulsion in a rotating shaker (120 rpm). After the equilibration, the samples were filtered to separate the fluid phase from the soil sample. Desorption of soil contaminant was investigated at four soil-to-microemulsion ratios (1/1, 1/2, 1/4 and 1/10) and different extraction times (30, 60, 120, 240, 360 min). The amount of residual contaminant adsorbed on sand was calculated from mass balance after determining the concentration by a UV/vis absorption technique. When the ratio and the extraction time were fixed, the effect of microemulsion composition was analyzed.

Results and discussion

Phase behavior of the microemulsions

The effects of the C/S ratio, the nature of the cosurfactant, the addition of a hydrotrope and the oil phase were studied by controlling the phase behavior of the microemulsion. Three C/S ratios (0.5, 1.0 and 1.5), using butyl alcohol as cosurfactant, were investigated as shown in Fig. 1a. Since the efficiency of the system to solubilize (DTC) decreased with increasing C/S ratio, its value was normalized at 1.0 in the subsequent experiments.

The effects of isopropyl, butyl and isoamyl alcohols as cosurfactants were compared in the system prepared with $C_{12}E_{10}$/water/DTC (Fig. 1a–c). The analysis of the results showed that butyl alcohol promoted a larger monophasic region, using $C/S = 1$, and it was chosen for the next experiments.

The effect of different hydrotrope solutions is shown in Fig. 2. The addition of the hydrotropes EBTEG caused a substantial increase in the monophasic region. The minimum water content to form the monophasic region was increased slightly, the minimum amount of surfactant to form the region was decreased and the oil solubility became greater, especially when triethylene glycol ether was used.

The effect of the nature of the oil was investigated by comparing light hydrocarbons (the mixture DTC) with the crude oil heavy fraction. These microemulsion systems were very effective in solubilizing low molar mass aromatic molecules and polynuclear aromatic hydrocarbons, i.e., the incorporation of the asphaltene residue did not result in important differences in the phase diagrams. This means that the asphaltene residue can be solubilized by this microemulsion.

Crude oil heavy fraction removal

The solubility of asphaltic residue (rasf) present in the contaminated sand was investigated using different extraction media, such as water, the surfactant solution and the same solution with the microemulsion precursor. The negative results showed the high hydrophobic nature of this material and underline the advantages of microemulsions for contaminant removal. The small amounts of rasf solubilized in some solutions at 296 K are presented in Table 1.

However, it can be observed in Fig. 3 that all the sand-to-microemulsion ratios investigated were efficient for the removal of the asphaltene residue. The contaminant extraction was also determined as a function of time. It can be noticed in Fig. 4 that the contaminant desorption and the subsequent solubilization occur within a short period of time owing to the change of the wetting behavior of the soil caused by the low

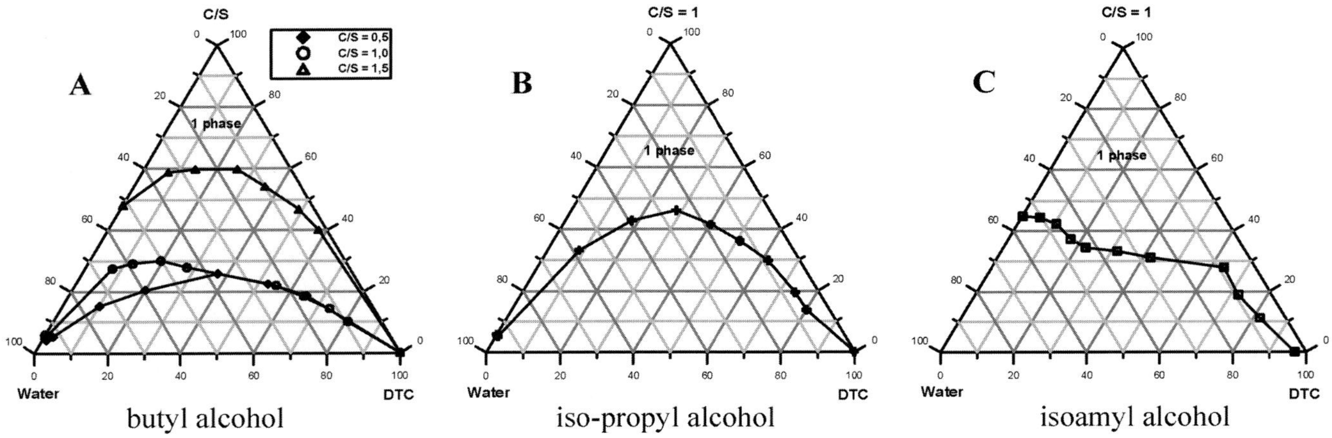

Fig. 1 Effect of cosurfactant-to-surfactant (*C*/*S*) ratio and cosurfactant on the pseudo-ternary phase diagram, using $C_{12}E_{10}$ as surfactant

interfacial tension of the microemulsions. After 1 h more than 90% of the contaminant was removed. Increasing the extraction time above 4 h did not result in any significant improvement in the removal. A ratio of 1:2 and an extraction time of 4 h were selected to analyze the effect of microemulsion composition.

The solubilization capacity of the microemulsions was studied (Table 2) for three different compositions, A (10% DTC, 15% $C_{12}E_{10}$, 15% butyl alcohol and 60% aqueous phase), B (30% DTC, 20% $C_{12}E_{10}$, 20% butyl alcohol and 30% aqueous phase) and C (60% DTC, 15% $C_{12}E_{10}$, 15% butyl alcohol and 10% aqueous phase), where the aqueous phases are water and different aqueous solutions of hydrotropes. The removal of the asphaltene residue by these microemulsions is shown in Table 2. It can be observed that more than 90% of asphaltene was removed in some cases. The extraction experiments showed the enhancing effect of the microemulsion as an extracting solvent for hydrophobic contaminants in the soil. Adding an oil phase (DTC) to the micellar surfactant solution enhanced the solubility overproportionally. The efficiency is higher when the oil phase increases. It is important to note that the microemulsion is achieved in composition A only with the presence of a hydrotrope although the extractions without the hydrostrope was higher (90%).

After washing the sand, the crude oil heavy fraction was incorporated to the organic phase (DTC) of the dispersion when compositions B and C were used, while composition A promoted free oil in equilibrium with the microemulsion. This result demonstrates that the quantity of the oil phase must be sufficiently high to form a one-phase solution.

These results show that the presence of DTC in the microemulsion was a very important variable for the washing process. Additional experiments were then carried out with natural solvents as an alternative microemulsion oil phase. The action of the DTC mixture was compared with the actions of terpene, orange oil and pine oil, all of them considered not to be harmful to the environment. These natural solvents, with the exception

Fig. 2 Effect of the hydrotropes triethylene glycol ether (*EBTEG*), sodium *p*-toluene sulfonate (*NaPTs*) and sodium cumene sulfonate (*NaCs*) on the pseudo-ternary diagram of $C_{12}E_{10}$-butyl alcohol/hydrotrope/DTC

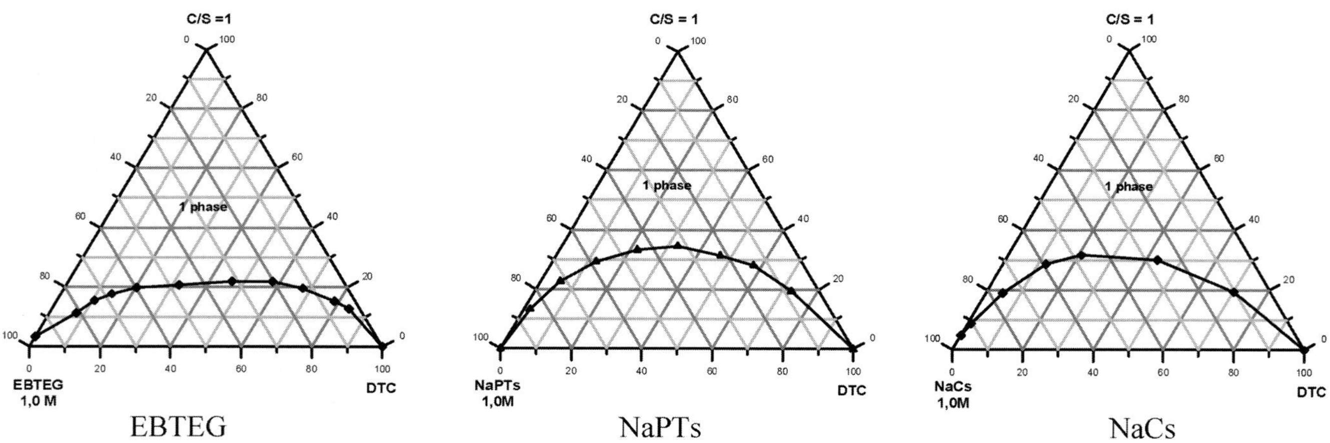

of pine oil, showed a solubilization ability very close to that of as DTC as can be seen in Table 3.

Conclusions

The pseudo-tenary diagrams showed that the addition of hydrotropes to microemulsions may cause a reduction of

Table 1 Solubilization of asphaltene residue, in polyethoxylene-10 lauryl ether (*$C_{12}E_{10}$*) solution, and after the addition of triethylene glycol ether (*EBTEG*) with butyl alcohol

Solution (% w/w)	Solubilization (%)
Water	0.0
10% $C_{12}E_{10}$ in water	0.24
10% $C_{12}E_{10}$ + 10% butyl alcohol + 80% 1 M EBTEG	1.18

Table 2 Extraction efficiency of asphaltene residue by hydrotrope-containing microemulsions. Sodium *p*-toluene sulfonate (*NaPTs*) and sodium cumene sulfonate (*NaCs*), quartz sand contaminated with 2,900 mg/kg of asphaltene residue (*rasf*)

Composition	Aqueous phase	Residual rasf (mg/L)	Removal efficiency (%)
A	Water	230.4	92.0
	NaPTs	241.2	91.6
	NaCs	542.0	81.2
	EBTEG	1,617.9	43.8
B	Water	208.7	92.7
	NaPTs	330.6	88.5
	NaCs	525.8	81.7
	EBTEG	582.7	79.7
C	Water	5.4	99.8
	NaPTs	13.6	99.4
	NaCs	16.3	99.5
	EBTEG	16.3	99.4

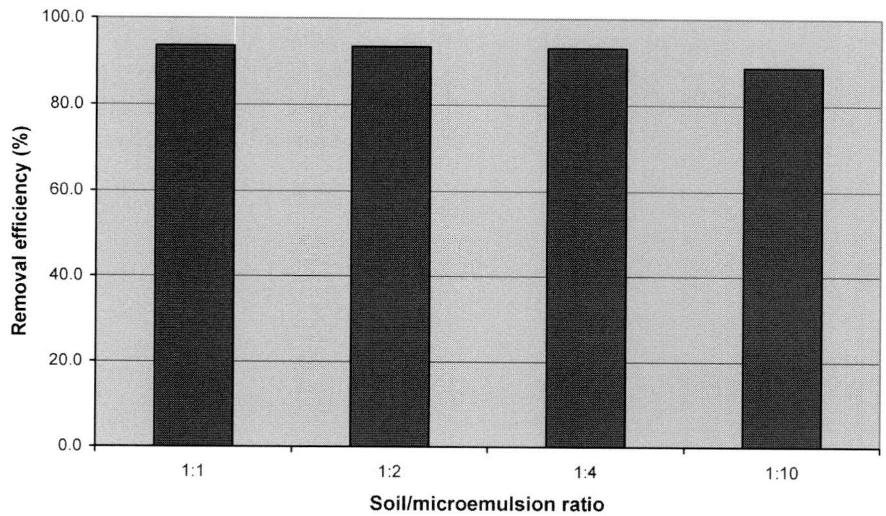

Fig. 3 Effect of soil-to-microemulsion ratio on the efficiency of contaminant removal

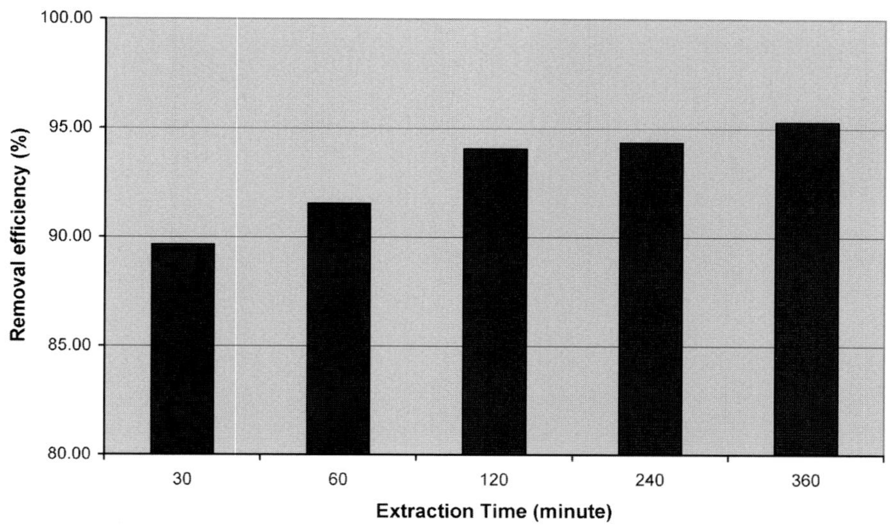

Fig. 4 Effect of extraction time on the efficiency of contaminant removal for a soil-to-microemulsion ratio of 1:2

Table 3 Effect of the microemulsion oil phase on the efficiency of contaminant removal

Composition	Oil phase	Residual rasf (mg/L)	Removal efficiency (%)
B	DTC mixture	208.7	92.7
	Terpene	208.7	92.8
	Orange oil	200.6	93.0
	Pine oil	2,848.3	1.1

the surfactant and cosurfactant minimum concentration necessary for the oil phase solubilization. It was also verified that the microemulsious containing DTC have a high capacity for removing crude oil heavy fractions from contaminated sands. The use of natural solvents as a microemulsion oil phase also showed high removal efficiency.

These results indicate the possibility of using microemulsions in the remediation of this type of contaminated soil and the selection of the microemulsion composition should obviously take into account the amount of contaminant present in the soil.

Acknowledgement The authors wish to acknowledge Petrobras for support of this research.

References

1. Bonkhoff K, Schwuger MJ, Subklew G (1997) Surfactant Sci Ser 66:355–374
2. Lowe DF, Oubre CL, Ward CH (1999) Surfactants and cosolvents for NAPL remediation: a technology practices manual. Lewis, FL
3. Desnoyers J, Quirion F, Hétu D, Perron G (1983) Can J Chem Eng 61:672–679
4. Dierkes F, Haegel H, Schwuger MJ (1998) Colloids Surf A 141:217–225
5. Sabatini DA, Knox RC, Harwell JH, Wu B (2000) J Contam Hydrol 45:99–121
6. Ouyang Y, Cho JS, Mansell RS (2002) Water Res 36:33–40
7. Ouyang Y, Mansell RS, Rhue RD (1996) J Hazard Mater 46:23–35

Progr Colloid Polym Sci (2004) 128: 293–297
DOI 10.1007/b97119

Roberto C.G. Oliveira
José F. Oliveira
Brij M. Moudgil

The effect of hydrophobic fine particles on the foam flushing remediation process

R.C.G. Oliveira
Petróleo Brasileiro S/A,
CENPES/Petrobras,
Rio de Janeiro, RJ, Brazil

J.F. Oliveira (✉)
Federal University of Rio de Janeiro,
COPPE, Centro de Tecnologia,
Bloco G, sala 106, Cidade
Universitária, Ilha do Fundão,
21949-900 Rio de Janeiro,
RJ, Brazil
e-mail: oliveira@metalmat.ufrj.br
Tel.: +55-21-25627030
Fax: +55-21-2290-6626

B.M. Moudgil
Engineering Research Center for Particle
Science & Technology,
Gainesville, FL, USA

Abstract Underground contamination with non-aqueous-phase liquids such as petroleum hydrocarbons is a major environmental problem. The organic phase immobilized by capillary trapping behaves as a continuing source of dissolved organic pollutants for the passing groundwater. Removal of these contaminants from the subsurface environment is essential to prevent the spreading of contamination. In situ foam flushing has proved to be an efficient process for non-aqueous-phase liquid soil remediation. Nevertheless, its efficiency can be affected by the surfactant's adsorption on the soil matrix and also by the presence of fine particles. These particles may interfere with the foam breakage process and also with the surfactant solution foaming ability. The present work shows the effect of hydrophobic fine particles on surfactant foaming ability and foam stability. The impact of the adsorption of surfactant molecules on the surface of the hydrophobic particles was also determined. The results showed that the reduction in the foam stability, due to the presence of hydrophobic fine particles, was much higher than the reduction in the surfactant foaming ability. It was observed that the rate of film breakage due to the presence of hydrophobic micro-sized particles is to some extent higher than that of the hydrophobic nano-sized particles. It was also verified that the adsorption of surfactant molecules on the surface of the hydrophobic particles reduces their ability in breaking foams.

Keywords Non-aqueous-phase liquid remediation · Foam flushing · Hydrophobic particle · Surfactant adsorption

Introduction

Around 50% of the drinkable water and 80% of the world's irrigation water reserves come from aquifers located in the subsurface environment [1]. Moreover, the solubility of non-aqueous-phase liquids in water commonly exceeds the drinking water standards by 3–4 orders of magnitude. These aspects clarify the importance of preventing groundwater contamination.

Different processes can be used for the removal of non-aqueous-phase liquids from the soil matrix. Unfortunately, the application of each process is limited to specific conditions. Bioremediation is the most widely applied process; however, it depends on the efficient delivery of oxygen, moisture and nutrients. This aspect restricts the bioremediation application to the superficial soil layers.

In situ soil flushing involves the use of an appropriate surfactant solution and can be used to mobilize, solubilize or emulsify the contaminant phase and to bring it to the surface for disposal or destruction. In soil flushing, a major concern is to restrain treatment fluids to the remediation zone [2]. The advantages of using foams as a remediation process are the better control of the injected

Table 1 Some characteristics of the Degussa hydrophobic particles

Particle	Average size (10^{-6} m)	Brunauer–Emmett-Teller surface area (m^2/g)	Contact angle (degrees) [9]
Sipernat D-11	5	85	134.8 ± 1.1
Aerosil R-972	0.016	110	137.6 ± 1.0

fluids, uniformity of contact, and their ability to contain the migration of contaminant-laden liquids. The use of foams can also be considered for delivering the necessary amendments for the bioremediation process [3].

An important parameter for the performance of foams is their stability. Foam stability refers to its ability to resist bubble breakdown. Foam stability under static conditions is quantified by the time required for the breakage of half of the initial volume of the foam.

Some of the materials present in the soil are able to adsorb surface-active molecules from the bubble's surface as the foam passes through. This leads to a reduction of the surfactant interfacial concentration and thus reduces the foam stability [4]. Some particles present in the soil can also act as antifoam agents. Most of them are hydrophobic and finely divided particles. When a hydrophobic particle comes in contact with a gas/aqueous film, it can enter the film and thus the local (Laplace) pressure around the particle increases. As a consequence, the internal aqueous phase flows away from the particle until a new equilibrium is reached or the film is broken [5, 6].

Dippenaar [7, 8] studied the effect of hydrophobic particles of different sizes on froth stability. He observed that large particles break films more quickly than smaller particles do, since a particle has to wait for a film to thin to less than its own dimensions before it can rupture the film.

Under actual field conditions the efficiency of the foam remediation process is affected by the presence of fine particles that may interfere with the foam breakage process and also with the surfactant solution foaming ability. Both parameters are crucial for the foam remediation process as they determine the type of flux and the pressure drop through the porous media.

The aim of this work was to determine the impact of hydrophobic fine particles on the surfactant foaming ability and foam stability, and how the particle's wettability is affected by the adsorption of surfactant molecules.

Experimental

Materials

The sodium dodecyl sulfate (SDS) used in this study was of 99% purity as obtained from the Aldrich Chemical Co, whereas the Sipernat D-11 and Aerosil R-972 hydrophobic particles were obtained from Degussa–Huls Co. Some data from the particles used are shown in Table 1.The SDS critical micellar concentration (cmc) was determined to be around 8.2 mM at 20 °C. The water used in this study was produced by a Millipore purification system.

Methods

Measurement of contact angles

The particles' contact angles were measured utilizing the capillary-rise technique [9]. This technique was also used to quantify the change in the particles' contact angles due to the adsorption of SDS molecules.

Particle dispersion

The hydrophobic particles and the SDS surfactant were mixed together as solid phases. Then, small volumes of water were added to the mixture until it reached the correct concentration. Finally, the solution prepared containing the hydrophobic particles was submitted to ultrasound waves for 5 min. This procedure proved to be effective in keeping the hydrophobic particles dispersed in the SDS solutions.

Apparatus

A batch column setup (Fig. 1) was built in order to quantify the effect of the hydrophobic particles on the foaming ability and foam stability of the SDS solutions. It was also used to show how the surfactant molecules adsorbed on the particles' surfaces can interfere with the foam stabilization process. These effects were studied using two different experimental procedures. The foaming chamber consists of a glass cylinder 1-m high and 5 cm in diameter.

The first procedure consists of flushing air at a controlled flow rate through a dry porous disc installed at the bottom of a glass column.

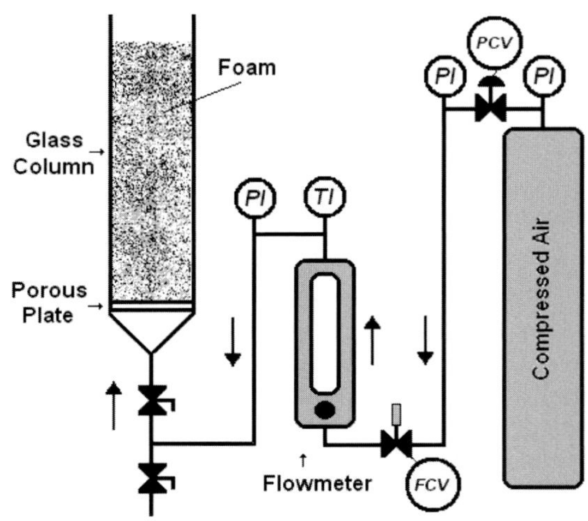

Fig. 1 The batch column setup

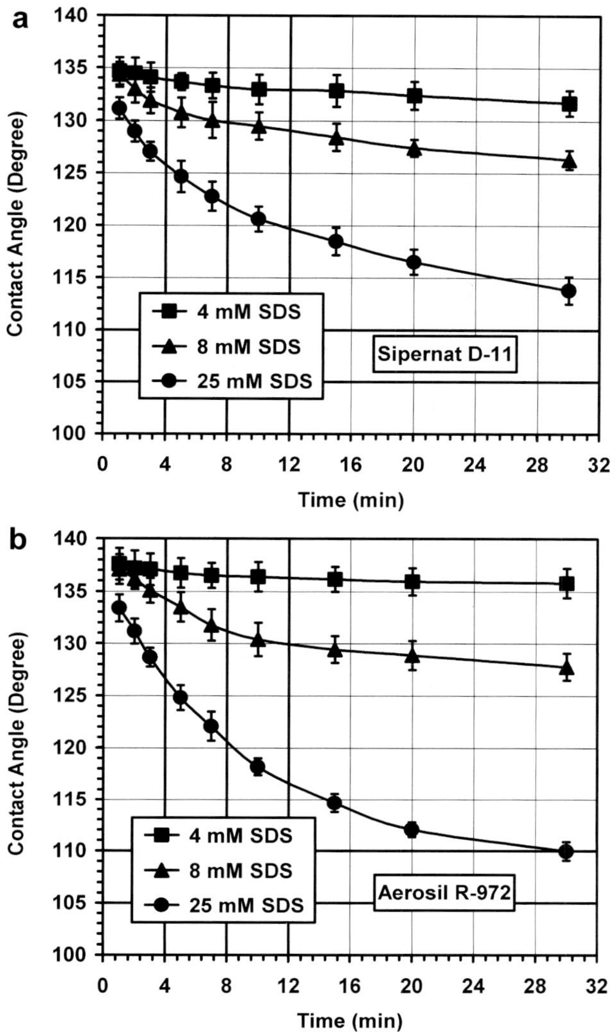

Fig. 2a,b The effect of sodium dodecyl sulfate (*SDS*) molecules adsorption on the particles' contact angles. **a** *Sipernat D-11* particles. **b** *Aerosil R-972* particles

When the process achieves equilibrium (constant pressure drop and air flow rate), 50 ml of a surfactant solution containing the dispersed particles is dropped through the top of the column and the foam starts to be generated. After 1 min, the foam generation process is interrupted and both the initial foam height (H_0) and the initial liquid level (L_0) are measured. The evolution of these levels is measured as a function of time and is used to determine the foam's half-life ($T_{1/2}$). The second procedure also consists of flushing air at a controlled flow rate through the porous disc but in this case the particles were placed into the column and the solution was poured onto them. Time control started at this point and the foam height was quantified as a function of SDS concentration.

Results and discussion

The influence of the hydrophobic particles on the foaming ability and foam stability of the SDS solutions was studied using the batch column setup described. The experiments were conducted at an air flow rate of 100 ml/min and using 500 mg/l of particles at three different SDS concentrations: 4, 8 and 25 mM.

The effect of the SDS adsorption on the particles' contact angles is shown in Fig. 2. As expected, the adsorption of the SDS species led to a reduction of the wettability of the particles. But, the important point is that the SDS adsorption in the particles' surfaces is a very slow process, especially when the SDS bulk concentration is below the cmc.

The effect of these hydrophobic particles on the initial foam height is presented in Fig. 3. The results show that the foaming ability of the SDS solutions is slightly reduced by the presence of hydrophobic particles, especially when the particles are not previously in the SDS solution. This destructive effect increases as the SDS concentration is reduced below the cmc.

The foaming process is essentially dependent on the concentration of monomers available in the surfactant solution, and the reduction in the monomer concentration tends to decrease the surfactant solution foaming ability. This effect can be observed for SDS concentrations below the cmc. However, it was not observed above the cmc as the reduction in the monomer concentration, due to the particle adsorption, was quickly compensated by the

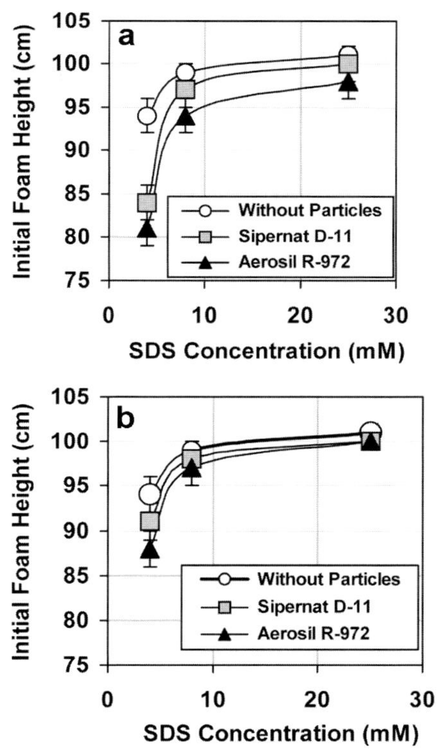

Fig. 3a,b The effect of hydrophobic particles on the foaming ability of SDS solutions. **a)** Particles not previously in contact with the SDS solution. **b)** Particles previously conditioned in the SDS solutions by ultrasound stirring

micellar relaxation mechanism [10] that tends to reestablish the monomer concentration equilibrium and to minimize the particle adsorption effect on the foaming ability.

The results presented in Fig. 4a show that the foam stability can be severely impaired by the presence of hydrophobic particles not previously dispersed in the surfactant solution. Owing to their hydrophobic behavior, Sipernat D-11 and Aerosil R-972 particles are able to anchor on the inner side of the bubble's film. This leads to a reduction in the surfactant packing and, as a consequence, increases the gas diffusion rate between bubbles. The presence of hydrophobic particles in the interfacial film also causes a localized reduction of the film thickness as the internal aqueous phase flows away

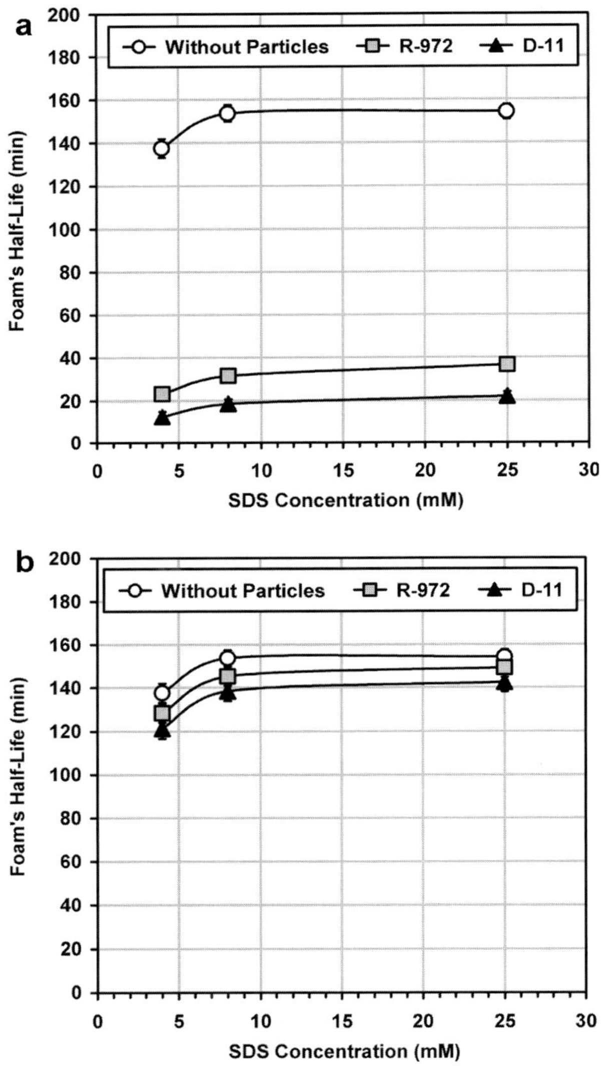

Fig. 4a,b The effect of hydrophobic particles on the foam's half-life. **a** Particles not previously in contact with the SDS solution. **b** Particles previously conditioned in the SDS solutions by ultrasound stirring

from the particles. Both effects reduce the foam stability and thus increase the foam's coalescence rate.

The adsorption of amphiphilic species reduces the hydrophobicity of the particles and thus their tendency to adsorb at the gas/liquid film. The results presented in Fig. 4b show that when the particles are previously treated with the surfactant solution their effect on the foam stability is drastically reduced.

Increasing the surfactant concentration also increases the adsorption of surfactant species on the particles' surfaces, rendering them less hydrophobic and, as mentioned earlier, reduces the particles' foam breaking ability. The results show that the surfactant adsorption effect on the particle behavior is more significant below the cmc. In this situation the concentration of SDS molecules adsorbed on the particle surface and at the bubble's gas/liquid interface is reduced.

It was also observed that the destructive effect of the Sipernat D-11 particles is to some extent stronger than that of the Aerosil R-972 particles. These results are in agreement with Dippenaar's [8] observations in which a particle has to wait for a film to thin to less than its own dimension before it can rupture the gas/liquid film. As the Sipernat D-11 particles are much bigger (5 μm) than the Aerosil R-972 particles (16 nm) the rate of film breakage of the Sipernat D-11 particles should be expected to be higher than that of the Aerosil R-972 particles.

Conclusion

The experimental results have shown that the foam stability can be severely impaired by the presence of hydrophobic fine particles. This brings another variable to the foam remediation process as the surfactant agent is generally selected on the basis of economical aspects and laboratory screen tests where the presence of hydrophobic particles is not considered.

It was observed that the rate of film breakage due to the presence of hydrophobic micro-sized particles is to some extent higher than that of the hydrophobic nano-sized particles. This result is in agreement with Dippenaar's [8] observations in which a particle has to wait for a film to thin to less than its own dimension before it can rupture the gas/liquid film.

It was also verified that the adsorption of surfactant molecules on the particle's surface reduces their ability in breaking foams. It may be supposed, therefore, that by increasing the affinity between the surfactant molecules and the particle's surface it is possible to reduce the destructive effect of the hydrophobic particles.

Acknowledgements The authors wish to acknowledge Petróleo Brasileiro, CENPES/Petrobras, the Federal University of Rio de Janeiro, COPPE/UFRJ, the Engineering Research Center (ERC) for Particle Science and Technology at the University of Florida, the National Science Foundation (grant EEC-94-02989), and the industrial partners of the ERC for their support of this research.

References

1. Rao PSC, Rao MP, Anderson BS (1987) Soil science, fact sheet SL 55. Institute of Food & Agricultural Science, University of Florida, Gainesville, FL, USA
2. Luthy RG, Dzombak DA, Peters CA, Roy SB, Romaswami A, Nakles DV, Nott BR (1994) Environ Sci Technol 28:266–276
3. Chowdiah P, Misra BR, Kilbane II, Srivastava VJ, Hayes TD (1998) J Hazard Mater 62:265–280
4. Kulkarni RD, Goddard ED, Kanner B (1977) Ind Eng Chem Fundam 16:472–474
5. Binks BP, Lumsdon SO (2000) Phys Chem Phys 2:2959–2967
6. Hedreul C, Frens G (2001) Colloids Surf A 186:73–82
7. Dippenaar A (1982) Int J Miner Process 9:1–14
8. Dippenaar A (1982) Int J Miner Process 9:15–27
9. US Provisional Patent Application 60/450,025, 04/03/03
10. Patist A, Axelberd T, Shah DO (1998) J Colloid Interface Sci 208:259–265

Progr Colloid Polym Sci (2004) 128: 298–302
DOI 10.1007/b97121

Roberto C.G. Oliveira
José F. Oliveira
Brij M. Moudgil

Optimizing micro-foam rheology for soil remediation

R.C.G. Oliveira
Petróleo Brasileiro S/A, CENPES/Petrobras, Rio de Janeiro, RJ, Brazil

J.F. Oliveira (✉)
Federal University of Rio de Janeiro, COPPE, Centro de Tecnologia, Bloco G, sala 106, Cidade Universitária, Ilha do Fundão, 21949-900
Rio de Janeiro, RJ, Brazil
e-mail: oliveira@metalmat.ufrj.br
Tel.: +55-21-25627030
Fax: +55-21-22906626

B.M. Moudgil
Engineering Research Center for Particle Science & Technology, Gainesville, FL, USA

Abstract In situ foam flushing has proved to be an efficient process for removing oil from contaminated soils. The drawback is that foams are thermodynamically unstable and also produce a significant pressure gradient as they pass through porous media. The use of micro-foams, usually called colloidal gas aphrons (CGAs), instead of regular foams minimizes the pressure gradient owing to their reduced size and increased stability. Furthermore, there is a lack of information as far as the micro-foam rheology is concerned. The present paper describes how the formulation and the generation conditions can affect the CGA flow properties. The effects of the surfactant concentration, the bubble size distribution and the gas volume fraction on the rheology of CGAs were quantified using a capillary viscometer. The results showed that CGAs display strong shear-thinning behavior, decreasing their apparent viscosity with increasing shear rate. It was also verified that the apparent viscosity of the CGAs increases as the surfactant concentration is enhanced. Finally, it was observed that there is a significant increment in the apparent viscosity, the yield stress and the mean bubble diameter of CGAs when the gas volume fraction is increased, especially above the critical volume fraction value ($\phi_c \sim 0.65$). This behavior confirms that CGAs are able to form packing structures.

Keywords Soil remediation · Colloidal gas aphron · Micro-foam rheology

Introduction

The petroleum industry is familiar with the use of foams for enhancing oil recovery. Foams are able to act as a mobility control agent helping to prevent channeling of the injected fluids in high-permeability zones. Nevertheless, there are huge differences between the uses of foams in oil recovery and their intended use in soil remediation. Oil recovery operation generally occurs in consolidated rocks where high injection pressures are permissible. Soil remediation is often conducted in unconsolidated soils, generally from 3 to 9 m in depth, where high injection pressures can cause soil heaving or fracturing. While displacement of mobile oil is the primary purpose of the process for enhancing oil recovery, the use of foams in soil remediation would be to dissolve and or to emulsify the soil contaminants in a soil flushing process.

Bioremediation has been widely used for the in situ removal of non-aqueous-phase liquids from contaminated soils. Nevertheless, it depends on the efficient delivery of oxygen, moisture and nutrients to the subsurface soils [1]. Foams are being considered for delivering the necessary components for this process. It can also be used to reduce soil heterogeneity and to improve soil wettability and contaminant desorption. Another advantage of using foams is their large surface area, which leads to a substantial reduction in the surfactant consumption [2].

Sebba [3, 4] was the first to study micro-foams, which he later termed as colloidal gas aphrons (CGAs). CGAs are stabler than regular foams and can be pumped from the point of generation to the point of application. CGAs also differ from regular foams in structure as well as in terms of their fluid mechanics properties. Regular foams are essentially a gas/liquid/gas system in which thin lamellae of a liquid form a polyhedral structure encapsulating the gas phase. By contrast, CGAs are aggregates of micron-sized spherical bubbles encapsulated in a thin shell of surfactant film and suspended in the bulk liquid, as shown in Fig. 1. CGA foams also exhibit a much higher liquid content than regular foams.

The stability of a CGA results from its ability to resist changes in bubble size, liquid content or degree of dispersion [5, 6]. This means that normally CGAs maintain an average bubble size and are uniformly distributed through the continuous liquid phase. This concept is very different from that of foam stability as related to the intrinsic nature of the films to resist rupture.

Foams have been studied for several decades but a considerable amount of research remains to be done, especially that related to CGA rheology. The dominant variables affecting foam rheology are the gas-phase volume fraction (ϕ), the bubble size distribution, and the interfacial tension. Other interfacial parameters, like surface viscosity, have a secondary effect on foam rheology although they are relevant to the mechanism of foam flowing through porous media. Kraynik [7] studied the flow of dry foams ($\phi > 0.7$) in large-diameter transparent tubes. He observed that at low shear stresses the foam moved as a solid body, slipping along the tube wall; however, at higher shear stresses the foam began to shear inside the tube, in addition to slipping along the tube wall. Kraynik's observations imply that dry foams have a yield stress, and flow experiments must be designed to measure or to eliminate wall slip.

Prud'homme and Khan [8] have developed a capillary viscometer for measuring foam viscosity (η_e) and slip velocity (V_s). It was shown that the slip velocity is affected by the wall roughness, the gas volume fraction, and the bubble size distribution. At high gas volume fractions, foams are structured materials and this structure influences their rheological properties. Foams are typically pseudo-plastic or "shear-thinning" fluids, which means that their apparent viscosity decreases with increasing shear rate. Khan and Armstrong [9] have shown that for very structured foams (dry foams) the yield stress (τ_0) value is inversely proportional to the bubble size and decreases with time owing to the foam's instability. For structured foams, the stress distribution across the measuring geometry is not uniform. There are regions of the foam that are below their yield stress and thus are in rigid-body motion while other regions are flowing. In addition, if the shear stress is below the yield stress everywhere, then the foam will slip at the boundaries rather than flow and the measured apparent viscosity will not reflect the fluid viscosity but the sliding friction of the foam across the solid surface.

The flow of foam in a capillary is shown schematically in Fig. 2. It should be noted that the foam slips at the solid wall and is in plug flow at the center of the capillary where the shear stress is below the yield stress. The shear stress is linear from the maximum value, τ_R, at the wall to zero at the center regardless of the fluid rheology.

The aim of the present study was to establish, using a capillary viscometer, how the formulation and the generation conditions can affect the CGA flow properties, and to design CGAs with optimal performance for soil remediation.

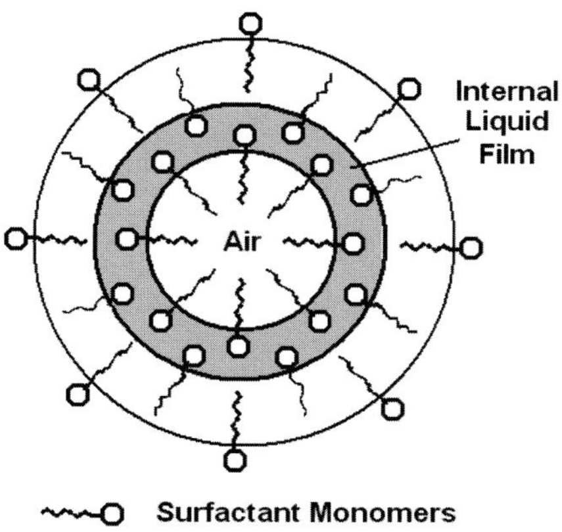

Fig. 1 Colloidal gas aphron (*CGA*) structure as proposed by Sebba [4]

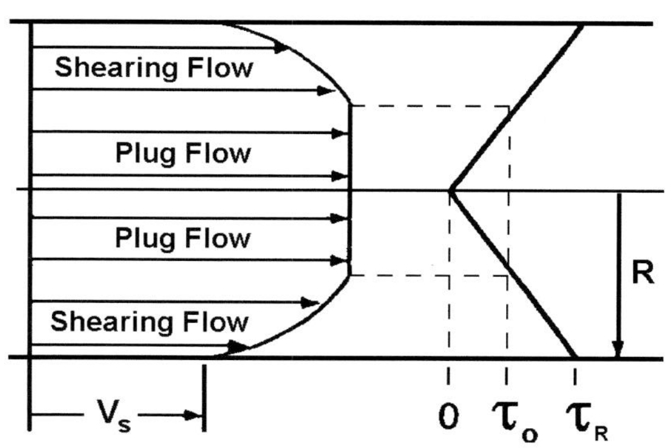

Fig. 2 Scheme of foam flow through a capillary of radius R and having a slip velocity V_s at the wall surface [8]

Experimental

Materials and apparatus

The sodium dodecyl sulfate (SDS) used in this study was of 99% purity and was obtained from Aldrich Chemical Co. The SDS critical micellar concentration (cmc) was determined to be around 8.2 mM at 20 °C. The water used was purified by a Millipore system. The capillary viscometer used for measuring the yield stress and the apparent viscosity of the CGAs is shown in Fig. 3. This viscometer was built on the basis of the concept proposed by Prud'homme and Khan [8]. The key elements of this viscometer are two long capillaries with different diameters, but the same length-to-diameter ratio (L/D). Each capillary is fed from the same central manifold and, consequently, is submitted to the same pressure drop and wall shear stress. Therefore, the CGA slip velocity in each capillary is also the same, since slip is only a function of wall shear stress [8]. The central manifold has an exit port connected to a valve that permits control of the flow of the CGA that will be delivered to the capillaries. So, the CGA flow rate through the capillaries is not varied by altering the gas and liquid flow rates through the CGA generator, but rather by controlling the CGA flow rate through the exit port. This setup ensures that data are taken at constant wall shear stress in both capillaries for the analysis of slip velocity and CGA viscosity.

Slip correction method

The apparent slip velocity (V_s) at the wall can be estimated by assuming that the observed flow rate (q_{obs}) is equal to the sum of the true flow rate (q_{true}) and the slip flow rate (q_{slip}):

$$q_{obs} = q_{true} + q_{slip}. \tag{1}$$

By multiplying Eq. (1) by ($8/D$) and dividing it by the cross-sectional area we obtain Eq. (2):

$$\left(\frac{8V}{D}\right)_{obs} = \left(\frac{8V}{D}\right)_{true} + \left(\frac{8V}{D}\right)_{slip}. \tag{2}$$

Jastrzebski [10] has shown that for fluids with macroscopic structure, such as foams, the apparent slip velocity depends on the wall stress (τ_w) and the tube diameter (D) as shown in Eq. (3), where β is the slip coefficient:

$$V_S = \frac{\beta\tau_W}{D}. \tag{3}$$

Thus, applying Eq. (3) in Eq. (2), we obtain Eq. (4):

$$\left(\frac{8V}{D}\right)_{obs} = \left(\frac{8V}{D}\right)_{true} + \frac{8\beta\tau_W}{D^2}. \tag{4}$$

Enzendorfer et al. [11] showed that the slip correction method proposed by Jastrzebski [10] provides a consistent means of apparent slip correction. They also proposed the use of a factor called the specific volume expansion ratio (ε) to equalize the rheological data, where ρ_l and ρ are, respectively, the liquid phase and foam densities:

$$\varepsilon = \frac{\rho_l}{\rho}. \tag{5}$$

Results and discussion

The capillary viscometer setup (Fig. 3) was used to estimate the CGA yield stress and apparent viscosity at different gas volume fractions. The experiments were conducted at different shear rates and SDS concentrations. The results were analyzed on the basis of the Jastrzebski [10] equation, Eq. (4), and the slip correction method proposed by Enzendorfer et al. [11].

The effect of the gas volume fraction on the true viscosity of CGAs for different shear rates is presented in Fig. 4. The results reveal that the true viscosity of CGAs increases substantially when the gas volume fraction exceeds 0.65. This value matched the maximum random packing of identical-size spheres (ϕ_c~0.65). Above this critical value, thin interfacial

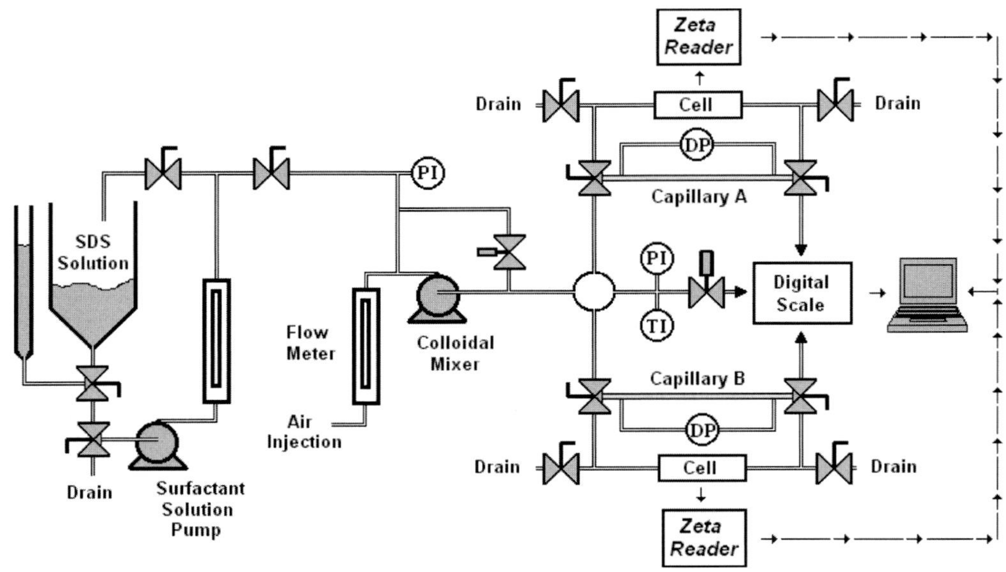

Fig. 3 Capillary viscometer setup

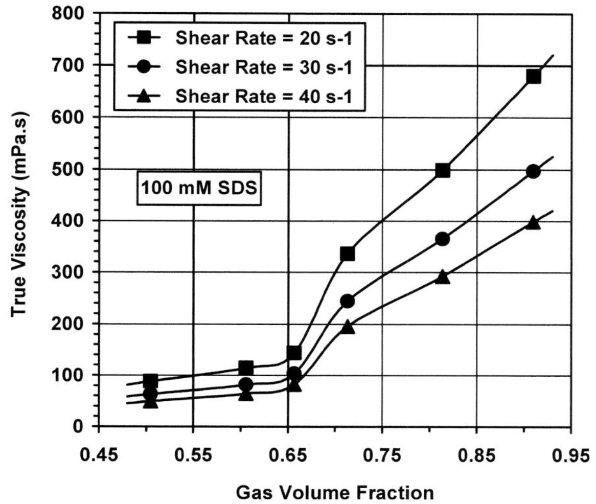

Fig. 4 True viscosity of CGAs as a function of the gas volume fraction for different shear rates

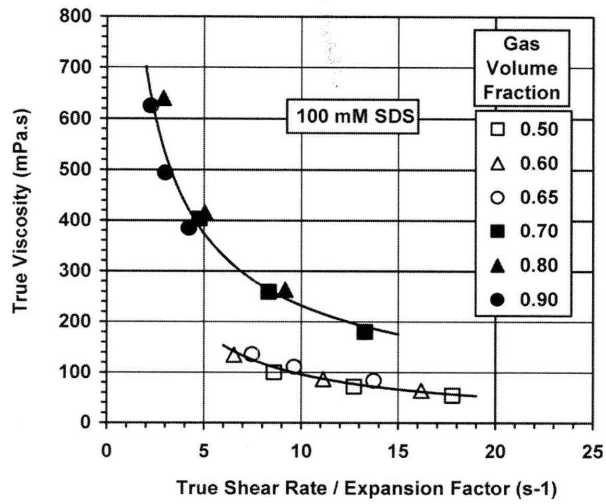

Fig. 6 The effect of the volume expansion factor on the true viscosity of CGAs for different gas volume fractions and a constant sodium dodecyl sulfate (*SDS*) concentration of 100 mM

films are formed between the bubbles, leading to large hydrodynamic forces and high viscosities. It was also verified that CGAs behave as shear-thinning fluids, especially above ϕ_c.

The effect of surfactant concentration on the CGA true viscosity is shown in Fig. 5. The results show that increasing the surfactant concentration above the cmc tends to slightly increase the CGA viscosity.

The effect of the volume equalization factor on the true viscosity curves of CGAs is shown in Fig. 6. It can be seen that the data tend to converge into two sets of curves when the true shear rate is normalized by the volume equalization factor. In part, this behavior corroborates the observations of Enzendorfer et al. [11]. Below the critical volume fraction value ($\phi_c \sim 0.65$) the CGAs' bubbles are spherical and do not form any packing structure. In this situation the apparent viscosity depends only on the interactions between the CGA bubbles. As the ϕ value is increased above ϕ_c the CGA bubbles start to deform each other, leading to a packing structure. This structure causes a huge influence on the viscosity of the CGAs, leading to the establishment of a different rheological curve.

The effect of the gas volume fraction on the yield stress of the CGAs is shown in Fig. 7. It can be verified that increasing the ϕ value of the CGAs above ϕ_c causes an

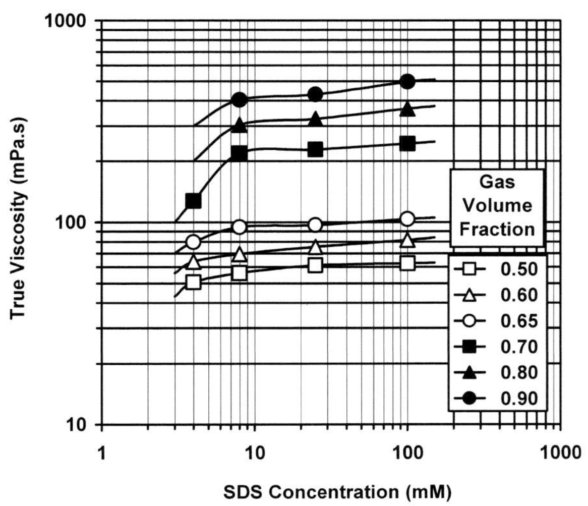

Fig. 5 The effect of surfactant concentration on the true viscosity of CGAs for several values of gas volume fraction

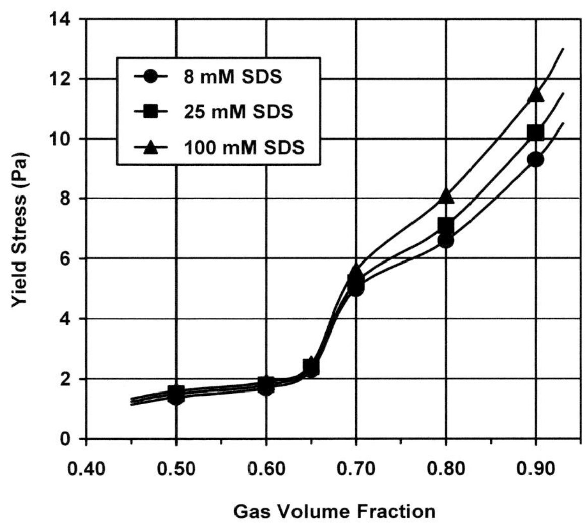

Fig. 7 The effect of the gas volume fraction on the yield stress of CGAs for several values of the SDS concentration

increase in the yield stress. This behavior is considered to be intrinsically related to the presence of a structured packing at ϕ values higher than ϕ_c.

Conclusion

It was found that CGAs display a strong shear-thinning behavior, decreasing their apparent viscosity with increasing shear rate. The slip correction method adopted showed consistent results when applied to the rheological measurements of CGAs using a capillary viscometer.

It was observed that there is a significant increment in the apparent viscosity and the yield stress of CGAs when the surfactant concentration and the gas volume fraction are increased, especially above the critical volume fraction ($\phi_c \sim 0.65$).

Finally, the use of the volume expansion factor lead to the rheological data of the CGAs fitting two distinct curves, one for ϕ values below ϕ_c and the other one for ϕ values higher than ϕ_c. This behavior confirms that CGAs are able to form packing structures above ϕ_c. Thus, in order to achieve trouble-free underground injection of CGAs, it is recommended to keep the CGA ϕ value below 0.65.

Acknowledgements The authors acknowledge Petróleo Brasileiro, CENPES/Petrobras, the Federal University of Rio de Janeiro, COPPE/UFRJ, the Engineering Research Center (ERC) for Particle Science and Technology at the University of Florida, the National Science Foundation (grant EEC–94–02989), and the industrial partners of the ERC for support of this research.

References

1. Chowdiah P, Misra BR, Kilbane II, Srivastava VJ, Hayes TD (1998) J Hazard Mater 62:265–280
2. Brockman FJ (1995) J Hazard Mater 41:287–298
3. Sebba F (1971) J Colloid Interface Sci 35:643–646
4. Sebba F (1982) OWRT/RU-82/10. US Department of the Interior, Office of the Water Resources and Technology
5. Longe TA (1989) PhD dissertation. Virginia Polytechnic Institute and State University, Blacksburg, VA
6. Vidrine WK, Willson CS, Valsaraj KT (2000) Colloids Surf A 175:277–289
7. Kraynik AM (1988) Annu Rev Fluid Mech 20:325–357
8. Prud'Homme RK, Khan SA (1996) Surfactant Sci Ser 57:217–241
9. Khan SA, Armstrong RC(1989) J Rheol 33:881–911
10. Jastrzebski ZD (1967) Ind Eng Chem Fundam 6:445–453
11. Enzendorfer C, Harris RA, Valko P, Economides MJ, Fokker PA, Davies DD (1995) J Rheol 39:345–358

Progr Colloid Polym Sci (2004) 128: 303–308
DOI 10.1007/b97123

Gabriela Piperata
Jorge M. Meichtry
Marta I. Litter

Photocatalytic reactions over TiO_2 supported on porcelain spheres

G. Piperata · J.M. Meichtry
M.I. Litter (✉)
Unidad de Actividad Química,
Centro Atómico Constituyentes,
Comisión Nacional de Energía Atómica,
Av. Gral. Paz 1499, 1650 San Martín,
Prov. de Buenos Aires, Buenos Aires,
Argentina
e-mail: litter@cnea.gov.ar
Tel.: +54-1-167727016
Fax: +54-1-167727886

Abstract Results of immobilisation of TiO_2 on low-cost porcelain beads and the resulting photocatalytic activities of the samples are presented. A very simple technique for immobilisation by immersion in a TiO_2 suspension, previously acidified to pH 2.5, followed by calcination at 450 °C, was used. After vigorous washing with a water jet, no detachment of the catalyst was observed at the end of the fixation process. Reasons for the stability of the material are discussed. The crystallographic nature of P-25 was not affected after attachment to the porcelain surface. The beads were submitted for 100 h to turbulent regimes corresponding to *Re* numbers of 100 and 400 in a recirculating system. The effect of the recirculation flow was carefully examined to evaluate the degree of detachment of the catalyst from the support and to assess how the treatment influences the activity of the supported catalyst. For this, photocatalytic tests were performed on two model compounds, 4-chlorophenol and potassium dichromate in the presence of ethylenediaminetetraacetic acid, and the results were compared with those of untreated beads. A good activity remained in the beads even after washing for more than 100 h in a turbulent regime (both $Re = 100$ and $Re = 400$). The activity of particles detached from the support by ultrasonication was also compared with that of the pure precursor.

Keywords Supported TiO_2 · Heterogeneous photocatalysis · Porcelain beads

Introduction

Heterogeneous photocatalysis for disinfection and detoxification of effluents is an emergent technology already applied in industrialised countries [1, 2]. By irradiation of a semiconductor with UV light (below 400 nm), electron–hole pairs are created that can react with acceptors and donors present in the suspension. Active species such as $HO^{\cdot}$ radicals are formed, provoking mineralisation of organic pollutants [3]. Even toxic ionic metals can be transformed into less noxious species [4].

So far, the most used photocatalyst is TiO_2, especially in its commercial Degussa P-25 powdered form. One of the most important drawbacks to achieve the successful implementation of heterogeneous photocatalysis is that the photocatalyst must be separated from the system after the treatment, and the cost of this separation stage may even invalidate economically the technique [5, 6]. To solve this problem, several attempts to immobilise TiO_2 have been performed. For the immobilisation, two techniques can be envisaged [6–9]: (1) "in situ" synthesis of the oxide on the support by a sol–gel technique, a chemical vapour deposition technique or others; (2) direct impregnation of a commercial or previously prepared TiO_2 powder on the surface of the support. The second one is the easiest and most economical procedure. The fixation is achieved starting from a TiO_2 suspension in an adequate dispersant followed by sinterisation, in some cases with the addition of ligand agents [10–12]. The process by which the particles are fixed is not totally clear. Probably, electrostatic interactions between particles and charged surfaces are involved, although covalent interactions cannot be discarded when the adherence is extremely strong (e.g., with glass). There is no agreement on which method produces films with better photocatalytic activity, and this depends generally on the nature and experimental conditions of the system [6].

There is an important variety of materials that can be used to support TiO_2. The most studied are the ones based on SiO_2, owing to their low cost and easy availability [6]. The success of these coatings is based on the large adherence between TiO_2 and glass, attributed to the occurrence of some sintering in the calcination process [13]. The thermal treatment is limited by the glass-softening temperature that, in the case of borosilicate, is about 500 °C. To overcome this limitation, more resistant and then more expensive glasses must be used. Other support materials include metals, ceramics, zeolites, plastics, cements, red bricks and inorganic fibres [5, 7, 14–18].

The first prototypes of photoreactors with supported TiO_2 were constructed by simply covering the internal wall of glass tubes with powdered TiO_2 [19, 20]. To have more surface area, beads, rings or glass fibres inside the tubes have been filled [5]. These materials have been successfully used in commercial devices for air purification [21]. In contrast to gas systems, the catalyst has been found to be mechanically unstable in aqueous systems, generally submitted to strong abrasion and turbulent regimes, which cause removal of the TiO_2 coating with the time of use. The geometrical shape of the support is a very important parameter for fixation: whereas the fixation to a flat or cylindrical surface yields rather good permanent adhesion, glass spherical shapes produce poor results, and TiO_2 is rapidly detached.

As is known, the surface area as well as the nature of the TiO_2 phase strongly influence photocatalytic activity. Changes in the electronic and crystalline structure of TiO_2 due to interaction with the support may result in alteration of the efficiency. The attachment of TiO_2 to a surface decreases considerably the available surface area and the amount of light reaching the particles. Also, mass-transfer limitations due to poorer access or contact of the pollutant with the catalyst are usually invoked. As a general behaviour, supported TiO_2 is always less active than suspended TiO_2.

In this work, we discuss results of immobilisation of TiO_2 on low-cost small porcelain spheres where TiO_2 was impregnated by a simple technique. The degree of detachment of the catalyst from the surface and the photocatalytic activity on model compounds such as 4-chlorophenol (4-CP) and potassium dichromate (Cr(VI)) were evaluated.

Experimental

Materials and equipment

As a support for TiO_2, commercial porcelain beads according to DIN standard 40685, group 100 (Del Morro, Argentina), of 6-mm diameter were used. Basically, this porcelain is a silicoaluminate containing vestiges of potassium.

TiO_2 Degussa P-25 was a commercial sample, used as provided. $K_2Cr_2O_7$ (Carlo Erba), 4-CP (Fluka), sodium ethylenediaminetetraacetate (Na_2EDTA) (Mallinkrodt) and all other chemicals were of analytical reagent grade. Doubly distilled water was used for the preparation of all solutions and for final washings. Dilute $HClO_4$ was used for pH adjustments.

X-ray diffraction (XRD) patterns were obtained at room temperature with a Philips PW-3710 diffractometer, using Cu Kα radiation. Spectrophotometric measurements were performed with a Hewlett-Packard diode-array UV–vis spectrophotometer, model HP 8453A. Ultrasonications were performed with a Cleanson (25 kHz) ultrasonicator, model CS-1109.

Impregnation procedure

Beads were impregnated with TiO_2 according to the following procedure. Spheres were introduced into a 0.1 M HNO_3 solution for 2 h, then washed four times with water and dried at 100 °C. This step was followed by immersion in a 1 M KOH solution for 48 h, washing four times with water and drying again at 100 °C. A 10% w/v TiO_2 suspension, previously adjusted to pH 2.5 with $HClO_4$, was ultrasonicated for 30 min and the spheres were immersed in the suspension for 15 min. The excess of suspension was drained and the material was dried at 100 °C and calcined at 450 °C for 2 h at a heating rate of 3.3 °C/min. At the end, the spheres were rinsed for 15–30 min with a vigorous water jet to eliminate loose particles, and dried at 100 °C. The impregnation procedure was repeated six times.

Evaluation of the abrasion resistance

The mechanical stability of TiO_2 on the porcelain spheres was tested by washing with bidistilled water for more than 100 h. For this sake, a recirculation setup was used, consisting of a 30-mm-diameter, 140-mm-length glass tube, in which 230 spheres were arranged as a fixed bed. From a reservoir, 500 mL distilled water was continuously recirculated by means of a peristaltic pump. Two recirculation rates, 0.6 and 2.4 L/min, corresponding to $Re = 100$ and $Re = 400$ regimes, were tested. The Reynolds number for spherical beads can be calculated according to [22]

$$Re = D_p \rho v_0 \mu^{-1}, \tag{1}$$

where D_p is the bead diameter, ρ is the fluid density, v_0 is the flow rate in an empty bed and μ is the fluid viscosity. For this geometry and configuration, a turbulent flow is defined for $Re > 40$; therefore, both recirculation rates correspond to turbulent regimes.

Periodically, samples of washings were taken and analysed spectrophotometrically against pure water to evaluate the bleeding of TiO_2 by changes in the UV region due to the scattering caused by TiO_2 particles. After around 100 h of recirculation, a 45-mL sample was filtered through a 0.2-μm Sartorius filter membrane and the membrane was analysed by XRD for a possible solid residue.

Hereafter in this work, beads will be named "uncoated" (without TiO_2), "untreated" (with TiO_2 but not submitted to the recirculation treatment), "Re 100" and "Re 400", respectively.

To estimate the amount of TiO_2 attached to the beads, 21 of them were ultrasonicated (25 kHz) for 30 min in 10-mL bidistilled water. Ultrasonication at this frequency did not provoke any deterioration of the beads. The beads were withdrawn from the suspension, rinsed and dried at 100 °C, and the mass of TiO_2 was calculated from differences in weight before and after the detachment. This procedure was repeated three times and the results were averaged. The amount of TiO_2 deposited on untreated beads was 2.1 mg TiO_2/g spheres (approximately 0.8 mg TiO_2/cm^2). For Re 100, 1.7 mg TiO_2/g spheres was calculated (approximately 0.7 mg TiO_2/cm^2) and for Re 400, 0.6 mg TiO_2/g spheres (approximately 0.3 mg TiO_2/cm^2). The suspensions obtained from the ultrasonication procedure were used for photocatalytic experiments, as described later.

Photocatalytic tests

Photocatalytic irradiations were carried out in a 30-mL cylindrical cell, open to air, irradiated from the top, using high-pressure xenon arc lamps (Osram XBO, 150 or 450 W). The setup was thermostatted at 25 °C. The cell was provided with a 300-nm cutoff filter, avoiding wavelengths lower than 300 nm. The IR fraction of the incident light was removed by a 50-mm cylindrical water filter.

The solution containing the model pollutant (4-CP or Cr(VI)) at a fixed concentration and pH, together with TiO_2, either supported or in suspension, was introduced into the cell. In the case of supported TiO_2, a layer of 21 beads was evenly arranged on the bottom of the cell. When suspended TiO_2 was used, the suspension was first ultrasonicated for 15 min and magnetically stirred during the run. Prior to irradiation, solutions or suspensions were kept in the dark for 30 min, to ensure substrate–surface equilibrium. No significant differences in concentrations were observed after this dark period. Once the irradiation was initiated, 500-μL samples were taken periodically and were diluted to 5 mL for analysis. Suspensions were filtered through a 0.45-μm MSF filter before analysis. At least duplicate runs were carried out for each condition, and the results were averaged. Photon flows per unit volume were determined with ferrioxalate [23], using the same volume of actinometric solution as used in the experiments to keep the same conditions.

For 4-CP, 25 mL (corresponding to a 2.8-cm height of the solution in the cell) of a 0.2 mM solution at initial pH 3 were irradiated. 4-CP degradation was followed by the decrease of the peak at 225 nm. During the photocatalytic reaction, some intermediates (phenol and hydroquinone, among others) can be formed. Under our conditions, the amount of them is very low and their absorption coefficient at 225 nm is lower than that of 4-CP [24, 25]. This validates our measurement methodology.

For Cr(VI) experiments, 20 mL (corresponding to a 2.5-cm height of the solution in the cell) 0.4 mM $K_2Cr_2O_7$, together with 0.9 mM Na_2EDTA as a sacrificial synergetic agent [4], was used. The initial pH was adjusted to 2. Cr(VI) measurement was performed spectrophotometrically at 349 nm [26].

The concentration of TiO_2 in the different experiments is indicated in the legends of the figures. For the case of supported TiO_2, the values correspond to equivalent concentrations in suspension, calculated from the estimated amount of TiO_2 in the beads.

Results and discussion

Preparation of the impregnated beads

A very simple procedure was used to impregnate the porcelain beads with TiO_2. A pretreatment with HNO_3 solution (pH 1) for 2 h allowed to eliminate any previous impurity of the surface to improve further treatments. As the surface of beads was very smooth, an alkaline treatment on the porcelain by immersion in KOH solution was done. A 10% w/v TiO_2 suspension, adjusted to pH 2.5 with $HClO_4$, was ultrasonicated to break possible aggregates. The supports were then introduced into this suspension and kept for a time as short as 15 min. At this pH value and TiO_2 concentration, the deposition of the catalyst is facilitated by electrostatic attraction. Use of a mineral acid other than $HClO_4$ could have left anions on the surface that would have reduced the photocatalytic activity of the semiconductor [12]. Calcination at 450 °C for 2 h at a controlled temperature ramp was performed. In this way, the beads were resistant to cracking. Finally, the material was rinsed vigorously with water to eliminate loose particles. Previous experiments in our laboratory had shown that the activity of the beads was higher when two TiO_2 layers were applied compared with only one impregnation. In this work, we decided to apply six layers, but this number should be still optimised.

The impregnation treatment did not affect the crystallographic composition of the TiO_2 coated onto the spheres, as it will be discussed later.

Evaluation of the abrasion resistance under different recirculation regimes

The adherence of TiO_2 to porcelain spheres was tested by washing with bidistilled water in two vigorous turbulent regimes for more than 100 h. After this treatment, the washings were analysed for the presence of TiO_2 particles, spectrophotometrically as well as by XRD after filtration. Concerning Re 100, no TiO_2 was observed in the washings or as a solid residue, but actually some detachment was detected from the difference in the weight before and after treatment in the turbulent regime (2.1 against 1.7 mg TiO_2/g spheres). In the case of Re 400, a much lower amount of TiO_2 remained after the treatment (0.6 mg TiO_2/g spheres) and a very weak but visible peak of anatase was detected by XRD as a solid residue in the membrane used for filtration.

It was verified that neither the turbulent recirculation nor the impregnation procedure had affected the crystallographic composition of the coated TiO_2. For this purpose, Re 100 beads were submitted to ultrasonication for 30 min and a XRD spectrum of the detached TiO_2 powder was obtained. Comparison of this spectrum with that of pure P-25 (not shown) indicated that both samples presented a similar anatase-to-rutile ratio (74% anatase), calculated from the intensity of the main diffraction peaks ((101) anatase and (110) rutile), according to Ref. [27].

On the other hand, it was proven that on this type of beads, made of porcelain, TiO_2 presented a higher adherence than on Pyrex glass spheres tested previously by us in preliminary experiments.

Photocatalytic experiments

To evaluate the activity of the coated beads, two photocatalytic model systems, tested previously in our and in other laboratories were used: (1) degradation of 4-CP; (2) reduction of Cr(VI) in the presence of an excess of ethylenediaminetetraacetic acid (EDTA). No attempts to compare the results of the different experiments after normalisation by TiO_2 mass were made.

The degradation of 4-CP with different TiO_2-containing beads at pH 3 is shown in Fig. 1. Untreated beads as well as beads submitted to both *Re* regimes are compared, together with TiO_2 in suspension at the same concentration as in the untreated beads (0.7 g/L). As expected, immobilised TiO_2 presented a lower efficiency than the suspended catalyst at the same concentration. As can be seen, although the activity is decreased after the turbulent recirculation treatment, it is worthwhile to remark that the treatment was prolonged for more than 100 h. The activity loss was higher with the highest *Re* number. Of course, the amount of TiO_2 in the reaction volume was also lower owing to the previous detachment. However, both Re 100 and Re 400 beads still show some efficiency.

As the photocatalytic degradation of 4-CP with immobilised TiO_2 was found to be slow for the comparison of the activities of the samples, another very well known and faster photocatalytic reaction, reduction of Cr(VI) in the presence of EDTA at pH 2 [4], was tested. The Results are shown in Fig. 2. In the absence of TiO_2, a minor photochemical reaction between Cr(VI) and EDTA takes place. The results were similar to those obtained for 4-CP; the Cr(VI) reduction efficiency decreased in immobilised TiO_2 compared with that in suspension and in Re 100 and Re 400 with respect to untreated beads.

The activities of Re 100 and a suspension prepared from TiO_2 detached from Re 100 by ultrasonication are compared in Fig. 3, together with a P-25 suspension of the same concentration. As expected, owing to a lower surface area, mass-transfer limitations and a lower amount of light reaching the beads only from the top,

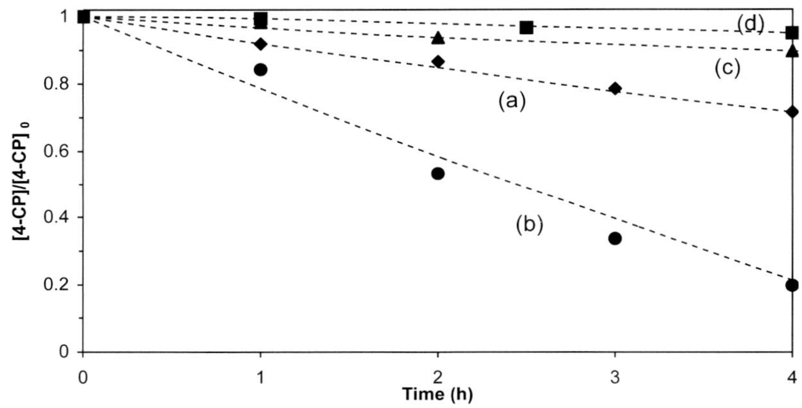

Fig. 1 4-Chlorophenol (4-*CP*) degradation with immobilised and suspended TiO_2. Conditions: [4-CP] = 0.2 mM; pH_0 3; T = 25 °C; open to air; P_0 = 12 μ einstein/s/L; $\lambda > 300$ nm. Untreated, [TiO_2] = 0.7 g/L (*a*); P-25 suspension, [TiO_2] = 0.7 g/L (*b*); Re 100, [TiO_2] = 0.5 g/L (*c*); Re 400, [TiO_2] = 0.15 g/L (*d*)

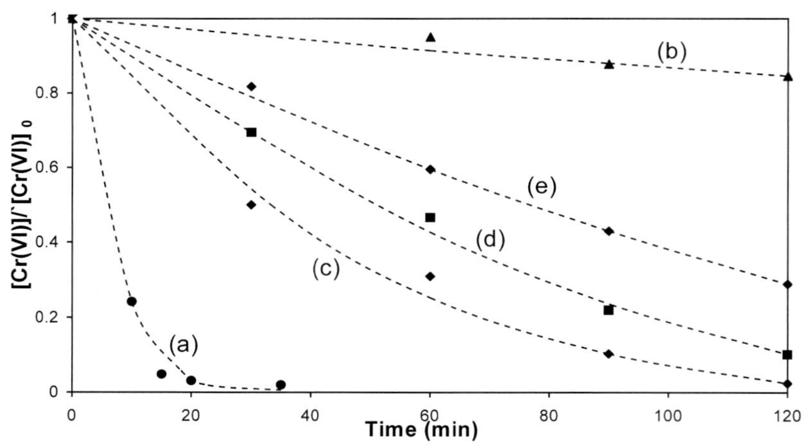

Fig. 2 Cr(VI) reduction with suspended TiO_2 and beads submitted to different treatments. Conditions: [Cr(VI)] = 0.8 mM; [EDTA] = 0.9 mM, where *EDTA* is ethylenediaminetetraacetic acid; pH_0 2; T = 25 °C; open to air; P_0 = 18 μ einstein/s/L; $\lambda > 300$ nm. P-25 suspension, [TiO_2] = 0.9 g/L (*a*); uncoated, [TiO_2] = 0 (*b*); untreated, [TiO_2] = 0.9 g/L (*c*); Re 100, [TiO_2] = 0.6 g/L (*d*); Re 400, [TiO_2] = 0.2 g/L (*e*)

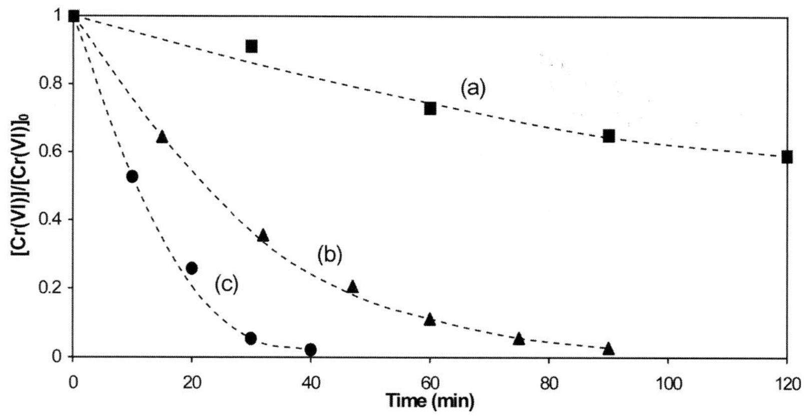

Fig. 3 Cr(VI) reduction efficiencies of immobilised and suspended TiO_2. The equivalent concentration of TiO_2 in all cases is 0.6 g/L. Conditions: [Cr(VI)] = 0.8 mM; [EDTA] = 0.9 mM; pH_0 2; T = 25 °C; open to air; P_0 = 3.6 μ einstein/s/L; λ >300 nm. Re 100 (*a*); TiO_2 detached from Re 100 after ultrasonication (*b*); P-25 suspension (*c*)

the efficiency was lower for the immobilised catalyst. It must also be taken into account that the suspensions were magnetically stirred in contrast with experiments done with beads. TiO_2 detached from Re 100 shows a lower activity compared with that of the precursor suspension. This lower activity can be explained by the presence of some impurities coming from the support [16].

The same experiments were performed using Re 400 and the results are presented in Fig. 4. As in the previous case, fixed TiO_2 presents a lower activity than TiO_2 in suspension. It is remarkable that here the activity of detached TiO_2 is almost identical to the activity of P-25. We think that a more profound washing than for Re 100 removed impurities, increasing the activity.

Conclusions

Our results indicate that a very good adherence of TiO_2 is obtained onto porcelain beads compared with Pyrex glass spheres. This aspect is important as this is one of the main drawbacks when spherical geometries are used to fix the catalyst by impregnation. We then obtained a cost-effective material by a very simple technique, in comparison with the sol–gel technique or other preparative synthesis of TiO_2 on the support.

We noticed that the adherence of TiO_2 to the support is affected by a turbulent regime in a recirculation system. However, a Reynolds number of 100 would be still suitable for application in a fixed-bed reactor without significant detachment of the catalyst. Further experiments are needed to determine the optimal Reynolds number that guarantees a turbulent regime without TiO_2 detachment.

Although immobilised TiO_2 on porcelain beads does not exhibit a very good photocatalytic activity for Cr(VI) reduction and 4-CP degradation under the conditions explored in this work compared with TiO_2 in suspension, the new material would be able to be applied in a fixed bed in which, with an optimum flow rate, geometry and other parameters, a better global performance of the system would be obtained.

Further studies are under way to optimise the fixation time and the number of TiO_2 layers to be applied on the support.

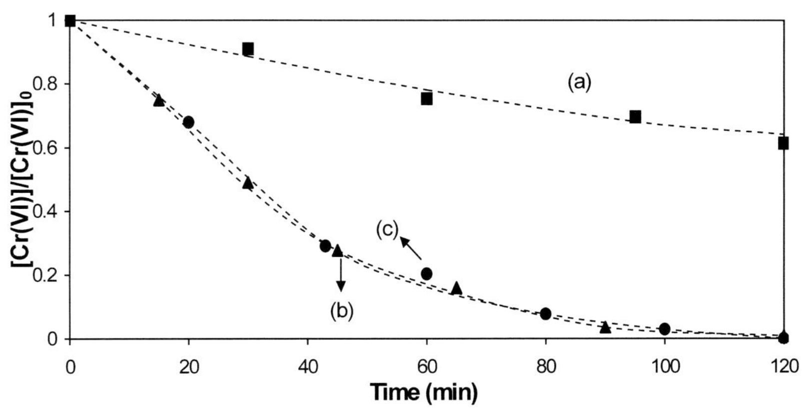

Fig. 4 Cr(VI) reduction efficiencies of immobilised and suspended TiO_2. The equivalent concentration of TiO_2 in all cases is 0.2 g/L. Conditions: [Cr(VI)] = 0.8 mM; [EDTA] = 0.9 mM; pH_0 2; T = 25 °C; open to air; P_0 = 3.6 μ einstein/s/L; λ> 300 nm. Re 400 (*a*); TiO_2 detached from Re 400 after ultrasonication (*b*); P-25 suspension (*c*)

Acknowledgements. We thank G. Leyva for taking XRD spectra. This work was performed as part of the Comisión Nacional de Energía Atómica CNEA P5-PID-36-4 programme, the Agencia Nacional de Promoción de la Ciencia y la Tecnología PICT98-13-03672 project and the Consejo Nacional de Investigaciones Científicas y Técnicas PIP662/98 project. G.P. thanks CNEA-CONICET for a fellowship to perform this work. M.I.L. is a member of CONICET.

References

1. Hoffmann MR, Martin ST, Choi W, Bahnemann DW (1995) Chem Rev 95:69, and references therein.
2. Domènech X, Jardim W, Litter M (2001) In: Blesa MA (ed) Eliminación de contaminantes por fotocatálisis heterogénea. Texto colectivo elaborado por la Red CYTED VIII-G. Digital Grafic, La Plata, pp 3–26, http://www.cnea.gov.ar/xxi/ambiental/cyted.asp, and references therein
3. Bahnemann D (1999) In: Boule P (ed) Handbook of environmental photochemistry. Springer, Berlin Heidelberg New York, p 285
4. Litter MI (1999) Appl Catal B 23:89
5. Candal RJ, Rodríguez J, Colón G, Gelover S, Vigil Santos E, Jiménez González A, Blesa MA (2001) In: Blesa MA (ed) Eliminación de contaminantes por fotocatálisis heterogénea. Texto colectivo elaborado por la Red CYTED VIII-G. Digital Grafic, La Plata, pp 143–166, http://www.cnea.gov.ar/xxi/ambiental/cyted.asp
6. Pozzo RL, Baltanás MA, Cassano AE (1997) Catal Today 39:219
7. Anandan S, Yoon M (2003) J Photochem Photobiol C 4:5
8. Hu C, Yizhong W, Hongxiao T (2001) Appl Catal B 30:277
9. Monneyron P, Manero M-H, Foussard J-N, Benoit-Marquisse F, Maurette M-T (2003) Chem Eng Sci 58:971
10. Barbé C, Arendse F, Comte P, Jirousek M, Lenzman F, Shklover V, Grätzel M (1997) J Am Ceram Soc 80:3157
11. Burnside SD, Shklover V, Barbé C, Comte P, Arendse F, Brooks K, Grätzel M (1998) Chem Mater 10:2419
12. Haarstrick A, Kut OM, Heinzle E (1996) Environ Sci Technol 30:817
13. Zeltner WA, Hill CG Jr, Anderson MA (1993) CHEMTECH 21
14. Candal RJ, Zeltner WA, Anderson MA (1999) J Environ Eng 125:906
15. Candal RJ, Zeltner WA, Anderson MA (1998) J Adv Oxid Technol 3:270
16. Fernández A, Lassaletta G, Jiménez VM, Justo A, González-Elipe AR, Herrmann JM, Tahiri H, Ait-Ichou Y (1995) App Catal B 7:49
17. Anpo M, Takeuchi M (2003) J Catal 216:505
18. Rachel A, Subrahmanyam M, Boule P (2002) Appl Catal B: Environ 37:301
19. Matthews R (1987) J Phys Chem 91:3328
20. Sabate J, Anderson MA, Kikkawa H, Edwards M, Hill CG Jr (1991) J Catal 127: 167
21. Zorn ME, Tompkins DT, Zeltner WA, Anderson MA (2000) Environ Sci Technol 34:5206
22. Perry JH (1979) Manual del Ingeniero Químico, 3rd edn. McGraw-Hill, México
23. Hatchard CG, Parker CA (1956) Proc R Soc Lond Ser A 235:518
24. Theurich J, Lindner M, Bahnemann DW (1996) Langmuir 12:6368
25. Hodak J, Quinteros C, Litter MI, San Román (1996) J Chem Soc Faraday Trans 92:5081
26. Wei C, German S, Basak SR, Rajeshwar K (1993) J Electrochem Soc 140:L60
27. Criado JM, Real C (1983) J Chem Soc Faraday Trans 1 2765

Progr Colloid Polym Sci (2004) 128: 309–311

AUTHOR/TITLE INDEX

Progr Colloid Polym Sci (2004) 128: 312–313

KEY WORD INDEX

Printing: Mercedes-Druck, Berlin
Binding: Stein + Lehmann, Berlin